The Northeast Shelf Ecosystem: Assessment, Sustainability, and Management

On the Cover

The image is from a scene taken by the National Oceanic and Atmospheric Administration's (NOAA) Nimbus 7 Satellite. It has been computer enhanced to depict spatial differences in surface chlorophyll. The reddish-brown to yellow gradient in pigment represents decreasing levels of chlorophyll concentrations in the highly productive coastal waters of the ecosystem. The blue coloration shows characteristically low values of chlorophyll in the less productive adjacent oceanic waters of the shelf and slope region of the Northeast Atlantic ocean. The satellite image depicts a warm core ring (WCR) that separated from the Gulf Stream. The diameter of the WCR is nearly equal to the distance across Massachusetts from Cape Cod to the Berkshires. Approximately ten WCRs are formed annually. They generally are moved by the offshore circulation westward along the shelf-slope boundary and dissipate in the vicinity of Cape Hatteras.

The Northeast Shelf Ecosystem: Assessment, Sustainability, and Management

Edited by

Kenneth Sherman

Director, Narragansett Laboratory, and Chief, Oceanography Branch, Northeast Fisheries Science Center, National Marine Fisheries Service, National Oceanic and Atmospheric Administration, Narragansett, Rhode Island; Adjunct Professor of Oceanography, Graduate School of Oceanography, University of Rhode Island, Narragansett, Rhode Island

Norbert A. Jaworski

Director, Environmental Research Laboratory, Office of Research and Development, Office of Environmental Processes and Effects Research, United States Environmental Protection Agency, Narragansett, Rhode Island

Theodore J. Smayda

Professor of Oceanography, Graduate School of Oceanography, University of Rhode Island, Narragansett, Rhode Island

Blackwell Science

Blackwell Science
Editorial offices:
238 Main Street, Cambridge, Massachusetts 02142, USA
Osney Mead, Oxford OX2 OEL, England
25 John Street, London WClN 2BL, England
23 Ainslie Place, Edinburgh EH3 6AJ, Scotland
54 University Street, Carlton, Victoria 3053, Australia
Arnette Blackwell SA, 224, Boulevard Saint Germain, 75007 Paris, France
Blackwell Wissenschafts-Verlag GmbH
Kurfürstendamm 57, 10707 Berlin, Germany
Zehetnergasse 6, A-1140 Vienna, Austria

Acquisitions: Jane Humphreys
Design and Production: Silverchair Science + Communications
Manufacturing: Lisa M. Flanagan

Printed and bound by Edwards Brothers, Inc.

Printed in the United States of America
96 97 98 99 5 4 3 2 1

Distributors
North America
Blackwell Science, Inc.
238 Main Street
Cambridge, Massachusetts 02142
(Telephone orders: 800-215-1000
or 617-876-7000)

Australia
Blackwell Science Pty Ltd.
54 University Street
Carlton, Victoria 3053
(Telephone orders: 03-347-0300;
fax: 03-349-3016)

Outside North America and Australia
Blackwell Science, Ltd.
c/o Marston Book Services, Ltd.
P.O. Box 269
Abingdon
OX14 4YN
England
(Telephone orders: 01235-465500; fax 01235-465555)

Library of Congress Cataloging-in-Publication Data

The northeast shelf ecosystem : assessment, sustainability, and management / edited by Kenneth Sherman, Norbert A. Jaworski, Theodore Smayda.
p. cm.
Includes bibliographical references and index.
ISBN 0-86542-468-3
1. Marine ecology--Atlantic Coast (U.S.) 2. Fishery resources--Atlantic Coast (U.S.)--Management. 3. Marine organisms--Effect of water pollution on--Atlantic Coast (U.S.) 4. Continental shelf--Atlantic Coast (U.S.) I. Sherman, Kenneth, 1932- II. Jaworski, Norbert A. III. Smayda, Theodore J.
QH104.5.A84N67 1996
333.95'2'091634--dc20 96-24468
CIP

Contents

Prologue

In this volume, one of the world's most productive large marine ecosystems—the United States Northeast Shelf—is examined in relation to stress on the system and the actions underway to relieve that stress. The most visible natural resource capital of the Northeast Shelf ecosystem is its rich biodiversity of fish, plankton, bird, and mammal species. The coastal states from Maine to North Carolina currently realize \$1 billion of economic benefits annually from the fisheries of the ecosystem. Management efforts are under way to rebuild the depleted condition of cod, haddock, and flounder stocks to recover the economic potential of these species.

The coastal zone draining into the Northeast Shelf ecosystem has an area of approximately 500,000 km^2 and supports a population of 55 million people, with an average density of 115 people per km^2. Preliminary estimates of waste-water flow into the system from municipal and industrial treatment facilities are about 7 billion gallons per day. The nitrate and phosphate loadings in several estuaries and embayments have exceeded the present "natural" capacity of the ecosystem to adequately recycle the nutrients, resulting in significant overproduction of phytoplankton ("eutrophication") and contributing to the increasing frequency and extent of harmful algal blooms in near-coastal waters. Control of the amount of nutrient loadings, coupled with adequate treatment, will reduce the threat of coastal eutrophication.

With appropriate management practices, the ecosystem should provide the necessary capital in natural productivity for full recovery of depleted fish stocks. Earlier severe declines of mackerel and herring populations due to overexploitation were fully reversed by limiting the fishery for these species through the use of licensing and other restrictions on foreign fishing.

The various jurisdictions responsible for the management of the ecosystem's rich biodiversity and other capital assets will need to develop more effective means for governance of the ecosystem at all levels, including municipal, state, regional, and national governments, if the large populations living in urban centers along the Northeast coast are to realize the long-term sustainable benefits to be derived from one of the most productive large marine ecosystems in the world.

Kenneth Sherman
Norbert A. Jaworski
Theodore J. Smayda

Background and Focus

The mountains of Vermont and New Hampshire, the rugged Maine coast, and the coastal plain of southern New England and the mid-Atlantic states are familiar sights in the Northeast. What is less familiar is how these topographic features extend into the coastal waters of the region. Offshore are the deep basins of the Gulf of Maine, large shoals on Georges Bank, and a gradual sloping of the continental shelf seaward off Massachusetts to North Carolina and into the deep waters of submarine canyons found 70–100 km offshore. The interaction of this submarine topography with currents, water temperatures, salinities, and nutrient runoff from regional rivers and estuaries contributes to the natural productivity of the coastal waters from the Gulf of Maine to Cape Hatteras. Observations of several key species in the waters of the Northeast Shelf—phytoplankton, zooplankton, fish, marine mammals, and marine birds—indicate that they have evolved into an interactive and changing ecosystem.

The Northeast Shelf ecosystem extends from the Gulf of Maine south to Cape Hatteras, and from the coast seaward to the edge of the continental shelf (Sherman et al., this volume). It is a distinct region that is one of the most productive of the 49 large marine ecosystems (LMEs) located around the coastal margins of the world's ocean basins. The boundaries of the LMEs encompass 95% of the world's fish yields, as well as areas of the world's oceans where human population pressures have increased pollution, habitat loss, and overexploitation of many fish and shellfish resources (see Chapter 1). The fish populations in U.S. waters of the Northeast Shelf ecosystem contribute about 25% of the value and 18% of the volume of the U.S. commercial fisheries. The ecosystem, however, is under significant stress.

During the past 30 years, human intervention has resulted in unprecedented biological changes in the Northeast Shelf ecosystem. A large number of marine species have been heavily fished in this region, depressing populations to levels causing significant economic losses. For example, in 1990, annual catch levels totaled 750,000 metric tons (mt) and were valued at $857 million.[1] Some economists use an economic multiplier—three times the landed value—to estimate the value of landed fisheries product to the gross national product of a region. Using that multiplier, the value of the fisheries is estimated at $2.57 billion annually. The long-term potential yield (LTPY) is a projected figure that represents the long-term average catch that can be achieved for the Northeast Shelf ecosystem annually. Assuming conscientious stewardship by controlling fishing mortality rates to maintain populations at sizes that would produce the highest sustainable yield, according to the National Marine Fisheries Service (NMFS), the LTPY for the Northeast Shelf ecosystem is estimated at 1.1 million metric tons (mmt). However, in the past several years, annual catches have been lower than the LTPY. The 1990 landings of 750,000 mt, for example, are 30% lower than the potential annual catch; the difference is estimated at approximately $200–300 million landed value and worth $600–900 million to the regional economy.

Marine fisheries research in the United States began when the Bureau of Fisheries Laboratory at Woods Hole, Massachusetts, was established in 1872. It is clear from the subsequent 123 years of data on Northeast Shelf ecosystem fisheries that annual fluctuations are to be expected. However, since the late 1960s and early 1970s, when large fleets of factory trawlers from eastern Europe landed catches that consistently exceeded LTPY levels, the principal species have been in serious decline (Murawski, this volume).

Human intervention has a significant impact on the biological cycles of the Northeast Shelf ecosystem. Scientists include important multispecies interactions when advising on management strategies. Other considerations are also necessary. In addition to the changing levels of fish stocks, other recent changes have been observed. Harmful blooms of planktonic algae have been increasing. Certain algal species have been implicated in oxygen depletions leading to shellfish mortalities. Other algal species have been responsible for massive die-offs of bay scallops and other molluscs in embayments along the southern New England coast. Whether the algal blooms are the result of excessive nitrates and phosphates entering coastal waters from waste water is being investigated by state and federal scientists.

From the studies reported in this volume, it appears that the Northeast Shelf ecosystem, as a whole, remains healthy. The full recovery of herring and mackerel stocks, when coupled with information on the system's zooplankton community, attests to the robustness of the ecosystem's health. Quantitative indices to measure the changing states of health of the ecosystem are given in Chapter 1. Warning signs have been detected. Nearshore waters are showing signs of stress. It is in these areas that algal blooms may respond to increased nutrient loadings from waste water from urban population centers. The U.S. Environmental Protection Agency (EPA) is considering requirements for the treatment of waste water to remove excess nitrates and phosphates as an option to further control increases in coastal eutrophication. Other actions to mitigate stress in relation to habitat degradation from oil spills and loss of wetlands have shown promise. The successful restoration of a highly oiled salt marsh on Staten Island is encouraging and indicative of how science and management strategies for mitigating ecosystem stress can be more tightly linked in the future.[2] The civil legal proceedings leading to the successful restoration of the wetlands area specifically directed the "governments" (United States—i.e., NOAA and the Department of the Interior; the states of New Jersey and New York; and the city of New York) to "administer those sums recovered from the Exxon Bayway oil spill to restore, rehabilitate, and acquire the equivalent of injured natural resources in the Arthur Kill and other portions of the greater New Jersey/New York harbor . . . "(Resolution 9; United States et al. v. Exxon Corp., Civil Action No. 91-1003, EDNY, Governments' Committee). Adaptive management techniques are being developed that will consider the effects of change on principal biological populations and their environments from the perspective of whole ecosystems. This approach will focus on the health of coastal ecosystems, their key components, and the economies that depend on them.

A significant milestone for strengthening links between science and ecosystem management occurred during August 1991, when 149 marine specialists came together for 4 days on the Bay Campus of the University of Rhode Island's Graduate School of Oceanography to deliberate the changing states of biodiversity and the sustainability of marine resources of the Northeast Shelf ecosystem. The focus of the symposium was an assessment of the effects of growing stress on the long-term yields of the system's living resources and a review of the conditions of the coastal environment.

The symposium was convened by the Northeast Fisheries Science Center of the National Marine Fisheries Service (NMFS) of the National Oceanic and Atmospheric Administration (NOAA). It was sponsored by the EPA, NMFS, the Marine Mammal Commission, the American Fisheries Society, the University of Rhode Island, and the Minerals Management Service of the Interior Department. This volume includes selected peer-reviewed papers from the symposium. These reports have been augmented with the results of other timely and pertinent studies related to the sustainability of the ecosystem, several of which have previously appeared in widely scattered literature.

This volume is the sixth in a series on LMEs. The first four LME conferences were held in the United States during annual meetings of the American Association for the Advancement of Science (1984, 1987, 1988, and 1990). The fifth volume was based on an international conference on LMEs held in Monaco in 1990. This sixth volume is a case study of the stressed Northeast Shelf of the United States, where efforts are under way to mitigate that stress. However, in the absence of an ecosystem-wide governing system, the sustainability of the marine resources of the ecosystem remains at risk. In each of the six volumes, emphasis is placed on the LMEs as a unifying concept for aiding in focusing science and management efforts within the boundaries of distinct, ecologically based units of ocean space. An initial step toward achieving this goal is the bringing together in a single volume the results of multidisciplinary studies to serve as a basis for assessing the changing states of the Northeast Shelf ecosystem, thereby providing a sound scientific foundation for a future governance regime.

The reports in this books are focused on the Northeast Shelf as a LME distinct from the Scotian Shelf to the north and the Southeast Shelf ecosystem south of Cape Hatteras (see Figure 1-1, Chapter 1). The volume brings together assessments of fisheries, pollution, and habitat from an integrated ecosystem perspective. This multidisciplinary focus on the ecosystem is a necessary precursor for any consideration of management of the ecosystem. The legislative authority for proceeding toward coastal ecosystem management can be found in several public laws. Foremost among these are the Magnuson Fishery Conservation and Management Act (P.L. 94-265), the Marine Mammal Protection Act (P.L. 103-238), the National Environmental Policy Act (P.L. 91-190), the Federal Water Pollution Control Act (P.L. 92-500), and the National Coastal Monitoring Act (H.R. 2130). Within each of these acts, the federal government is directed, as the nation's steward of natural resources, to collaborate closely with the coastal states in moving forward to implement more coherent and integrated marine resource management. This theme is explored in the introductory papers to the volume.

The next section of the book discusses the results of studies focused on physical and biological dynamics of the Northeast Shelf ecosystem, ecosystem stressors, and mitigation actions for improving its health. These issues, critical to the resource sustainability of the Northeast Shelf ecosystem, recognize the need to reduce stress on fish, water and benthic quality, and habitat. Actions required to mitigate stress should be based on (1) quantitative science, (2) consensus of user groups, and (3) a framework for governance. Substantial progress has already been made in addressing two of these issues: (1) Federal agencies have acknowledged their stewardship responsibilities and initiated action to mitigate stress on ecosystem components; and (2) an extensive information base on the resources of the ecosystem is available to user groups on electronic networks. Considerable scientific effort by federal and state governments is funded and under way in the region, along with pertinent research activities in academia and the private sector.

An important unfinished task is the development of a governing mechanism for ensuring that future management is planned and executed with the participation and concurrence of the principal constituent groups throughout the area. It is clear from recent reports that the long-term productivity of marine resources will be contingent on the implementation of a governing system that recognizes the spatial and temporal extent of marine resources and the capacity of the ecosystem to support these resources, within the framework of sustainable economic practices.

The level of primary production required to support marine fish stocks on a sustainable basis has been examined on a global basis. It appears from the results that the present annual fishery yield, approximately 90 million metric tons, has reached the limit imposed by primary production. Any large-scale increase in annual yield of living marine resources could only be achieved by harvesting nontraditional species at lower trophic levels than are currently being exploited. This strongly suggests that long-term sustainability of traditional fish species will require more highly regulated access to fishery resources than presently exists.[3,4] Conflicts over straddling stocks and national and international borders can be overcome, in part, by recognizing the ecological coherence of biological communities and ecosystem dynamics (e.g., Canada vs. Spain over disputed stocks within the Newfoundland Shelf LME). The integrity of ecosystem and management prospects are, for the most part, different in spatial extent than the designated 200-mile Law of the Sea coastal nation boundary. This issue is addressed in Chapter 29 by Lawrence Juda. Evolving models of governance for marine ecosystem management can be found in the reports of the Commission for the Conservation of Antarctic Marine Living Resources,[5] the Ministerial Declaration for the Black Sea,[6] and the Ministerial Declaration for the North Sea.[7] These topics are reviewed in Chapter 1.

The symposium provided a forum for the examination of the utility of the LME approach to the mitigation of problems related to resource sustainability within the U.S. Exclusive Economic Zone (EEZ) and in other LMEs where efforts are under way to reduce stress on marine ecosystems from human interventions and natural perturbations. The symposium participants reached consensus on the desirability of reviewing the status of resource sustainability of the Northeast Shelf ecosystem at intervals of about 3 years.

John Knauss, former Undersecretary for Oceans and Atmosphere in the U.S. Department of Commerce, indicates in Chapter 2 that the holistic approach inherent in the LME concept encourages agencies to address issues of overfishing, habitat loss, pollution, and recreation needs from a multidisciplinary ecosystems perspective. He stresses the importance of organizing a set of regional programs, each designed for a specific LME, with the goal of monitoring it to determine how humans may be perturbing the system, thereby providing a basis for appropriate mitigating action.

Charles Peterson of the University of North Carolina reviews in Chapter 3 the sustainable biosphere initiative (SBI) of the Ecological Society of America. Dr. Peterson emphasizes the positive relationships in the SBI initiative and the application of the Global Ocean Ecosystem Dynamics Program (GLOBEC) of the National Science Foundation (NSF) and NOAA, presently under way within the Northeast Shelf ecosystem. He describes the Northeast Shelf ecosystem with emphasis on seven points being addressed by government and academic scientists: (1) Species do not exist, and cannot be managed, in isolation; (2) Process-based ecosystem models must be created, incorporating trophic interactions and physical dynamics as a substitute for the lack of capacity for replication of a specific LME; (3) Ecosystem models must include key human interactions relating to exploitation, nutrient and pollutant inputs, and climate change; (4) Rigorous experimental and analytic methodologies of manipulative ecology need to be applied in "adaptive management"; (5) Modelers must be charged with, and agencies made responsive to, the need for identifying precisely those state variables that require monitoring to test model predictions and assess ecosystem health; (6) Enhanced attention to inclusion of prey dynamics, such as forage fish and zooplankton, is critical to fisheries management; and (7) Studies and models must evaluate the roles of biodiversity and indirect interactions (those other than direct predator-prey relations) in ocean ecosystem dynamics to allow for the simplification of a diverse system.

The physical forces influencing productivity and sustainability of the Northeast Shelf ecosystem are reviewed by several investigators who consider the environmental forces influencing ecosystem structure and function, including weather and climate (Hertzman, Chapter 5); water mass structure (Brooks, Chapter 4). Following the presentations of trends in coastal physics, a summary of trends in populations is given for principal components of the ecosystem (Sherman et al., Chapter 7); zooplankton (Durbin and Durbin, Chapter 8); fish (Anthony, Chapter 9); cetaceans (Kenney et al., Chapter 10); and marine birds (Schneider and Heinemann, Chapter 11).

Based on the information presented to the symposium, river basin drainage into the ecosystem is undergoing some degree of improvement in the removal of toxins, including heavy metals (Windom, Chapter 16). Most of the waste-water plants discharging into the Northeast Shelf ecosystem are providing secondary treatment, and recent regulations have begun to address overflows from combined sewage and stormwater discharge systems that introduce inadequately treated waste-water into the ecosystem. It was concluded, based on NOAA's Status and Trends Program information, that large-scale biological consequences from existing levels of toxins were unlikely. Levels of lead and

DDT have declined. Elevated levels of toxins are limited to the vicinities of large urban outfalls, with a few relatively small areas that can be described as biological "hot spots" (O'Connor, Chapter 13). Evidence of neoplasms in fish and shellfish has been found, and periodic advisories have been issued to limit the human consumption of striped bass, bluefish, and blue crabs because of potential health risks from chlorinated hydrocarbons that have been detected in the tissues of these species. However, no linkage has, as yet, been made to allow for extrapolation from individual specimens with relatively high levels of toxins to any significant biofeedback at the population level related to increased levels of contamination. Diagnostics are improving, and states continue to issue advisories on red tides if they present any risk to human health. Also, research is under way to improve surveys of pollution effects on individual organisms and populations (White and Robertson, Chapter 14).

The results of an investigation conducted by the EPA (Godshall and Walker, Chapter 6) suggest that Northeast coastal waters are in a warming trend. In recognition of the need for increasing funds available to academic institutions for conducting retrospective and prospective long-term studies in relation to global climate changes, NSF and NOAA are supporting the GLOBEC program. This program is designed to assist scientists in academia and the federal government to support research on the physical and biological factors influencing the productivity and dynamics of zooplankton production and the survival of early life stages of codfish.[8]

With regard to eutrophication, Malone and Conley (Chapter 17) indicate, based on case-history studies in the Hudson River and Chesapeake Bay systems, that in the former there is no consistent pattern of any surplus production attributable to measured increased eutrophication. In the offing of the Hudson, in some years it is possible to recognize the effects of eutrophication in the discharge plume that contributed periodically to anoxia, significant benthic mortalities, and economic losses of shellfish in the 1970s and 1980s. Phytoplankton blooms in the Chesapeake seem to vary in extent and intensity, but at the present time, no evidence of encroachment out onto the relatively open waters of the mid-Atlantic Bight subarea of the Northeast Shelf ecosystem is apparent.

Within the Northeast Shelf ecosystem, there is evidence of increasing frequency of biotoxin-related mortality of marine mammals, along with recent closures of offshore shellfish beds because of increased levels of paralytic shellfish poisoning (White, Chapter 22). Heavy-metal concentrations in demersal fish, crustaceans, and bivalve molluscs continue to be monitored as biological indicators (Schwartz et al., Chapter 15) in relation to development of strategies for mitigation of pollution stress on the ecosystem. General trends of pollutant loading for heavy metals appear to be decreasing, while the introduction of excessive nutrient loadings is increasing (Jaworski and Howarth, Chapter 18). Actions to investigate and protect important marine habitats are considered in the papers by Bigford (Chapter 19) and Thomas (Chapter 20). Other habitat observations focus on coastal wetlands, habitat stressors on fish (Peters and Cross, Chapter 21), and public health (Epstein, Chapter 23).

Coastal sites used to dump sludge from New York City waste treatment plants are showing signs of repopulation and improved water quality (Ingham, Chapter 24), indicating that mitigating actions that result in dump closures will result in a degree of recovery of stressed benthic communities. Sites contaminated with PCBs, however, do not show rapid recovery following mitigation. Residence times of PCBs in sediments of New Bedford Harbor are quite long, and adverse effects on shellfish species continue to be detected several years after banning the disposal of PCBs in coastal waters (Capuzzo, Chapter 25).

Highly migratory species of marine mammals and apex predators are important components of the Northeast Shelf ecosystem. They frequent the coastal waters seasonally, principally on feeding migrations. A strategy to improve estimates of population sizes and better define migratory patterns is proposed in Chapter 26 (Smith et al.). For the offshore waters of the Northeast Shelf ecosystem, a major shift among dominant species was reported by Murawski (Chapter 27). Over the past two decades, dominant species have shifted from predominantly gadoids to a state in which elasmobranch and pelagic species dominate the fish community. This shift in ecosystems state reduces the turnover rates of fish, as elasmobranchs are longer lived than gadoids, thereby imposing a long-term reduction in the production levels of the fish community in relation to long-term resource sustainability.

The Northeast Shelf ecosystem continues to be highly productive at the lower end of the food chain. Model simulations based on biological and economic data suggest that fishing effort will need to be reduced by 50–70%, if long-term sustainability of preferred high-demand and high-priced species is to be realized (Edwards and Murawski, Chapter 28). Options for implementing an ecosystem-based governance regime under the terms of the Law of the Sea (LOS) is of interest to marine managers who recognize the need for improving present management regimes. The possibilities of developing a LOS-based ecosystem management regime are explored in Chapter 29 (Juda).

Kenneth Sherman
Norbert A. Jaworski
Theodore J. Smayda
Narragansett, Rhode Island

References

1. Our Living Oceans: Report on the Status of U.S. Living Marine Resources. 1992. NOAA Tech. Mem NMFS-F/SPD2. NOAA, U.S. Department of Commerce. 148 pp.
2. Matsil, M. A., and Bergen, A. 1995. On the evolution of an oil spill remediation plan: Facts and fiction of restoration, research, and monitoring of an urban intertidal salt marsh. Presented at Coastal Zone 95, 9th International Symposium on Coastal and Ocean Management. Tampa, Florida.
3. Beddington, J. The primary requirements. Nature 374:213–214, 1995.
4. Pauly, D., and Christensen, V. Primary production required to sustain global fisheries. Nature 374:255, 1995.
5. Scully, R. T. 1993. Convention on the Conservation of Antarctic Marine Living Resources. In K. Sherman, L. M. Alexander, and B. D. Gold (eds.), Large Marine Ecosystems: Stress, Mitigation, and Sustainability. Washington, D.C.: AAAS Press, pp. 242–251.
6. Hey, E., and Mee, L. D. 1993. Black Sea—The ministerial declaration: An important step. Environ. Pol. Law 2315:215–217, 235–236.
7. North Sea Quality Status Report. 1993. North Sea Task Force, Oslo and Paris Commissions, International Council for the Exploration of the Sea, London.
8. 1995 U.S. GLOBEC News No. 8, U.S. GLOBEC Steering Committee Office; Department of Integrative Biology, Berkeley, California.

Acknowledgments

The editors are indebted to the contributors for their willingness to take time out of busy schedules to develop the expert syntheses and reviews needed to move toward ecosystem governing regimes. We are pleased to acknowledge the interest and financial support of the National Oceanic and Atmospheric Administration (NOAA) and the National Marine Fisheries Service (NMFS). We gratefully acknowledge support for the Northeast Shelf LME conference, from which many of the presentations in this volume were drawn. Organizations providing support include the Marine Mammal Commission, the Minerals Management Service of the Department of the Interior, and the U.S. Environmental Protection Agency. The LME symposium would not have been possible without the willing and capable cooperation of many people who gave unselfishly of their time and effort. We are indebted to Jeanne Burns of NMFS in Narragansett, Rhode Island, for her extraordinary dedication, care, and expertise in technically editing the entire volume. We thank Jane Humphreys and Debra Lance of Blackwell Science for their care in the final production of the volume. We also thank Janet Kelly of NMFS in Narragansett, whose contribution in the early stages of production of the volume was invaluable. A special debt of thanks is extended to Dr. John Knauss, who, as Undersecretary of Commerce for Oceans and Atmosphere, provided a challenging and inspiring opening address to the participants.

Kenneth Sherman
Norbert A. Jaworski
Theodore J. Smayda

Contributors

Vaughn C. Anthony
Boothbay, Maine

Carol A. Batdorf
Cat Cove Marine Laboratory, Division of Marine Fisheries, Salem, Massachusetts

Thomas E. Bigford
National Oceanic and Atmospheric Administration/National Marine Fisheries Service, Silver Spring, Maryland

David A. Brooks
Department of Oceanography, Texas A & M University, College Station, Texas

Donna Busch
National Oceanic and Atmospheric Administration, Narragansett Laboratory, Narragansett, Rhode Island

Judith E. McDowell Capuzzo
Woods Hole Oceanic Institution, Woods Hole, Massachusetts

John G. Casey
National Oceanic and Atmospheric Administration/National Marine Fisheries Service, Narragansett, Rhode Island

Daniel J. Conley
National Environmental Research Institute, Frederiksborgvej, Denmark

Ford A Cross
National Oceanic and Atmospheric Administration/National Marine Fisheries Service, Beaufort, North Carolina

Ann G. Durbin
Graduate School of Oceanography, University of Rhode Island, Narragansett, Rhode Island

Edward G. Durbin
Graduate School of Oceanography, University of Rhode Island, Narragansett, Rhode Island

Nina M. Duston
Cat Cove Marine Laboratory, Division of Marine Fisheries, Salem, Massachusetts

Steven F. Edwards
National Oceanic and Atmospheric Administration/National Marine Fisheries Service, Woods Hole, Massachusetts

Paul R. Epstein
Harvard School of Public Health, Cambridge, Massachusetts

Fredric A. Godshall
Science Applications International Corporation, Narragansett, Rhode Island

John Green
National Oceanic and Atmospheric Administration/National Marine Fisheries Service, Narragansett, Rhode Island

Robert B. Griffin
Graduate School of Oceanography, University of Rhode Island, Narragansett, Rhode Island

Marvin Grosslein
National Marine Fisheries Service, Woods Hole Laboratory, Woods Hole, Massachusetts

Dennis Heinemann
Manomet Observatory for Conservation Sciences, Manomet, Massachusetts

Owen Hertzman
Department of Oceanography, Dalhousie University, Halifax, Nova Scotia, Canada

Robert W. Howarth
Cornell University, Ithaca, New York

Merton C. Ingham
South Dennis, Massachusetts

Norbert A. Jaworski
Environmental Protection Agency, Narragansett, Rhode Island

Jack Jossi
National Oceanic and Atmospheric Administration/National Marine Fisheries Service, Narragansett, Rhode Island

Lawrence Juda
Department of Marine Affairs, University of Rhode Island, Kingston, Rhode Island

Joseph Kane
National Oceanic and Atmospheric Administration/National Marine Fisheries Service, Narragansett, Rhode Island

Robert D. Kenney
Graduate School of Oceanography, University of Rhode Island, Narragansett, Rhode Island

John A. Knauss
Graduate School of Oceanography, University of Rhode Island, Narragansett, Rhode Island

Thomas C. Malone
University of Maryland System, Horn Point Environmental Laboratory, Cambridge, Maryland

David Mountain
National Oceanic and Atmospheric Administration/National Marine Fisheries Service, Woods Hole, Massachusetts

Steven A. Murawski
National Oceanic and Atmospheric Administration/National Marine Fisheries Service, Woods Hole, Massachusetts

Thomas P. O'Connor
National Oceanic and Atmospheric Administration, National Status and Trends Program, Silver Spring, Maryland

Jay O'Reilly
National Oceanic and Atmospheric Administration/National Marine Fisheries Service, Naragansett, Rhode Island

P. Michael Payne
National Oceanic and Atmospheric Administration/National Marine Fisheries Service, Silver Spring, Maryland

David S. Peters
National Oceanic and Atmospheric Administration/National Marine Fisheries Service, Beaufort, North Carolina

Charles H. Peterson
University of North Carolina at Chapel Hill, Institute of Marine Sciences, Morehead City, North Carolina

Andrew Robertson
Coastal Monitoring and Bioeffects Assessment Division, National Oceanic and Atmospheric Administration, Silver Spring, Maryland

David C. Schneider
Ocean Sciences Center, Memorial University, St. John's, Newfoundland, Canada

Jack P. Schwartz
Cat Cove Marine Laboratory, Division of Marine Fisheries, Salem, Massachusetts

Kenneth Sherman
National Oceanic and Atmospheric Administration/National Marine Fisheries Service, Narragansett, Rhode Island

Theodore J. Smayda
Graduate School of Oceanography, University of Rhode Island, Narragansett, Rhode Island

Tim D. Smith
National Oceanic and Atmospheric Administration/National Marine Fisheries Service, Woods Hole, Massachusetts

Wallace Smith
National Marine Fisheries Service/Northeast Fisheries Science Center, James J. Howard Marine Sciences Laboratory, Highlands, New Jersey

Andrew Solow
Marine Policy Center
Woods Hole Oceanographic Institute
Woods Hole, Massachusetts

Roger Theroux
East Falmouth, Massachusetts

James Postles Thomas
National Oceanic and Atmospheric Administration/National Marine Fisheries Service, Silver Spring, Maryland

Henry A. Walker
United States Environmental Protection Agency, Narragansett, Rhode Island

Gordon T. Waring
National Oceanic and Atmospheric Administration/National Marine Fisheries Service, Woods Hole, Massachusetts

Alan W. White
Massachusetts Maritime Academy, Buzzards Bay, Massachusetts 02532

Harris H. White
National Oceanic and Atmospheric Administration, Silver Spring, Maryland

Herbert L. Windom
Graduate School of Oceanography, University of Rhode Island, Narragansett, Rhode Island

Howard E. Winn
Graduate School of Oceanography, University of Rhode Island, Narragansett, Rhode Island

The Northeast Shelf Ecosystem: Assessment, Sustainability, and Management

I

Ecosystem Sustainability

1

Marine Ecosystem Sustainability*

Kenneth Sherman

Background

For the past ten years, the concept of large marine ecosystems (LMEs) as geographic units for improving the assessment and management of marine resources has been examined (AAAS 1986, 1989, 1990, 1991, 1993). Large marine ecosystems are regions of ocean space encompassing coastal areas from river basins and estuaries out to the oceanic margins of continental shelves and the seaward extent of coastal current systems (Figure 1-1). They are relatively large regions, on the order of 200,000 km^2 or larger, characterized by distinct bathymetry, hydrography, productivity, and trophically dependent populations. The theory, measurement, and modeling relevant to monitoring the changing states of LMEs are imbedded in reports on ecosystems with multiple steady states and on the pattern formation and spatial diffusion within ecosystems (Holling 1973, 1986, 1993; Pimm 1984; Beddington 1986; Mangel 1991; Levin 1993; Sherman 1994). Since 1984, the National Oceanic and Atmospheric Administration (NOAA) has convened five symposia, published several volumes, and obtained expert syntheses of the principal forces controlling the biomass yields of 29 LMEs, including five within the U.S. Exclusive Economic Zone (EEZ)—i.e., the Northeast U.S. shelf, the Southeast U.S. shelf, the Gulf of Mexico, the California Current, and the Eastern Bering Sea (Table 1-1). On a global basis, 49 LMEs account for 95% of the annual yield of the global fisheries (Table 1-2).

---

*Parts of this chapter appeared in K. Sherman. 1994. Sustainability, Biomass Yields, and Health of Coastal Ecosystems: An Ecological Perspective. Marine Ecology Progress Series 112:277–301.

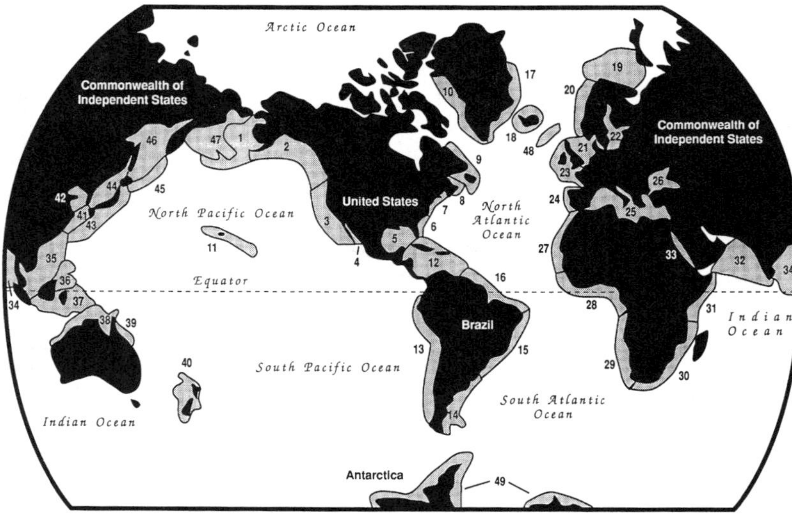

Figure 1-1. Boundaries of 49 large marine ecosystems.

1. Eastern Bering Sea
2. Gulf of Alaska
3. California Current
4. Gulf of California
5. Gulf of Mexico
6. Southeast U.S. Continental Shelf
7. Northeast U.S. Continental Shelf
8. Scotian Shelf
9. Newfoundland Shelf
10. West Greenland Shelf
11. Insular Pacific—Hawaiian
12. Caribbean Sea
13. Humboldt Current
14. Patagonian Shelf
15. Brazil Current
16. Northeast Brazil Shelf
17. East Greenland Shelf
18. Iceland Shelf
19. Barents Sea
20. Norwegian Shelf
21. North Sea
22. Baltic Sea
23. Celtic-Biscay Shelf
24. Iberian Coastal
25. Mediterranean Sea
26. Black Sea
27. Canary Current
28. Guinea Current
29. Benguela Current
30. Agulhas Current
31. Somali Coastal Current
32. Arabian Sea
33. Red Sea
34. Bay of Bengal
35. South China Sea
36. Sulu-Celebes Seas
37. Indonesian Seas
38. Northern Australian Shelf
39. Great Barrier Reef
40. New Zealand Shelf
41. East China Sea
42. Yellow Sea
43. Kuroshio Current
44. Sea of Japan
45. Oyashio Current
46. Sea of Okhotsk
47. West Bering Sea
48. Faeroe Plateau
49. Antarctic

Table 1-1. List of 29 large marine ecosystems and subsystems for which syntheses relating to principal, secondary, or tertiary driving forces controlling variability in biomass yields have been completed as of February 1993

Large marine ecosystem	Source	Authors
U.S. Northeast Continental Shelf	AAAS, 1986	Sissenwine
	AAAS, 1991	Falkowski
U.S. Southeast Continental Shelf	AAAS, 1991	Yoder
Gulf of Mexico	AAAS, 1989	Richards and McGowan
	AAAS, 1991	Brown et al.
California Current	AAAS, 1986	MacCall
	AAAS, 1991	Mullin
	AAAS, 1993	Bottom
Eastern Bering Shelf	AAAS, 1986	Incze and Schumacher
West Greenland Shelf	AAAS, 1990	Hovgaard and Buch
Norwegian Sea	AAAS, 1990	Ellertsen et al.
Barents Sea	AAAS, 1989	Skjoldal and Rey
	AAAS, 1991	Borisov
North Sea	AAAS, 1986	Daan
Baltic Sea	AAAS, 1986	Kullenberg
Iberian Coastal	AAAS, 1989	Wyatt and Perez-Gandaras
Mediterranean-Adriatic Sea	AAAS, 1993	Bombace
Canary Current	AAAS, 1993	Bas
Gulf of Guinea	AAAS, 1993	Binet and Marchal
Benguela Current	AAAS, 1989	Crawford et al.
Patagonian Shelf	AAAS, 1993	Bakun
Caribbean Sea	AAAS, 1990	Richards and Bohnsack
South China Sea-Gulf of Thailand	AAAS, 1989	Piyakarnchana
Yellow Sea	AAAS, 1989	Tang
Sea of Okhotsk	AAAS, 1993	Kusnetsov
Humboldt Current	AAAS, 1993	Alheit and Bernal
Indonesia Seas-Banda Sea	AAAS, 1990	Zijlstra and Baars
Bay of Bengal	AAAS, 1993	Dwivedi
Antarctic	AAAS, 1986, 1993	Scully et al.
Weddell Sea	AAAS, 1990	Hempel
Kuroshio Current	AAAS, 1989	Terazaki
Oyashio Current	AAAS, 1989	Minoda
Great Barrier Reef	AAAS, 1989	Bradbury and Mundy
	AAAS, 1993	Kelleher
South China Sea	AAAS, 1993	Pauly and Christensen

Sustainability: An Ecosystems Approach

Human intervention and climate change are sources of variability in the natural productivity of coastal marine ecosystems. Within the nearshore areas and extending seaward around the margins of the global land masses, coastal ecosystems are being subjected to increased stress from toxic effluents, habitat

Table 1-2. Contributions to global fisheries yield, by country and large marine ecosystem (LME), representing 95% of the annual global catch in 1990

Country	Percentage of world marine nominal catch[a]	LMEs producing annual biomass yield	Cumulative percentage
Japan	12.25	Oyashio Current, Kuroshio Current, Sea of Okhotsk, Sea of Japan, Yellow Sea, East China Sea, W. Bering Sea, E. Bering Sea, and Scotia Sea	
USSR	11.37	Sea of Okhotsk, Barents Sea, Norwegian Shelf, W. Bering Sea, E. Bering Sea, and Scotia Sea	
China	8.28	W. Bering Sea, Yellow Sea, E. China Sea, and S. China Sea	
Peru	8.27	Humboldt Current	
USA	6.76	Northeast U.S. Shelf, Southeast U.S. Shelf, Gulf of Mexico, California Current, Gulf of Alaska, and E. Bering Sea	
Chile	5.98	Humboldt Current	52.91
Korea Republic	3.28[b]	Yellow Sea, Sea of Japan, E. China Sea, and Kuroshio Current	
Thailand	2.96[b]	South China Sea and Indonesian Seas	
India	2.78	Bay of Bengal and Arabian Sea	
Indonesia	2.76	Indonesian Seas	
Norway	2.11	Norwegian Shelf and Barents Sea	
Korea (Democratic People's Republic)	1.98[b]	Sea of Japan and Yellow Sea	
Philippines	1.96	S. China Sea, Sulu-Celebes Sea	
Canada	1.90	Scotian Shelf, Northeast U.S. Shelf, Newfoundland Shelf	
Iceland	1.82	Icelandic Shelf	
Denmark	2.07	Baltic Sea and North Sea	76.25
Spain	1.73	Iberian Coastal Current and Canary Current	
Mexico	1.46	Gulf of California, Gulf of Mexico, and California Current	
France	1.03[b]	North Sea, Biscay-Celtic Shelf, Mediterranean Sea	80.47
Vietnam	0.74	South China Sea	
Myanmar	0.72	Bay of Bengal, Andaman Sea	
Brazil	0.71[b]	Patagonian Shelf and Brazil Current	
Malaysia	0.71[b]	Gulf of Thailand, Andaman Sea, Indonesian Seas, and S. China Sea	
UK, Scotland	0.70	North Sea	
New Zealand	0.68	New Zealand Shelf Ecosystem	
Morocco	0.68	Canary Current	
Argentina	0.66	Patagonian Shelf	
Italy	0.57	Mediterranean Sea	
Netherlands	0.52	North Sea	

Poland	0.52	Baltic Sea	
Ecuador	0.47	Humboldt Current	
Pakistan	0.44	Bay of Bengal	
Turkey	0.41	Black Sea, Mediterranean Sea	
Germany	0.41	Baltic Sea and Scotia Sea	
Ghana	0.40	Gulf of Guinea	
Portugal	0.39	Iberian Shelf and Canary Current	90.20
Venezuela	0.38	Caribbean Sea	
Namibia	0.35	Benguela Current	
Faeroe Islands	0.34	Faeroe Plateau	
Senegal	0.34	Gulf of Guinea and Canary Current	
Sweden	0.31	Baltic Sea	
Bangladesh	0.31	Bay of Bengal	
Ireland	0.28	Biscay-Celtic Shelf	
Hong Kong	0.28	S. China Sea	
Nigeria	0.26	Gulf of Guinea	
Australia	0.25	N. Australian Shelf and Great Barrier Reef	
Iran	0.24[b]	Arabian Sea	
UK, England and Wales	0.21	North Sea	
Cuba	0.20	Caribbean Sea	
Panama	0.19	California Current and Caribbean Sea	
Greenland	0.17	East Greenland Shelf, West Greenland Shelf	
Sri Lanka	0.19	Bay of Bengal	
Greece	0.16	Mediterranean Sea	
Oman	0.15	Arabian Sea	
Angola	0.12	Guinea Current, Angola Basin	
United Arab Emirates	0.11	Arabian Sea	95.01

[a]Percentages based on fish catch statistics from FAO (1992). Some fraction of the catch made by fishing operations of long-distance trawlers may actually be caught in other LMEs.
[b]Percentage calculated using FAO estimate from available sources of information.

degradation, excessive nutrient loadings, harmful algal blooms, emergent diseases, fallout of aerosol contaminants, and episodic losses of living marine resources from the effects of pollution and overexploitation. The long-term sustainability of coastal ecosystems as sources for healthy economies appears to be diminishing. A growing awareness that multiple driving forces are having an adverse impact on the quality of the global coastal ecosystems has accelerated efforts to assess, monitor, and mitigate coastal stressors from an ecosystem perspective. The Intergovernmental Oceanographic Commission (IOC) of the United Nations Educational, Scientific, and Cultural Organization (UNESCO) is encouraging coastal nations to establish national programs for assessing and monitoring coastal ecosystems to enhance the ability of national and regional management organizations to develop and implement effective remedial programs for improving the quality of degraded ecosystems (IOC 1992). This encouragement follows from the significant milestone achieved in July 1992 with the adoption by a majority of coastal countries of follow-on actions to the United Nations Conference on Environment and Development (UNCED). The UNCED declarations on the ocean explicitly recommended that nations of the globe (1) prevent, reduce, and control degradation of the marine environment so as to maintain and improve its life-support and productive capacities; (2) develop and increase the potential of marine living resources to meet human nutritional needs as well as social, economic, and development goals; and (3) promote the integrated management and sustainable development of coastal areas and the marine environment.

The concept of sustainability as applied to marine resources and ecosystems has been the subject of debate. Ludwig et al. (1993) argued that the basic research in ecology being conducted in support of the Sustainable Biosphere Initiative of the Ecological Society of America (Lubchenco et al. 1991) may be leading to complacency, because the most pressing issues being addressed to avoid further degradation of global renewable resources are the increasing size of human populations and excessive use of resources, not necessarily more scientific research. They supported their arguments with citations of (1) the apparent failures in fisheries management to maintain ecologically balanced stock viability in the face of increasing fishing effort, and (2) the difficulty of obtaining scientific consensus on resource status as populations are overexploited, making claims of sustainability suspect. Rosenberg et al. (1993) disputed the assertions of Ludwig et al. (1993) and maintained that "fisheries management provides positive examples of sustainable resource use and lessons for future improvements." They argued that "there is a sound theoretical and empirical basis for sustainable use of marine resources, that overexploitation is not inevitable or the result of inadequate scientific advice, and that sustainable use of renewable resources is an obtainable management objective." In the development of a sustainable harvesting management regime, Rosenberg et al. (1993) emphasized the need for taking into account environmental conditions and exploitation rates by implementing appropriate monitoring programs to identify and track the shifts in population responses. They concluded that sustainable development of marine resources is achievable if scientific advice based on biologic, social, and economic considerations is an integral part of the development of policies for use of renewable resources. This conclusion is in agreement with that of Mangel et al. (1993), who argued that "sustainable use" of renewable resources is achievable when humans use living components of ecosystems in ways that allow natural processes to replace what is used. Under these conditions, the ecosystem "will renew itself indefinitely and human use will be sustainable."

Large-Scale Programs

An ecosystems approach for ecologists interested in contributing to a scientifically based strategy for resource sustainability was given by Holling (1993). He emphasized that there is an emerging science that is multidisciplinary and focused on populations and ecosystems of large spatial scales that includes socioeconomic considerations in the planning and implementation appropriate to the issue of resource sustainability. The more traditional, but nonetheless important, discipline-oriented ecologic studies can contribute more to resource sustainability when they are conducted within a framework of science at a multidisciplinary level of organization focused on issues affecting populations within an ecosystem. The international GLOBEC regional programs (Skjoldal et al. 1993), the International Geosphere–Biosphere Program on the Land–Oceans Interaction in the Coastal Zone Program (IGBP 1994), and the Joint Global Ocean Flux Study (IGBP 1994) are examples of this emerging large-scale multidisciplinary marine ecosystem science. Linkages between these process-oriented studies and the more applied studies that support the sustainability of marine resources are being encouraged by national and international funding agencies concerned with the sustainability of natural resources. In this regard, the definition of sustainability used by Holling and carried forward in this perspective focuses on ecosystem studies in support of "the social and economic development of a region with the goals to invest in the maintenance and restoration of critical ecosystem functions, to synthesize and make accessible knowledge and understanding for economies, and to develop and communicate the understanding that provides a foundation of trust for citizens" (Holling 1993).

Governance

In practice, it is important to establish institutional arrangements that ensure that appropriate socioeconomic considerations are exercised when scientific regimes aimed at the sustainability of renewable resources are applied. Regional examples of this approach to ecosystem sustainability can be found in the objectives of the Convention for the Conservation of Antarctic Living Marine Resources (Scully 1993) and the ministerial declarations for the protection of the Black Sea (Hey and Mee 1993) and the North Sea (NSTF 1991). The Black Sea Declaration refers specifically to the objectives in UNCED Agenda 21, Chapter 17. These objectives call for integrated management and sustainable development of coastal areas, marine environmental protection, sustainable use and conservation of living resources under national jurisdiction, the need for addressing critical uncertainties in the management of the marine environment, and the strengthening of international and regional cooperation and coordination (Hey and Mee 1993). An effort to develop a management system for the Barents Sea Ecosystem that provides stronger links between science and socioeconomic considerations from an ecosystems perspective is under consideration (Eikeland 1992).

An essential component of an ecosystem management regime is the inclusion of a scientifically based strategy that monitors and assesses the changing states and health of the ecosystem by tracking key biologic and environmental variables. From this perspective, marine ecosystem assessment and monitoring is a component of a management system that includes regulatory, institutional, and decision-making aspects relative to marine ecosystems, and therefore would include a range of activities needed to provide management information about ecosystem conditions, contaminants, and resources at risk. Based on experiences in North America, Europe, and elsewhere, a comprehensive coastal ecosystem assessment and monitoring system whose core components include conceptual and numerical modeling capabilities; laboratory and field research; time–series measurements; data analysis, synthesis, and interpretation; and a capacity for initiating the effort with preliminary or scoping studies is the most likely to be successful (NRC 1990). The principal characteristic of a comprehensive ecosystem assessment and monitoring program is the integration and coordination of the components of the effort into a total ecosystems approach designed to produce scientific information in support of coastal resources management. This strategy is consistent with the conclusion that monitoring efforts at the regional scale need to be strengthened to improve understanding of broader-scale trends in marine ecosystem quality.

Actions to mitigate stress on marine ecosystems are required to ensure the long-term sustainability of marine resources. The principles adopted by coastal states under the terms of the United Nations Convention for the Law of the Sea (UNCLOS) have been interpreted as supportive of the management of living marine resources and coastal habitats from an ecosystems perspective (Belsky 1986, 1989). At present, however, no single international institution has been empowered to monitor the changing ecologic states of marine ecosystems and to reconcile the needs of individual nations with those of the community of nations in taking appropriate mitigating actions (IUCN 1990; Myers 1990). With regard to the development and sustainability of marine resources, the strategic importance of research implementation, monitoring, and stress mitigation at less than a global level has been recognized (Taylor and Groom 1989; Malone 1991; Hey 1992). Achievement of UNCED goals will require the implementation of a new paradigm that will provide greater integration of the highly sectorized approach to solving the problems of coastal habitat degradation, marine pollution, and the overexploitation of fisheries than has been practiced by coastal nations during most of this century. It will also require a working partnership between the developed and developing nations of the world. Such an approach should be based on principles of ecology and sustainable development.

An ecologic framework that may be useful in achieving the UNCED objectives is the large marine ecosystem (LME) concept. Large marine ecosystems are areas that are being subjected to increasing stress from growing exploitation of fish and other renewable resources, coastal zone damage, habitat losses, river basin runoff, dumping of urban wastes, and fallout of aerosol contaminants. The Northeast U.S. shelf ecosystem is being subjected to these stressors. Mitigation of these stressors will require the collaboration of national agencies working in close cooperation with regional and local governments and other stakeholders in an effort to ensure the long-term sustainability of the resources and economies at risk because of a lack of an overarching governance regime.

Application of the Large Marine Ecosystem Approach to Resource Management

The empirical and theoretical aspects of yield models for LMEs have been reviewed by several ecologists. According to Beddington (1986), Daan (1986), Levin (1990), and Mangel (1991), the dynamic models of marine ecosystems published thus far offer little guidance on the detailed behavior of communities. However, these authors concur on the need for covering the common ground between observation and theory by implementing monitoring efforts on the large spatial and long temporal (decadal) scales of "key" components of the LMEs. The sequence for improving understanding of the possible mechanisms underlying observed patterns in LMEs as described by Levin (1990) is (1) examination of statistical analyses of observed distributional patterns of physical and biological variables; (2) construction of competing models of variability and patchiness based on statistical analyses and natural scales of variability of critical processes; (3) evaluation of competing models through experimental and theoretical studies of component systems; and (4) integration of validated component models to provide predictive models for population dynamics and redistribution. The approach suggested by Levin is consistent with the observation by Mangel (1991) that empirical support for the currently used models of LMEs is relatively weak and that a new generation of models that enhance the linkage between theory and empirical results is needed.

There is a growing awareness among marine scientists, geographers, economists, government representatives, and lawyers of the utility of a more holistic ecosystem approach to resource management (Byrne 1986; Christy 1986; Alexander 1989; Belsky 1989; Crawford et al. 1989; Morgan 1989; Prescott 1989). On a global scale, the loss of sustained biomass yields from LMEs as a result of mismanagement and overexploitation has not been fully investigated but is likely very large (Gulland 1984). Effective management strategies for LMEs will be contingent on the identification of the major forces causing large-scale changes in biomass yields. Management of species responding to strong environmental signals will be enhanced by improving the understanding of the physical factors forcing biologic change, thereby enhancing forecasts of El Niño-type events. In other LMEs, where the driving force is predation, options can be explored for implementing adaptive management strategies (Collie 1991). Remedial actions are required to ensure that "pollution" of the coastal zones of LMEs is reduced and does not become a driving force in any LME.

A distinguishing feature of the LME approach is its emphasis on practical applications of its scientific findings in managing the LME and on the explicit integration of economic analysis with scientific research to ensure that prospective management measures are cost-effective. Economists and policy analysts will need to work closely with ecologists and other scientists to identify and evaluate management options that are both scientifically credible and economically practical.

The topics of change and persistence in marine communities and the need for multispecies and ecosystem perspectives in the sustainability of living marine resources were reviewed at the Dahlem Conference on Exploitation of Marine Communities in May 1984. Significant multispecies ecosystem-oriented studies in support of the management of marine resources have not been practiced until recently. In 1991, the Advisory Committee on Fisheries Management of the International Council for the Exploration of the Sea (ICES) reached an important milestone by including guidance for providing "the advice necessary to maintain viable fisheries within sustainable ecosystems." The designation and management of LMEs is, at present, an evolving scientific and geopolitical process (Morgan 1988; Alexander 1989; AAAS 1993). Sufficient progress has been made to allow for useful comparisons among different processes influencing large-scale changes in the biomass yields of LMEs (Bax and Laevastu 1990; Bakun 1993). Among the more recent findings is the marked decline in biodiversity among the principal fish stocks of the Black Sea, based in part on the introduction of a "foreign" species of ctenophore that has undergone a population explosion, resulting in severe predation on the early life-stages of both fish and their prey (Mee 1992). The influence of changes in the gene pool of wild stocks from inadvertent releases of cultured stocks is another concern of scientists engaged in LME studies.

Among the ecosystems being managed from a more holistic perspective are the Yellow Sea Ecosystem, where the People's Republic of China is providing the principal effort (Tang 1989); the multispecies fisheries of the Benguela Current Ecosystem, under the management of the government of South Africa (Crawford et al. 1989); the Great Barrier Reef Ecosystem (Bradbury and Mundy 1989; Kelleher 1993) and the Northwest Australian Continental Shelf Ecosystem (Sainsbury 1988), under management of the state and federal governments of Australia; and the Antarctic Marine Ecosystem, under management of the Commission for the Conservation of Antarctic Marine Living Resources (CCAMLR) and its 21-nation membership (Scully et al. 1986; Sherman and Ryan 1988). Movement toward ecosystems management is emerging for the North Sea (NSTF 1991), the Barents Sea (Eikeland 1992), and the Black Sea (Hey and Mee 1993). Within the EEZ of the United States, the state governments of Washington and Oregon have developed a comprehensive plan for the management of marine resources within the Northern California Current Ecosystem (Bottom et al. 1989).

Socioeconomic Considerations

Economic and management research will be closely integrated with science throughout and is intentionally designed to respond adaptively to enhanced scientific information. This component of the LME approach to marine resources management was developed by James Broadus, former Director of the Marine Policy Center, Woods Hole Oceanographic Institution. It consists of six interrelated elements.

Human Forcing Functions

The natural starting point is a generalized characterization of the ways in which human activities affect the natural marine system and the expected sensitivity of these forcing functions to various types and levels of human activity. Population dynamics, coastal development, and land use practices in the system's drainage basin are clear examples. Work integrating the efforts of natural and social scientists should concentrate further on resolving apparent effects (such as eutrophication-associated red tide events or changing fish population structures) that are confounded by cycles or complex dynamics in the natural system itself. Progress also is possible in achieving better characterizations of the way in which human forcing is mediated by alternate management options. Emphasis should be on isolating and quantifying those forcing activities (sewage discharge, agricultural runoff, fishing effort) likely to be expressed most prominently in effects on the natural system.

Assessing Impacts

Another natural element in the systemic approach is to estimate and even predict the economic impacts of unmanaged degradation in the natural system and, obversely, the expected benefits of management measures. Such assessment is a form of standard benefit–cost analysis, but it requires scientific information to describe the effects of human forcing so they may be quantified in economic terms. Initial analysis should focus on the social and economic sectors likely to experience the largest impacts: fishing, aquaculture, public health, recreation, and tourism.

Feedbacks

Collaborative effort should also be devoted to identifying and estimating the feedbacks of economic impacts into the human forcing function. Extensive coastal eutrophication, for example, associated with coastal development and runoff, might reduce the suitability of coastal areas for aquaculture production and increase its exposure to red tide damage, thereby putting a premium on capture fishery and increasing pressure on wild stocks. Similar feedbacks, both negative and positive, will be addressed and expressed in economic terms for all the major sectors.

Ecosystem Service and the Value of Biodiversity

Special consideration should be given to improved knowledge of how the natural system generates economic values. Many valuable services provided by natural systems are not traded in markets or included in planning evaluations, so extra care must be taken to ensure that they are not sacrificed through ignorance. The services provided by coastal wetlands as nurseries for fisheries, natural pollution filters, and storm buffers is a well-known example that has particular relevance to coastal reclamation activities. Other examples are more subtle, including the importance of predator–prey relationships and the possibility of losing unrecognized "keystone" species in a valuable ecosystem. Experience suggests that the growing economic value of aesthetic and recreational/tourism amenities may be expected in the LME setting as well. A variety of sources of economic value arising from the natural diversity of the LME should be identified and assessed with regard to existing uses and potential management innovations.

Environmental Economics

Many of the elements described in this section constitute topics in "environmental economics." Specialists in that field attempt to estimate the economic values (both use and non-use) associated with environmental resources and to identify the conditions associated with their optimal management (to derive the greatest net benefits for society). An important element is the collaboration between scholars from developing nations and those from the developed countries to transfer and adapt to the needs and techniques of environmental economics.

Integrated Assessment

The ultimate objective is the integration of all the results achieved above with scientific characterizations of the LME into a comprehensive analytic framework (decision-support environment) that will permit integrated assessment of human practices, effects, and management options in the region. Such work is at the forefront of recent research on the human dimensions of global environmental change as well as research on human interactions with natural coastal/marine systems.

Future Applications of the Large Marine Ecosystem Approach

The broad-spectrum approach to LME research and monitoring provides a conceptual framework for collaboration in process-oriented studies conducted by the National Science Foundation (NSF)/NOAA-sponsored Global Ocean Ecosystems Dynamics (GLOBEC 1991) Program in the United States and the International GLOBEC Program. Developing LME monitoring strategies are compatible with the proposed Global Ocean Observing System (GOOS) of the IOC (1992) and those modules to be focused on living marine resources and ecosystem health (IOC 1993).

Efforts are under way to place greater emphasis on the linkage between scientific and societal needs and the utility of long-term, broad-area coastal ocean assessment and monitoring studies aimed at enhancing the long-term sustainability of marine resources. If the proposition for time–series monitoring of changing ecosystem states is to be realized in this period of shrinking budgets, it would be in the best interests of science and socioeconomic entities to be tightly linked in the endeavor. The basis for the linkage was emphasized not only in the UNCED declarations on the oceans, but also in a series of recent developments revolving around (1) global climate change; (2) legal precedent for international cooperation implicit in the Law of the Sea; (3) a growing interest in marine ecosystems as regional units for marine research, monitoring, and management; (4) the effort of the IOC to encourage the implementation of a GOOS; and (5) renewed national interests in improving the health of degraded coastal ecosystems. In the United States, this interest has resulted in the enactment of legislation mandating the establishment of a national coastal monitoring program for assessing the changing states of "coastal ecosystem health" and reporting the findings to the U.S. Congress as a recurring biannual responsibility of NOAA and the EPA (NCMA 1992).

A more holistic approach to coastal ecosystems assessment and monitoring as a means for fostering international cooperation in achieving sustainability objectives for marine resources between the more developed and less developed countries is presently under way. The 49 LMEs that have been identified for comparative sustainability studies are located around the margins of the ocean basins and extend over the coastlines of several countries. They are in regions of the world ocean most affected by overexploitation, pollution, and habitat degradation, and collectively represent target areas for mitigation efforts, particularly in the stressed coastal ecosystems adjacent to dense centers of population in developing countries. The Global Environment Facility (GEF) of the World Bank, in collaboration with NOAA, IOC, UNEP, FAO, the Natural Environment Research Council (NERC), the Sir Alister Hardy Foundation for Ocean Science, and scientists from national marine resource agencies of several of the more developed countries (e.g., Belgium, Canada, Denmark, France, Germany, the Netherlands, Norway, and the United Kingdom) are prepared to assist developing nations in implementing coastal ecosystem assessment, monitoring, and mitigation programs aimed at providing a scientific basis for improving prospects for the long-term sustainable development of marine resources (Sherman et al. 1992). Two of these programs are in the advanced planning stage. One, for the Gulf of Guinea Ecosystem, brings together into a single program five nations of the region—Ivory Coast, Ghana, Nigeria, Benin, and Cameroon. The other program is being developed jointly by marine specialists from China and Korea for the Yellow Sea

Ecosystem (Wu and Qiu 1993). The first of these projects was implemented in the Gulf of Guinea LME in the summer of 1995.

A comprehensive regional project to assess, monitor, and mitigate stresses on the Black Sea Ecosystem is being supported by the GEF (Mee 1992). It appears that marine resource managers and scientists are being responsive to management needs by reversing the cancellation of monitoring programs described recently by Duarte et al. (1992), and that monitoring initiatives related to the support of a more comprehensive systems approach to ensuring the long-term sustainability of marine resources are likely to be under way once again in marine ecosystems of Europe, North America, Africa, and Asia.

The growing partnership among funding agencies, marine ecologists, and socioeconomic interests marks an important step toward realization of the UNCED declaration aimed at reversing the declining condition of coastal ecosystems and enhancing the long-term sustainability of marine resources.

References

Alexander, L. M. 1989. Large Marine Ecosystems as Global Management Units. In American Association for the Advancement of Science (AAAS), Biomass Yields and Geography of Large Marine Ecosystems. AAAS Selected Symposium 111. Boulder, CO: Westview Press, pp. 339–344.

American Association for the Advancement of Science (AAAS). 1986. Variability and Management of Large Marine Ecosystems. AAAS Selected Symposium 99. Boulder, CO: Westview Press.

American Association for the Advancement of Science (AAAS). 1989. Biomass Yields and Geography of Large Marine Ecosystems. AAAS Selected Symposium 111. Boulder, CO: Westview Press.

American Association for the Advancement of Science (AAAS). 1990. Large Marine Ecosystems: Patterns, Processes, and Yields. Washington, DC: AAAS Press.

American Association for the Advancement of Science (AAAS). 1991. Food Chains, Yields, Models, and Management of Large Marine Ecosystems. Boulder, CO: Westview Press.

American Association for the Advancement of Science (AAAS). 1993. Large Marine Ecosystems: Stress, Mitigation, and Sustainability. Washington, DC: AAAS Press.

Bakun, A. 1993. The California Current, Benguela Current, and Southwestern Atlantic Shelf Ecosystems: A Comparative Approach to Identifying Factors Regulating Biomass Yields. In Large Marine Ecosystems: Stress, Mitigation, and Sustainability. Washington, DC: AAAS Press, pp. 199–221.

Bax, N. J., and Laevastu, T. 1990. Biomass Potential of Large Marine Ecosystems: A Systems Approach. In Large Marine Ecosystems: Patterns, Processes, and Yields. Washington, DC: AAAS Press, pp. 188–205.

Beddington, J. R. 1986. Shifts in Resource Populations in Large Marine Ecosystems. In Variability and Management of Large Marine Ecosystems. AAAS Selected Symposium 99. Boulder, CO: Westview Press, pp. 9–18.

Belsky, M. H. 1986. Legal Constraints and Options for Total Ecosystem Management of Large Marine Ecosystems. In Variability and Management of Large Marine Ecosystems. AAAS Selected Symposium 99. Boulder, CO: Westview Press, pp. 241–261.

Belsky, M. H. 1989. The ecosystem model mandate for a comprehensive United States ocean policy and Law of the Sea. San Diego L. Rev. 26;3:417–495.

Bottom, D. L., Jones, K. K., Rodgers, J. D., and Brown, R. F. 1989. Management of living resources: A research plan for the Washington and Oregon continental margin. NCRI-T-89-004. Newport, OR: National Coastal Resources Research Development Institute.

Bradbury, R. H., and Mundy, C. N. 1989. Large-Scale Shifts in Biomass of the Great Barrier Reef Ecosystem. In K. Sherman and L. M. Alexander (eds.), Biomass Yields and Geography of Large Marine Ecosystems. AAAS Selected Symposium 111. Boulder, CO: Westview Press, pp. 143–167.

Byrne, J. 1986. Large Marine Ecosystems and the Future of Ocean Studies. In K. Sherman and L. M. Alexander (eds.), Variability and Management of Large Marine Ecosystems. AAAS Selected Symposium 99. Boulder, CO: Westview Press, pp. 299–308.

Christy, F. T. Jr. 1986. Can Large Marine Ecosystems Be Managed for Optimum Yields? In Variability and Management of Large Marine Ecosystems. AAAS Selected Symposium 99. Boulder, CO: Westview Press, pp. 263–267.

Collie, J. S. 1991. Adaptive Strategies for Management of Fisheries Resources in Large Marine Ecosystems. In Food Chains, Yields, Models, and Management of Large Marine Ecosystems. Boulder, CO: Westview Press, pp. 225–242.

Crawford, R. J. M., Shannon, L. V., and Shelton, P. A. 1989. Characteristics and Management of the Benguela as a Large Marine Ecosystem. In Biomass Yields and Geography of Large Marine Ecosystems. AAAS Selected Symposium 111. Boulder, CO: Westview Press, pp. 169–219.

Daan, N. 1986. Results of Recent Time-Series Observations for Monitoring Trends in Large Marine Ecosystems with a Focus on the North Sea. In Variability and Management of Large Marine Ecosystems. AAAS Selected Symposium 99. Boulder, CO: Westview Press, pp. 145–174.

Duarte, C. M., Cebrian, J., and Marba, N. 1992. Uncertainty of detecting sea change. Nature 356:190.

Eikeland, P. O. 1992. Multispecies Management of the Barents Sea Large Marine Ecosystem: A Framework for Discussing Future Challenges. Lysaker, Norway: The Fridtjof Nansen Institute.

Food and Agriculture Organization of the United Nations (FAO). 1992. FAO Yearbook of Fishery Statistics, Vol. 70 (for 1990). Rome: FAO.

Global Ocean Ecosystems Dynamics (GLOBEC). 1991. Report Number 1. Initial Science Plan. February 1991. Washington, DC: Joint Oceanographic Institutions Inc.

Gulland, J. A. 1984. Epilogue. In R. M. May (ed.), Exploitation of Marine Communities. Berlin: Springer-Verlag, pp. 335–337.

Hey, E. 1992. A healthy North Sea ecosystem and a healthy North Sea fishery: Two sides of the same regulation? Ocean Develop. Int. Law 23:217–238.

Hey, E., Mee, L. D. 1993. Black Sea—The ministerial declaration: An important step. Environ. Pol. Law 2315:215–217; 235–236.

Holling, C. S. 1973. Resilience and Stability of Ecological Systems. Vancouver: Institute of Resource Ecology, University of British Columbia.

Holling, C. S. 1986. The Resilience of Terrestrial Ecosystems: Local Surprise and Global Change. In W. C. Clark and R. E. Munn (eds.), Sustainable Development of the Biosphere. London: Cambridge University Press, pp. 292–317.

Holling, C. S. 1993. Investing in research for sustainability. Ecol. Applic. 3:552–555.

Intergovernmental Oceanic Commission (IOC). 1992. The Use of the Large Marine Ecosystem Concept in the Global Ocean Observing System (GOOS). Presented at the Twenty-fifth Session of the IOC Executive Council, Paris, 10–18 March 1992. IOC/EC-XXV/Inf.7.

Intergovernmental Oceanic Commission (IOC). 1993. Report of the IOC Blue Ribbon Panel for a Global Ocean Observing System (GOOS): The Case for GOOS. IOC/INF-915 Corr. Paris, 23 February 1993. SC-93/WS3.

International Geosphere–Biosphere Programme (IGBP). 1994. IGBP in Action: Work Plan 1994–1998. Global Change Report No. 28. The International Geosphere–Biosphere Programme (IGBP): A Study of Global Change of the International Council of Scientific Unions (ICSU). Stockholm: IGBP.

International Union for Conservation of Nature and Natural Resources (IUCN). 1990. Caring for the World: A Strategy for Sustainability. Second Draft, June 1990. Gland, Switzerland: World Conservation Union (IUCN), the United Nations Environment Programme (UNEP), and the World Wide Fund for Nature (WWF).

Kelleher, G. 1993. Sustainable Development of the Great Barrier Reef as a Large Marine Ecosystem. In Large Marine Ecosystems: Stress, Mitigation, and Sustainability. Washington, DC: AAAS Press, pp. 272–279.

Levin, S. A. 1990. Physical and Biological Scales and Modelling of Predator–Prey Interactions in Large Marine Ecosystems. In Large Marine Ecosystems: Patterns, Processes, and Yields. Washington, DC: AAAS Press, pp. 179–187.

Levin, S. A. 1993. Approaches to Forecasting Biomass Yields in Large Marine Ecosystems. In Large Marine Ecosystems: Stress, Mitigation, and Sustainability. Washington, DC: AAAS Press, pp. 36–39.

Lubchenco, J., Olson, A. M., Brubaker, L. B., et al. 1991. The sustainable biosphere initiative: An ecological research agenda. Ecology 72:371–412.

Ludwig, D., Hilborn, R., and Walters, C. 1993. Uncertainty, resource exploitation, and conservation: Lessons from history. Science 260:17–36.

Malone, T. C. 1991. River Flow, Phytoplankton Production and Oxygen Depletion in Chesapeake Bay. In R. V. Tyson and T. H. Pearson (eds.), Modern and Ancient Continental Shelf Anoxia. Geological Society Spec. Publ. No. 58. pp. 83–93.

Mangel, M. 1991. Empirical and Theoretical Aspects of Fisheries Yield Models for Large Marine Ecosystems. In Food Chains, Yields, Models, and Management of Large Marine Ecosystems. Boulder, CO: Westview Press, pp. 243–261.

Mangel, M., Hofman, R. J., Norse, E. A., Twiss, J. R. Jr. 1993. Sustainability and ecological research. Ecol. Appl. 3(4):573–575.

Mee, L. 1992. The Black Sea in crisis: A need for concerted international action. Ambio 21(4):1278–1286.

Morgan, J. R. 1988. Large marine ecosystems: An emerging concept of regional management. Environment 29(10):4–9; 26–34.

Morgan, J. R. 1989. Large Marine Ecosystems in the Pacific Ocean. In Biomass Yields and Geography of Large Marine Ecosystems. AAAS Selected Symposium 111. Boulder, CO: Westview Press, pp. 377–394.

Myers, N. 1990. Working towards one world. [Book review.] Nature 344(6266):499–500.

National Coastal Monitoring Act (NCMA). 1992. 102nd Congress, 2nd Session, H.R. 2130 (29 October 1992). The National Oceanic and Atmospheric Administration Authorization Act of 1992.

National Research Council (NRC). 1990. Managing Troubled Waters: The Role of Marine Environmental Monitoring. Washington, DC: National Academy Press.

North Sea Task Force (NSTF). 1991. Scientific Activities in the Framework of the North Sea Task Force. North Sea Environment Report No. 4. London: North Sea Task Force, Oslo and Paris Commissions, International Council for the Exploration of the Sea.

Pimm, S. L. 1984. The complexity and stability of ecosystems. Nature 307:321–326.

Prescott, J. R. V. 1989. The Political Division of Large Marine Ecosystems in the Atlantic Ocean and Some Associated Seas. In Biomass Yields and Geography of Large Marine Ecosystems. AAAS Selected Symposium 111. Boulder, CO: Westview Press, pp. 395–442.

Rosenberg, A. A., Fogarty, M. J., Sissenwine, M. P., et al. 1993. Achieving sustainable use of renewable resources. Science 262:828–829.

Sainsbury, K. J. 1988. The Ecological Basis of Multispecies Fisheries, and Management of a Demersal Fishery in Tropical Australia. In J. A. Gulland (ed.), Fish Population Dynamics (2nd ed.). New York: John Wiley & Sons, pp. 349–382.

Scully, R. T. 1993. Convention on the Conservation of Antarctic Marine Living Resources. In Large Marine Ecosystems: Stress, Mitigation, and Sustainability. Washington, DC: AAAS Press, pp. 242–251.

Scully, R. T., Brown, W. Y., and Manheim, B. S. 1986. The Convention for the Conservation of Antarctic Marine Living Resources: A Model for Large Marine Ecosystem Management. In Variability and Management of Large Marine Ecosystems. AAAS Selected Symposium 99. Boulder, CO: Westview Press, pp. 281–286.

Sherman, K. 1994. Sustainability, biomass yields, and health of coastal ecosystems: An ecological perspective. Mar. Ecol. Prog. Ser. 112:277–301.

Sherman, K., and Ryan, A. F. 1988. Antarctic marine living resources. Oceanus 31(2):59–63.

Sherman, K., Jaworski, N., and Smayda, T. 1992. The Northeast Shelf Ecosystem: Stress, Mitigation, and Sustainability, 12–15 August 1991, Symposium Summary. NOAA Tech. Mem. NMFS-F/NEC-94. Washington, DC: U.S. Department of Commerce.

Skjoldal, H. R., Noji, T. T., Giske, J., et al. 1993. Mare Cognitum. Science Plan for Research on Marine Ecology of the Nordic Seas (Greenland, Norwegian, Iceland Seas) 1993–2000. A regional GLOBEC program with contributions also to WOCE and JGOFS. Bergen, Norway: Institute of Marine Research.

Tang, Q. 1989. Changes in the Biomass of the Yellow Sea Ecosystems. In Biomass Yields and Geography of Large Marine Ecosystems. AAAS Selected Symposium 111. Boulder, CO: Westview Press, pp. 7–35.

Taylor, P., and Groom, A. J. R. (eds.). 1989. Global Issues in the United Nation's Framework. London: Macmillan.

Wu, B., and Qiu, J. 1993. Yellow Sea fisheries: From single and multi-species management towards ecosystems management. J. Oceanogr. Huanghai and Bohai Seas 11(1):13–17.

2

The Northeast Shelf Ecosystem: Stress, Mitigation, and Sustainability Symposium —Keynote Address

John A. Knauss

I am delighted to be here today to stir up discussion about large marine ecosystems (LMEs). The discussion of LMEs, I believe, merits further stirring, which is what this symposium is all about. And I believe LMEs may be ready for more than discussion. Maybe the time has come to think about implementation. My intention today is to help generate some discussion and to get you thinking about implementation.

The topic for this symposium, "The Northeast Shelf Ecosystem: Stress, Mitigation, and Sustainability" is a large and timely subject. We know that our environment, at least our coastal environment, is stressed, and that in spite of our best efforts, that stress is likely to increase. Certainly, mitigation is a concept we are trying hard to apply in many areas, more successfully in some applications than in others, in large part depending on how well we understand the situation.

Sustainability of natural resources as an achievable goal is likely just over the horizon, just beyond our view. Nevertheless, increasing numbers of us are convinced that we must set our course now for sustainability, because there are most assuredly other, far less palatable, options lurking over the horizon as well. We are a population of an incredible five billion people and growing, almost all with boundless aspirations for increasing our standard of living. We have been operating under the paradigm that growth is good and necessary—the faster the better. In doing so, we have tended to function with a view limited to short-term gratification, while ducking the growing pressure to consider long-term sustainability. I would venture to presume that most of you here share with me the uneasy sense that human-induced environmental stresses have the potential to reduce the long-term carrying capacities of the very systems that support us.

We are here to explore our role with respect to the health of the ecosystems. I am not so much concerned with the distant open ocean, the present health of which I believe is relatively good, if only because of its enormous volume and capacity for dilution. It is the systems on the edges of the world's oceans about which I am concerned and about that I suspect most of you are concerned, for they are experiencing the lion's share of our disruptive impacts—our consumptive appetites, our wastes, and our penchant for physically altering the land–sea interface in the name of economic and social comfort.

Many of us wonder what is the most intelligent way to approach the question of the health of our coastal oceans and the role we might play in helping to maintain it. According to *Webster's Dictionary*, health is defined as that "condition of being sound in body, mind, or spirit." That sounds pretty good. In the field of medicine, we have some experience dealing with health which may provide some useful ideas. Medicine is defined by *Webster's* as "the science and art of dealing with the maintenance of health and the prevention, alleviation, or cure of disease." Pushing that analogy a bit further, perhaps we can equate medicine for the oceans with good stewardship of the oceans and their resources.

Imagine that an unconscious patient is wheeled into an emergency room unable to communicate what is wrong. The first actions taken by nurses, technicians, and physicians involve hooking up a multitude of monitors and taking dozens of samples of every body fluid imaginable. They measure heart rate, blood pressure, temperature, blood oxygen levels, brain waves, a whole suite of biochemical indicators related to the functioning of specific organ systems. They test for bacteria, viruses, and toxic substances. They also solicit anecdotal information regarding the activities and functioning of the patient from those familiar with the patient. They attempt to obtain a family history as well in an effort to apprise themselves of the patient's normal idiosyncrasies or unique hereditary vulnerabilities.

Armed with information gleaned from this process, the physician is able to respond with what an ecologist must view as an enviable degree of confidence. In most instances in which a human body has received some sort of insult, the physician can go about placing the patient on the path to health. It is the physician's intensive monitoring that supplies the information making it possible for him or her to diagnose which problem, or array of problems, require immediate attention. It is this knowledge that enables the physician to fulfill his or her role of furnishing remedial influence and promoting health. This is a role we would like to play with regard to promoting the health of our coastal oceans, and exerting a remedial influence where and whenever necessary. The difficulty for us lies in the fact that we do not have some of the critical information we need to do this properly.

What the physician has that we do not have is a good understanding of what is normal. Decades, even centuries, of experience have provided physicians with a highly reliable understanding of normal parameters, including the parameters of normal variation. We simply do not have a comparable knowledge base for ocean ecosystems. Many times we do not know what is normal. Just as important, we seldom know the range of normal variability. This makes it extremely difficult for us to be the confident and accurate diagnosticians we would like to be.

Fifteen or 20 years ago we experienced a scare about the mercury content in the flesh of swordfish. Our initial assumption was that these levels were abnormal and were the result of human introduction of mercury into the oceans. As it turned out, it was a question of our not knowing that swordfish normally concentrate the mercury that occurs naturally in the oceans. We made a false-positive diagnosis because we lacked sufficient understanding of the normal parameters. If it is dangerous to eat swordfish, then it has probably always been dangerous to eat swordfish. It is simply that previously we had never gotten around to measuring the mercury content of swordfish flesh.

Another example of our inability to distinguish between natural and unnatural, or anthropogenic, processes in our coastal oceans lies in the increased incidence of paralytic shellfish poisoning that has intermittently shut down the clam industry in the Georges Bank area. Are the responsible blooms of toxic dinoflagellates caused by natural ocean variability such as changing current patterns, or are they the result of nearshore to offshore movement of nutrients introduced from anthropogenic sources? We simply do not know. We need to know these things. Until we can distinguish between that which is natural, or normal—whatever that means—and perturbations that are anthropogenic, we can hardly fulfill the roles of diagnosticians, suppliers of remediation, or stewards of ocean health.

I am particularly concerned about two important ways in which we differ from our counterparts in the field of human medicine. The first is the question of time scales. The physician appropriately thinks in terms of a human lifetime. We know that human lives are finite, and we have a very good idea of what a normal life span is likely to be. Most individuals are monitored and assessed periodically during their lives, beginning when they are born, thereby providing their physicians with an excellent basis on which to judge changes in health. This is to say that they have a time-series of data that is virtually complete.

But over what time scale do we need to monitor our coastal oceans? Decadal scales at a bare minimum. Our efforts to determine whether changes we observe over a few years are the result of pollution, overfishing, or natural environmental changes require longer-term time-series monitoring than we presently have accomplished—very likely a good deal longer than the average life span of a person.

A second critical difference between us and our colleagues in medicine is that for physicians, the choice of a convenient and appropriate unit of study is obvious—the human body. Things are not so simple for those of us trying to get a better grasp of ocean systems. We have a long way to go before we in our field of study have in our black bags the wealth of background knowledge and the understanding of the normal state and its natural variability that physicians have in theirs. But while we may not have done a very good job of observing ocean variability thus far, we are not completely ignorant. More important, I believe we have a much better understanding of what we need to know and are busily formulating strategies that will give us that knowledge.

Let me be very clear, however, about the limitations of monitoring. Monitoring is no substitute for understanding the physical and biochemical processes that occur in a human body or in an ecosystem. Monitoring does, however, alert you to possible problems, and long time-series monitoring can often put some bounds on the changes that are likely to be within the normal range of an ecosystem and those that are beyond.

That brings me to the subject of large marine ecosystems. The development and application of the concept of LMEs represents one strategy for monitoring and understanding the health of the coastal oceans. A point I continue to make to those concerned about the health of the world ocean is that we need to concentrate on the ocean edges, the coastal oceans. The effects of humankind on the ocean will first and most intensively be seen along the coasts and in the near offshore. To the extent that the coastal oceans are healthy, I believe we can be relatively sanguine about the health of the vast central ocean regions.

Distinctive physical, chemical, and oceanographic features; productivity; and community trophodynamics characterize a large marine ecosystem. The LME gives us a well-defined regional unit for research, monitoring, and management, allowing us to focus on the health of entire marine ecosystems. This is a critical first step. We in government, both state and federal, have much to answer for. Traditionally, coastal zones and their resources have been studied and managed by a wide range of single-function agencies and institutions concerned with fisheries, or transportation, or conservation, or water quality, or waste disposal, or recreation, or minerals management and development, and more. This practice of working independently, within agency boundaries, has led and can lead to significant progress, but it is often an inefficient approach for addressing the interrelated, multidisciplinary issues facing our coastal oceans.

I believe the LME concept has much to offer in this respect. LMEs are relatively large areas of 200,000 square kilometers or more and are typically located in waters adjacent to land masses, therefore encompassing the areas under greatest stress from overexploitation, pollution, and habitat alteration. Taking an ecosystem approach highlights the interrelatedness of the different variables of each system and encourages cooperative dialogues across traditional disciplinary boundaries. I believe that this is not only a good idea, it is also essential if we are sincere in our desire to address this increasingly complex suite of coastal ocean issues—issues such as coastal zone management, pollution reduction, fisheries productivity and sustainability, and habitat protection. I do not want to suggest there is no room for the individual specialist any more than there is no room for the individual agency requirement. A LME approach to understanding and managing the coastal ocean is no panacea, but I do believe it can help. The problems here are seldom single-issue, single-answer problems. In this respect particularly, the holistic approach inherent in the LME concept encourages us in the right direction.

In an address at the Massachusetts Institute of Technology last fall, I made a proposition. (I had tried it out previously in Monaco with representatives from a number of different European countries. Some of you may have heard it already, but let me reiterate it here again.) If one set out to design a coastal ocean monitoring system to monitor the health of the ocean, are LMEs an appropriate geographic unit? If they are, would it be useful to organize a set of regional programs, each designed for a specific LME? Each nation, or set of nations, bordering on an LME would be responsible for the design and implementation of the program. The goal of the programs would be to monitor the system and understand how the system works, what its normal parameters are, and how humans are perturbing the system.

Those responsible for the program of each LME could meet locally on a regular basis. Perhaps every few years representatives from each region could come together internationally to compare notes and report on the health of all of the LMEs. By doing this, they would, in effect, be reporting on the health of the ocean.

I went on, in Monaco, to suggest that such a program could be organized through one, or some combination of, United Nations agencies and, further, that the development of such a program should be considered at the United Nations Conference on Environment and Development (UNCED) in Brazil. The LME approach lends itself to international application well. Focusing fisheries, pollution, and coastal zone studies on some 49 LMEs in which pertinent studies are already under way, and financed by national interests, will promote more coherent and effective national and U.N. ocean research and monitoring programs.

Today there is clear interest on the part of developing countries, as well as industrialized nations, to develop some form of coastal zone management to improve the prognosis for sustainable development of their coastal resources. LMEs provide an approach that is flexible enough to account for the fact that the economic values of the coastal zone within LMEs differ from nation to nation. In some areas, such as in the deltas of Egypt and Bangladesh, coastal agriculture is of primary importance. In Peru and Chile, fishing may be more important, and in other areas, such as in the Mediterranean, the Pacific Islands, or the Caribbean Islands, marine-related tourism is critically important. Regional LME management plans can be tailored to meet the multiple-use needs of the bordering nation or nations.

Around the world, management efforts have been initiated that embrace this ecosystem approach. These include the Yellow Sea ecosystem, where a principal effort is under way by China, the multispecies fisheries of the Benguela Current ecosystem under the management of the government of South Africa, the Great Barrier Reef ecosystem and the Northwest Australian Continental Shelf ecosystem under management of the state and federal governments of Australia, and the Antarctic Marine Ecosystem under management by the Commission for the Conservation of Antarctic Living Marine Resources and its 21-nation membership. By comparing results of research among the LMEs, it should be possible to accelerate an understanding of how the systems work and how they respond to, and recover from, stress. Such comparisons will allow us to narrow the context of unresolved problems and to capitalize on research efforts under way in different ecosystems. But the global perspective will be only as good as the regional and local efforts that go into studying and understanding each of the LMEs.

This gathering represents an effort to bring together those federal and state agencies, academic institutions, and individuals who are contributing to, or simply interested in, a better understanding of the Northeast Shelf ecosystem. I believe it is especially appropriate that this first symposium on the health of a particular LME is focusing on this particular LME. A great deal of work has focused on the Northeast Shelf ecosystem in the past, which is well-suited to the LME approach, particularly with respect to efforts to obtain long time-series monitoring data in terms of interdisciplinary and intergovernmental collaboration.

The scientists here at the Graduate School of Oceanography of the University of Rhode Island have some of the longest time-series monitoring studies for the coastal waters in the United States. Professor Ted Smayda has concluded from his analysis of a 25-year time-series monitoring of phytoplankton collections in Narragansett Bay that major changes have occurred that may be implicated in the greater frequency and extent of unusual plankton blooms. Professor Perry Jeffries has concluded from his analysis of a 25-year time-series monitoring of bottom fish collections that significant changes have taken place. The once-dominant flounder are being replaced by rock crab and other large benthic invertebrates. These are significant findings, and they underscore the importance of time-series monitoring efforts. Continuing such efforts will enable us to determine whether these changes are the result of pollution, overfishing, or natural environmental variability so that appropriate mitigating actions can be taken.

Other long time-series monitoring studies are available for analysis in other parts of the Northeast Shelf ecosystem as well, including Long Island Sound, Massachusetts Bay, and, more recently, the Gulf of Maine. The findings from these studies will lead to actions designed to reduce the input of pollutants into the Shelf ecosystem. One of the longest time-series monitoring studies of all is NOAA's National Marine Fisheries Service trawling survey for the Georges Bank. The data are being used to develop strategies for aiding in the recovery of depleted groundfish resources of the Northeast Shelf. And our NOAA National Marine Fisheries Service will be re-instituting the 11-year Marine Resources Monitoring Assessment and Prediction Program (MARMAP).

This is also a region in which there is a relatively long history of interdisciplinary and intergovernmental cooperation and collaboration in dealing with issues related to the coastal zone. The Sea Grant College Program was born here in Rhode Island some 25 years ago. The idea, initially suggested by Athelstan Spilhaus at the University of Minnesota, was picked up by those of us here at the University of Rhode Island. Congressional support was championed by Rhode Island's Senator Claiborne Pell, and the National Sea Grant College Act was signed into law by President Lyndon Johnson in 1966. From the very beginning, the Sea Grant Program has combined the best of what a university–federal partnership should be: research into problems, education about issues, and outreach to those affected and concerned. Sea Grant has contributed enormously to increasing our understanding of ocean and coastal environments over the past 25 years. I also believe Sea Grant can take considerable credit for ensuring that much of this understanding has been used for public benefit through Sea Grant's education and outreach programs. Sea Grant, and other groups such as Save the Bay and programs such as the Volunteer Salt Pond Watchers, provide models for other regions, lending insight into the benefits of establishing a wide circle of participation in addressing the issues facing our coasts.

Further evidence of coordination and cooperation are found in the Environmental Protection Agency's national estuary programs and NOAA's Coastal Zone Management program, where federal, state, and local governments as well as academic institutions, industry, and environmental interest groups, are brought together on technical and nontechnical advisory committees to help formulate regional and local management plans. NOAA and the EPA are also joining forces and combining NOAA's Status and Trends and its MARMAP time-series monitoring programs with the pollution monitoring efforts of EPA's new EMAP program.

What all this means to the Northeast Shelf ecosystem is that here we need not start from scratch in our efforts to embrace the LME approach. Much is already in place in terms of long-term monitoring and the breaking down of institutional boundaries in recognition of the benefits of collaboration. Adjacent to what has historically been, and continues to be, the most densely populated region of the United States, studies of this LME can provide a model for other regions that may not have had the same head-start.

Let me close with a few words about the urgency of this effort. We are all aware of recent examples of environmental changes that we believe are a consequence of ecological stress, such as decreasing water quality, declining fisheries stocks, and alarming losses of wetlands, coral reefs, or sea grasses—stress that we humans are imposing on our support systems.

The oceans are not constant. They have certainly changed over geologic time, as has the life within them. Slow changes allow for the necessary compliance by the systems and life forms affected. Mankind's changes have been virtually instantaneous in terms of geologic time. As NOAA's Chief Scientist, Sylvia Earle, is fond of saying, "We humans have hit the fast forward button, and risk the fate of bacteria in a culture dish, or locusts, who consume themselves out of business."

Let me return briefly to the medical analogy. Remember that physicians have a good idea of to what extent a system, such as the cardiopulmonary system, can compensate for a chronic or acute insult before reaching its limit and collapsing. Experience has taught them where the thresholds lie. This experience has come as the result of crossing such thresholds and of people dying.

An ecosystem analogy is an overfertilized pond or estuary. A small amount of extra phosphate or nitrates causes little harm. In many cases, it enhances the local productivity. But too much, and the system "dies." Or to be more explicit, that system dies and another one takes its place. Our knowledge of the tolerance level of marine ecosystems is very limited. The lessons to be learned from exceeding the analogous ecosystem thresholds are not pleasant to contemplate. The collapse of the systems that sustain us would have catastrophic consequences. It is time now to take stock, to determine what we have, how it works, how it does not work, and how to manage it for real, long-term health and benefits.

Considering the human resources represented here in this room, I believe that the prognosis can be a good one. I applaud you for being here, and I applaud those of you whose dedication and hard work have made it possible. Thank you.

3

Ecological Research for a Sustainable Northeast Shelf Ecosystem and Biosphere

Charles H. Peterson

Abstract

In 1991, the Ecological Society of America published recommendations for research priorities in the 1990s. The outcome of this two-year-long, grassroots planning process called for establishment of the Sustainable Biosphere Initiative (SBI). This research initiative contains three priority research foci: (1) global change, (2) biological diversity, and (3) sustainable ecological systems. For ocean ecosystems, the federal research program that is most responsive to this agenda is Global Ocean Ecosystems Dynamics (GLOBEC). The GLOBEC program is part of the U.S. Global Change Program, which is supported by multiple agencies and is part of an international GLOBEC program. It combines all three research priorities identified within the SBI to address how changing global climate might be expected to alter ecosystems dynamics and especially secondary production in the sea. The initial field system identified for study by the GLOBEC steering committee is the Northwest Atlantic, which is essentially equivalent to the Northeast Shelf ecosystem. This paper applies the principles extracted from the SBI and GLOBEC planning processes to the Northeast Shelf ecosystem. Specifically, the following lessons are included: (1) species do not exist and cannot be managed in isolation; (2) process-based ecosystem models incorporating trophic interactions and physical dynamics must be created as an antidote to the lack of replication of any specific large marine ecosystem (LME); (3) ecosystem models must include key human interactions of exploitation, nutrient and pollutant inputs, and climate change; (4) rigorous experimental and analytical methodologies of manipulative ecology need to be applied in "adaptive management"; (5) modelers must be charged with, and agencies made responsive to, the need for identifying precisely those state variables that require monitoring to test model predictions and assess ecosystem health; (6) enhanced attention to inclusion of prey dynamics, such as forage fish and zooplankton, is critical to fisheries management; and (7) studies and models must evaluate the roles of biodiversity and indirect and higher-order interactions (those other than direct predator–prey relations) in ocean ecosystems dynamics to allow tractable simplification of a diverse system.

The intent of this paper is to apply the general recommendations of the Ecological Society of America (ESA) resulting from a review of research priorities for the 1990s (Lubchenco et al. 1991) to the specific case of developing a research agenda for understanding and managing the Northeast Shelf ecosystem. The Northeast Shelf ecosystem is the coastal ecosystem stretching from the Gulf of Maine to Cape Hatteras (as defined by Sherman et al. 1988), a system whose important biological resources would doubtless benefit from ecologically informed management of the entire ecosystem as a large marine ecosystem, or LME (Sherman and Alexander 1986). I first describe briefly the major conclusions and research priorities put forward by the Ecological Society of America. I then demonstrate how the Initial Science Plan of the Global Oceans Ecosystems Dynamics (GLOBEC) program (as described in JOI 1991a), the most appropriate multiagency federal research initiative to respond in the marine environment to the appeal of the ESA, is indeed responsive to the ESA's guidance. Finally, I go beyond the previously published general documents to apply their broad recommendations to the development of some specific guidelines for research in and management of the Northeast Shelf ecosystem.

The Sustainable Biosphere Initiative

In response to a recognition that resources for scientific research are necessarily limited and that not all the excellent research now proposed by scientists can be funded, the ESA established in 1989 a broad-based committee of its members to review the discipline of ecology and to identify the priority areas for research emphasis in the 1990s. The committee met frequently and worked intensively for over a year to create a draft document describing priorities and an agenda for ecological research in the coming decade. The draft document was then subjected to the review and scrutiny of the entire society through presentation at the ESA annual meeting in 1990 and through solicitation of written reviews from many individuals in allied disciplines who were not members of the ESA. Throughout the process, community input and criticism were also sought through workshops and publication of announcements in the *ESA Bulletin* and newsletter. The input from numerous ecologists was ultimately incorporated into the final document, which was accepted and published by the ESA (Lubchenco et al. 1991). This document presents the blueprint for a Sustainable Biosphere Initiative, the result of a successful grassroots process of research prioritization carried out by practicing scientists in the discipline to guide the allocation of resources for research in the 1990s.

The Sustainable Biosphere Initiative (SBI) reflects a recognition that the human population and its influence have grown to a point where the sustainability of ecological systems cannot be assumed. Additional basic research is needed to enhance our understanding of ecosystems so that their important functions can be sustained through wise management. Public understanding of the mechanisms by which they function is critical to ensuring that ecological principles and knowledge are used appropriately in managing the biosphere, so the SBI calls for increased public education and the transfer of ecological information. Most important, the SBI contends that proper management of our natural resources demands that sound ecological knowledge be incorporated into management and policy (Risser et al. 1991).

The SBI is composed of three research priorities: global change, biological diversity, and sustainable ecological systems.

Global Change

The SBI strongly endorses the ongoing global change research program (U.S. Committee on Global Change 1990). Understanding the ecological causes and consequences of global change is critical to furthering an understanding of ecosystem processes at all scales (local and regional as well as global) and vital to wise development of national policy and informed management of ecosystems that deliver goods and services. Despite support for the presently formulated U.S. Global Change Program, the SBI also identifies one area of global change research that is not adequately addressed by current initiatives: the broad issue of how ecological conditions and processes control global processes. Quite simply, the composition, structure, and functioning of an ecosystem determine the biogeochemical fluxes, transfers, transformations, and storage of materials that occur within it, and thereby contribute to global element cycles and global change. For example, the degree of success in reducing atmospheric CO_2 through stimulation of phytoplankton production by application of iron, a limiting micronutrient, to the southern oceans depends on the extent to which any enhanced primary production is respired back into the atmosphere by penguins, whales, and other consumers in the Antarctic food webs (Huntley et al. 1991). The question of how geochemical cycling is altered by changes in ecological community composition and function needs to be the target of explicit research in the global change program.

Biological Diversity

Like the first research priority identified in the SBI, global change, the second priority, biological diversity, is also the focus of a large ongoing research program (Committee on International Science's Task Force on Global Biodiversity

1989). The SBI endorses these efforts, directed largely toward cataloguing the species composition of various habitats and protecting areas of high biological diversity. This effort, however, must be expanded in scope to address two additional issues. First, basic research on ecosystems is necessary to develop an understanding of how different patterns of biodiversity influence the functioning of ecological systems. For example, do simple ecosystems with short food chains (e.g., estuarine benthos) respond differently to interventions such as pollution stress than do more diverse ecosystems with longer food chains (e.g., pelagic water-column ecosystems)? Second, we need to develop greater fundamental appreciation for the ecological processes that influence biodiversity. Without this improved understanding of the relationships between ecosystem function and biodiversity, reliable strategies to manage and maintain the biological diversity required to support the biosphere cannot be developed.

Sustainable Ecological Systems

In contrast to the first two research priorities identified by the ESA's SBI, the third priority, sustainable ecological systems, is new, and not the focus of any direct integrated federal research program. The thrust of the recommended research effort is to develop the rigorous understanding of underlying ecological processes required for successful management and restoration of natural, exploited, and artificial ecosystems (e.g., timberlands, agricultural systems, aquaculture systems). There are, of course, numerous research activities designed to develop information required to manage particular exploited resources, such as fish stocks, crops, and timber. However, these efforts are fragmented, rather than united through the basic ecological problems held in common, and often inappropriately directed toward single resources without incorporating adequately the entire ecosystem context. Effective management must include an appreciation for function of the ecosystem in which the target resource is embedded. This task requires a large multidisciplinary effort and a strong foundation built on ecological principles to produce effective predictions and to avoid costly ad hoc repetition of efforts on every individual system. The development of this third priority of the SBI is crucial to sustain human activities, lifestyles, and societies.

Global Ocean Ecosystems Dynamics

Global Ocean Ecosystems Dynamics (GLOBEC) is a component research initiative of the U.S. Global Change Program. An international GLOBEC program has also been established. The core goal of this research initiative is to develop the necessary understanding of how physical processes affect, both directly and indirectly, marine ecosystems dynamics, so as to predict how global change might be expected to alter the structure, dynamics, and productivity of ocean ecosystems (JOI 1991a). The heaviest emphasis of the program is on secondary production, especially fisheries resources. A scientific steering committee for GLOBEC was named in the spring of 1989, after numerous workshops in the preceding years on marine fish ecology, zooplankton ecology, benthic ecology, physical-biological coupling, and recruitment dynamics. In the spring of 1991, the GLOBEC steering committee published its Initial Science Plan (JOI 1991a).

The U.S. GLOBEC Program is designed as a multiagency research initiative, with the National Science Foundation (NSF), the National Oceanic and Atmospheric Administration (NOAA), and the Office of Naval Research (ONR) providing most of the present support. Initial projects funded under the GLOBEC umbrella included modeling efforts supported jointly by ONR and NSF Ocean Sciences, retrospective analyses of particular ecosystems, biotechnology development, research on the effects of ultraviolet light on larvae, and preparatory planning for field programs in the Northeast Shelf ecosystem by the NOAA Marine Ecosystem Response (MER) program. A NSF panel met in August 1991 to recommend decisions on proposed projects submitted in response to a call for proposals to develop biotechnological tools for assessing the physiologic status of zooplankters and for mechanizing counting of taxa of zooplankters. In addition to these specific projects, relevant laboratories of the National Marine Fisheries Service of NOAA are using internal resources to promote GLOBEC field projects within their purview. The first large field program supported by GLOBEC has been an interdisciplinary study of ecosystems dynamics and physical forcing around the Georges Bank region of the Northeast Shelf, with a special, ultimate focus on cod and haddock production and dynamics. Despite this present level of GLOBEC activity within the United States, the GLOBEC-related projects abroad greatly exceed it, with Canadian, Norwegian, and French programs being especially active.

The strategy adopted by the scientific steering committee for GLOBEC is to precede major field programs with efforts to (1) produce appropriate models of relevant physical dynamics, population dynamics, and ecosystem processes that can serve to guide later field experiments; and (2) develop new instrumentation that will allow biological sampling in the sea to occur more rapidly and over larger areas such that it can more closely match the scales of physical data collection and contributions of physical processes. With physical-biological models in hand to guide hypothesis formation and with new, more powerful biological sampling tools available, the GLOBEC steering committee plans to mount major future field efforts in a small number of LMEs. These field programs will be designed to understand how physical processes, including those sensitive to change on local, regional, and global scales, contribute directly and indirectly to ecosystems dynamics.

The initial field system selected for intensive study and hypothesis testing by the GLOBEC steering committee is the Northwest Atlantic ecosystem (JOI 1991b), which is essentially equivalent to the Northeast Shelf ecosystem of Sherman et al. (1988). The intent is for efforts in the United States to focus on the Georges Bank area, while analogous Canadian efforts focus on other banks (e.g., Browns Bank and the Flemish Cap) and the more physically open Scotian Shelf, and European efforts consider the shelf ecosystems of the Northeast Atlantic. Comparison of these systems, physically different but with similar biota, can yield important insights into the processes controlling marine ecosystems dynamics and their universality. This particular LME was chosen for the initial intensive field program for several reasons. Because the food chains in this system are topped by gadoid fishes of commercial importance, cod and haddock, there is practical importance to understanding this system, and historical time-series data on key system variables are available. These data include not only fisheries catch statistics, but also a large amount of information on physical circulation, transport, and mixing processes. For the Georges Bank, there is clear evidence of overfishing and a consequent shift from a dominance of gadoid to relatively undesirable elasmobranch top predators (Clark and Brown 1977; Sissenwine 1986), implying the need for an ecosystems-oriented management approach. In addition, several lines of argument suggest that this region may be highly sensitive to global change, and therefore an appropriate choice for detection of a signal and testing hypotheses on ecosystem response. For example, general circulation models (GCMs) suggest that the North Atlantic is a region expected to experience a strong set of physical signals from global warming. Changing precipitation patterns and ice melt during warming would be expected to alter the buoyancy-driven currents that strongly affect the circulation patterns in this coastal, moderately high-latitude system. Finally, biogeographic changes in response to warming sea temperatures might be expected to alter the Georges Bank ecosystem dramatically, given the narrow latitudinal range of this shallow coastal feature and the present biogeographic boundaries.

The Initial Science Plan for GLOBEC (JOI 1991a) correlates remarkably well with the relevant appeals of the ESA for an SBI (Sustainable Biosphere Initiative in Lubchenco et al. 1991). All three priority research objectives of the SBI come together in the science plan for GLOBEC. First, GLOBEC is part of the U.S. Global Change Program, which has the responsibility for assessing how marine ecosystems are expected to be altered by various processes of global change. As such, it responds (for ocean ecosystems) to the SBI's call for research on how ecosystems dynamics will react to global change. Unless in partnership with the Joint Global Ocean Flux Study (JGOFS), however, GLOBEC will probably not have the resources to address the other key issue of how different ecosystems contribute differently to global materials budgets. Furthermore, although the long-range plans for the GLOBEC research program include an intention to assess different types of ocean ecosystems, it is unclear how much assessment will be permitted by ultimate funding levels. Second, the GLOBEC program necessarily and explicitly addresses important aspects of the biodiversity problem. In particular, the ecosystem models to be tested by GLOBEC must answer the question of the importance of species diversity to the functioning and dynamics of marine ecosystems because, as a practical matter, not all ecosystem variables can be continuously monitored. Thus, one must address, in up-front modeling, the importance of the rarer components of ecosystems in understanding and predicting systems behavior and, conversely, the degree to which the ecosystem can be conceptually simplified without loss of ability to predict its dynamics (JOI 1991c). Finally, the SBI's call for research on sustainable ecological systems is served explicitly by the GLOBEC intent to focus on exploited marine ecosystems. The need to further develop an ecosystems approach to model and understand the dynamics of important fisheries resources is a major motivation of GLOBEC. The novelty of incorporating the roles of physical dynamics, including variables responding to global change, into ecosystems models that themselves include species under exploitative harvest is a promising approach and perhaps the most important justification for support of GLOBEC. For ocean ecosystems, the science plan of GLOBEC depicts a process to bring together fisheries biologists, biological oceanographers, physical oceanographers, technologists, and ecosystems modelers to address explicitly the most serious void identified by the ESA in its call for an SBI, namely the need to develop the fundamental scientific basis for managing and sustaining ecological systems.

Application of the Sustainable Biosphere Initiative: Recommendations for the Northeast Shelf Ecosystem

1. *Do not model or manage species in isolation from their ecosystem context.* Although the classic quantitative modeling of exploitation effects on individual species populations (e.g., Ricker 1958) has led to much insight into

fisheries management and the impact of harvest on fish stocks, the ESA's SBI document would argue that modeling and management that ignore the effects of ecosystems dynamics are grossly incomplete and unlikely to be successful. This message came clearly from the Dahlem Conference on Exploitation of Marine Communities (May 1984) and has been embraced well in the concept of management at the level of the LME (Sherman and Alexander 1986; Sherman 1991). For the Northeast Shelf ecosystem, this approach has several implications. Specifically, it implies that codfish and haddock should not be managed outside the context of the ecosystem in which they are found. Moreover, since the elasmobranch fishes that have come to dominate the demersal predator trophic level in the wake of overexploitation of gadoid stocks on the Georges Bank share prey resources with, and even prey on, the juvenile life stages of the cod and haddock (Fogarty et al. 1987; Sherman et al. 1988), there are reasonable grounds to postulate that the recovery of the gadoid stocks may be strongly inhibited by interactions with elasmobranchs in the ecosystem. Basic research on the dynamics of the Northeast Shelf ecosystem is required to address this and other issues of how the gadoid dynamics are tied to other components of the ecosystem.

2. *Develop ecosystems models based on understanding of processes of population changes in component species.* Although much insight can be gained by comparisons of LMEs, each individual LME is essentially unique. Consequently, the ability to use replicated experiments and truly independent observations as the means of making rigorous inferences is limited. The only viable antidote to this problem of lack of replication is to develop an excellent understanding of the mechanisms by which populations change within the ecosystem. This mechanistic appreciation for process must encompass the complete range of relevant scales across which population processes, as well as controlling physical dynamics, operate (Steele 1988). The implications of this recommendation for the Northeast Shelf LME are numerous, including, notably, the need to incorporate physical dynamics directly and indirectly into models of ecosystems dynamics. Description of the circulation patterns of the currents from the Gulf of Maine through Georges Bank to the southern end of the Northeast Shelf LME, in combination with measurements of population processes, is required to establish the proper scales that define component populations, especially for the important zooplankton species such as *Calanus finmarchicus.* The importance of the seasonal stratification of the water column over Georges Bank and the timing of wind-driven mixing to the feeding success of zooplankters, as well as larval and juvenile fishes,

apparently needs to be included in models of population dynamics in this system. The ecosystem models developed to predict dynamics in this ecosystem must achieve an accurate balance of the potentially conflicting impacts of physical mixing, which may alter feeding success of juvenile gadoids while simultaneously changing exposure and risk to their own consumers. Logic might suggest that the significance of mortality through immediate predation would dominate over effects on growth, but slower growth implies a longer period of exposure to size-dependent predators in this water-column environment. A proper ecosystem dynamics model is required to answer such critical questions.

3. *Include anthropogenic influences explicitly in models of ecosystems dynamics.* Traditionally, academic models of ecosystems dynamics have tended to address natural systems in the absence of human influence, whereas models of the consequences of human activities, especially exploitation of resources, have tended to neglect the full range of important natural, physical, and biological variables. Sustaining important ecosystems demands that appropriate contributions of anthropogenic influences, in addition to the effects of natural processes, be explicitly included in our formalizations and studies of ecosystems dynamics. In general, the human influences that need consideration are harvest pressure (exploitation), nutrient and pollutant inputs, and signals of global change (warming, increases in ultraviolet radiation, changes in ocean circulation patterns, effects on upwelling intensity and turbulent mixing, CO_2 increase, etc.). For the Georges Bank region of the Northeast Shelf LME, it is evident that the impacts of exploitation must be included in ecosystem dynamics models, whereas nutrient loading and pollution seem comparatively unimportant except in the estuarine areas (Sherman 1991). Impacts of global change on this system are probably also subtle at this time but require incorporation to ensure that the goods and services provided by this important system will be sustained indefinitely into the future. At the southern margin of the Northeast Shelf LME, the consequences of nutrient loading are evident where the discharge of the Chesapeake Bay enters the continental shelf north of Cape Hatteras; for that region, ecosystem modeling should not ignore eutrophication effects. If these recommended analyses and models of the joint impact of anthropogenic influences and natural variables are successfully incorporated into an integrative understanding of ecosystem dynamics for this LME, one very significant outcome could be derived—namely, the mechanistic understanding of the consequences of by-catch from relatively nonselective trawling gear. This would address an urgent management need.

4. *Apply rigorous approaches of experimental ecology to "adaptive management" of exploited systems.* Fishing is a heavily regulated process. Such regulation can be treated as an ecosystem manipulation conceptually, such that the methodologies and analytical techniques of experimental ecology can be applied to test various impacts of fishing practices. These manipulations of fishing pressure rarely represent a controlled experiment because ecosystems differ sufficiently over space such that control sites are never identical to the sites where exploitation is being manipulated. Nevertheless, this use of regulation of exploitation of natural ecosystems is a powerful technique and can be cleverly used to enhance our understanding of how specific exploited ecosystems function, thereby leading to improved management (Sugihara et al. 1984).

   The design and use of regulation as an experiment has been termed "adaptive management" by Walters (1986). This approach is an excellent example of the application of ecology to the development of a management plan for a sustainable ecosystem, as urged by the ESA's SBI (Lubchenco et al. 1991). The National Research Council's panel on sea turtle conservation demonstrated how effectively and convincingly the consequences of manipulation of fishing regulations could be used to test an impact of fishing by correlating strandings of dead sea turtles in several localities with the precise period of shrimp trawling (NRC 1990). New methods for analysis of such environmental manipulations are continually being developed by ecologists and biostatisticians; a very significant recent example is the Before-After-Control-Impact (BACI) analysis of Stewart-Oaten et al. (1986), which helps deal with the problem of non-identity between treatment and control sites, as well as the lack of replication in many environmental "experiments."

   Application of the adaptive management approach to the Northeast Shelf ecosystem represents an important implication of the SBI for this exploited and clearly affected system. Such a recommendation is consistent with the suggestion of Sissenwine and Cohen (1991), who similarly advocated consideration of a cautious experimental approach to regulation and management of the Georges Bank system to test those species interactions critical to improving management and enhancing important overfished gadoid stocks. The actual experiments used must be carefully designed and are constrained by economics and many other practical considerations. Nevertheless, the SBI recommendations imply strong advocacy for this approach.
5. *Charge modelers to identify key biological, chemical, and physical variables to be monitored to assess ecosystem status and health and to test the predic-*

tions of the ecosystem dynamics models, and incorporate these monitoring needs into agency monitoring programs. One fundamental challenge faced by modelers of ecosystems is the development of a defensible set of criteria that accurately define the status and health of an ecosystem. Despite substantial attention to this question, we still lack rigorous indices of ecosystem stress, status, and health (Lubchenco et al. 1991). Furthermore, the ecosystem models required to understand the functional effects of anthropogenic influences on natural and exploited ecosystems must be tested, modified, and refined through interactive field and modeling programs. It is unrealistic to envision complete monitoring of all important biotic and abiotic components of the ecosystem, making identification of key monitoring variables a critical task for ecosystem modelers.

Modelers also need to communicate with the federal and state agencies charged with environmental monitoring. Federal agencies with important monitoring responsibilities and capabilities, such as NOAA, the Environmental Protection Agency (EPA), and the National Aeronautics and Space Administration (NASA), should be encouraged to approach the leaders of the necessarily large research programs designed to develop ecological understanding and management strategies for specific LMEs to initiate development of a monitoring package that is feasible, yet appropriate and telling. The complexity of processes that jointly control ecosystem dynamics necessarily implies a challenge for monitoring. The task includes a need not only to identify what to monitor, but also to identify the spatial and temporal scales (e.g., Steele 1988). Without this partnership between ecosystems modelers and responsible agencies, models will not be adequately tested, and monitoring will not yield the information required for sustainable management. Although application of SBI recommendations may not immediately result in modification of monitoring procedures for the Northeast Shelf LME, a long lead time will nevertheless be required for interaction between ecosystems modelers and leaders of monitoring programs. Now is the time to start.

6. *Devote substantial effort to understanding controls on prey population dynamics, especially for zooplankton and forage fish in LMEs.* The majority of past research efforts in the field of fisheries has been directed toward developing an understanding of the population dynamics of exploited species (see Rothschild 1986). Recently, fisheries research has broadened in scope, with the term *fisheries oceanography* describing this new focus. Fisheries oceanographic studies are intended to incorporate an integrative vision of exploited populations in a context of the complete physical and biological environment. Understanding the dynamics of exploited stocks

in the context of the entire ecosystem is a key feature of the ESA's SBI recommendations, as already indicated. Nevertheless, one specific aspect of this approach deserves special mention because of its importance and our lack of attention to it in past programs: the understanding of the controls on population dynamics of critical prey organisms, especially zooplankton and forage fish.

The key prey organism that requires explicit attention in the Northeast Shelf LME appears to be the zooplankter *Calanus finmarchicus* (JOI 1991b). This species of calanoid copepod is the primary prey for codfish and haddock during a critical early stage in their life history, when food limitation could well influence the recruitment success of both species. As a general rule, much more must be known about the determinants of population fluctuations and variability in production of copepods to be able to advance further our understanding of how ecosystems dynamics help determine the year-class strength of exploited fish populations (JOI 1991a). The abundance of *Calanus* can influence the rate of starvation of young gadoids, their growth rate, and thus, indirectly, their survivorship by altering the length of time required in the small size class so vulnerable to predation, and the strength of many possible indirect interactions affecting gadoid abundance through a multitude of food chain interactions. Hypotheses involving prey species dynamics lie at the cutting edge of fisheries oceanographic research, and for the Northeast Shelf LME, demand a focused study of zooplankton dynamics, especially *Calanus finmarchicus*. In addition, key forage fishes, such as sand lance and capelin, deserve similar directed efforts to assess the processes controlling their population dynamics and their role in affecting the dynamics and yields of exploited bottom fish.

7. *Evaluate the significance of biological diversity to the functioning of marine ecosystems so that we know the extent to which ecosystem models and studies can be simplified yet remain accurate.* Perhaps the most critical decision that must be made before initiation of any ecosystem study is the components to be measured. The answer requires an understanding of the role of biotic diversity in ecosystem and community dynamics, which is a question at the cutting edge of basic ecological research (Lubchenco et al. 1991). For example, it seems clear that any study of ecosystems dynamics should measure the population sizes of the most abundant species, but to how many trophic levels does that level of specific attention need to be carried? If the ultimate goal of an LME study is to understand the dynamics of an exploited fish species that primarily occupies a specific predatory trophic level, is it sufficient to merely measure cell densities of major

taxa of phytoplankton? What level of information on decomposers is needed? Also, how important are indirect and higher-order species interactions (those not represented by direct predation or consumption) in ecosystem dynamics?

These questions are generic problems of ecosystems ecology, but they must be addressed specifically for the Northeast Shelf LME as part of any effective ecosystem study. Modeling and field study will need to be closely coordinated for efficiency. For example, marine planktonic ecosystems are probably well characterized by food web interactions, whereas marine benthic ecosystem models would require the addition of more complex indirect interactions involving habitat modification. Unfortunately, marine planktonic food chains tend to be long compared with other food chains (Schoener 1989), implying the need to measure more populations to understand the behavior of any marine planktonic ecosystem. These issues need immediate attention to allow effective design of a study of the Northeast Shelf LME dynamics, so that the important functions of that vital ecosystem can be sustained in the face of obvious human impacts.

ACKNOWLEDGMENTS

I thank the National Science Foundation for support required to prepare this paper and NOAA/NMFS and EPA for support to attend the "Northeast Shelf Ecosystem: Stress, Mitigation, and Sustainability" Conference to present this paper.

References

Clark, S. H., and Brown, B. E. 1977. Changes of biomass of finfishes and squids from the Gulf of Maine to Cape Hatteras, 1963–74, as determined from research vessel survey data. Fish. Bull., U.S. 75:1–21.

Committee on International Science's Task Force on Global Diversity. 1989. Loss of biological diversity: A global crisis requiring international solutions. A Report to the National Science Board. Washington, DC: National Science Foundation.

Fogarty, M. J., Sissenwine, M. P., and Grosslein, M.D. 1987. Fish population dynamics. In R. H. Backus (ed.), Georges Bank. Cambridge, MA: MIT Press, pp. 494–509.

Huntley, M. E., Lopez, M. D. G., and Karl, D. M. 1991. Top predators in the Southern Ocean: A major leak in the biological carbon pump. Science 253:64–66.

Joint Oceanographic Institutions (JOI). 1991a. GLOBEC initial science plan. Washington, DC: Joint Oceanographic Institutions.

Joint Oceanographic Institutions (JOI). 1991b. GLOBEC Northwest Atlantic Program. Washington, DC: Joint Oceanographic Institutions.

Joint Oceanographic Institutions (JOI). 1991c. Theory and modeling in GLOBEC. Washington, DC: Joint Oceanographic Institutions.

Lubchenco, J., Olson, A. M., Brubaker, L. B., et al. 1991. The sustainable biosphere initiative: An ecological research agenda. Ecology 72:371–412.

May, R. M. (ed.). 1984. Exploitation of marine communities. Berlin: Springer-Verlag.

National Research Council (NRC). 1990. Decline of the sea turtles: Causes and prevention. Washington, DC: National Academy Press.

Ricker, W. E. 1958. Handbook of computations for biological statistics of fish populations. Bull. Fish. Res. Board Can. 119:1–300.

Risser, P. G., Lubchenco, J., and Levin, S. A. 1991. Biological research priorities: A sustainable biosphere. Bioscience 9:625–627.

Rothschild, B. J. 1986. Dynamics of Marine Fish Populations. Cambridge, MA: Harvard University Press.

Schoener, T. W. 1989. Food webs from the small to the large. Ecology 70:1559–1589.

Sherman, K. 1991. The large marine ecosystem concept: A research and management strategy for living marine resources. Ecol. Applic. 1(4):349–360.

Sherman, K., and Alexander, L.M. (eds.). 1986. Variability and Management of Large Marine Ecosystems. AAAS Selected Symposium 99. Boulder, CO: Westview Press.

Sherman, K., Grosslein, M., Mountain, D., et al. 1988. The Continental Shelf Ecosystem Off the Northeast Coast of the United States. In H. Postma and J. J. Zijlstra (eds.), Ecosystems of the World 27—Continental shelves. Amsterdam: Elsevier, pp. 279–337.

Sissenwine, M. P. 1986. Perturbation of a Predator-controlled Continental Shelf Ecosystem, pp. 55–85. In K. Sherman and L. M. Alexander (eds.), Variability and Management of Large Marine Ecosystems. AAAS Selected Symposium 99. Boulder, CO: Westview Press, pp. 55–85.

Sissenwine, M. P., and Cohen, E. B. 1991. Resource Productivity and Fisheries Management: Northeast Shelf Ecosystem. In K. Sherman, L. M. Alexander, and B. D. Gold (eds.), Food Chains, Yields, Models, and Management of Large Marine Ecosystems. Boulder, CO: Westview Press, pp. 107–123.

Steele, J. H. 1988. Scale Selection for Biodynamic Theories. In B. J. Rothschild (ed.), Toward a Theory on Biological-Physical Interactions in the World Ocean. NATO ASI Series. Series C: Mathematical and Physical Sciences. Dordrecht, The Netherlands: Kluwer Academic, Vol. 239, pp. 513–526.

Stewart-Oaten, A., Murdoch, W.W., and Parker, K. 1986. Environmental impact assessment: "Pseudoreplication" in time? Ecology 67:929–940.

Sugihara, G., Garcia, S., Guilland, J. A., et al. 1984. Ecosystems Dynamics: Group Report. In R. M. May (ed.), Exploitation of Marine Communities. Berlin: Springer-Verlag, pp. 130–153.

U.S. Committee on Global Change Research. 1990. Research Strategies for the U.S. Global Change Research Program. Washington, DC: National Academy Press.

Walters, C. J. 1986. Adaptive management of renewable resources. New York: Macmillan.

II

Physical and Biological Characteristics

4

Physical Oceanography of the Shelf and Slope Seas from Cape Hatteras to Georges Bank: A Brief Overview

David A. Brooks

Abstract

The United States Northeast Shelf ecosystem is conveniently defined by the ocean region from Cape Hatteras to Georges Bank, including the Gulf of Maine and the Slope Sea, which extends offshore to the edge of the Gulf Stream. The region is influenced on a broad range of time and space scales by winds, climate changes, river runoff, estuarine exchange, tides, and Gulf Stream meanders and rings. These forcing processes individually and jointly produce a complex circulation regime with distinctive features in various parts of the ecosystem. To wisely husband (i.e., "manage") the resources of such a complex region, it is first necessary to understand the basic physical processes that control its natural dynamics; only then can we sensibly assess the influence of anthropogenic stresses on the ecosystem and propose mitigation for them.

This paper reviews some of the important physical mechanisms active in the Northeast Shelf ecosystem, beginning on the largest scale with an overview of the circulation in the Slope Sea and then moving to the coupled local subsystems of Georges Bank and the Gulf of Maine. Finally, on a smaller scale, some fresh insights are offered about an isolated benthic community that may be sustained by a remote energy source, and a bay ecosystem that has at various times been considered as a site for tidal power development.

The brief review reaffirms that an adequate level of physical understanding should form the foundation of rational management and monitoring strategies for ecosystems on all scales.

To this end, it may be convenient to define the scope of an ecosystem by artificial or real geographic boundaries, but it is important to remember that the significant physical processes may not be confined by those choices.

The United States Northeast Shelf ecosystem has been defined as the ocean region from Cape Hatteras on the south to Georges Bank on the north, including the waters of the Gulf of Maine and extending offshore to the Gulf Stream. The region, shown in Figure 4-1, is subject to a wide variety of physical forcing mechanisms, including the winds, tides, atmospheric cyclogenesis, surface buoyancy modifications from seasonal heating and cooling, estuarine exchanges, and meanders and rings from the Gulf Stream. In the eastern portion of the region, Georges Bank and the Gulf of Maine form a coupled ecosystem with prominent characteristics of its own.

This paper gives a brief overview of some of the major physical oceanographic features of the Northeast Shelf ecosystem, beginning with the largest scale of the Slope Sea and focusing on aspects of the Georges Bank–Gulf of Maine region. In any ocean region so large and complex, a thorough review of significant processes would require many volumes, so this brief overview is necessarily limited and certainly incomplete. A broader review of physical processes on U.S. continental shelves is given by Allen et al. (1983), and a detailed description of the oceanography of Georges Bank can be found in the excellent review book edited by Backus (1987). For the Gulf of Maine, the comprehensive volume by Bigelow (1927) still stands as the most extensive overview of fisheries and physical oceanography, although the large amount of new information gained since then points to the need for a modern review volume. The intent here is simply to highlight some of the salient physical mechanisms and to give an introductory background to stimulate further reading in the rich literature for this region.

When considering how best to "manage" an ecosystem—by which is meant how best to husband the ecosystem's resources while using them wisely—it is natural to divide the ocean into regions that seem to have well-defined geographic limits. In the case of the Northeast Shelf ecosystem, for example, it is sensible to choose the Gulf Stream front as the outer boundary, but the locations of the "ends" of the system are not so obvious. For example, the confluence of Virginia coastal water with the Gulf Stream as it leaves Cape Hatteras forms a convenient southwestern vertex of the Slope Sea, but it does not give an explicit appreciation of the fact that much of what happens in the southern part of that region depends on the meandering of the Gulf Stream

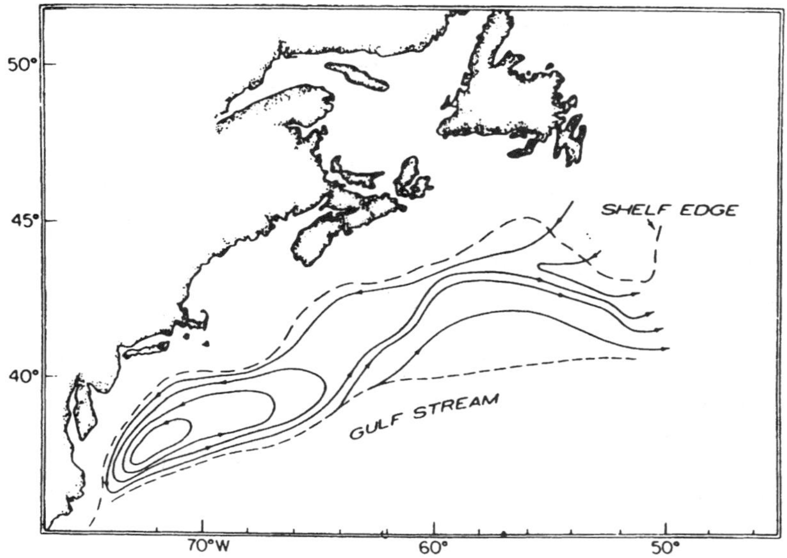

Figure 4-1. Schematic of horizontal circulation in the Slope Sea, inferred from empirical evidence. Solid curves are streamlines, showing the cyclonic gyre in the southern end of the region. (Reprinted from G. T. Csanady and P. Hamilton. 1988. Circulation of slope water. Cont. Shelf Res. 8:565–624. With permission from Elsevier Science Ltd, The Boulevard, Langford Lane, Kidlington 0X5 1GB, UK.)

between the Florida Straits and Cape Hatteras. The northeastern boundary of the ecosystem is even more amorphous because there is no similarly convenient geographic delimiter formed by the land or the ocean bottom, although the Northeast Channel separating the Scotian Shelf from the Gulf of Maine has been suggested by Sherman et al. (1988). In this regard, it is important to note that the recently determined United States–Canadian boundary across the eastern part of Georges Bank (the "Hague line") is most definitely *not* a dynamically sensible ecosystem boundary, because the international line is transparent to the physical processes that prominently influence the circulation in the region.

Wise management also supposes that mitigative actions are possible and desirable in a stressed ecosystem—for example, actions that tend to reduce degrading anthropogenic influences such as oil spills or other forms of pollution. Such strategies presume that influences of anthropogenic activities can be distinguished from natural variations of the ecosystem, which occur on a wide variety of time and space scales and with varying intensity. Thus, before a sensible management strategy can be devised for a particular ecosystem, it is necessary to understand the natural dynamics that influence it. The purpose of this paper is to introduce some of the prominent physical factors that control the natural variability of the Northeast Shelf ecosystem, and to point out that, although much is known about natural variability in this region, in many cases there is just beginning to emerge an adequate understanding of why things are the way they are.

To cite just one obvious example, the precipitous decline of finfish stocks on Georges Bank in recent years focuses attention on the mechanisms that control nutrient and larvae influxes to the well-mixed waters over the bank. These mechanisms involve the complex movement of dissolved and drifting quantities across prominent thermohaline and tidal fronts in which are imbedded strong currents. A satisfactory understanding of frontal dynamics, particularly the ways in which mass, momentum, and other properties can be transferred across fronts, is therefore needed to "manage" the Georges Bank fishery, but such understanding has only recently begun to come into focus. Adequate understanding awaits additional knowledge about the seasonal development and occasional disruption of the thermohaline fronts around the bank, and these processes are in turn related to the seasonal formation, import, and export of water masses in the Gulf of Maine, as well as the influence of Gulf Stream rings on the offshore side of the bank. Ultimately, to predict the impact of natural variability on the productivity of Georges Bank, sophisticated numerical models capable of simulating the complexities of strong stratification over steep bottom slopes will be required. All of these are active areas of contemporary research.

The Slope Sea

The Slope Sea is that body of water lying approximately between the shelf "break," near the 100-m isobath, and the inner edge or front of the Gulf Stream, shown between the dashed lines in Figure 4-1. The sea extends in the along-shelf direction from Cape Hatteras, where the Gulf Stream departs from the coastline, to the eastern end of the Grand Banks of Newfoundland, where it "ends" only as a matter of convenience.

The surface circulation in the Slope Sea, shown schematically in the mean in Figure 4-1, consists of a general southwestward movement over the outer shelf, augmented by outflow from the Gulf of Maine, rivers, and the major estuaries such as Delaware Bay and Chesapeake Bay (cf. Bumpus 1973). In the inner shelf region, low-salinity estuarine outflows form persistent southward coastal jets (Chao and Boicourt 1986). The off-

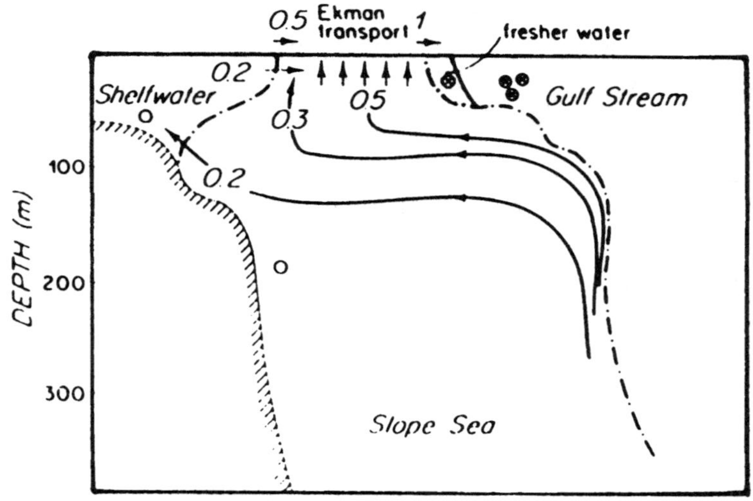

Figure 4-2. Schematic circulation in a vertical-plane cross section through the Slope Sea. Numbers give transport per unit length perpendicular to the transect in m^2s^{-1}. The circled crosses indicate eastward flow and the circled dots indicate westward flow. (Reprinted from G. T. Csanady and P. Hamilton. 1988. Circulation of slope water. Cont. Shelf Res. 8:565–624. With permission from Elsevier Science Ltd, The Boulevard, Langford Lane, Kidlington 0X5 1GB, UK.)

shore limb of the Slope Sea circulation consists of intense eastward or northeastward water movement along the inner edge of the Gulf Stream farther offshore. At the southern end of the sea, where the shelf width narrows and essentially vanishes, the southward moving shelf water (locally, Virginia Coastal Water) encounters the Gulf Stream and is mostly deflected by it or entrained in it, as suggested by the streamlines in Figure 4-1. The wind stress curl over the region is cyclonic (Leetma and Bunker 1978), which also contributes to the tendency for the surface flow to retroflect at Cape Hatteras and form a cyclonic gyre south of the New York Bight. The entrained coastal water often produces a relatively fresh and cool near-surface band in the innermost part of the surface Gulf Stream flow, first noted by Ford et al. (1952) and shown schematically in the cross section view of Figure 4-2. Relatively cool entrained water also is apparent in surface infrared maps obtained from satellites, as shown clearly by Garfield and Evans (1987). According to Csanady and Hamilton (1988), the cyclonic gyre in the western Slope Sea carries a westward transport of about 3×10^6 m^3s^{-1} in a broad boundary current off New Jersey. They also point out that the surface waters of the Slope Sea drain eastward along the inner edge of the Gulf Stream, a process perhaps aided by the formation of cold-core rings from cyclonic meanders of the Stream (Figure 3 in Richardson et al. 1978).

Csanady and Hamilton (1988) also noted that a seaward increase of the eastward wind stress over the Slope Sea produces a divergence of the surface Ekman transport, which must be compensated by an upwelling of water mostly drawn from isopycnals in the Gulf Stream front, as suggested in Figure 4-2. From the wind stress distribution, they calculate that an upwelling velocity of approximately 2.5×10^{-6} ms^{-1} (about 0.2 md^{-1}) is required at the base of the surface Ekman layer, a process that may be important in supplying deep nutrients to the upper layers in the Slope Sea.

In the winter, the loss of heat from the Slope Sea to the atmosphere is substantial (Bunker and Worthington 1976) due to the frontal outbreaks of cold, dry continental air. Intense air–sea heat exchange also plays an important role in sudden storm cyclogenesis as the low pressure center associated with the seaward moving front moves over the warm surface waters of the Gulf Stream.

The sketch shown in Figure 4-1 attempts to represent the mean circulation, defined over a suitably long period, in the Slope Sea. On a shorter time scale, the intense variability of the Gulf Stream dominates, and this is particularly evident at the surface in the form of meanders and warm-core rings (Figure 4-3). Warm-core rings are pinched-off eddies that form from an anticyclonic meander of the Gulf Stream front, which results in a separated deep pool of anomalously warm water with an energetic clockwise circulation (Joyce 1984). Warm-core rings can have lifetimes of many months, during which they generally drift slowly westward in the Slope Sea, bringing Gulf Stream water close to the edge of the continental shelf (Richardson 1983). When rings perturb the shelf–slope front, which normally separates shelf and slope waters, important cross-shelf exchanges of nutrients, heat, salt, and momentum can occur. For example, warm-core rings can significantly influence upwelling and nutrient exchange on the Scotian Shelf (Smith 1978), affect the water masses entering the Gulf of Maine (Ramp et al. 1983; Brooks 1987), and distort or rupture the shelf–slope front south and west of Georges Bank, causing large exchanges of shelf and slope water (Houghton et al. 1986; Butman and Beardsley 1987). At times, warm-core rings can nearly fill the surface area of the Slope Sea, whereas at other times, when the Gulf Stream front is quiet, they can be entirely absent (Garfield and Evans 1987).

The meandering process that sometimes spawns warm-core rings also directly influences the westward surface flow in the inner part of the Slope

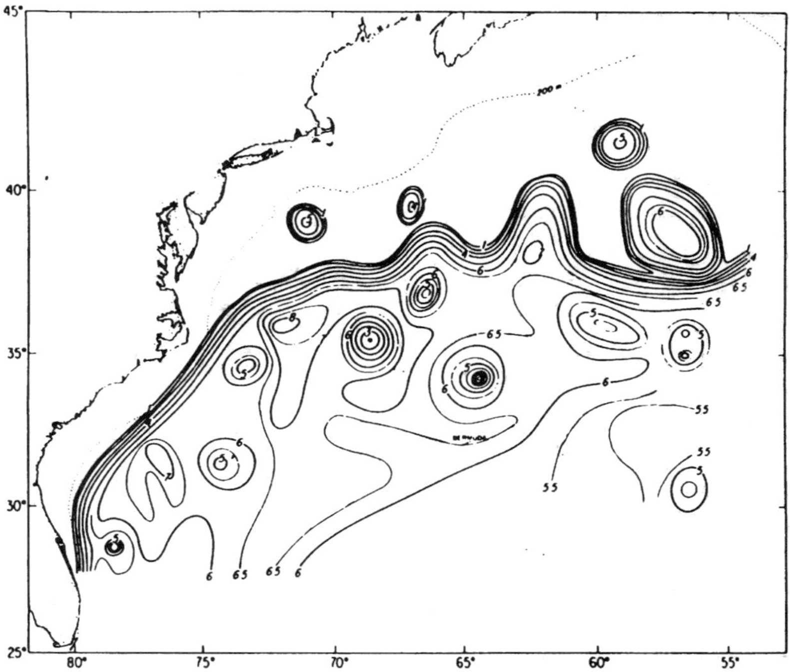

Figure 4-3. Depth topography (100s of meters) of the 15°C isothermal surface in the western North Atlantic Ocean, showing the Gulf Stream front, nine cyclonic rings on the shoreward side of the Stream, and three anticyclonic rings on the offshore side of the Stream. (Reprinted from P. L. Richardson, R. E. Cheney, and L. V. Worthington. 1978. A census of Gulf Stream rings, Spring 1975. J. Geophys. Res. 83:6136–6144. Copyright by the American Geophysical Union.)

Sea (Figure 4-4; Bane et al. 1988). The 2-year mean upper level currents shown in Figure 4-4A reveal the counterclockwise circulation in the western Slope Sea gyre, with the strongest flow northeastward in the Gulf Stream front. Figure 4-4B shows that the southwestward current between the 200-m and 2,000-m isobaths increases significantly during an onshore movement of the Gulf Stream front, which can meander in the cross-shelf direction as much as a few hundred kilometers in a month. The correlation suggests that the increased southwestward flow over the shelf results because the Slope Sea water entrained along the inner edge of the Gulf Stream must be supplied from a narrower width of shelf when the Gulf Stream front is in its inshore position. Figure 4-4B shows that the mechanism may account for a large part of shelf current variability in the Slope Sea on time scales of a few months to a year.

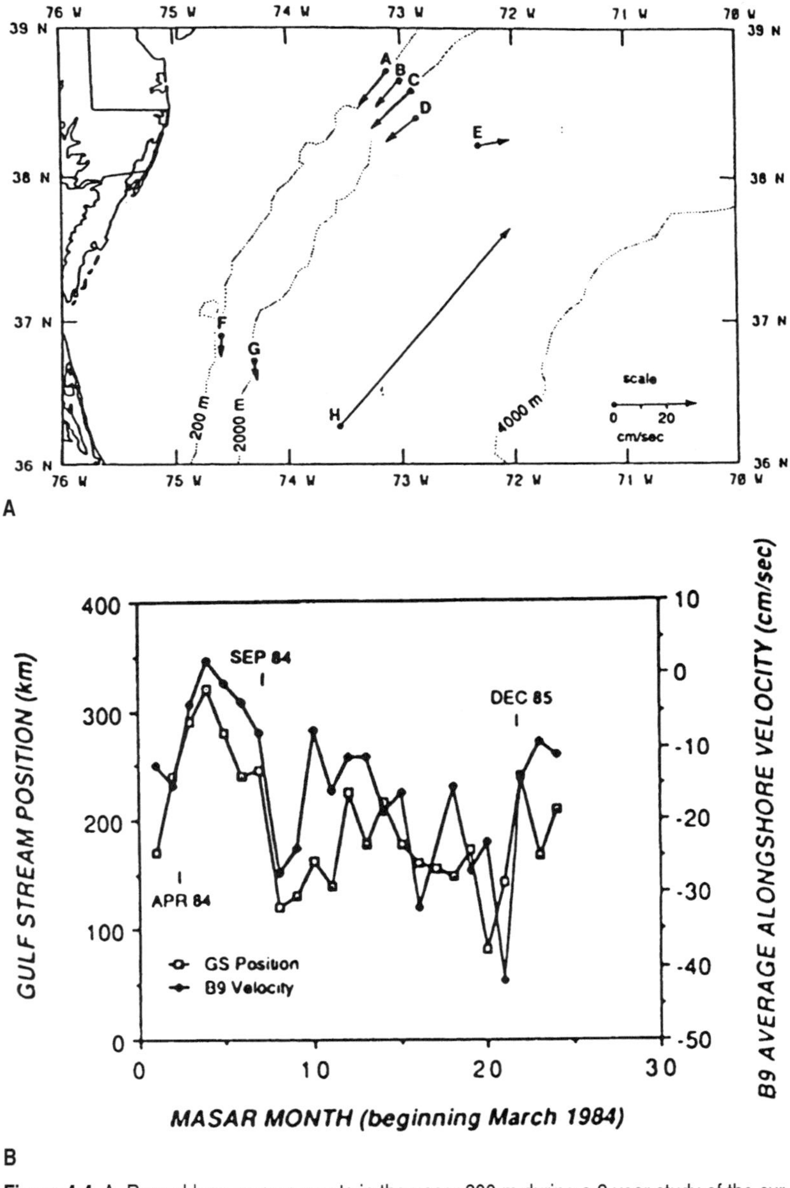

Figure 4-4. A. Record-long mean currents in the upper 200 m during a 2-year study of the currents in the Slope Sea region. B. Time series of monthly averaged Gulf Stream position and monthly averaged alongshore flow at the "B" mooring in Figure 4-4A, showing that the strength of the southwestward alongshore flow is correlated with a shoreward displacement of the Gulf Stream. (Reprinted from J. M. Bane, Jr., O. M. Brown, R. H. Evans, and P. Hamilton. 1988. Gulf Stream remote forcing of shelfbreak current in the Mid-Atlantic Bight. Geophys. Res. Lett. 15:405–407. Copyright by the American Geophysical Union.)

The Shelf–Slope Front and Georges Bank

The inner boundary of the Slope Sea is formed by the shelf–slope front, which separates relatively cool and fresh water of coastal or shelf origin from warmer and saltier water of the continental slope (Beardsley and Flagg 1976; Flagg 1987). The front is a highly variable but permanent feature that extends at least from Nova Scotia to Cape Hatteras (Horne 1978). In the mean, the shelf–slope front intersects the bottom near the 100-m isobath and extends upward and offshore, reaching the surface approximately over the 200-m isobath, as shown for example in the New York Bight region in Figure 4-5 and schematically over Georges Bank in Figure 4-6. In the summer, surface heating and stratification may mask the thermal front and seal off a near-bottom water mass known as the "cold band," which displays a prominent temperature minimum (Figure 4-5). Near Georges Bank, cold band waters are often continuous with a similarly formed Gulf of Maine water mass known as Maine Intermediate Water (MIW), which normally exits the Gulf through the Northeast Channel, following the topography of the eastern end of Georges Bank.

The shelf–slope front is a permanent but highly variable feature. The front meanders laterally with amplitudes of about 50 km on time scales of days to weeks (Halliwell and Mooers 1979). Warm-core rings from the Gulf Stream sometimes rupture the surface front, and the ring's clockwise circulation can bring warm and salty offshore water shoreward while exporting shelf water seaward (Ramp et al. 1983). When rings approach the south side of Georges Bank, shelf water and sometimes bank water can be drawn off the shelf in "streamers" that appear to be entrained in the peripheral clockwise circulation of a ring (Figure 4-7; Butman and Beardsley 1987). The streamer entrainment process plays an important role in the transport of biological substances between slope and ocean waters off Georges Bank, and the interaction of warm-core rings with the southern flank of Georges Bank may bring deep, nutrient-rich waters from the Gulf Stream front into the photic zone near the top of the bank (Yentsch and Phinney 1986). Such ring-induced exchange of shelf and slope waters is common along the shelf break from Nova Scotia to Cape Hatteras (Garfield and Evans 1987). When warm-core rings approach the mouth of the Northeast Channel, which is the primary conduit for deep water entry into the Gulf of Maine, ring-modified slope water sometimes enters the channel and disrupts or even reverses the usual outflow of intermediate water adjacent to Georges Bank (Brooks 1987).

Tidal currents over the shallow top of Georges Bank are very strong, typically with speeds of about 1 ms^{-1}, a result of the nearly resonant semidiurnal

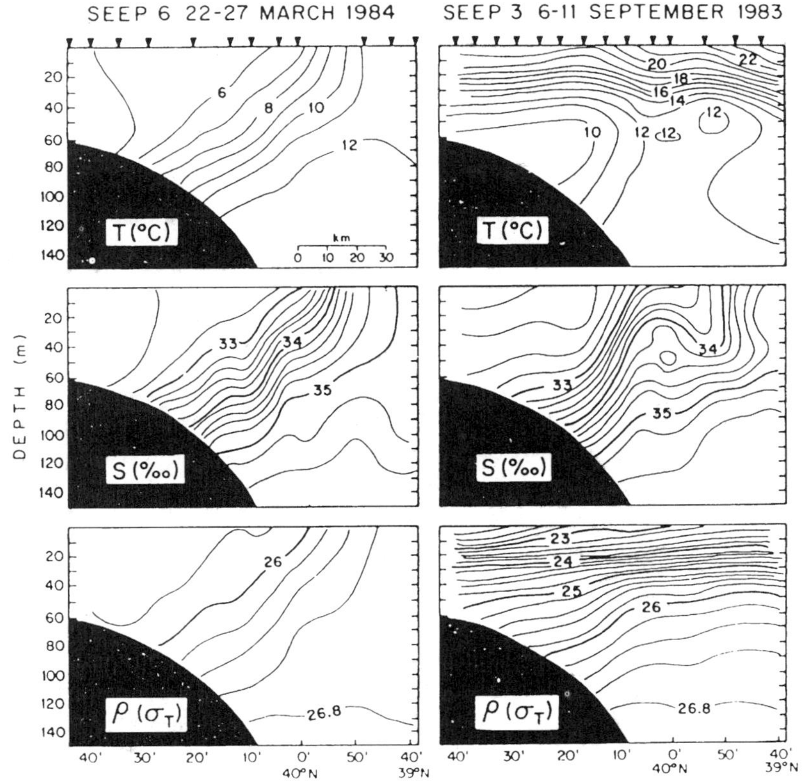

Figure 4-5. Hydrographic sections (temperature, salinity, and sigma-t) over the shelf break in the Slope Sea off southern New England for summer and winter conditions. (Reprinted from R. W. Houghton, F. Aikman, III, and H. W. Ou. 1988. Shelf-slope frontal structure and cross-shelf exchange at the New England shelf break. Cont. Shelf Res. 8:687–710. With permission from Elsevier Science Ltd, The Boulevard, Langford Lane, Kidlington 0X5 1GB, UK.)

response of the Gulf of Maine and Bay of Fundy to the Atlantic tidal wave (Garrett 1972). The tidal currents account for more than 80% of the total current variance near and over the bank (Moody et al. 1984). Over the top of the bank, within a large area shallower than about 60 m (Figures 4-6, 4-7, and 4-8), vigorous tidal currents keep the waters well mixed vertically, with the result that nearly homogenous conditions with anomalously cold surface temperatures are maintained year-round inside that isobath (Loder and Greenberg 1986). Consequently, the "cap" of Georges Bank is surrounded by a tidal front that separates the cool waters of the well-mixed shallow area from the warmer, seasonally stratified shelf waters on the seaward and shoreward sides of the bank, as shown schematically in Figure 4-6.

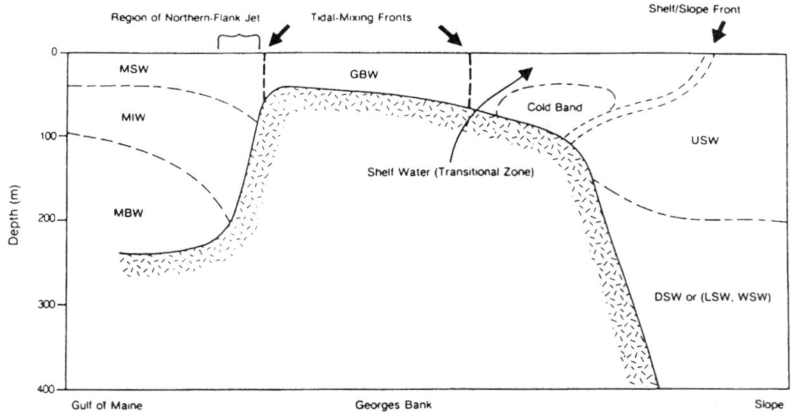

Figure 4-6. Sketch showing the principal water masses near Georges Bank in summer (MSW, Maine Surface Water; MIW, Maine Intermediate Water; MBW, Maine Bottom Water; GBW, Georges Bank Water; USW, Upper Slope Water; DSW, Deep Slope Water; LSW, Labrador Slope Water; WSW, Western Slope Water.) (Reprinted from C. N. Flagg. 1987. Hydrographic Structure and Variability. In R. H. Backus [ed.], Georges Bank. Cambridge, MA: MIT Press, pp. 108–124. With permission.)

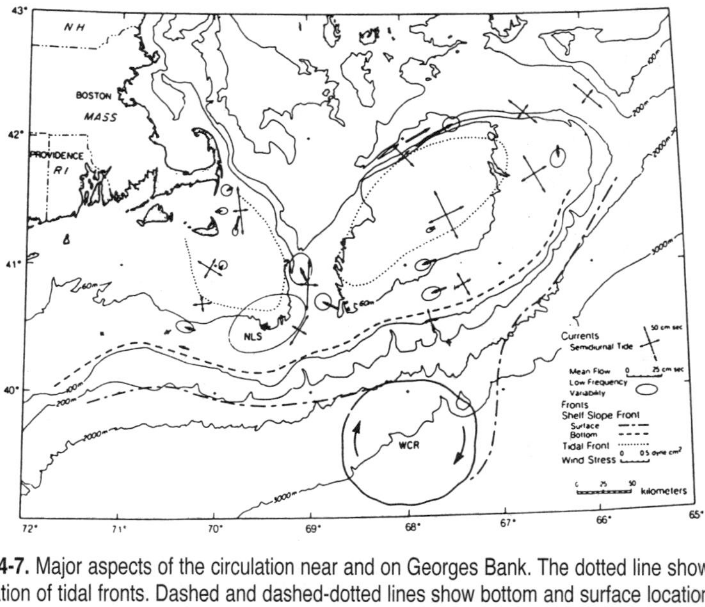

Figure 4-7. Major aspects of the circulation near and on Georges Bank. The dotted line shows the location of tidal fronts. Dashed and dashed-dotted lines show bottom and surface locations, respectively, of the shelf–slope front; the latter is perturbed by a warm-core ring (WCR) from the Gulf Stream. Rotary tidal currents are shown by crossed arrows, which define tidal ellipses. Monthly mean currents typical of summer conditions are shown by bold arrows. (Reprinted from B. R. Butman and R. C. Beardsley. 1987. Physical Oceanography. In R. H. Backus (ed.) Georges Bank. Cambridge, MA: MIT Press, pp. 88–98.)

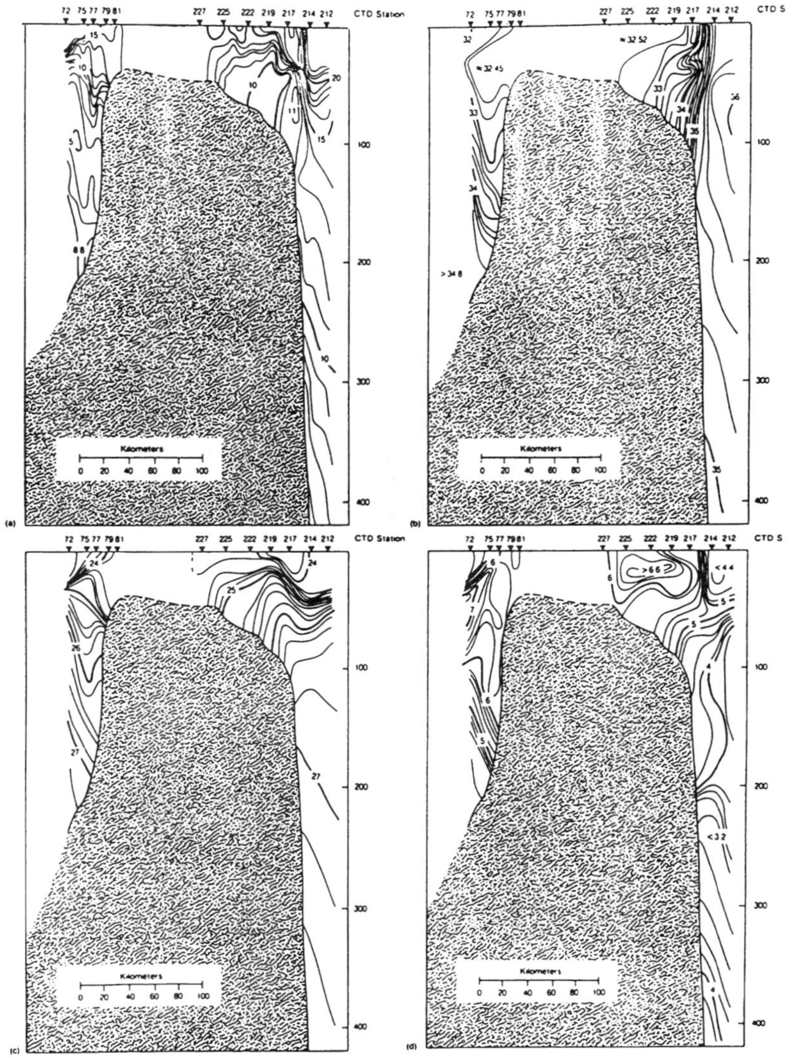

Figure 4-8. Hydrographic transects of (a) temperature, (b) salinity, (c) sigma-t, and (d) dissolved oxygen across Georges Bank in summer, showing the structure of tidal fronts on both sides of the bank and the well-mixed zone on top of the bank. (Reprinted from C. N. Flagg. 1987. Hydrographic Structure and Variability. In R. H. Backus [ed.], Georges Bank. Cambridge, MA: MIT Press, pp. 108–124. With permission.)

On the Gulf side, the seasonal thermohaline front is associated with a strong eastward jet that carries MIW and surface water toward the Northeast Channel. The eastward flow is also associated with nonseasonal rectification of the tidal currents over the steep edge of the bank (Loder and Wright 1985). A summer example of a detailed hydrographic section across the bank is shown in Figure 4-8 (Flagg 1987). The temperature minimum of the MIW is evident inside the Gulf, and the cold band, with MIW properties, can be seen on the offshore side near the bottom. The divergent isopycnals on the Gulf side of the bank are consistent with a deep inflow (out of the page) and a narrow eastward jet crowded against the bank with a subsurface maximum speed; this interpretation is generally consistent with the larger scale flow patterns within the Gulf (e.g., Brooks 1985). The oxygen contours suggest upwelling in the thermohaline front on the inner side of the bank, which is consistent with theoretically predicted upwelling near the 50-m isobath driven by tidal rectification over the steep edge of the bank (Loder and Wright 1985). The inner thermohaline front weakens or disappears in winter, when intense surface cooling erases much of the density contrast between surface waters inside the Gulf and on top of the bank (Flagg 1987) and the large-scale interior circulation in the Gulf relaxes (e.g., Bigelow 1927; Bumpus and Lauzier 1965).

Averaged over seasons, the residual flow reveals a near-surface clockwise movement around the edge of Georges Bank, as shown for example in Figure 4-7 (Butman and Beardsley 1987). The "Georges Bank gyre" is one of the often-cited features of the mean circulation in the region (e.g., Bigelow 1927; Butman et al. 1982), but it should be kept in mind that the seasonal mean flow speed around the bank is typically only about one-third the instantaneous current speed (Butman and Beardsley 1987; NAS 1991). Lagrangian drifters generally reveal the around-bank water movement, particularly in summer, but most drifters leave the bank without completing a circuit of the bank, indicating the variability in the flow patterns (Flagg et al. 1982). The low-frequency variability, with periods of days to months, is associated with wind fluctuations, seasonal variation in thermohaline structures, baroclinic instabilities in the around-bank flow, spring-to-neap tidal variations (and thus variation in the nonlinear rectification mechanism), and, on the offshore and eastern sides of the bank, the perturbing influence of warm-core rings, as noted earlier. Noble et al. (1983) and Brink et al. (1987) give more information about the complex low-frequency motions that characterize the region.

Gulf of Maine

The nontidal surface currents in the Gulf of Maine produce a general counterclockwise circulation, as originally described by Bigelow (1927). In spring and summer, partially closed counterclockwise gyres tend to form in the deep basins, particularly in Georges Basin at the inner end of the Northeast Channel and in Jordan Basin in the inner part of the eastern Gulf, where accumulations of dense bottom water entering from the Atlantic continental slope encourage cyclonic upper level flows (Brooks 1985). Surface salinity in June of 1983, for example, indicates that the southwestward-flowing eastern Maine coastal current turns offshore near Mt. Desert Island and then back toward the east as part of a counterclockwise flow in Jordan Basin (Figure 4-9; Brooks 1985); the surface temperature (not shown) hints at a similar pattern in Georges Basin, with water moving westward off the Scotian Shelf. In both basins, cyclonic upper level movement is associated with an accumulation of dense slope water in the bottom of the basins. There is less evidence to indicate a surface recirculation in Wilkinson Basin in the western Gulf; in the summer most of the surface waters there appear to move generally southward or southeastward, feeding an outflow past Cape Cod and Nantucket Shoals and also feeding into the eastward jet along the northern edge of Georges Bank, as suggested in Figure 4-9. Ten-week means of currents measured at a depth of 25 m, shown by the heavy arrows in the figure, provide some support for the surface water movement inferred from the property distributions.

The circulation pattern just described strengthens in spring and persists into the late summer, but with the onset of winter cooling, the surface gyre in the Gulf weakens and becomes more strongly influenced by winds (e.g., Bumpus and Lauzier 1965). Intense heat loss from the surface occurs in winter, with vigorous convection at least to mid-depth occurring in much of the Gulf (Brown and Beardsley 1978). In the subsequent spring and summer, restratification of the surface layer seals off a mid-depth layer that preserves the low temperature and salinities of the winter water. The prominent characteristic of this Maine Intermediate Water is a pronounced temperature minimum, which can be traced throughout most of the Gulf in summer (Hopkins and Garfield 1979). In late winter and early spring, a significant inflow of relatively fresh near-surface water occurs from the Scotian Shelf, which is in turn influenced by river runoff from the Laurentian Basin in the previous year (Smith 1983). In addition, warm but relatively saline (and therefore dense) water from the Atlantic slope flows into the Gulf through the lower part of the Northeast Channel (Ramp et al. 1985). The slope water inflow is pulselike and correlated with the winter winds but in summer seems to be more controlled

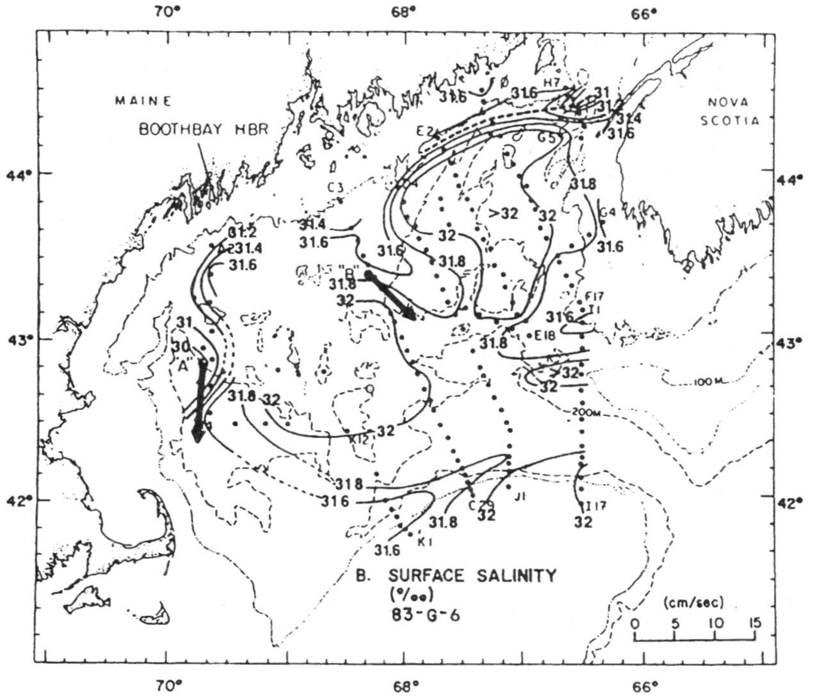

Figure 4-9. Surface salinity map for the Gulf of Maine in June, 1983, showing the tendency of surface waters to move cyclonically around the deep basins in the eastern Gulf. Bold arrows show 10-week mean currents at a depth of 25 m. The heavy dashed line traces the salinity minimum associated with outflow from the Saint John River in New Brunswick. (Reprinted from D. A. Brooks. 1985. Vernal circulation in the Gulf of Maine. J. Geophys. Res. 90(C5):4697–4705. Copyright by the American Geophysical Union.)

by thermohaline differences within and without the Gulf. Maximum deep inflow in the channel seems to occur in July or August, and the properties of inflowing water are sometimes influenced by warm-core rings (Brooks 1987). Since the distribution and character of slope water inside the Gulf play an important role in establishing the seasonal surface circulation, a knowledge of the factors that control the entry and spreading of slope water is needed in order to understand or predict the circulation and its interannual variation.

The contrast between warm and salty Atlantic bottom water in the basins against fresher and cooler surface waters of coastal or shelf origin results in a three-layer "sandwich" structure of the summer water masses in the Gulf, as shown schematically in Figure 4-10. The figure, which shows a vertical section from the Maine coast extending seaward across Georges Bank, also indicates

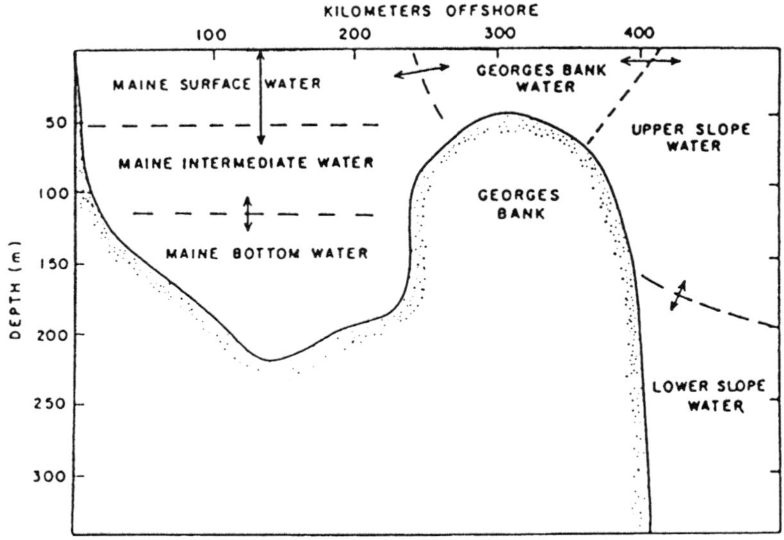

Figure 4-10. Schematic cross section of the Gulf of Maine, showing layering of the principal water types in summer. (Reprinted from T. S. Hopkins and N. Garfield, III. 1979. Gulf of Maine intermediate water. J. Mar. Res. 37:103–139. With permission.)

the vertical extent of seasonal movement of the various water mass boundaries; for example, the interface between Maine Intermediate Water and bottom water varies little seasonally, but the warm Maine Surface Water disappears entirely under the influence of winter cooling and convection.

A hydrographic section from the Maine coast to the inner edge of Georges Bank clearly reveals the three-layer summer structure (Figure 4-11). The section crosses the Jordan and Georges basins near the locations of their greatest depths. The MIW temperature minimum centered at a depth of about 75 m spans the section except in the northern part of Georges Basin, where warm, westward moving slope water from the Northeast Channel divides the MIW. A thin (ca. 25 m) surface layer of relatively warm water indicates the cumulative effect of summer restratification as of the June 1984 date of the section. The accompanying salinity section shows the high salinity (>34.5 ppt) of the slope water domed up in Georges Basin and to a slightly lesser extent in Jordan Basin, which is indicative of the tendency to produce separate cyclonic upper level currents, as noted earlier. Low salinities near the coast, which originate from the inflow of Scotian Shelf water and from the major coastal rivers (in the section shown, primarily the Saint John River in New Brunswick) produce

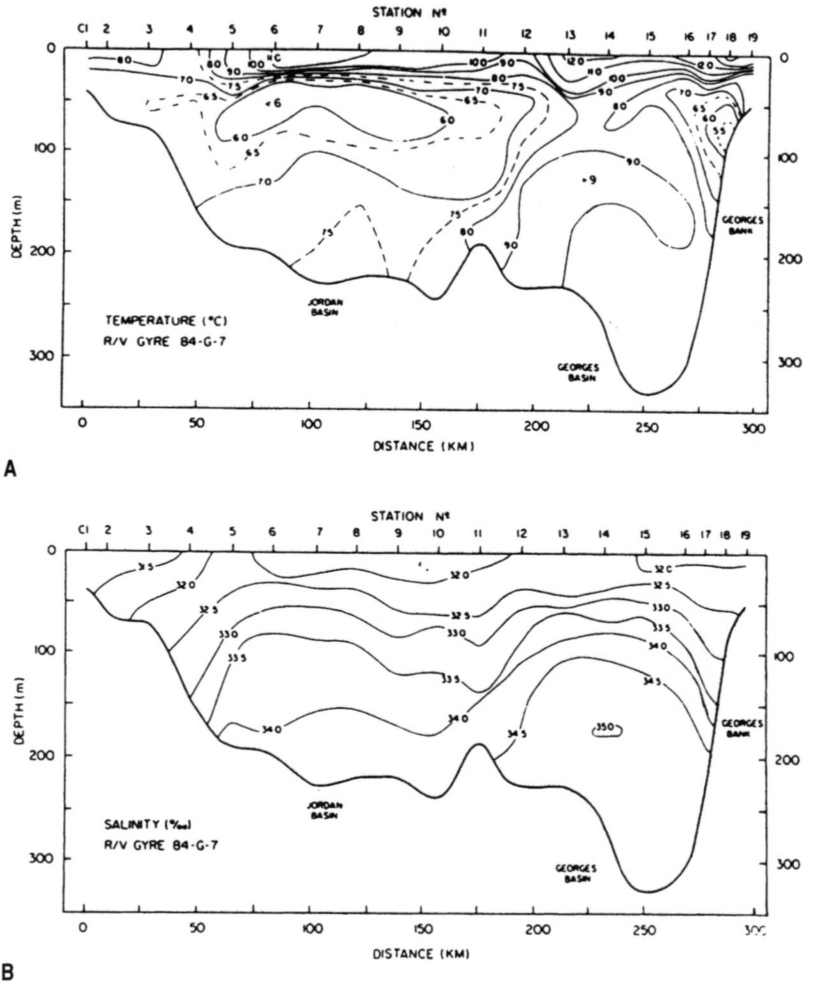

Figure 4-11. Transects from the Maine coast to Georges Bank in the eastern Gulf of Maine, showing (A) temperature and (B) salinity in June, 1984. The temperature minimum of MIW is evident, and the up-domed isohalines are consistent with separate cyclonic gyres in the Jordan and Georges Basins. (Reprinted from D. A. Brooks. 1985. Vernal circulation in the Gulf of Maine. J. Geophys. Res. 90(C5):4697–4705. Copyright by the American Geophysical Union.)

a nearshore downwarping of isohalines, also associated with the eastern Maine coastal current mentioned previously.

Generally, the three-layer hydrographic structure can be found throughout the central and western regions of the Gulf in the summer. However, in the eastern Gulf, the water mass contrasts begin to fade, and as one approaches the western tip of Nova Scotia and the entrance to the Bay of Fundy, where tidal currents are very strong, vertical mixing tends to erode most vertical structure of the hydrographic fields, although the low salinity (ca. 31.5 ppt) of Scotian Shelf water can readily be traced into the northeastern corner of the Gulf.

The obvious significance of the import of slope water and export of MIW points to the Northeast Channel as the "front door" of the Gulf of Maine. To illustrate, a temperature section across the channel from Browns Bank to Georges Bank (Figure 4-12) shows warm slope water entering on the north side, with the warmest core waters (>11°C) located near 130-m depth (cf. Ramp et al. 1985). Above the slope water, relatively cool and fresh (salinity not shown) Scotian Shelf water enters the Gulf as an upper level flow adjacent to and inshore of Browns Bank. On the other side of the channel, adjacent to Georges Bank, the low temperatures of exiting MIW span the depth range of about 35–150 m, with minimum core temperature found near a depth of 75 m.

Tides

Parts of the Gulf of Maine and the Bay of Fundy experience an unusually large semidiurnal tide, as mentioned earlier. The tidal sea level amplitude (half-range) increases from about half a meter in the Slope Sea offshore Georges Bank to more than 5 m in the upper reaches of the Bay of Fundy, a result of the near-resonance of the Gulf–Bay system with the principal lunar semidiurnal (M-2) tide in the Atlantic Ocean (Figure 4-13). The rapid inward increase in amplitude and relatively small phase variation are indications of a quarter-wave resonant response, sometimes referred to as a Helmholtz mode (and more colloquially compared to the response of an organ pipe). The associated tidal currents are intensified over the shallow top of Georges Bank, near Nantucket Shoals, on the continental shelf off the southwestern end of Nova Scotia, and in the Bay of Fundy, confirmed by computer simulations of the M-2 tide, which show that the current speed may exceed 1 ms^{-1} over the bank near times of maximum flood and ebb (Greenberg 1979, 1983). The tidal currents are rotary in nature, with clockwise rotation of the velocity vectors except in the western Gulf, where amplitudes are considerably reduced (Brown 1984). Elsewhere, the tidal velocity vectors form elliptical hodographs, with relatively small ellipticity over Georges Bank (Figure 4-7) and large ellipticity near coasts and in the Bay of Fundy, where the currents are more nearly rectilinear (Brown 1984).

The cotidal chart in Figure 4-13 shows the maximum amplitude of the M-2 tide at each location. However, there is a small but significant phase varia-

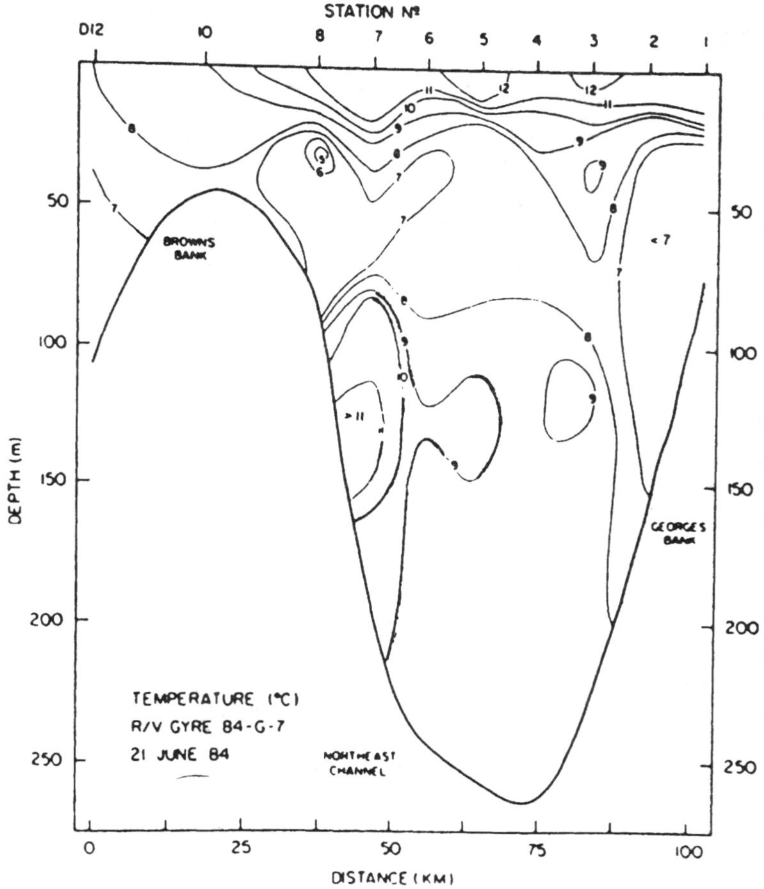

Figure 4-12. Temperature section across the Northeast Channel from Browns Bank to Georges Bank ("looking" out of the Gulf of Maine), showing cold MIW leaving the Gulf next to Georges Bank and much warmer water from the continental slope entering the Gulf at depth next to Browns Bank. A very cold near-surface inflow from the Scotian Shelf is also evident adjacent to Browns Bank. (Reprinted from D. A. Brooks. 1985. Vernal circulation in the Gulf of Maine. J. Geophys. Res. 90(C5):4697–4705. Copyright by the American Geophysical Union.)

tion so that the times of high and low water are not the same at all locations inside the Gulf. Three-dimensional computer simulations of the tide show that the semidiurnal tidal wave enters the Gulf around the southwestern tip of Nova Scotia and proceeds up the eastern side of the Bay of Fundy, compressed against the western shore of Nova Scotia in the manner of a Kelvin wave (Figure 4-14; Brooks 1992a). On reaching the head of the Bay, the wave reflects

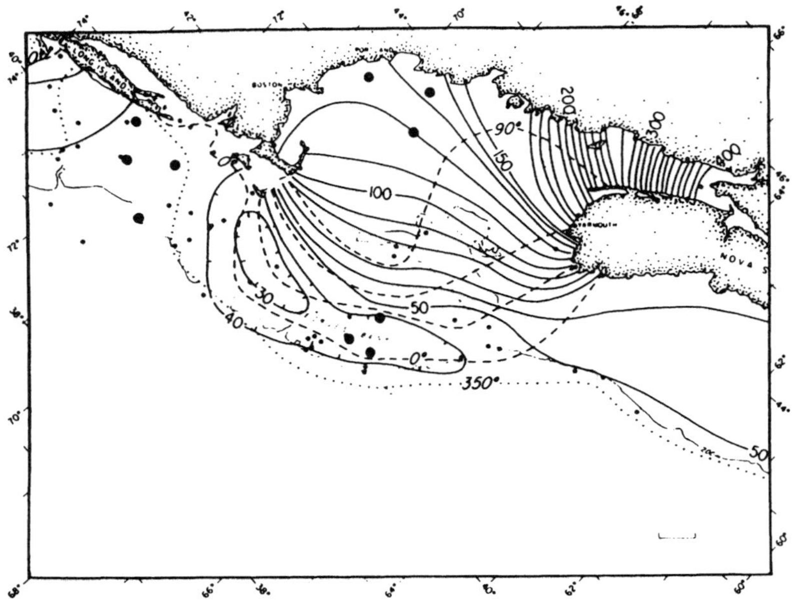

Figure 4-13. Cotidal chart for the principal semidiurnal (M-2) tide in the Gulf of Maine region, showing the large amplitude increase in the inner gulf and the Bay of Fundy as a result of the near-resonance of the tide in the region. Solid lines show M-2 amplitude (half range) in centimeters; dashed lines show Greenwich phase in degrees. (Reprinted from W. S. Brown. 1984. A comparison of Georges Bank, Gulf of Maine, and New England Shelf tidal dynamics. J. Phys. Oceanogr. 14:145–167. With permission.)

and travels southwestward along the New Brunswick shore with slightly diminished amplitude. Thus the tide arrives slightly later on the west side of the Bay, and the tidal range is slightly greater on the Nova Scotia side as a result of the energy lost to bottom friction. Sea level contours show that the rising tide wave sweeps southwestward along the New England coast. The speed of advance of the wave is such that high water at Boston lags behind that at the eastern end of the Maine coast (Eastport) by only about 30 minutes.

Computer simulations also indicate that the tide floods most strongly around the tip of Nova Scotia, over Georges Bank, and near Nantucket Shoals off Cape Cod at about 3 hours after low water in the Bay of Fundy. Tidal currents flood inward and northeastward in most of the Gulf, except for a westward component toward Boston Harbor in the region between Cape Cod and Cape Ann, and near major coastal bays and river mouths, where there may be significant local deviation. With regard to the tides in the western Gulf, theoretical calculations based on measured currents suggest that the tidal response there has some of the character of a standing Kelvin wave in a closed channel, whereas over Georges Bank and in the eastern Gulf the wave travels more like a frictionally retarded rotational gravity wave (Brown 1984). The computer simulation also shows that the tide first begins to turn near the tip of Cape Cod, while the flood continues over the banks, so that a weak counterclockwise eddy appears over Nantucket Shoals about an hour before high water in the Bay of Fundy (Figure 4-14). At about the same time, the ebb begins off southeastern Nova Scotia while the flood continues into the Bay of Fundy, so that a momentary divergence is established over the western Scotian Shelf. The current vectors continue to rotate clockwise over Georges Bank, so that near the time of slack water in the upper Bay of Fundy there is a strong ebb established around the tip of Nova Scotia, southwestward over Nantucket Shoals, and generally eastward off Georges Bank.

Two Coupled Subsystems

The purpose of this brief review has been to bring attention to some of the important physical processes operating in the Northeast Shelf ecosystem, arguing that an effective ecosystem management strategy presupposes an adequate understanding of the local dynamics. To illustrate the physical–biological coupling implicit in this argument, summary examples from two small-scale subsystems in the Gulf of Maine are presented. Many other examples on other scales could be chosen. The two briefly described here offer insight into an unexpected coupling mechanism in one case, and provide a renewed look at old concerns in the other.

Cashes Ledge: Remote Tidal Forcing

Cashes Ledge is a small, isolated, shoal outcropping of rock located in the western-central part of the Gulf of Maine, about 100 km north of the inner edge of Georges Bank. Ammen Rock Pinnacle, on the southern part of the ledge, rises to about 28 m below the surface. The upper part of the ledge, which reaches a shallowest depth of less than 6 m, is well known for its biological productivity, a fact the fishing community has exploited for generations. Unlike other shoal areas in the eastern Gulf and over the offshore banks, where local tidal currents are strong enough to keep the water column well mixed vertically, the tidal currents at Cashes Ledge are too weak (ca. 20–30 cm s^{-1}) to break down the summer stratification, which is primarily manifested by

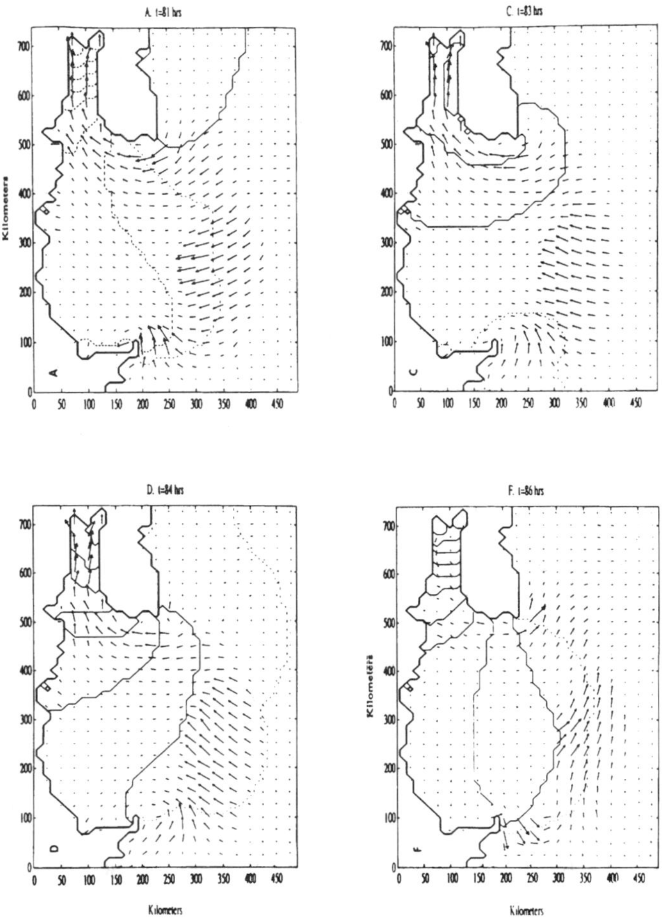

Figure 4-14. Model-predicted surface currents and surface elevations at four stages of one tidal cycle in the Gulf of Maine. Calculations were carried out in a 40 × 60 grid with horizontal resolution about 12 km and with ten levels in the vertical; for clarity, only alternate vectors are shown. Maximum vector magnitudes are 1.58, 2.00, 1.50, and 1.34 ms^{-1} in panels A, C, D, and F, respectively. The sea level contour interval is 0.5 m, with the zero contour by a dot-dash line; negative contours are dashed and positive contours are solid. The only forcing was an M-2 tide with 0.5-m amplitude at the open ocean boundary. The rising tide enters the Gulf like a Kelvin wave, hugging the coast of Nova Scotia. After reflecting from the head of the Bay of Fundy, the tide wave spreads southwestward along the Maine coast, filling the rest of the Gulf. The surface currents first begin to ebb at the tip of Nova Scotia and off Cape Cod, while the flood continues into the upper Bay of Fundy and over Georges Bank, where the large vectors outline the shape of the shoal topography. Near the time of high water in the upper Bay of Fundy (panel F), the sea level elevation is >5 m above its mean value. (Reprinted from D. A. Brooks, 1992. An Overview of the Physical Oceanography of the Gulf of Maine. In J. Wiggin and C. N. K. Mooers [eds.], Proceedings of the Gulf of Maine Scientific Workshop. Boston: Urban Harbors Institute, University of Massachusetts at Boston. With permission.)

a shallow thermocline found in the depth range of 10–20 m. Hence, local tidal mixing over the ledge may be inadequate to account for the rich benthic population found there. Other agents, such as lateral advection from the eastern Maine coastal current (Brooks and Townsend 1989), may bring near-surface nutrients into the region of Cashes Ledge and help stimulate primary productivity and provide some near-bottom mixing there (Townsend et al. 1989), but an additional process may be required to make the phytoplankton stock (i.e., the subsurface chlorophyll maximum just below the summer thermocline) available to the filter-feeding benthic community (Witman et al. 1993).

Preliminary results indicate that short-period internal waves generated remotely by the tides over the steep inner edge of Georges Bank, roughly 100 km distant, may contribute to the high productivity of Cashes Ledge. Recent visible-band photographs from a U.S. Space Shuttle flight that passed over the southwestern tip of Nova Scotia clearly show arcuate groups or "packets" of wavelike disturbances that appear to emanate from the inner edge of the bank and propagate inward into the Gulf, where they pass over and interact with the shallow topography of Cashes Ledge (La Violette et al. 1990). The wave groups evidently are internal soliton packets generated on each semidiurnal cycle, a result of internal hydraulic jumps caused by the ebbing tide as the thermocline is pushed up over the steep edge of the bank; a similar process is well documented over a sill in the Sulu Sea (Apel et al. 1985). By analogy with the Sulu Sea case, when the ebb slackens, the jump, which has the form of a prominent thermocline depression, evidently is released to propagate into the Gulf; nonlinearity and dispersion then combine to convert the original depression into an inward-propagating wave train or "packet," typically consisting of four to six internal solitons. Within each packet, the thermocline depression associated with each passing soliton can be of order 50 m (Apel et al. 1985), with strong associated subsurface currents and a correlated pattern of surface convergence and divergence zones. The convergences, sometimes called "rips," occur over and travel with the thermocline depressions. The resulting modulation of small surface wave amplitudes produces a traveling roughness pattern that is visible in sun-glint photographs such as those taken from the Space Shuttle (La Violette et al. 1990). The importance of internal waves produced by the tides has long been recognized near Georges Bank (e.g., Marsden 1986) and over Stellwagen Bank in Massachusetts Bay (Halpern 1971), but the widespread significance in the inner Gulf of short-period soliton packets generated along the northern edge of Georges Bank has only recently been appreciated.

Over Cashes Ledge, measurements indicate that the traveling sequence of thermocline depressions associated with each soliton packet brings the waters of the thermocline region down to the level of the top of the ledge, intermittently bathing the benthic community in warm, plankton-rich near-surface waters for perhaps a total of an hour or two during each semidiurnal cycle (Witman et al. 1993). Preliminary measurements show an irregular sequence of thermocline depressions bringing water temperatures of 10–13°C down to Ammen Rock Pinnacle, where the typical summer bottom temperature is about 6–7°C. The high temperature events are also correlated with high levels of chlorophyll fluorescence determined by a bottom-mounted fluorometer. Thus it appears that the wave mechanism indirectly supplies food to the benthic community. Further field work is needed to test this hypothesis, but it appears that the Cashes Ledge ecosystem is partially supported by internal waves generated near Georges Bank, some 100 km away. In view of the fact that internal waves are commonly observed over the continental margins and shelves of the world's oceans (Gasparovic et al. 1985), a similar process may be important in many other ecosystems.

Passamaquoddy Bay: Consequences of Tidal Power Development

At the eastern end of the Maine coast, at the border with Canada, a prominent coastal indentation is formed by Passamaquoddy Bay and the adjacent, smaller Cobscook Bay, where the mean semidiurnal tidal range is 5.4 m. The region is marked by many islands and narrow passages that define a body of water nearly cut off from the rest of the Gulf of Maine. Because of its large tidal range, the Passamaquoddy region has several times been seriously considered as a site for the development of tidal power, most recently in the late 1950s (e.g., Trites 1961). The most ambitious version of the tidal project would have produced about 1,000 MW of peak power, comparable with that produced by a modern nuclear plant. Although construction actually began on a version of the project in 1935, the work was never completed. Subsequent studies recommended against construction of the tidal project, largely because of environmental concerns about potential damage to the rich fishery of the bays, and because it was expected that reduced tidal mixing in the presence of the required dams would result in increased winter freezing of harbors inside the bays. The latter concerns were based on qualitative arguments and engineering comparisons with other embayments that had been impounded for other reasons. At the time, computer technology was inadequate to simulate the circulation in the bays.

Recently, the tidal and wind-driven circulation in the Passamaquoddy Bay region has been numerically simulated using a sigma-coordinate primitive-equation hydrodynamic model (Hess 1989) to integrate the equations of motion on a 60-by-40 grid with 600-m horizontal resolution and ten levels in the vertical (Brooks 1992b). Current computer technology makes such simulations feasible with desktop workstations. After establishing that the tidal motions were reasonably reproduced in the natural domain, the model was modified to represent the tidal structures as they were proposed in the most recent version of the project, so the modified circulation regime could be compared against the natural one. With regard to ecosystem concerns, the modeling strategy also allowed the surface freezing issue to be quantitatively addressed by comparing surface temperature in the modified domain against that in the natural domain.

A central result of the experiment was that, under severe but not extreme winter conditions, the tidal power project had a slight *moderating* influence on surface freezing in the bays compared with the extent of freezing in the natural system. This result, which is contrary to earlier expectations, follows because the circulation in the modified system includes a tidal residual flow that brings slightly warmer, more saline water into Passamaquoddy Bay; the influence of this water appears to be (barely) sufficient to overcome the increased surface freezing tendency caused by reduced vertical mixing, particularly in the St. Croix River estuary in the western part of the bay, where the impact was expected to be most severe. Earlier studies recognized that the modification would result in a significant one-way tidal flow through the bay, but the outcome of the competing influence of lateral versus vertical mixing could not have been anticipated. Mixing effectiveness in models depends on incompletely known turbulence processes and their parameterization, but in any case the model indicates that the state of freezing of the bays is far more sensitive to external environmental variables such as wind speed than it is to whether or not the tidal power structures are in place. Other potential impacts on the bay ecosystem, such as interference with the fishery, remain unresolved, but a fresh look at the physical factors controlling the ecosystem indicates that the impact of the tidal project might not be as severe as once thought.

Conclusion

In an ocean region as large and complex as the Northeast Shelf ecosystem, a brief review such as this can only begin to point out the many physical factors that determine the circulation, mixing, and exchange processes that are fundamental to a knowledge of how the ecosystem "works." Yet, a sensible (meaning publicly defensible) ecosystem management strategy presupposes a certain level of understanding of the important physical processes; otherwise, mitigative actions must be based on empirical strategies that may fail for unknown reasons at critical times. For example, a strategy to confine an oil spill on Georges Bank based on the assumption of a continuous clockwise flow around the bank (a common view) might be successful, but it is very unlikely that such an approach would produce expected results if a warm-core ring from the Gulf Stream were disrupting the offshore limb of the bank gyre at the time of the spill. Many other examples could be cited, but the important point is that advances in dynamic understanding should proceed apace with efforts to "manage" ecosystems, and pressure to engage in the latter should also extend to the former. A combined, unified approach to such complex oceanic issues certainly will, in the long run, produce benefits to society far in excess of those likely to be returned by independent efforts.

References

Allen, J. S., Beardsley, R. C., Blanton, J. O., et al. 1983. Physical oceanography of continental shelves. Rev. Geophys. Space Phys. 21(5):1149–1181.

Apel, J. R., Holbrook, J. R., Liu, A. K., and Tsai, J. J. 1985. The Sulu Sea internal soliton experiment. J. Phys. Oceanogr. 15:1625–1651.

Backus, R. H. (ed.). 1987. Georges Bank. Cambridge, MA: MIT Press.

Bane, J. M., Jr., Brown, O. M., Evans, R. H., and Hamilton, P. 1988. Gulf Stream remote forcing of shelfbreak currents in the Mid-Atlantic Bight. Geophys. Res. Lett. 15: 405–407.

Beardsley, R. C., and Flagg, C. N. 1976. The water structure, mean currents and shelf water/slope water front on the New England continental shelf. Mem. Soc. R. Sci. Liege 6:209–225.

Bigelow, H. B. 1927. Physical oceanography of the Gulf of Maine. Fish. Bull., U.S. 40:511–1027.

Brink, K. H., Magnell, B. A., and Noble, M. A. 1987. Low Frequency Current and Bottom Pressure Variability. In R. H. Backus (ed.), Georges Bank. Cambridge, MA: MIT Press, pp. 140–146.

Brooks, D. A. 1985. Vernal circulation in the Gulf of Maine. J. Geophys. Res. 90(C5): 4687–4705.

Brooks, D. A. 1987. The influence of warm-core rings on slope water entering the Gulf of Maine. J. Geophys. Res. 92(C8):8183–8196.

Brooks, D. A. 1992a. An Overview of the Physical Oceanography of the Gulf of Maine. In J. Wiggin and C. N. K. Mooers (eds.), Proceedings of the Gulf of Maine Scientific Workshop. Boston: Urban Harbors Institute, Univ. of Mass.

Brooks, D. A. 1992b. Tides and tidal power in Passamaquoddy Bay: A numerical simulation. Cont. Shelf Res. 12:675–716.

Brooks, D. A., and Townsend, D. W. 1989. Variability of the coastal current and nutrient pathways in the eastern Gulf of Maine. J. Mar. Res. 47:303–321.

Brown, W. S. 1984. A comparison of Georges Bank, Gulf of Maine and New England Shelf tidal dynamics. J. Phys. Oceanogr. 14:145–167.

Brown, W. S., and Beardsley, R. C. 1978. Winter circulation in the western Gulf of Maine: 1. Cooling and water mass formation. J. Phys. Oceanogr. 8:265–277.

Bumpus, D. F. 1973. A description of the circulation on the continental shelf of the east coast of the United States. Prog. Oceanogr. 6:111–157.

Bumpus D. F., and Lauzier, L. M. 1965. Surface circulation on the continental shelf of eastern North America between Newfoundland and Florida. Ser. Atlas Mar. Environ. Folio 7, 4 pl., 8 pp. New York: American Geographic Society.

Bunker, A. F., and Worthington, L. V. 1976. Energy exchange charts of the North Atlantic Ocean. Bull. Am. Meteor. Soc. 57:670–678.

Butman, B. R., and Beardsley, R. C. 1987. Long-term observations on the southern flank of Georges Bank. Part I. A description of the seasonal cycle of currents, temperature, stratification and wind stress. J. Phys. Oceanogr. 17:367–384.

Butman, B. R., Beardsley, R. C., Magnell, B., et al. 1982. Recent observations of the mean circulation on Georges Bank. J. Phys. Oceanogr. 12:569–591.

Chao, S.-Y., and Boicourt, W. C. 1986. Onset of estuarine plumes. J. Phys. Oceanogr. 16:2137–2149.

Csanady, G. T., and Hamilton, P. 1988. Circulation of slope water. Cont. Shelf Res. 8:565–624.

Flagg, C. N. 1987. Hydrographic Structure and Variability. In R. H. Backus (ed.), Georges Bank. Cambridge, MA: MIT Press, pp. 108–124.

Flagg, C. N., Magnell, B. A., Frye, D., et al. 1982. Interpretation of the physical oceanography of Georges Bank. Report prepared under contract AA851-CT1-39, Bureau of Land Management. Waltham MA: EG&G Environmental Consultants.

Ford, W. L., Longard, J. R., and Banks, R. E. 1952. On the nature, occurrence and origin of cold low-salinity water along the edge of the Gulf Stream. J. Mar. Res. 11:281–293.

Garfield, N., III, and Evans, D. L. 1987. Shelf water entrainment by Gulf Stream warm-core rings. J. Geophys. Res. 92:13003–13012.

Garrett, C. 1972. Tidal resonance in the Bay of Fundy and Gulf of Maine. Nature 238:441–443.

Gasparovic, R. F., Apel, J. R., Brandt, A., and Kaischke, E. S. 1985. Synthetic aperture radar imaging of internal waves. Johns Hopkins APL. Tech. Dig. 6(4):338–345.

Greenberg, D. A. 1979. A numerical model investigation of tidal phenomena in the Bay of Fundy and Gulf of Maine. Mar. Geod. 2:161–187.

Greenberg, D. A. 1983. Modeling the mean barotropic circulation in the Bay of Fundy and the Gulf of Maine. J. Phys. Oceanogr. 13:886–904.

Halliwell, G. R., and Mooers, C. N. K. 1979. The space-time structure and variability of the shelf water–slope water and Gulf Stream surface temperature fronts and associated warm-core eddies. J. Geophys. Res. 84:7707–7725.

Halpern, D. 1971. Observations of short-period internal waves in Massachusetts Bay. J. Mar. Res. 29:116–132.

Hess, K. W. 1989. MECCA Program Documentation. NOAA Tech. Rep. NESDIS 46. Washington, DC: NOAA.

Hopkins, T. S., and Garfield, N., III. 1979. Gulf of Maine intermediate water. J. Mar. Res. 37:103–139.

Horne, E. P. W. 1978. Interleaving at the subsurface front in the slope water off Nova Scotia. J. Geophys. Res. 83(C7):3659–3671.

Houghton, R. W., Olson, D. B., and Celone, P. J. 1986. Observation of an anticyclonic eddy near the continental shelf break south of New England. J. Phys. Oceanogr. 16:60–71.

Joyce, T. M. 1984. Velocity and hydrographic structure of a Gulf Stream warm-core ring. J. Phys. Oceanogr. 14:936–947.

La Violette, P. E., Johnson, D. R., and Brooks, D. A. 1990. Sun-glitter photographs of Georges Bank and the Gulf of Maine from the space shuttle. Oceanogr. Mag. 3(1):43–49.

Leetma, A., and Bunker, A. 1978. Updated charts of the mean annual wind stress, convergences in the Ekman layers, and Sverdrup transports in the North Atlantic. J. Mar. Res. 36:311–322.

Loder, J. W., and Greenberg, D. A. 1986. Predicted positions of tidal fronts in the Gulf of Maine region. Cont. Shelf Res. 6(3):397–414.

Loder, J. W., and Wright, D. G. 1985. Tidal rectification and frontal circulation on the sides of Georges Bank. J. Mar. Res. 43:581–604.

Marsden, R. F. 1986. The internal tide on Georges Bank. J. Mar. Res. 44:35–50.

Moody, J. A., Butman, B., Beardsley, R. C., et al. 1984. Atlas of tidal elevation and current observations on the Northeast American continental shelf and slope. U.S. Geol. Surv. Bull. 1611. Washington, DC: USGS.

National Academy of Sciences (NAS). 1991. The adequacy of environmental information for outer continental shelf oil and gas decisions: Georges Bank. Washington, DC: National Academy Press.

Noble, M., Butman, B., and Williams, E. 1983. On the longshelf structure and dynamics of subtidal currents on the eastern United States continental shelf. J. Phys. Oceanogr. 13:2125–2147.

Ramp, S. R., Beardsley, R. C., and Legeckis, R. 1983. An observation of frontal wave development on a shelf-slope/warm-core ring front near the shelf break south of New England. J. Phys. Oceanogr. 13:907–912.

Ramp, S. R., Schlitz, R. J., and Wright, W. R. 1985. The deep flow through the Northeast Channel, Gulf of Maine. J. Phys. Oceanogr. 15:1790–1808.

Richardson, P. L. 1983. Gulf Stream Rings. In A. R. Robinson (ed.), Eddies in Marine Science. New York: Springer-Verlag, pp. 19–64.

Richardson, P. L., Cheney, R. E., and Worthington, L. V. 1978. A census of Gulf Stream rings, spring 1975. J. Geophys. Res. 83:6136–6144.

Sherman, K., Grosslein, M., Mountain, D., et al. 1988. The continental shelf ecosystem off the northeast coast of the United States. pp. 279–337. In H. Postma and J. J. Zilstra (eds.), Ecosystems of the World. 27: Continental Shelves. Amsterdam: Elsevier, pp. 279–337.

Smith, P. C. 1978. Low-frequency fluxes of momentum, heat, salt, and nutrients at the edge of the Scotian Shelf. J. Geophys. Res. 83(C8):4079–4096.

Smith, P. C. 1983. The mean and seasonal circulation off southwest Nova Scotia. J. Phys. Oceanogr. 13:1034–1054.

Townsend, D. W., Corwin, S., Ackleson, S. G., et al. 1989. A hydrographic survey of the north-central Gulf of Maine: Results of R/V *Argo Maine* Cruise, 14-16 August, 1989. Bigelow Laboratory for Ocean Sciences, Tech. Rep. No. 70. W. Boothbay Harbor, ME: Bigelow Laboratory for Ocean Sciences.

Trites, R. W. 1961. Probable effects of proposed Passamaquoddy power project on oceanographic conditions. J. Fish. Res. Board Can. 18(2):163–201.

Witman, J. D., Leichter, J. J., Genovese, S. J., and Brooks, D. A. 1993. Pulsed phytoplankton supply to the rocky subtidal zone: Influence of internal waves. Proc. Natl. Acad. Sci. USA 90:1686–1690.

Yentsch, C. S., and Phinney, D. A. 1986. The role of streamers associated with mesoscale eddies in the transport of biological substances between slope and ocean waters. p. 42. In J. C. J. Nihoul (ed.), Marine Interfaces Ecohydrodynamics. New York: Elsevier, pp. 42.

5

Meteorology of the Northeast Shelf

Owen Hertzman

Abstract

There are five main influences on the weather and climate of the Northeast Shelf region of the U.S. east coast: (1) the warm waters of the Gulf Stream, (2) the cold water flowing southwestward from the Scotian Shelf, (3) the winter cold air from central North America, (4) the warm moist air from the Gulf of Mexico, and (5) the position of the predominant westerly jet across eastern North America. These five factors produce the primary weather features that influence the heat budget and vertical mixing in the shelf region.

Climate in the region is affected by variability in the strength and location of the Gulf Stream and by changes in the low–wave-number hemispheric wave patterns in the atmosphere that modulate the input airflows to the region on weekly and longer time scales. The climate exhibits a strong annual cycle of temperature and a rather uniform precipitation rate. The region's climate would respond to global increases in greenhouse gases with a modest warming, rather small compared with current seasonal changes.

The strongest heat losses from the ocean accompany winter outbreaks of very cold, dry continental air, which can induce upward latent and sensible heat fluxes over 1,000 W m^{-2}. The dominant mixing events are associated with the gradients of wind stresses accompanying cold season cyclones. These storms can pass through the region on a number of tracks but appear to be steered most often along the coast and the north wall of the Gulf Stream, the regions of strongest surface temperature gradient. Occasional hurricanes spawned at lower latitudes also influence the region's weather. However, the occurrence of strong hurricanes in the region is rare and should be treated as a rather minor part of the climatology of the Northeast Shelf.

The Region

The region under consideration here extends from Cape Hatteras, North Carolina to the Gulf of Maine. The latitude range (35° N to 45° N) puts the region clearly within the middle latitudes. The dominant meteorological elements at these latitudes generally come from the west (southwest and northwest), steered by the dominant westerly flow in the middle and upper troposphere. This basic flow is modified by upstream topography and other regional and local features. Here, the major upstream topography is the Appalachian mountain range. In winter, the mountains can enhance thermal gradients near the coast by trapping cold air between the mountains and the sea when an anticyclone lies over south-central Canada. The temperature contrast between this cold air and warmer air from the Gulf of Mexico and from over the Gulf Stream (southeast of Georgia and the Carolinas) provides the potential energy for several major storm generation events in a typical cold season. The important local and regional features include the shape of the coastline, the temperature contrast across the coast in certain seasons, and the temporal and spatial distribution of sea surface temperature (SST), which is dominated here by the Gulf Stream.

Data Coverage

Surface Platforms

The region is reasonably well covered by coastal meteorological stations. The concave shape of the coast, the network of coastal radars, heavy shipping traffic, and the number of NOAA and other buoys provide reasonable upstream and in situ surface meteorological data coverage. Upper air data are available from near coastal stations at Hatteras, North Carolina; Dulles Airport; Chatham, Massachusetts; Albany, New York; Portland, Maine; and Yarmouth, Nova Scotia. These stations provide reasonable upstream coverage, but the region lacks an on-the-shelf site such as Sable Island, Nova Scotia, on the Scotian Shelf. The recently installed NEXRAD radars have improved both qualitative and quantitative in situ coverage of precipitation events in this region. The quality of the information decreases beyond about 100 km because of Earth curvature effects.

Satellite Data

Satellite information, both geostationary (GOES) and polar orbiting, is essential for meteorological analysis in this region, at both the synoptic and local scales. Of particular importance are the SST maps, which indicate the position of the northwestern edge of the Gulf Stream and the associated regions of strong SST gradients, and the presence of warm-core or cold-core rings. Though the geostationary satellites fail from time to time, new platforms are provided by both U.S. and European sources. The data from both GOES and METEOSAT are more than adequate for most offshore oceanographic and meteorological applications, which require time intervals of about 1 hour. These data are often not as detailed as the research quality data available from polar orbiting platforms.

Research Data

In addition to the routine observations, several groups of workers have examined certain aspects of the weather in the region in concentrated, limited time research projects. Scientists from the Massachusetts Institute of Technology (Sanders, et al.), the State University of New York (Bosart, et al.), Drexel University (Kreitzberg), and collaborative group experiments involving many scientists from government laboratories and universities (e.g., Genesis of Atlantic Lows Experiment [GALE] 1986 and Experiment on Rapidly Intensifying Cyclones over the Atlantic [ERICA] 1988–1989) have examined such features as precipitation distributions and dynamics, explosive cyclogenesis, the role of the coastal front and cold air damming, and air–sea interaction. The conceptual models of the processes that have been deduced by the various workers are the starting point for future work at the air–ocean interface. Additional useful information about winter storms along the U.S. east coast is contained in Kocin and Ucellini (1990).

Climate

It is quite difficult to unambiguously characterize the climate of a large marine ecosystem the size of the Northeast Shelf. The weather and climate are clearly quite different in the various land areas adjacent to the shelf: lowland Virginia,

Long Island, and coastal Maine. None of these areas is actually on the shelf; however, they provide a good indication of the magnitude of the variability along the shelf.

The basic climatology for the coastal region is rather straightforward. The region has a pronounced annual thermal cycle, abundant precipitation well distributed throughout the year, and generally increasing fractions of frozen precipitation and subfreezing temperatures as latitude increases. The climate in the northern regions, near Georges Bank and the Gulf of Maine, is affected by the cold waters of the St. Lawrence basin and the ice that forms seasonally in the Gulf of St. Lawrence. The climate may be classified as modified continental, clearly more maritime than Illinois or Missouri, yet more continental than Oregon or California. Continental here refers primarily to strong winter-to-summer thermal contrasts and the history of the predominant air masses. In an earlier examination of a part of this region, the climate of the Gulf of Maine was classified as a "cold-type" marine climate (Hertzman 1992). It is clear that the climate of this region is a hybrid between the low-dewpoint, rather unchangeable west coast marine climate and the rapidly changing continental climate in the middle of the continent. The important point is that both continental and maritime influences are present.

Temperature

MEANS Using data from Quayle and Presnell (1991), Figure 5-1A shows a comparison of monthly average daily mean temperatures for three stations in the region: Norfolk, Virginia; New York City; and Portland, Maine. These stations represent, to a first approximation, the range of conditions on the coast next to the Northeast Shelf. The difference in mean temperatures along the coast is greater in winter than in summer by about 50% (19°F or 12°C versus 13°F or 8°C). This reflects the greater importance of offshore airflow in winter compared with the more general southwesterly airflow during the warm season. The winter north–south thermal contrast inland is typically larger than that on the coast. Hence, the air flowing from the west or northwest is often much colder at Portland than at Norfolk.

The amplitudes of the annual cycles at each station are two to four times the range of temperatures among the stations in a given month, indicating the importance of seasons in the region. Note also that Portland has four months with mean temperatures below freezing, New York, two, and Norfolk, none. Those figures are important when considering the phase of precipitation and the effects on subsequent river runoff on the shelf.

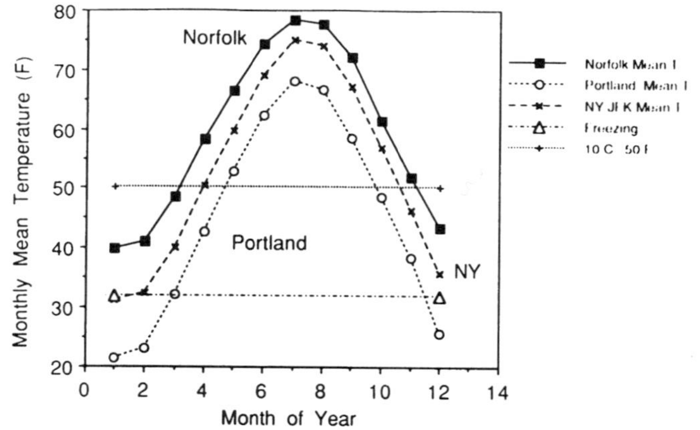

A

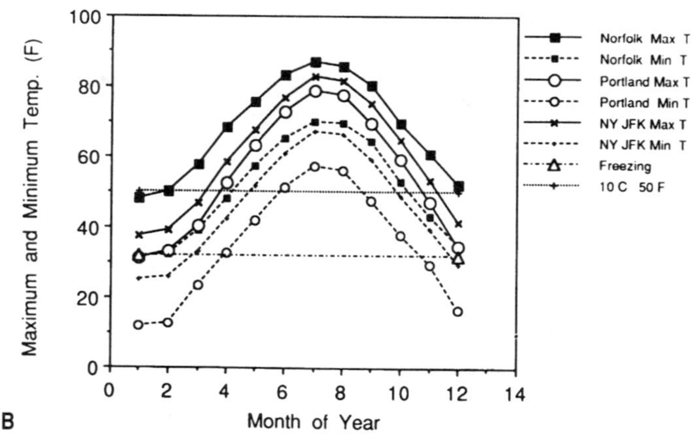

B

Figure 5-1. A. Monthly mean temperatures (degrees F) at Norfolk, Virginia (solid squares), New York City, JFK Airport, NY (crosses), and Portland, Maine (open circles). (Data from Quayle and Presnell [1991].) Freezing (0°C) is denoted by a dash-dot line, 10°C (50°F) by a dotted line. B. Same as A but for the monthly mean of daily maximum and minimum temperatures, with the smaller symbols indicating minima.

EXTREMES Figure 5-1B shows the monthly average daily maximum and minimum temperatures for the same stations. The average minimum temperatures are a surrogate for the dewpoint temperature, an indicator of the mean moisture status for the station. Summer dewpoint temperatures above 59°F (15°C) are not common in the region, indicating rather humid condi-

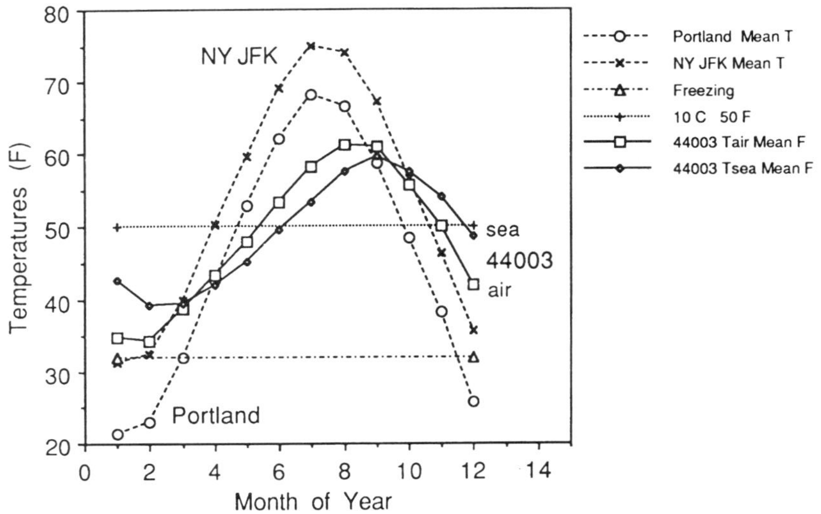

Figure 5-2. Monthly mean air temperatures at Portland and New York (as in Figure 5-1A) and offshore buoy 44003 (open squares). The sea temperature at 44003 is denoted by open diamonds.

tions and relatively weak average transfers of latent heat between the ocean and the free atmosphere in this season. Winter average minimum temperatures are at or below freezing, indicating substantial average upward fluxes of both latent and sensible heat. Average maximum temperatures exceed 59°F (15°C) in the warmest five months everywhere. When compared with the sea surface temperatures (such as those from buoy 44003 in Figure 5-2, after Hopkins and Raman 1987), these data indicate summer daytime transfers of heat from the atmosphere to the ocean during periods when air from over the near coastal areas is advected over the coastal waters. The average summer minimum air temperatures indicate little sensible heat transfer during typical nights. During particularly warm periods early in the warm season, substantial heating can occur, especially in the northern portions of the shelf.

Figure 5-2 shows the mean air and sea temperatures from buoy 44003 (latitude 41.8° N, longitude 68.5° W, southeast of Cape Cod). Note the expected decrease in the amplitude of the annual temperature cycle as one moves offshore. Also, both the air temperature (open squares) and the sea surface temperature (small open diamonds) show temperature maxima delayed by one or two months compared with New York or Portland.

1991 There is anecdotal evidence from the summer of 1991 that the coastal waters in Maine have been abnormally warm. A component of this event may have been the anomalously warm air, and downward sensible heat transfer, associated with the prolonged late spring anticyclone over eastern North America. In addition, the Gulf Stream SSTs have exceeded 30°C for extended periods. The origin of this warm surface water is still unclear. This abnormally warm surface water was associated with substantial convection along the Gulf Stream north wall during July, including Tropical Storm Ana in the early part of the month. Horizontal advection of some of this warm water may also have contributed to the warm water off the Maine coast. The water at the location of buoy 44003 appears to be several degrees Kelvin above normal this year.

Precipitation

MEANS The precipitation in the region, on average, is rather well distributed throughout the year. Annual precipitation is 1,000–1,100 mm (40–44 inches). Figure 5-3A shows climatologic precipitation values from the three stations above. Portland has a winter maximum, Norfolk a summer maximum, whereas the New York signal shows no annual cycle. This indicates that as one moves north through the region the total annual precipitation is more dominated by winter cyclones than by summer convection. The mean number of days with significant precipitation in the region is about 120, with monthly values ranging between 7.5 and 12.6 for the three stations. Combined with the statistics above, these numbers indicate that long periods with little precipitation are likely to be rare. However, a full study of this aspect of the coastal climate is beyond the scope of this report.

VARIABILITY In Figure 5-3B, the extreme monthly precipitation totals at New York are compared. It is clear that, despite the average values summarized above, there are some rather large excursions in particular years. An examination of the data in Quayle and Presnell (1991) indicates that occurrence of these extreme conditions does not appear to have increased in recent years.

SNOW AND RUNOFF Snow is rather rare in the southern part of the region, yet common in the north. Average yearly days with significant snow (>2.5 mm water equivalent) for the three stations are 2, 7, and 17 for Norfolk, New York, and Portland, respectively. Extreme snowy months produce about 60 mm of water equivalent in Norfolk and New York and 150 mm in Portland. Combined with the temperature information above, it is clear that runoff characteristics (which are not reviewed here) should have a substantially stronger spring peak in the northern shelf compared with the southern shelf.

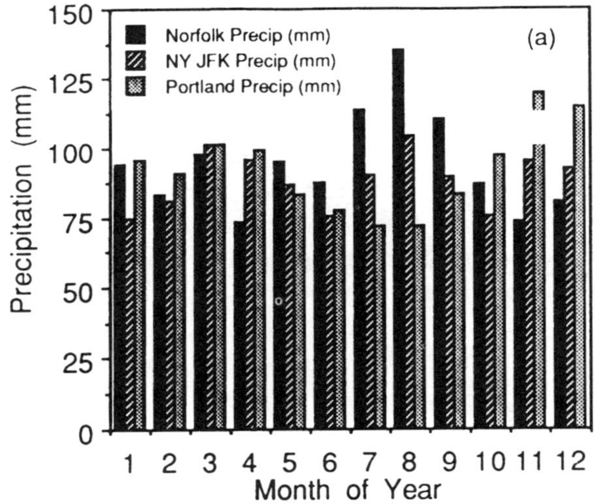

A

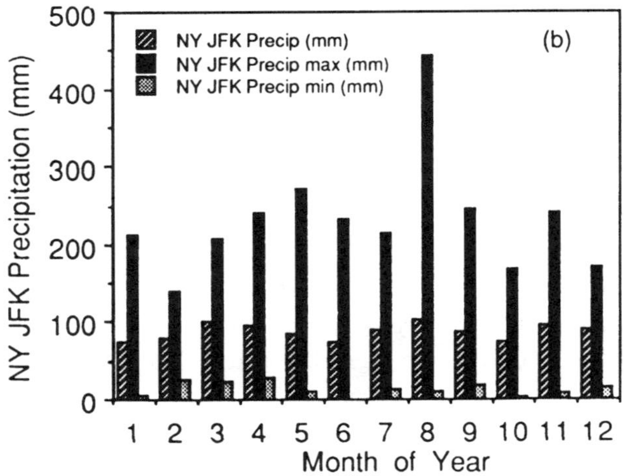

B

Figure 5-3. A. Monthly mean precipitation (millimeters) for Norfolk (solid), New York (heavy hatching), and Portland (light hatching). Data from Quayle and Presnell (1991). B. Precipitation statistics for New York, JFK. Solid color indicates the maximum monthly value in the last 43 years, light hatching the minimum, and heavy hatching the climatologic mean from A.

The net effect of this frozen precipitation is to enhance spring and summer fresh water input to the northern shelf at the expense of that during the winter. Combined with the winter maximum in precipitation in this region (Figure 5-3), the net horizontal fresh water input rate to the shelf is rather constant in both space and time compared with regions farther northeast, which are strongly influenced by the St. Lawrence River outflow (J. Umoh, personal communication, June 1990).

OFFSHORE Estimates of precipitation amounts offshore are rather questionable. Some of the Canadian experiences with drillship data are summarized in Bursey and Shaw (1990). Work done on both Canadian coasts indicated precipitation amounts from weatherships may also be seriously in error (Knox, personal communication, 1987). The strong evidence of cyclonic storm formation and intensification in the area would lead one to estimate higher precipitation amounts offshore. However, most investigators feel that there is some coastal enhancement of precipitation resulting from increased convergence at the land–ocean boundary associated with the coastal front. This shallow feature, the boundary between cold air dammed by the Appalachians and warmer marine air offshore, has been documented from New England (Bosart 1975) to North Carolina (several cases in GALE). It is often a precursor of intense storm development (e.g., Bosart 1981). These two effects on the precipitation allow the use of coastal information as a reasonable first approximation for the offshore areas.

CONVECTION The number of thunderstorm days per year varies from 17 at Portland to 37 at Norfolk. Some of this convection occurs in cyclonic storms, either in the forced convection region along the cold front or associated with the leading region of warm advection. Most of the summer convection is associated with thunderstorms, either isolated or organized in mesoscale lines or systems. These features are rather small and can interact with the ocean on spatial scales close to the observed Rossby radii of a few tens of kilometers. Occasionally, pre-existing convective features contribute to strong cyclonic storm development, as in the September 1978 "*Queen Elizabeth II* (QE II) storm" (Gyakum 1983), although there is still some debate about the relative importance of the convection compared with other factors (Gyakum 1983; Ucellini 1986). During ERICA, very strong convection was observed several times during the intensifying and mature stages of rapidly intensifying winter cyclones. The role of thunderstorms in both the lower-atmospheric preconditioning and the actual intensification of larger scale cyclonic storms in the shelf region is still a matter of active research.

GULF STREAM RAINBAND There is a region of banded convection along and southeast of the northwest wall of the Gulf Stream, east and northeast of Cape Hatteras (Hobbs 1987). It is quasi-continuous, being observed many days in most seasons, except when there are strongly unfavorable large scale synoptic conditions, such as post-frontal subsidence, in the region. Sweet et al. (1981) noted that this convection is often associated with changes in visibility along the Gulf Stream north wall. In addition, they noted "ghost" radar echoes or chaff corridors in earlier naval documents. From the field work in GALE, these echoes are likely those associated with the Gulf Stream rainband (Hobbs 1987), a shallow convective feature that can actually help dissipate pre-existing winter storms passing through the region (Sienkiewicz et al. 1989). The cloud tops in this type of rainband can be below 60 kPa (4 km), yet the band can contain intense cloud-to-cloud lightning and precipitation regions. In the warm season these rainbands can be deeper, much deeper, much like usual lines of thunderstorms, and can contribute to cyclogenesis as noted above.

Winds

The climatologic wind information indicates mean flow from the south or southwest at Norfolk most of the year and at all stations from June to August, from the northwest quadrant at the northern stations in winter, and from the northeast occasionally (Figure 5-4). When the mean monthly stress at buoy 44003 (Hopkins and Raman 1987) is compared with the winds at the land stations, the directional agreement is excellent, whereas the annual variation of the magnitudes is in phase as expected.

Despite the good agreement from the buoy, the believable offshore wind record is rather sparse. The microwave scatterometer aboard the ERS-1 satellite is providing state-of-the-art winds for marine applications. Though launched in 1991, the data are still under evaluation in 1996. The archive data from these platforms would appear to be of most use for oceanographic applications. Details on scatterometers and the ERS-1 can be found in Ikeda and Dobson (1996). However, high-quality continuous, calibrated surface wind measurements well offshore are still needed to derive useful numbers from the scatterometer data.

Air–Sea Exchange

The air–sea exchange part of the regional climatology is perhaps the most difficult to assess. Knowing the characteristics of the air over different parts of the shelf at a given time is part of the problem. Latent heat transfer, associated with evaporation, dominates air–sea energy exchange in regions such as Station P in

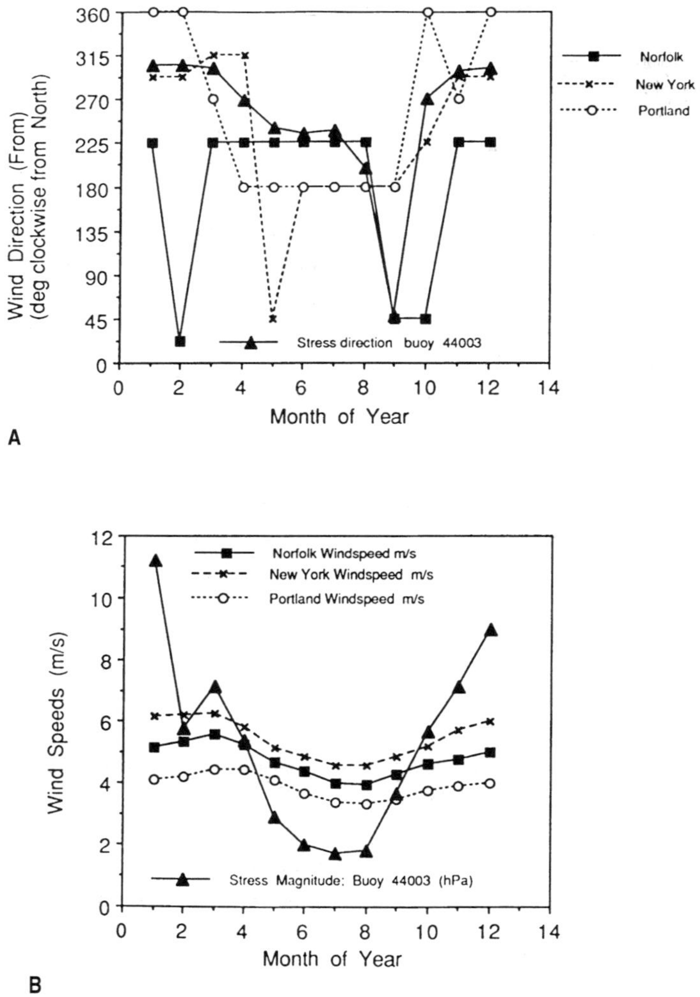

Figure 5-4. A. Monthly mean wind direction for the stations in Figure 5-1A and for buoy 44003 (solid triangles). B. Monthly mean wind speeds for the stations in Figure 5-1A. Monthly mean stress magnitude (hPa) at buoy 44003 (after Hopkins and Raman, 1987), indicated by solid triangles, is superimposed.

the North Pacific, both climatologically and on a per event basis (Hertzman et al. 1974). However, on the Northeast Shelf, although we have a reasonable knowledge of the air temperature and SST, our lack of knowledge of the evolution of surface dewpoint temperatures as air moves offshore decreases our ability to accurately partition the energy transfer over the region.

OBSERVATIONS Close to the shore, during particular winter cold-air outbreaks, total upward energy flux can exceed 1,000 W m^{-2} (Bane and Osgood 1989). The best summary of the climatologic importance of these events was done by Grossman and Betts (1990), using aircraft eddy-flux observations of the same GALE case:

> . . . we have the following hierarchy of total energy exchange from ocean to atmosphere: in the latitude belt 30°–40°N, an annual average of 70 W m^{-2}; within the same latitude belt but just off the eastern coast of the United States, an annual average of 380 W m^{-2}; for the same area a January monthly average of 550 W m^{-2}; within a much smaller area over the Gulf Stream during a January cold air outbreak . . . a peak total flux of about 1000 W m^{-2}. . . . Assuming two cold air outbreaks a winter month, each lasting about two days, and that one-third of the area is associated with a total flux of 1,000 W m^{-2} while the rest is associated with about 700 W m^{-2} (slightly less than the lowest level flux measured over the 20°C water of the eastern Gulf Stream front) then extreme cold air outbreaks could account for about 20% of the monthly average loss of energy from ocean to atmosphere during winter (in the region).

The partitioning of the largest fluxes was about 30% sensible and 70% latent heat transfer.

The Northeast Shelf intersects that zone but also includes regions where mean winter air temperatures and SSTs are both much colder. Because of nonlinearities in the saturation water vapor pressure as a function of temperature, sensible heat fluxes grow in relative importance as temperature decreases. So, for similar events over the shelf, perhaps 35–45% of the total energy transferred is via sensible heat. Thus, both sensible and latent heat fluxes are important in both the climatology and in the evolution of particular weather events.

ROLE IN CYCLOGENESIS The role of these fluxes in cyclogenesis is another area of active research. Kuo et al. (1991), in a series of numerical experiments, emphasized the key role of fluxes in preconditioning the environment for storm development. These conclusions are largely being confirmed by recent observational studies as well.

Weather

In general, the predominant weather features are cold season cyclonic storms, which either are swept eastward through the region or form near Cape Hatteras, and warm season convection and weaker cyclones.

Winter Storms

In winter, the Northeast Shelf area is a region of surface air temperature intermediate between generally subfreezing daily mean temperatures over land (Figure 5-1) and greater than 20°C temperatures over the Gulf Stream. Because of this strong temperature contrast, there is a chronic atmospheric baroclinic zone in the lower half of the atmosphere between the northwestern wall of the Gulf Stream and the St. Lawrence Valley. The winter weather on the Northeast Shelf is largely determined by how this baroclinic zone interacts with the vorticity maxima swept along in the westerly jetstream (which typically crosses the coast nearby) to generate, intensify, or dissipate mid-latitude cyclones.

Many of the usual "normal strength" winter storms are steered across eastern North America by generally zonal (west-to-east) jets. During winter these jets often cross the coast of the Canadian Maritimes or New England, steering storms generated further west through the northern part of the region. In some years (such as late 1990), this jet remains further north than usual. This tends to decrease the number and strength of cold air outbreaks while enhancing warm, moist surface flow from over the Gulf Stream and the southeastern United States. The speed of this jet (10–20 m s^{-1} in the middle atmosphere) combined with a global wave number of 7–14 gives an advective time scale of 1.5–5 days. The time scale is shorter for a more intense jet and smaller scale storms. If the SST gradient is strong on the Northeast Shelf, this additional baroclinic forcing can act to intensify these pre-existing storms locally and downstream.

Both types of cyclones (those advected in and those that form along the coast) appear to preferentially track northeastward along the coast or along the north wall of the Gulf Stream, although control by the 50 kPa flow is also important.

Storms of the first type are often disrupted by the Appalachians. In a GALE study, O'Handley (1988) noted that most cold fronts slow down and are deformed approaching the mountains, with considerable low-level blocking occurring. On passing the mountains, about half the cases he examined showed redevelopment, with about 70% of the organized redevelopment occurring to the southeast. The majority of these went on to become strong cyclones. Hence, the region of interest here is an area of storm redevelopment and intensification.

Storms of the second type, with an offshore track, often give rise to strong northeasterly winds to the northwest of the track ("nor'easters"). It is important to realize that large portions of the shelf may in fact experience no northeasterly winds during such events. The strong cold air outbreaks alluded to above usually occur behind the cold fronts of these types of storms. Coastal tracking storms also bring cold air south but not over the shelf.

Often coincident with a cyclonic disturbance decaying west of the Appalachians, a secondary development occurs along the coast, to the southeast of the original center. This secondary cyclone often tracks along the coast. Although it may deepen rapidly, it may be of rather ordinary strength if one or more of the ingredients required for rapid deepening, such as some support by the upper atmospheric vorticity and divergence fields (Ucellini 1986), is missing.

The particular geography of the eastern United States is conducive to the generation of rapidly (explosively) deepening cyclones under certain conditions. There is well documented evidence of this maximum in synoptic data (Sanders and Gyakum 1980). The width of the coastal plain in North and South Carolina seems to provide enough cold air to allow intense cyclogenesis to occur just downstream. The mean alignment of the Gulf Stream in the region is close to that of the mean middle atmospheric flow with a cyclonic center approaching from the west. The particular necessary and sufficient conditions for these explosive cyclones are still the subject of research.

EXPLOSIVE CYCLOGENESIS If an anticyclone over central Canada is (quasi-) stationary long enough to pump a large amount of cold air southward, east of the Appalachians (damming cold air against the mountains), then a thermal contrast is set up between the air over the mid-Atlantic (or other coastal) states and the air offshore over the shelf and the Gulf Stream. This contrast is often expressed as a shallow quasi-stationary coastal front, with rather sharp horizontal convergence and cyclonic vorticity associated with it. Temperature gradients across these fronts are perhaps as much as 20°C per 50 km.

If the upper-level atmospheric flow is such that a baroclinic wave and the associated upper level divergence and mesoscale updraft locks into the lower level convergence and vorticity along the coastal front, with optimum upstream tilting of similar wave phases, then a rather rapid cyclonic development often occurs. If, in addition, the sea surface temperature (SST) distribution ahead of the developing system is such that the boundary layer air has been warmed and moistened enough by surface sensible and latent heat fluxes to allow convection to be triggered rather easily as the disturbance moves over water, then there is a high probability that the upper-level disturbance will deepen rapidly (explosively) into a rather large and intense cyclonic storm. Deepening rates of 20 mb (2 kPa) in 12 hours and 40 mb in 24 hours are often observed in these situations. If the track of the storm center is along the edge of the Gulf Stream or just to the north of it, its sector of northeasterly surface flow often crosses the northern part of the Northeast Shelf region.

Depending on the strength and trajectories of the cold air outflow from a storm such as this, and the new position of the jet stream after the storm has gone by, one of these storms can enhance the intensification of the next upper-level disturbance. The Canadian Atlantic Storms Program (CASP) and ERICA field projects have provided additional evidence for the scenario of two preferred storm tracks for rapidly deepening east coast cyclones (and perhaps for all cyclones, although the evidence is less clear).

The baroclinicity associated with the temperature contrast between the Northeast Shelf and the continent to the west provides additional available potential energy for cyclones forming in the region or passing through from the west. The relatively warm waters strengthen pre-existing cold fronts and coastal fronts, aid in the secondary deepening of coastal lows, and help to steer systems up into the Bay of Fundy. The baroclinicity associated with the contrast between the colder inshore waters and the Gulf Stream enhances storm development by providing a long, moist fetch (for northwesterly cold inflow) where the water temperature is well below 20°C. Despite the strong surface fluxes, air can exit the Gulf of Maine and enter a cyclone developing along the north wall of the Gulf Stream 15°C cooler than the air entering on the warm side.

OCEANIC EFFECTS Storms on these two tracks should provide quite different wind forcing for the Northeast Shelf. Coastal track storms will produce clockwise wind turning in the shear zones associated with the warm and cold waves. Gulf Stream storms will produce more gradual counterclockwise turning, fewer frontal passages, and longer surface wave fetches. Inertial wave generation would be enhanced more by coastal storms (e.g., Anderson and Smith 1989) than by Gulf Stream storms.

Storms that cross the New England coast when a strong stationary anticyclone is over the eastern Maritimes and Newfoundland will tend to dissipate over the Gulf of Maine. From anecdotal evidence, this appears to be an event most common in the warm season, although occasional winter anticyclones also cause this to occur.

Rather weak disturbances crossing the mid-Atlantic coast can often generate quite intense convection, especially in the transition seasons as seen in the latter part of GALE in March 1986.

Summer Storms

Hertzman (1992) noted that the conditions in the Gulf of Maine in summer are rather more benign than in winter. Because the mean jet stream and its associated baroclinic zone tend to be well to the north in the warm season, cyclonic storms tend to be fewer and weaker. This conclusion also holds for the Northeast Shelf as a whole. However, convective activity is much more important for the Northeast Shelf as a whole than for the northern areas alone (see earlier section on climate).

The strong wind and heavy precipitation events that occur in the eastern United States during the summer are usually associated with convective systems. Some of these systems are locally generated thunderstorms, which last an hour or two then disappear. Some of the systems are more intense and longer lasting (lifetimes up to 36–48 hours). These systems, known variously as pre-frontal squall lines or systems, mesoscale convective complexes, or mesoscale convective systems (MCSs) consist of groups of thunderstorms that grow and decay as the group moves together.

Mesoscale convective systems are primarily spawned on the plains, although they can also form east of the Mississippi River. Several times each summer, the mid-Atlantic states experience the passage of a MCS. They rarely reach downstream areas such as Nova Scotia. Hence, it appears that the coastal and shelf waters (especially the Gulf of Maine) are a zone of dissipation for these systems. Presumably the wet-bulb temperature of the air over the coastal waters is not generally high enough to sustain the convective systems for tens of hours. Wind magnitudes associated with these systems are of the same order as those from strong cyclones, and the directional variability is larger, in both space and time. Because they are generally smaller than a full-blown cyclone, their mixing strength would be somewhat lower as a whole. Locally, however, these are strong events and may cause substantial changes in the surface water properties.

Hurricanes

Atlantic hurricanes occasionally pass along or east of the Northeast Shelf. Strong hurricanes very rarely pass through it. Heavy precipitation can occur even 1,500 km from the center of these storms. These effects can be felt even if the storm itself is not a factor in the local dynamics. The main dynamic effect of distant hurricanes is on the swell field and on the water levels, especially near the coast. During the infrequent nearby hurricane occurrences, the wind stress forcing on the system can be treated in much the same way as an intense winter cyclone.

By the time hurricanes reach latitude 35°N, they are usually speeding up and beginning to acquire some mid-latitude storm characteristics. Most of the published hurricane tracks that reach these latitudes (e.g., Gray 1990) indicate that there is a pronounced turning to the east (right, looking downstream) as the hurricanes track northward up the east coast. Because hurricanes dissipate rapidly when their centers are over land, and moderately fast when they are over cool water (<20°C), any hurricane in the Gulf of Maine must be dissipating. In the region near Cape Hatteras, tropical storms can still maintain themselves (Tropical Storm Ana in 1991 is an example). In this case, however, steering by both the Gulf Stream and the upper flow rapidly removes the active storm from the shelf.

Because of the increased oceanic stratification in the summer, the effectiveness of the subsurface mixing from a convective storm or hurricane is lower than that for a winter storm of similar wind stress or wind stress curl.

Climate Change

Attempting to predict changes in regional climate from scenarios with global scale forcing is always risky. However, because of the focus of this symposium, I have a few remarks to share. There is very little question that a mean global warming of 1–4°K over the next several decades would produce changes in the climate of the region. As I have attempted to show above, the weather and climate of the region are dependent on influences far removed from the region itself. If the climate in those areas changes, then some effects of the changes will reach this region advectively, if not by other means.

Qualitatively, the tropically generated components of the climate system, the Gulf Stream and the Gulf of Mexico warm air, would be expected to warm rather modestly (perhaps a third to a half of the global average air temperature increase). The continental cold air, especially in winter, would also warm, with the effect increasing with latitude. This would certainly decrease the strength of the cold air outbreaks discussed and the subsequent buoyancy-driven mixing. The strength of the mean westerly jet across North America would decrease slightly because of a decrease in meridional mean temperature gradient. This effect would be largest in winter, according to most projections. Finally, freshwater outflow from the Arctic would increase, cooling the climate in eastern Canada and perhaps also cooling the freshwater input to the region.

The net effect of these changes could be quite substantial if the actual mean climate change reaches the high end of current predictions (4°K increases in global temperature for an equivalent carbon dioxide doubling). However, if the actual global changes are on the low end of the predictions, the small resulting changes in the mean conditions on the Northeast Shelf would have relatively little importance when superimposed on the rather large seasonal changes in the region.

Summary

This overview of the meteorology of the Northeast Shelf provides the background to both thermal and dynamic mixing mechanisms for the ocean in the Northeast Shelf region. Both mechanisms are stronger in winter, although the forcing in the warm season is on a smaller scale, closer to the oceanic synoptic scale. The region is an important one for ocean-to-atmosphere heating and moistening and cyclonic storm generation and intensification.

References

Anderson, C., and Smith, P. C. 1989. Oceanographic observations on the Scotian Shelf during CASP. Atmosphere–Ocean 27:130–156.

Bane, J. M., and Osgood, K. E. 1989. Wintertime air-sea interaction processes across the Gulf Stream. J. Geophys. Res. (Oceans) 94:10755–10772.

Bosart, L. F. 1975. New England coastal frontogenesis. Q. J. Roy. Meteor. Soc. 101:957–978.

Bursey, J. O., and Shaw, R. W. 1990. The Representativeness of Precipitation Measurements Taken on East Coast Drilling Platforms. Atlantic Region. Report MAES 1-90, 26 pp. Ottawa: Environment Canada Atmospheric Environment Service.

Gray, W. M. 1990. Strong association between West African rainfall and U.S. landfall of intense hurricanes. Science 249:1251–1256.

Grossman, R. L., and Betts, A. K. 1990. Air-sea interaction during an extreme cold air outbreak from the eastern coast of the United States. Month. Weather Rev. 118:324–342.

Gyakum, J. R. 1983. On the evolution of the QE II storm. II: Dynamic and thermodynamic structure. Month. Weather Rev. 11:1156–1173.

Gyakum, J. R. 1981. Meteorologic precursors to the explosive intensification of the QE II Storm. Month. Weather Rev. 119:1105–1131.

Hertzman, O. 1992. Meteorology of the Gulf of Maine. In J. Wiggins and C. N. K. Mooers (eds.), Proceedings of the Gulf of Maine Workshop. Boston: Gulf of Maine Council on the Marine Environment, Urban Harbors Inst. of U. Mass. Boston, pp. 39–50.

Hertzman, O., Miyake, M., and Pond, S. 1974. Ten years of meteorologic data at ocean weather station Papa. Institute of Oceanography (UBC) Manuscript report No. 29, UBC, Vancouver, British Columbia, Canada.

Hobbs, P. V. 1987. The Gulf Stream rainband. Geophys. Res. Lett. 14:1142–1145.

Hopkins, T. S., and Raman S. 1987. Atmospheric variables and patterns. In R. H. Backus (ed.), Georges Bank. Cambridge, MA: MIT Press, pp. 66–73.

Ikeda, M., and Dobson, F. 1996. Remote Sensing Applications for Oceanography. New York: CRC Press.

Kocin, J., and Ucellini, L. 1990. Snowstorms along the Northeastern Coast of the United States: 1955 to 1985. Boston: American Meteorological Society.

Kuo, Y. -H., Reed, R. J., and Low-Nam, S. 1991. Effects of surface energy fluxes during the early development and rapid intensification stages of seven explosive cyclones in the western Atlantic. Month. Weather Rev. 119:457–476.

O'Handley, C. 1988. Orographic effects during GALE. GALE/CASP Workshop Reports, December 1988, p. 215. (GALE Project Office, NCAR, P.O. Box 3000, Boulder, CO)

Quayle, R., and Presnell, W. 1991. Climatic averages and extremes for U.S. Cities. NOAA Historical Climatology Series 6-3. Asheville, NC: U.S. Dept. Commerce, NOAA/NCDC.

Sanders, F., and Gyakum, J. R. 1980. Synoptic-dynamic climatology of the bomb. Month. Weather Rev. 108:1589–1606.

Sienkiewicz, J. M., Locatelli, J. D., Hobbs, P. V., and Geerts, B. 1989. Organization and structure of clouds and precipitation on the mid-Atlantic Coast of the United States. Part II: The mesoscale and microscale structures of some frontal rainbands. J. Atmos. Sci. 46:1349–1364.

Sweet, W., Fett, R., Kerling, J., and LaViolette, P. 1981. Air-sea interaction effects in the lower troposphere across the north wall of the Gulf Stream. Month. Weather Rev. 109:1042–1052.

Ucellini, L. W. 1986. Possible influence of upstream upper-level baroclinic processes on the development of the QE II Storm. Month. Weather Rev. 114:1019–1027.

6

Climate Change and Winter Coastal Marine Circulation in the Mid-Atlantic Region of the United States

Fredric A. Godshall and Henry A. Walker

Abstract

Over the past century, the climate warmed in most regions of the northern hemisphere but with variations that included a period of cooling during the 1950s. This cooling period ended during the late 1960s or early 1970s, when the warming trend resumed. Decadal summaries of ship-reported winds and wind observations from coastal meteorologic stations show systematic changes of wind within this period of climate cooling. Salinity changes at Montauk, New York, and Woods Hole, Massachusetts, coincidental with wind direction changes are explained by coastal circulation changes that are expected with wind changes.

Air temperature is one of the variables in the ensemble considered to represent climate; therefore, temperature changes are used here as indications of climate change. Air temperatures in the northern hemisphere (Hansen and Lebedeff 1987) have generally increased since the late 1800s. Short-period variations in temperature, such as a cooling period from the late 1940s to the late 1960s, also occurred. This period of cooling exhibited persistent blocking in the large-scale upper-air circulation over eastern North America (Andrews 1956; Klien 1956) and in Europe (Dickson et al. 1975). Low-index flow was associated with the blocking, and northerly winds increased over Europe and the North Atlantic.

A long record of observations from New Haven, Connecticut (Ingham 1982), shows that temperature variations in the mid-Atlantic region were generally in phase with the temperature changes over the hemisphere reported by Hansen and Lebedeff (1987). The period of increased northerly winds over the North Atlantic and Europe during the period of climate cooling is also anticipated in the mid-Atlantic region, since the period of cooling in the region is coincidental with cooling over the hemisphere.

Coastal orientation relative to wind direction affects wind influences on coastal circulation. The mid-Atlantic region (Figure 6-1) is geographically bounded in the north by the New England coast, Long Island, and Cape Cod, which are generally oriented west and east. The contours of bathymetry on the continental shelf also follow the coastline; in southern parts of the mid-Atlantic region, however, the coast and bathymetry are generally oriented north and south. Because of these differences, wind effects on coastal circulation will be different in different areas of the region.

Analysis

The Comprehensive Ocean–Atmosphere Data Set (COADS) (NOAA 1985) summary of ship-reported wind observations was used to compute decadal averages of winter (January, February, and March) wind direction and speed in a 2° square ocean area centered at 41° N, 71° W. Decadal averages of monthly wind observation summaries were computed from archives of wind data from island stations on Nantucket, Massachusetts, and Block Island, Rhode Island. Regional wind variations are illustrated by these wind vectors. The wind direction was generally from the west in the offshore area and at the island stations before the 1950s, but during the 1950s and 1960s, the summaries show that winds in the mid-Atlantic region were more northerly (Figure 6-2).

Coastal tide measuring stations at Montauk, New York, on Long Island and the tide-measuring station at Woods Hole, Massachusetts, reported measurements of surface water temperature and density. Salinity was estimated from the archives of the water temperature and density data using equations of state for water (Millero et al. 1980). A graph of winter averages of these estimates shows that salinity was lower at these stations during the 1950s and 1960s (Figure 6-3). A graph of gauged flow from the Hudson River watershed (Figure 6-4) shows a flow variation similar to that of most major rivers in the northeastern United States during the period of the mid-1930s through 1980. A plot of surface salinity estimated at a Sandy Hook, New Jersey, tide-measuring station (see Figure 6-4) shows how salinity is reduced in some areas because of river outflow. However, salinity changes estimated at Montauk and

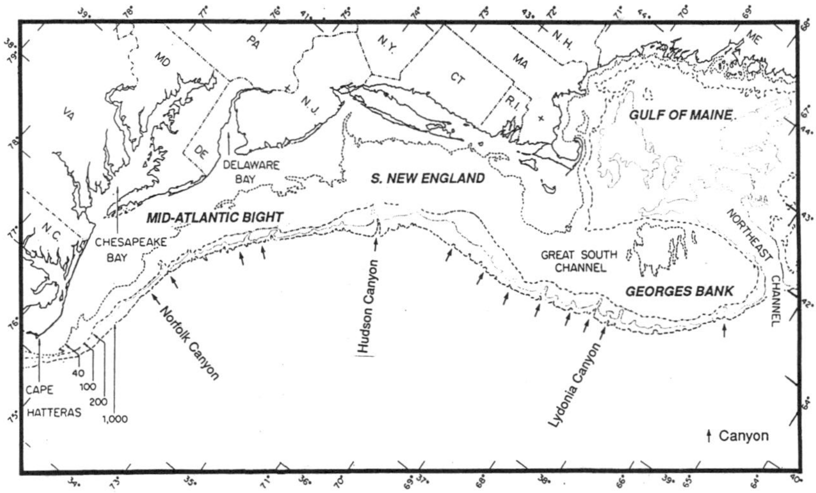

Figure 6-1. The four geographic areas of the northwest Atlantic—Gulf of Maine, Georges Bank, southern New England, and Middle Atlantic Bight—and stations sampled for zooplankton during MARMAP operations.

Woods Hole during the 1950s and 1960s tend to decrease with decreases in regional land runoff.

Decadal winter season salinity estimates and surface water temperature observations reported from the Woods Hole tide-measuring station are presented in a temperature/salinity (T/S) diagram (Figure 6-5). Comparison of the distribution of these coastal T/S characteristics to an area on the graph enclosing the winter T/S characteristics of a nearshore, shelf water mass of this mid-Atlantic region (as reported by Williams and Godshall 1977) shows that the range of T/S characteristics for New England near-coast shelf water overlaps the T/S characteristics from Woods Hole during the 1970s and 1980s but not during the 1950s or 1960s.

Conclusions

Considering the orientation of Long Island and the bathymetry of the coastal areas, upwelling along the coast is expected when wind directions are approximately parallel to shorelines and bathymetry. During the 1970s and 1980s, winds tended to be oriented along coastlines in the northern part of the mid-Atlantic region, and upwelling would be expected to force shelf water shore-

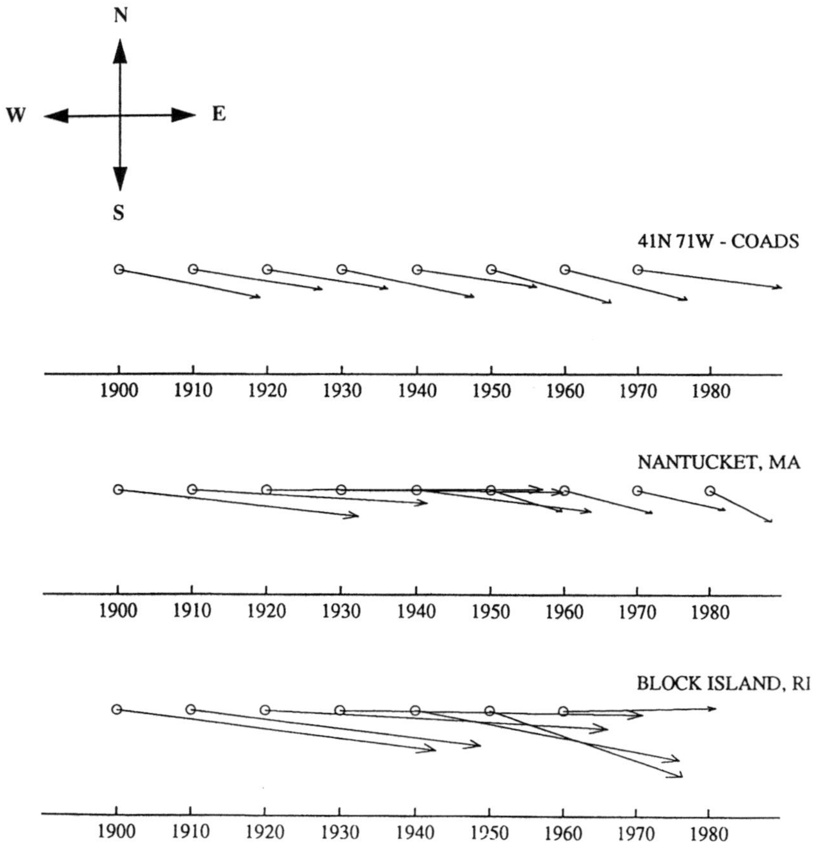

Figure 6-2. Decadal vector summaries of ship-reported winter wind from a COADS 2° area and island stations in the mid-Atlantic region. (No data at Block Island for 1970s or 1980s; no COADS data for 1980s.)

ward. During the 1950s and 1960s, winds tended to have less orientation along the coastlines, and relatively saline shelf water would have had less effect on salinity at the tide-measuring station locations. Therefore, the temporal salinity changes estimated from observations at Montauk, New York and Woods Hole, Massachusetts may be caused by the wind changes during the 1950s and 1960s. Salinity changes at the Montauk and Woods Hole tide-measuring stations could not be caused by a change in land runoff and dilution, since the variation in runoff is generally with the salinity changes.

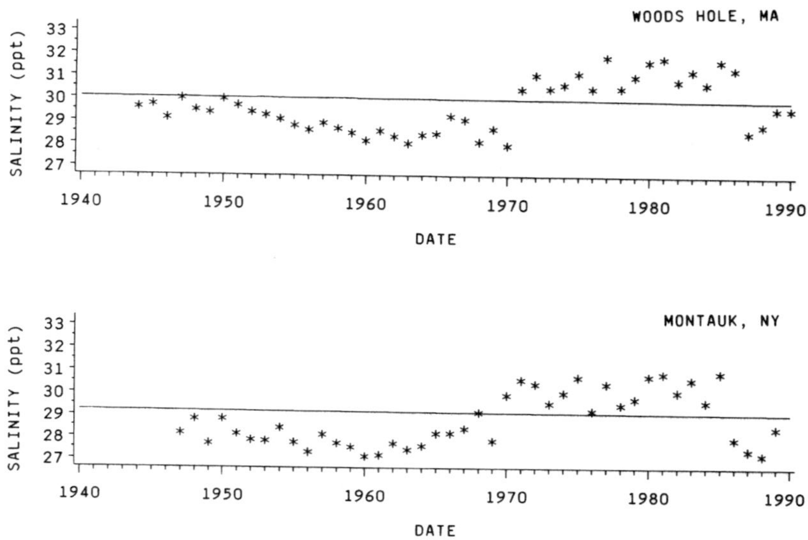

Figure 6-3. Surface salinity estimates from measured temperature and surface water density during winter at Woods Hole, Massachusetts, and Montauk, New York, tide-measuring stations.

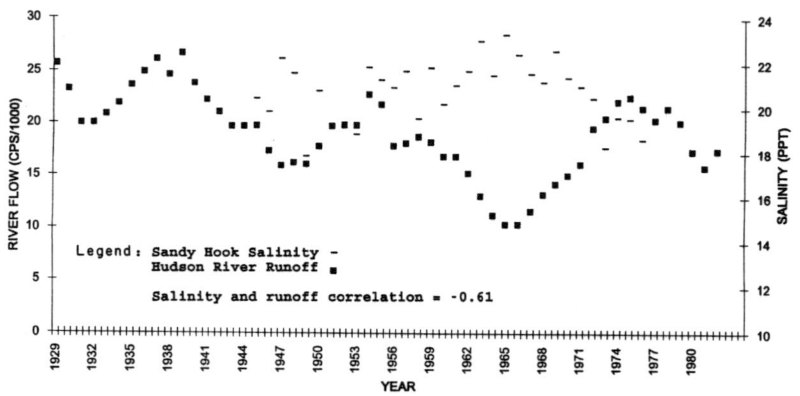

Figure 6-4. Outflow from the Hudson River watershed through Rondout Creek (a measured surrogate to the main flow of the Hudson), smoothed over 6 months; surface salinity estimated from measurements of surface water temperature and density at the Sandy Hook, New Jersey, tide-measuring station.

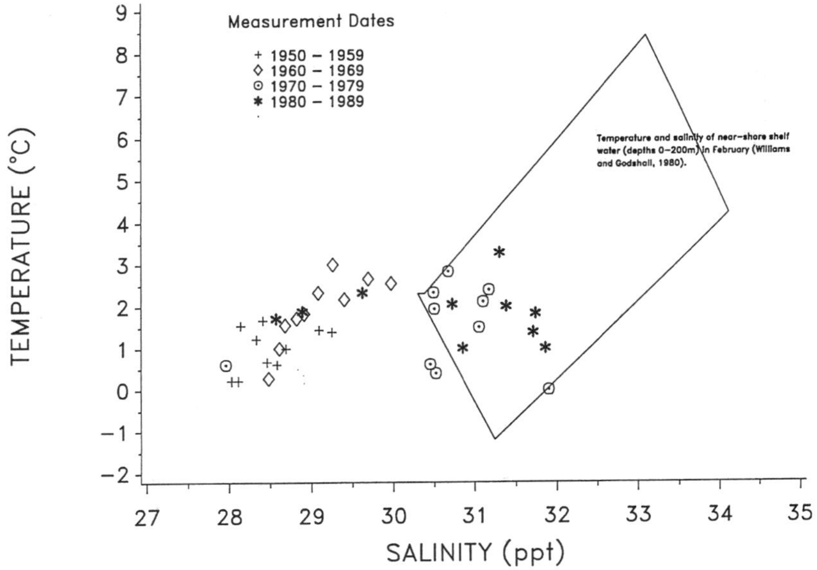

Figure 6-5. Surface water temperature and salinity decadal summaries from Woods Hole, Massachusetts, and temperature–salinity characteristics of the mid-Atlantic region continental shelf water mass.

The wind changes in the mid-Atlantic region occur coincidentally with the wind changes over the North Atlantic, and lower air temperatures in the mid-Atlantic region are coincidental with broad-scale climate cooling over the northern hemisphere during these decades. This association of changes provides a chain of relationships from the broad-scale hemispheric processes to local scales of change in the mid-Atlantic region.

References

Andrews, J. F. 1956. The weather and circulation of February 1956. Month. Weather Rev. 84:66–74.

Dickson, R. R., Lamb, H. H., Malmberg, S.-A., and Colebrook, J. M. 1975. Climate reversal in northern North Atlantic. Nature 256:479–481.

Hansen, J., and Lebedeff, S. 1987. Global trends of surface air temperature. J. Geophys. Res. (D11) 92:13345–13372.

Ingham, M. C. 1982. Weather Conditions and Trends in the Maine–Virginia Coastal and Offshore Areas During 1970–79. Northwest Atlantic Fisheries Organization Scientific Council Studies, Number 5, pp. 33–37.

Klein, W. H. 1956. The weather and circulation of January 1956: A month with a record low index. Month. Weather Rev. 84:25–34.

Millero, F. J., Chen, C.-T., Bradshaw, A. B., and Schleicher, K. 1980. A new high pressure equation of state for seawater. Deep-Sea Res. 27A:255–264.

National Oceanic and Atmospheric Administration (NOAA). 1985. Comprehensive ocean-atmosphere data set: Release 1. Boulder, CO: Cooperative Institute for Research in Environmental Sciences, Univ. of Colorado.

Williams, R. G., and Godshall, F. A. 1977. Summarization and interpretation of historical physical oceanographic and meteorological information for the mid-Atlantic region. Washington, DC: U.S. Department of Commerce.

Editors' Note

We have included a reference below on the North Sea ecosystem because of its similarity to the Northeast Shelf ecosystem in terms of species diversity, resource utilization, the advanced technological development of its adjacent coastal countries, and the resultant urban and industrial inputs to the ecosystem. Like the Northeast Shelf ecosystem, the fish community of the North Sea is heavily harvested, with some evidence of overexploitation. A comparison of the North Sea and Northeast Shelf ecosystems reveals more severely depressed fish stock levels in the latter. By contrast, coastal eutrophication from excessive nutrient loadings is in a more advanced state in the North Sea and appears to be related to an increasing frequency and extent of harmful algal blooms there. The environmental ministers of the North Sea coastal states have initiated governance control, having reached an agreement to limit nutrient inputs by the coastal states in an effort to control eutrophication. Although it is not practical to conduct large-scale experiments within large marine ecosystems (LMEs), comparisons of their changing states and the driving forces behind those changes can be most instructive when developing mitigation strategies; ecosystem degradation is more easily minimized when its causes are known from previous LME assessments. The North Sea report is particularly instructive in this regard. It is published as the North Sea Task Force. 1993. North Sea Quality Status Report, 1993. Oslo and Paris Commissions, London. Fredensborg, Denmark: Olsen & Olsen. It is available from the publishers.

7

The Northeast Shelf Ecosystem: An Initial Perspective

Kenneth Sherman, Marvin Grosslein, David Mountain, Donna Busch, Jay O'Reilly, and Roger Theroux

Editors' Note

Before 1988, published reports on the marine populations, resources, and dominant physical features of the Northeast Continental Shelf were rather specialized. Topics receiving attention in various publications were the fish stocks and fisheries, the plankton, the benthos, and the oceanography of particular areas of the shelf (e.g., coastal bays, Georges Bank, Gulf of Maine, Long Island Sound, Chesapeake Bay, Narragansett Bay, Boston Harbor). Separate monographs by Bigelow and others treated the oceanography of the Gulf of Maine and the plankton of the Gulf of Maine. The first integrative report to focus on the Northeast Shelf as an interconnected large ecosystem was published as part of a series on continental shelves of the world. It provides an overview of the ecosystem's richness in species biodiversity, the importance of the resources produced by the system's natural capital, and the considerable level of understanding gained by the scientific community during the past several decades of the ecosystem's major physical characteristics and rather high level of annual biomass productivity. Lack of space in the present volume precludes the reproduction here of the entire report. An abridged and annotated version is presented here to provide an overview of the principal findings of the original 1988 paper published by Elsevier Press. Annotations bring the text forward to the present state of knowledge where appropriate. The full reference to the work is: K. Sherman, M. Grosslein, D. Mountain, D. Busch, J. O'Reilly, and R. Theroux of the Northern Fisheries Center, NMFS-NOAA, Woods Hole, MA. 1988. The Continental Shelf Ecosystem off the Northeast Coast of the United States. In H. Potsma and J. J. Zijlstra (eds.), Ecosystems of the World 27: Continental Shelves. Amsterdam: Elsevier, pp. 279–337.

The continental shelf off the northeast coast of the United States is one of the most intensively studied regions of the North Atlantic. Results of early studies of the shelf, its physics, chemistry, geology, and biology can be found in monographs of the U.S. Fish Commission (now the National Marine Fisheries Service), the U.S. Geological Survey, the Woods Hole Oceanographic Institution, and the Bingham Oceanographic Collection of Yale University. An excellent synthesis of the early work from the turn of the century through 1970 is given by Emery and Uchupi (1972). The early research on shelf productivity and biology ranks among the best in the world, and can be found in the papers of Bigelow (1926, 1927), Fish (1936a, b, c), Redfield et al. (1937), Bigelow and Sears (1939), Bigelow et al. (1940), Riley (1941, 1947), Redfield (1941), Clarke (1946), Bigelow and Schroeder (1953), and Ketchum and Keen (1955). More recent summaries of shelf ecosystem productivity are given in Colton (1964), Cohen (1976), Sherman et al. (1978), Sherman (1980), Grosslein et al. (1980), Falkowski et al. (1980), Walsh (1981), Cohen et al. (1982), O'Reilly and Busch (1984), O'Reilly et al. (1987), and Backus (1987).

For three and a half centuries, the ecosystem has supported large and important common-resource fisheries, extending from the export trade in salted cod of the early colonists in the late seventeenth century to the heavy exploitation of the total finfish biomass in the late 1960s and early 1970s (see Chapters 9 and 27). Heavy fishing mortality imposed on the resources by European factory fleets precipitated the passage of legislation by the United States and the creation of a Fishery Management Zone (FMZ) in 1976. Off the northeastern United States, this zone encompasses 260,000 km^2 extending from the Gulf of Maine in the north to Cape Hatteras in the south (Figure 7-1). The shelf of this region is among the most productive in the world. Fishery yields support a harvesting and processing industry that contributes $1 billion annually to the economies of the coastal states from Maine to North Carolina.

Uses of the shelf as a potential source of petrogenic hydrocarbons and formerly as a repository for wastes have heightened concerns about the "health" of the ecosystem. Efforts to obtain energy resources accelerated the exploration for offshore gas and petroleum resources in the late 1970s. The waste products of a sprawling industrial urban megalopolis, from Baltimore to Boston, are transported onto the shelf in increasing volume through the natural forces of runoff, atmospheric fallout, and, until 1992, by the barge-dumping of dredge spoils, urban sludge, and wastes containing toxic chemicals, radionuclides, and heavy metals. Based on legal proceedings, this dumping has since been discontinued. Although no large-scale, irreversible changes caused by this dumping have been observed, the evidence of coastal eutrophication of the shelf is symptomatic of a growing problem. Also, although the effects of

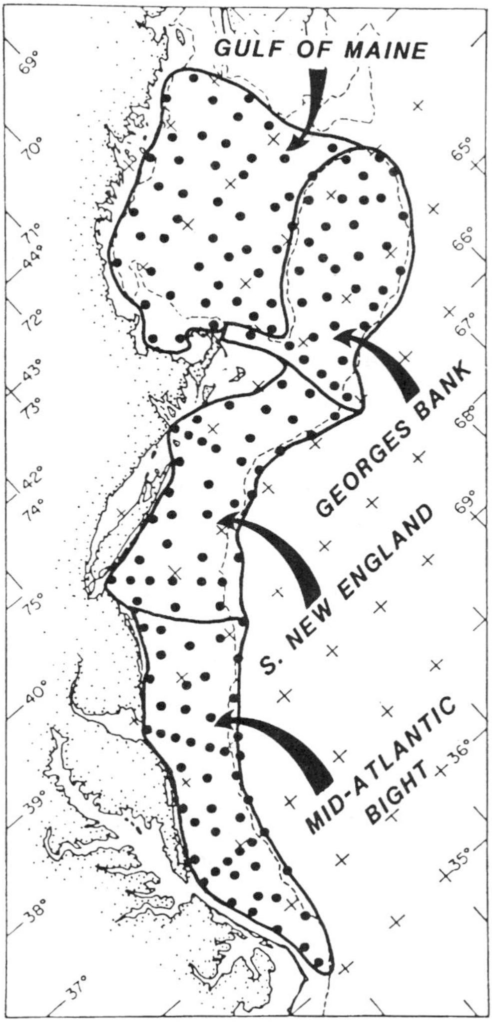

Figure 7-1. The four geographic areas of the Northwest Atlantic—the Gulf of Maine, Georges Bank, southern New England, and the Middle Atlantic Bight—and stations sampled for zooplankton during MARMAP operations.

an increasing pollution load are not, as yet, readily detectable in the open waters of the shelf, it is clear that the fish stocks have recently been under the stress of intensive fishing. During the period 1968 through 1975, the total finfish biomass of the shelf was reduced by several million metric tons of mid-sized predators (cod, haddock, mackerel, herring, and other species), representing just over 50% of the fish biomass, which resulted in a detectable perturbation in the structure of the shelf ecosystem. The predatory fish displaced through heavy fishing mortality appear to have been partially replaced by small elasmobranches (spiny dogfish and skates). During the late 1970s, the herring and mackerel biomass declined under the pressures of an exessive foreign fishing effort. Since 1980, the numbers of mackerel and herring have increased as a result of much reduced fishing effort to reach an unprecedented high biomass level of approximately 5 million metric tons (Northeast Fisheries Science Center 1995). Management actions are now beng implemented to reduce fishing effort on the depressed cod, haddock, and flounder stocks in an effort to enhance stock recovery.

To provide the information base from which ecosystem perturbation can be monitored and forecast, the National Marine Fisheries Service of the National Oceanic and Atmospheric Administration (NOAA) initiated a long-term study of the northeast shelf to measure changes in key structural components of the shelf ecosystem. This recurrent sampling effort has covered 260,000 km^2 several times a year from the late 1970s to the present. The results of these studies can be found in several reports on the changing states of the Northeast Shelf ecosystem, that are included in the reference section of this chapter (Cohen and Grosslein 1987; Cohen and Wright 1978; Cohen et al. 1982; Grosslein et al. 1980; Northeast Fisheries Science Center 1995; O'Reilly and Busch 1984; Sherman 1980; Sherman et al. 1983; Sherman et al. 1984; Sherman et al. 1987; Sissenwine et al. 1984).

Hydrography

The first major description of the hydrography of the Gulf of Maine including Georges Bank was by Bigelow (1927) based on observations made between 1912 and 1922. Despite the passage of half a century and a large increase in oceanographic sampling in recent decades, Bigelow's conclusions remain largely valid and still represent the most comprehensive work on the area. For the shelf south of New England and the Middle Atlantic Bight, the first comprehensive studies were by Bigelow (1933) and Bigelow and Sears (1935). These descriptions also portray the hydrographic regime as it is generally understood today. (See Brooks, Chapter 4, this volume; Hopkins and Garfield 1979; Colton et al. 1968; Wright 1976; Butman et al. 1982; Beardsley et al. 1976.)

Circulation

The mean circulation in the shelf region was first described in a comprehensive manner by Bumpus (1973) and Bumpus and Lauzier (1965) on the basis of data from the return of drift bottles and seabed drifters. The flow is generally toward the west-southwest (Figure 7-2), although it may reverse at times, particularly in summer (Bumpus 1969). Direct current measurements also show a mean westerly flow at 5 to 10 cm sec^{-1} that approximately parallels the isobaths (Beardsley et al. 1976). The same measurements, however, show that the variation is considerably larger than the mean, so that over short periods of time the flow may be in any direction.

The mass balance for the shelf region includes the mixing of about 157 km^3 of fresh coastal runoff with slope water to form 2.4 to 3.4×10^3 km^3 of shelf water annually (Wright 1976; Bush 1981). In addition, water enters the region from south of Nantucket as part of the general westward flow on the shelf. As a result, a substantial volume of the shelf water must be removed from the shelf each year. A continuous loss of water occurs near Cape Hatteras. Also, some exchange probably occurs continuously at the shelf–slope front. In addition, detached parcels of shelf water have been observed in the slope-water region (Ford et al. 1952; Cresswell 1967), and Wright (1976) has suggested that these parcels may occur frequently. A mechanism that may remove large volumes of water from the shelf is the passage of warm-core rings. These clockwise-rotating features originate from meanders in the Gulf Stream that pinch off and then move westward through the slope region. On the east side of the rings, a tongue of cooler shelf water is often seen extending off the shelf and around the ring.

Primary Productivity

Phytoplankton primary productivity and chlorophyll *a* concentrations on the northeast United States continental shelf have been monitored routinely by the Northeast Fisheries Center on shelf-wide cruises from Cape Hatteras to Nova Scotia since March 1977. The following summary is based on measurements made on those cruises and reports by O'Reilly and Busch (1984) and O'Reilly et al. (1987).

Annual Primary Production

Estimates of annual total phytoplankton primary production from Cape Hatteras to Nova Scotia are presented in Figure 7-3 by region using data from 1,089 productivity stations. The regions selected are based on the recurring seasonal patterns of chlorophyll distribution along the continental shelf. Annual production on the shelf ranges from 10,834 to 21,043 kJ m^{-2} yr^{-1}

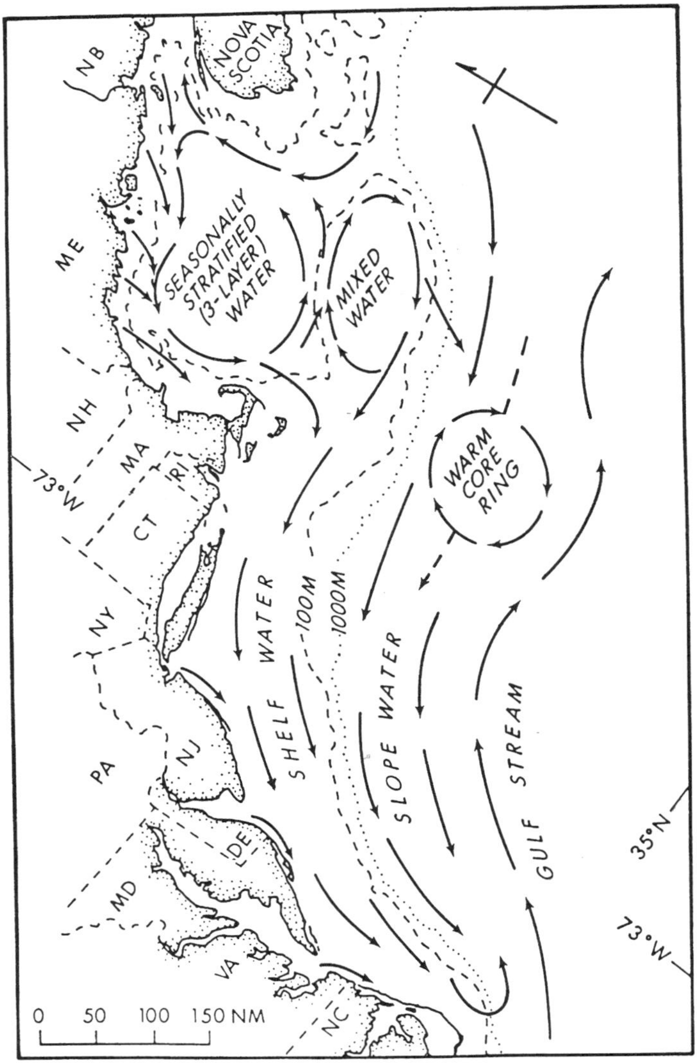

Figure 7-2. Mean nontidal surface circulation in the Gulf of Maine, Georges Bank, southern New England, and the Middle Atlantic Bight. The northern sector of the shelf ecosystem is characterized by a cyclonic gyre and a seasonally stratified three-layered water-mass system over the deep basins of the Gulf of Maine and mixed water with an anticyclonic gyre over the shoal bottom of Georges Bank. Further south, the waters move southwesterly along the broad shelf of southern New England to the narrower, gently sloping shelf plain of the Middle Atlantic Bight. (Reprinted from M. C. Ingham, R. S. Armstrong, J. L. Chamberlin, S. K. Cook, D. G. Mountain, R. J. Schlitz, J. P. Thomas, J. J. Bisagni, J. F. Paul, and C. E. Walsh. 1982. Summary of the Physical Oceanographic Processes and Features Pertinent to Pollution Distribution in the Coastal Waters of the Northeastern United States, Virginia to Maine. NEMP Oceanography Summary Rept. NEMP-IV-82-C-004. Washington, DC: U. S. Department of Commerce.)

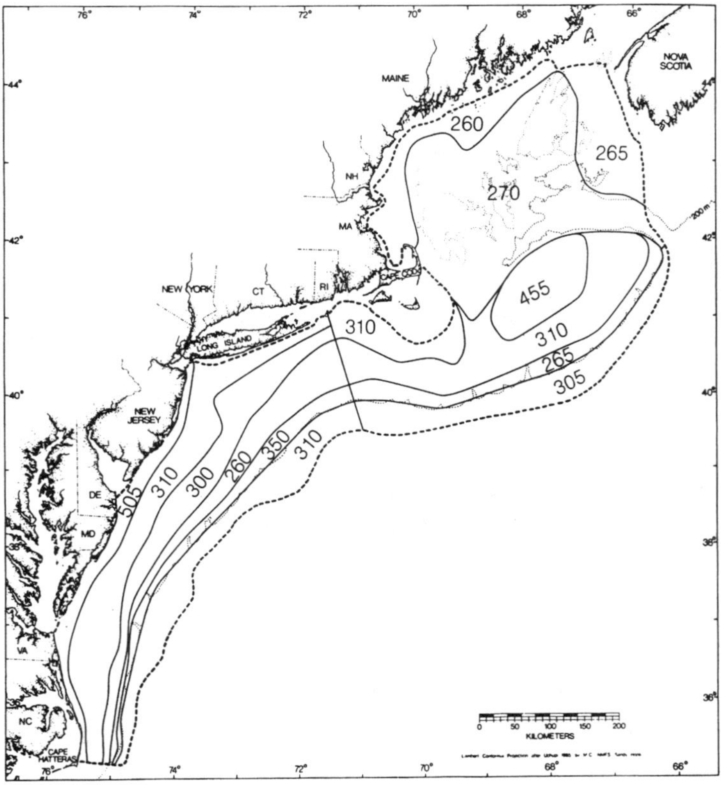

Figure 7-3. Estimated annual total primary production (particulate and dissolved organic carbon) from Cape Hatteras to Nova Scotia by subarea [g C m^{-2} yr^{-1} (1 g C = 10 kcal = 41.67 kJ)].

(260–505 g C m^{-2} yr^{-1}). The areas of highest estimated production on the shelf occur on the central, shallow portion of Georges Bank [18,960 kJ m^{-2} yr^{-1} (445 g C m^{-2} yr^{-1})] and along the coast between New Jersey and North Carolina [21,043 kJ m^{-2} yr^{-1} (505 g C m^{-2} yr^{-1})], which correspond to the areas with consistently high chlorophyll *a* concentrations. These productivity values indicate that the northeast United States continental shelf is one of the most productive areas of the world. Riley's (1941) work on Georges Bank produced estimates of net production lower than the values presented here. His results, however, were based on data collected using the dissolved-oxygen method with surface water, and therefore would be expected to yield lower estimates than the ^{14}C method. The areas of the shelf with the lowest estimated annual production include the mid-shelf area between Long Island and Cape Hatteras, the Gulf of Maine, and the southern edge of Georges Bank (Figure 7-3).

Nutrients

Georges Bank

The annual cycle of nitrate concentrations on Georges Bank has a peak in February, with concentrations declining in spring due to utilization by phytoplankton, then falling to a low in summer (Pastuszak et al. 1981). As autumn begins, concentrations increase and continue to do so until a winter maximum is again reached. Water on both sides of Georges Bank at and below 150 m contains high nutrient levels because of influx of slope water into the Gulf of Maine (Ramp and Vermersch 1978). It has been suggested by Sigaev (1978), from analysis of a ten-year data base, that upwelling of the deeper water occurs.

During February, nitrate is high at the surface on southeastern Georges Bank (9 µg-at l^{-1}) and in the vicinity of Nantucket Shoals (10 µg-at l^{-1}). In spring, nitrate concentrations are very low on top of the Bank, but increase to the north and south (Cohen and Wright 1978). During April and May, all nutrient concentrations generally are lower than during February. Nitrate concentrations range from 0.5 to 4 µg-at l^{-1} in the near-surface water and from 2 to 4 µg-at l^{-1} at 50 m. Phosphate and silicate concentrations are low (0.1–0.3 µg-at l^{-1}) to a depth of 100 m (Pastuszak et al. 1981). In September and October, concentrations are higher than in summer, but are still low (0–1 µg-at l^{-1}) in the shallow portion of the Bank. High concentrations occur in the eastern portion (3–10 µg-at l^{-1} nitrate and silicate) and at depths of 30–50 m on the northern and southern flanks (1–11 µg-at l^{-1}). From November to January nutrient concentrations begin to increase, with high concentrations in deeper waters to the north and the south. Maximum concentrations are reached during January and February.

Middle Atlantic Area

In the Middle Atlantic area in March, Walsh et al. (1978) measured more than 5 µg-at l^{-1} nitrate in well-mixed waters off New York. By summer, when stratified conditions exist in the water column, nitrate diminishes to undetectable concentrations in the euphotic zone. Walsh et al. (1978) suggested that nutrient concentrations are limiting to the phytoplankton in the Middle Atlantic area during summer, because the nitrogen must be derived from recycled sources including benthos, zooplankton, and bacterioplankton. Forty percent of the nitrogen demand for production of phytoplankton is thought to come from regeneration mechanisms, even though there is evidence of cross-shelf advective input of nutrients during spring and summer when upwelling winds are present [see Walsh (1981) for a more complete discussion of pelagic-demersal coupling and alongshore cross-shelf patterns from the Middle Atlantic Bight to Georges Bank].

O'Reilly and Busch (1984) stressed the importance of ammonium nitrogen in the sewage sludge that is discharged into the waters of the New York Bight apex in addition to ammonium from the sources listed above. Eppley (1980) and O'Reilly and Busch (1984), as well as Walsh (1981), imply that regeneration plays an important role in providing nitrogen in shelf ecosystems. During March, there are strong inshore-offshore gradients in surface waters, then in April the water column becomes well mixed, with ammonium nitrogen isopleths generally being vertical. Concentrations decrease from 4.8 μmol l^{-1} in the Raritan-Hudson estuary to 0.4 μmol l^{-1} at mid- shelf, with low levels (<0.2 μmol l^{-1}) in deep water. As succession of the phytoplankton from spring to summer communities occurs, concentrations of ammonium nitrogen continue to decrease. In May, surface concentrations are low and reflect vertical stratification, with highest concentrations at 40–80 m in water about 100–140 km from the coast. In June, concentrations are similar to those in May, and highest near the seabed at the inshore apex area (4–9 μmol l^{-1}), but by July there are strong inshore-offshore gradients, with very low concentrations in offshore water (0.2–1.0 μmol l^{-1}); however, by October there are again high levels in the surface waters near the estuary (Waldhauer et al. 1981).

Zooplankton

The zooplankton of the shelf have been described in comprehensive early works by Bigelow (1926) and Bigelow and Sears (1939). Other accounts have been limited to smaller areas of the shelf for relatively short periods of time (Fish 1936a, b; Clarke and Zinn 1937; Clarke 1940; Clarke et al. 1943; Deevey 1952, 1956, 1960; Grice and Hart 1962; Sherman 1968, 1970, 1976; Malone 1977; Judkins et al. 1980; Maurer and Toner 1983). Since the early work of Bigelow, little attention has been given to a systematic assessment of the changes in the biomass, species composition, and production of the zooplankton of the shelf. As recently as 1976, information on seasonal changes in the standing stock of zooplankton was limited to a rather confused series of data compiled by Cohen (1976) from available literature, in which both spring and autumn were depicted by different authors as annual peak periods of zooplankton abundance. More recent results are based on 32 surveys of zooplankton conducted by the United States, Poland, the Soviet Union, and the German Democratic Republic as part of a joint MARMAP study of shelf productivity. Between six and eight surveys were done per year. Sampling was done in four subareas: the Gulf of Maine, Georges Bank, southern New England, and the Middle Atlantic Bight, each characterized by distinct bathymetry and circulation (Ingham et al. 1982). Samples were collected with paired bongo nets (333-μm mesh netting) at an average of 129 locations per survey situated 25–35 km apart, resulting in a total of 3,568 samples. Details of the methods used in sampling are given in Sherman et al. (1984).

Biomass

In the northern half of the shelf, the zooplankton standing stock is greatest in spring. On Georges Bank, the spring peak is followed by a sharp decline from spring through early summer and a more gradual reduction from summer through autumn. In the Gulf of Maine the seasonal decline is more gradual, from an annual maximum in spring to a minimum in the autumn. Zooplankton in the southern half of the shelf undergoes a two-step increase in southern New England, with the most rapid growth from late winter to early spring, followed by a secondary but less precipitous increase from spring through late summer and a decline from summer through autumn. In the Middle Atlantic Bight, zooplankton standing stock gradually increases from late winter to a single peak in late summer (Figure 7-4). The biomass expressed in the seasonal curves represents minimal values, because the data are based on collections of macrozooplankton retained by 333-μm mesh apertures. It has been estimated by Sushin (unpublished manuscript, AtlantNIRO, Kaliningrad) based on measurements made during MARMAP surveys of zooplankton collected by 165-μm and 53-μm mesh apertures that the microzooplankton constituents of the northeast shelf represent approximately 30% of the biomass retained by a 333-μm net. The predominant microzooplankton on the shelf are copepod nauplii, copepodites, and small copepod species of the genus *Oithona*, which are undersampled in the coarser 333-μm meshes.

The measure of dominance used for the shelf zooplankton is based on the Fager and McGowan (1963) index of dominance. A large number of taxa (394) are represented in the shelf zooplankton but only 12 taxa, all copepods, constitute 85% of the dominance—*Acartia clausi, Acartia longiremis, Acartia tonsa, Calanus* spp., *Calanus finmarchicus, Centropages hamatus, Centropages typicus, Metridia lucens, Oithona* spp., *Paracalanus parvus, Pseudocalanus minutus,* and *Temora longicornis.* Among these twelve taxa, three—*Calanus finmarchicus, Pseudocalanus minutus,* and *Centropages typicus*—account for 75% of the total dominance. An important missing component is the cyclopoid *Oithona,* which is too small to be sampled adequately with 333-μm mesh nets.

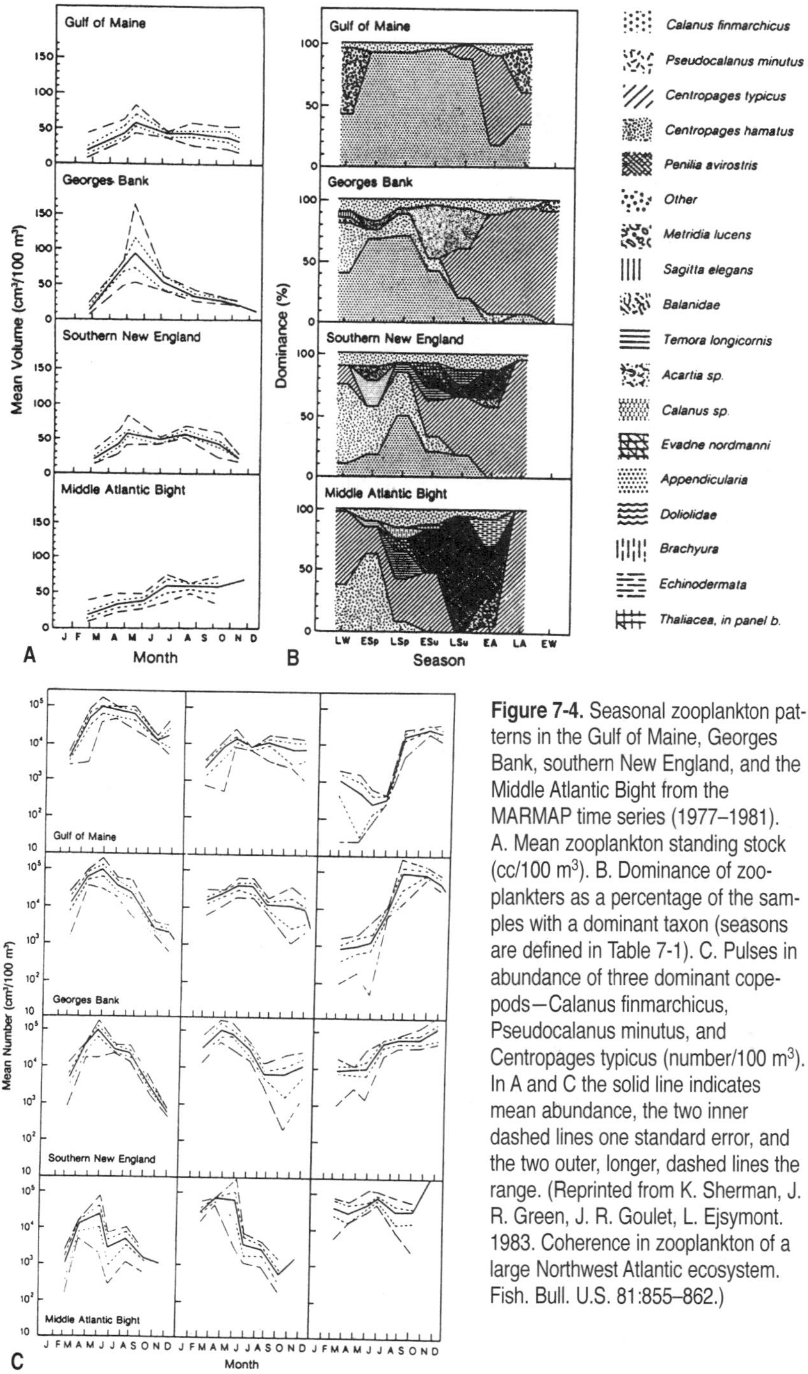

Figure 7-4. Seasonal zooplankton patterns in the Gulf of Maine, Georges Bank, southern New England, and the Middle Atlantic Bight from the MARMAP time series (1977–1981). A. Mean zooplankton standing stock (cc/100 m^3). B. Dominance of zooplankters as a percentage of the samples with a dominant taxon (seasons are defined in Table 7-1). C. Pulses in abundance of three dominant copepods—Calanus finmarchicus, Pseudocalanus minutus, and Centropages typicus (number/100 m^3). In A and C the solid line indicates mean abundance, the two inner dashed lines one standard error, and the two outer, longer, dashed lines the range. (Reprinted from K. Sherman, J. R. Green, J. R. Goulet, L. Ejsymont. 1983. Coherence in zooplankton of a large Northwest Atlantic ecosystem. Fish. Bull. U.S. 81:855–862.)

Although the three species co-occur on the shelf, their temporal and spatial centers of abundance differ. In the early spring *Calanus finmarchicus* is the predominant copepod in the northern part of the shelf (Gulf of Maine, Georges Bank). It is succeeded later in the year by *Centropages typicus.* In southern New England, *Pseudocalanus minutus* is the dominant copepod early in the year, succeeded by *C. typicus* later. In the Middle Atlantic Bight, *P. minutus* and *C. typicus* are predominant in winter and spring. By summer, *C. typicus* exceeds *P. minutus* in dominance (Figure 7-4).

Production Estimates

The relationship of zooplankton numbers to biomass and productivity rates is under investigation. Available information, unfortunately, is fragmentary. Secondary production based on older stages of *Calanus finmarchicus* in spring was estimated by Green et al. (1977) at 3.31 kJ m^{-2} d^{-1} (79.46 mg C m^{-2} d^{-1}). The instantaneous growth method was used in the calculation from the formula:

$$P = [(W_B - W_A)/t][(N_1 + N_2)/2]$$

where W_A is the weight at the beginning of a stage, W_B is the weight at the end of a stage, t is the stage duration in days, N_1 is the count of organisms in a stage, and N_2 is the count of organisms in the succeeding stage.

The production rates for each of five surveys made during the spring generations of *C. finmarchicus* populations were used to estimate total production for the 100-day period of the surveys. As was pointed out by Green et al. (1977) the values were of the same order of magnitude as the production rate of 2 kJ m^{-2} d^{-1} (46 mg C m^{-2} d^{-1}) estimated for *C. finmarchicus* in the North Sea (Mullin 1969) and the rate of 3 kJ m^{-2} d^{-1} (77 mg C m^{-2} d^{-1}) reported for *Acartia tonsa* in the Patuxent River estuary of Chesapeake Bay by Heinle (1966).

Sherman et al. (1987) estimated production by two methods: production to biomass ratios (P:B) and transfer efficiencies applied to particulate primary production measurements. Based on a five-year mean macrozooplankton biomass of 4.76 g dry wt. m^{-2} the microzooplankton biomass was calculated to be 2.05 g dry wt. m^{-2} (43% of the macrozooplankton biomass; Beers and Stewart 1969). Using P:B ratios of 7 for macrozooplankton (Steele 1974; Crisp 1975) and 25 for microzooplankton (Banse and Mosher 1980; Tremblay and Roff 1983), estimated secondary production was 1.8 kJ m^{-2} d^{-1} (42 mg C m^{-2} d^{-1}) for macrozooplankton and 2.7 kJ m^{-2} d^{-1} (64 mg C m^{-2} d^{-1}) for microzooplankton, totaling 4.5 kJ m^{-2} d^{-1} (106 mg C m^{-2} d^{-1}).

In the second estimate the partition of macro- and microzooplankton production was made by assuming that the macrozooplankton consumed particles larger than 20 μm and the microzooplankton consumed particles smaller than 20 μm. Assuming that 50% of the particulate primary production is grazed by zooplankton (Dagg and Turner 1982; Walsh 1983), that transfer efficiency is 32% (Steele 1974), and that 57% of the particulate primary production of 12,501 kJ m^{-2} d^{-1} (300 g C m^{-2} d^{-1}) (O'Reilly and Busch 1984) is caused by nanoplankton, macrozooplankton production is 2.4 kJ m^{-2} d^{-1} (57 mg C m^{-2} d^{-1}) and microzooplankton 3.1 kJ m^{-2} d^{-1} (75 mg C m^{-2} d^{-1}), totaling 5.5 kJ m^{-2} d^{-1} (132 mg C m^{-2} d^{-1}).

The lower macrozooplankton production estimate presented in Sherman et al. (1981) is reasonable in relation to the production estimate by Green et al. (1977), which was calculated for a 100-day period in the spring when *Calanus finmarchicus* growth is at a maximum. At this time it is the dominant species on Georges Bank, increasing in abundance 100-fold from the winter minimum (Sherman et al. 1983). The production estimates calculated were derived from yearly means of biomass and productivity levels and as such were probably more representative of the seasonally varying high-amplitude areas of the North Atlantic, including Georges Bank.

Production

The lack of knowledge on growth, fecundity, mortality rates, and feeding and conversion efficiencies for the principal components of benthic populations of the shelf has made it difficult to obtain accurate estimates of secondary productivity by the benthos. Recently, several investigators have attempted to estimate secondary production by benthos from estimates of standing crop biomass on Georges Bank and in the Middle Atlantic Bight. Cohen et al. (1980, 1982) gave a conservative estimate of yearly productivity as 750 kJ m^{-2} d^{-1} (180 kcal m^{-2} d^{-1}) for the macrobenthos of Georges Bank based on an average wet weight biomass of 240 g m^{-2}, and a meiobenthos annual productivity range between 26 and 63 kJ m^{-2} d^{-1} (6.25 and 15 kcal m^{-2} d^{-1}) based on data extrapolated from similar bottom areas on the southern New England Shelf. They speculated that this estimate was low, and expected that, when suitable data were available, a value approaching 83 kJ m^{-2} d^{-1} (20 kcal m^{-2} d^{-1}) would be more realistic.

Steimle (1987), using estimates of average wet weight biomass from two separate surveys on Georges Bank (1956–1965 and 1977), applied two methods in his attempt to estimate secondary production there. The total community production/biomass (P:B) method yielded macrofauna production estimates between 379 and 333 kJ m^{-2} d^{-1} (91 and 80 kcal m^{-2} d^{-1}); whereas, combining individual ratios, calculated for each major taxonomic group, yielded estimates of 425 kJ m^{-2} d^{-1} (102 kcal m^{-2} d^{-1}) for the first (1956–1965) survey, and 388 kJ m^{-2} d^{-1} (93 kcal m^{-2} d^{-1}) for the second (1977). Estimates of meiofaunal production for the two surveys were 46 and 64 kJ m^{-2} d^{-1} (11.2 and 15.4 kcal m^{-2} d^{-1}).

Walsh (1981) and Walsh et al. (unpublished data) have estimated annual benthic production, again based on estimates of the mean wet-weight standing-crop biomass, for the Middle Atlantic Bight. Their estimate based on 100 g m^{-2} mean wet-weight biomass, is about 463 kJ m^{-2} d^{-1} (111 kcal m^{-2} d^{-1}). They did not give separate estimates for macrobenthos and meiobenthos.

Fish Fauna of the New England Shelf Region

The shelf region from southwest Nova Scotia to Cape Hatteras, particularly the area from Georges Bank and Cape Cod to Cape Hatteras, represents an area of rapid transition from cold-temperate to warm-temperate fauna. The region from Cape Cod to Cape Hatteras is considered by some zoogeographers to represent the southern limit of a cold-temperate (boreal) faunal province that extends to the north of Cape Cod (Briggs 1974). Others consider the region between these two capes to be an overlap area between cold-temperate fauna and warm-temperate fauna (south of Cape Hatteras), or even a separate faunal province (Virginian) with the capes marking southern and northern boundaries (Hazel 1970). Common to all of these faunal classifications is the recognition that a high proportion of the fishes in the area from Cape Cod to Cape Hatteras are seasonal migrants, and only a few are truly endemic (Grosslein and Azarovitz 1982).

There are about 300 species of fish found in the entire shelf region from Cape Hatteras to western Nova Scotia, of which about 15% are cold-temperate (or boreal) species with principal abundance north of Cape Cod. Most of the remaining fish species are warm-temperate or tropical species, at least three-fourths of which range north of Cape Hatteras only in the summer (Briggs 1974).

Intensive research-vessel trawling surveys from Cape Hatteras to western Nova Scotia have been conducted each spring and autumn since the late 1960s, and these surveys have added a new quantitative dimension to zoogeographic data on fishes of the region. For example, the spring and autumn surveys of 1974 caught a total of 180 species in 739 bottom-trawl hauls representing 18 orders and 82 families: 37% of the species were cold temperate, 5% were residents of the Middle Atlantic Bight area, and the remainder (58%) were warm-temperate species that migrate north of Cape Hatteras during the spring and summer. The percentage of southern forms would be much higher if the many nearshore and estuarine species (not caught in the offshore trawl survey) had been included. Although the surveys do not provide a complete list of shelf species, they represent all of the major taxonomic groups.

The transitional nature of this faunal region and its seasonal character can be illustrated by comparing the spring and autumn surveys for each of the four subareas in terms of the frequency of occurrence of species classified as northern (cold-temperate or boreal), Middle Atlantic (residents of the Middle Atlantic Bight), and southern (warm-temperate and tropical). The total number of species decreases from south to north, with only about one-half as many in the Gulf of Maine as in the Middle Atlantic. The largest number of southern forms occurs in the southernmost Middle Atlantic subareas, and declines steadily northward. The maximum contribution of southern species (77%) occurs in the Middle Atlantic in the autumn, which corresponds there to the latter part of the warm period, and the minimum catch of southern species (10%) occurs in the Gulf of Maine in the spring, which corresponds to about the seasonal minimum in temperature for the subarea. The reverse pattern is present for northern species, the largest numbers occurring in the northern areas and diminishing southward. The maximum proportion of northern species (88%) occurs in the northernmost Gulf of Maine subarea during spring, and the minimum (17%) occurs in the southernmost Middle Atlantic subarea in autumn. These shifts are a result of the seasonal migrations forced by the large seasonal temperature changes that occur in the area, particularly from Cape Hatteras to Georges Bank (Parr 1933). When water temperatures rise in spring, there is a large influx of warm-water species (for example, drums, bluefish, jacks) into the Middle Atlantic area from the south (Grosslein and Azarovitz 1982). At the same time a number of cold-water species (for instance, cod, Atlantic herring, Atlantic mackerel, spiny dogfish, and shad), which had migrated into the Middle Atlantic from the north during the previous winter, then begin moving north. When water temperatures drop in autumn the north-south pattern reverses.

In addition to north-south migratory patterns, many species exhibit seasonal inshore-offshore migrations correlated with the strong temperature cycle. This is particularly true of the region between Cape Hatteras and Cape Cod, where most of the fish species have temperature tolerances that are exceeded by the normal seasonal temperature cycle of <1–26°C in inshore waters (Grosslein and Azarovitz 1982). Therefore, most resident Middle

Atlantic species (for example, scup, summer flounder, longfin squid, butterfish) are forced to migrate offshore in winter, where they can find temperatures ranging from about 5°C at midshelf to a maximum of about 13°C at the shelf edge. Southern species have higher minimum temperature tolerances and preferences, and they migrate south as well as offshore in autumn.

Fish Biomass and Yield

(*Editors' Note:* See Chapters 9 and 27 for an update of the fisheries biomass and yield through the early 1990s.)

Estimates of total finfish biomass are available (see Chapter 9) for the past two decades in the region from Cape Hatteras to the Gulf of Maine based on joint analysis of research-vessel survey data and commercial fishery statistics. The total standing crop of finfish in the International Commission for the Northwest Atlantic Fisheries (ICNAF) Subareas 5 and 6 declined from a maximum of about 7.0×10^6 MT(t) before 1964 to a minimum of about 2.0×10^6 t in 1975, as a result of very heavy fishing (Brown et al. 1976; Clark and Brown 1977). A major part of this decline in biomass was the result of the drop in herring stocks that began in the late 1960s, but virtually all other species showed significant declines as well, and the total finfish biomass for the region as a whole declined by about 65% between the early 1960s and 1975 (Clark and Brown 1977). The low level of 2×10^6 t in 1975 was estimated to be about one-fourth the virgin (maximum) biomass level, and about one-half the level of biomass (4×10^6 t) corresponding to the maximum sustainable (fishery) yield (MSY) of about 950,000 t (Brown et al. 1976; Clark and Brown 1977).

The finfish biomass began gradually increasing again after 1975, initially as a result of quotas on total catch (all species combined) first imposed in 1974 by ICNAF, and later as a result of further reductions in fishing by distant-water fleets following establishment of the United States 200-mile (320 km) limit in 1977. The annual rate of increase was estimated to be 32% between 1975–1976 and 1976–1977; but the increases occurred primarily in silver hake and squids, whereas the formerly large herring and mackerel stocks remained at a low level (Clark and Brown 1979). No further increases occurred after 1977, with the biomass leveling off around $3.0–3.5 \times 10^6$ t during the period 1978 to 1981 and even some suggestion of a decrease in 1982 (NOAA 1983). Despite the large decline in abundance observed for most species up until 1975, overall it appeared that there had been no drastic shift in species composition of the finfish biomass, except that sand lance (*Ammodytes* spp.) increased dramatically after 1975 coincident with the low biomass of virtually all fish predators and potential competitors (Grosslein et al. 1980; Sherman et al. 1981). Estimates of the size of the sand lance population based on back-calculations from larval abundance and fecundity data range from 0.25–5.00 × 10^6 t during the peak abundance in 1978 depending upon assumptions of egg mortality rate (Morse 1982). Our best guess is around 2 × 10^6 t corresponding with an egg mortality of about 90%. At the level of 2 × 10^6 t, the total finfish biomass would be about 5 × 10^6 t, which is roughly the level of biomass corresponding to MSY for the region.

In conjunction with the declines in abundance of most species after the mid-1960s, there was a reduction in mean size of fish, with average weights dropping from 0.5 kg in 1965–1969 to about 0.3 kg in the early 1970s, followed by a sharp rise from 0.25 kg in 1974 to 0.45 kg in 1977 (Grosslein et al. 1980). This pattern was consistent with the observed reductions in mean age by the early 1970s as a result of heavy fishing, and with the rebuilding of traditional stocks after 1975.

Average total yield (landings) to the fisheries of the entire region during the period 1969 to 1978, was about 900,000 t; there were 26 species (including two species of squid and one species of shrimp) with average landings of 3,000 t or more (McBride and Brown 1980). Mackerel and sea herring represented nearly one-half the total yield during this period, and the next largest yield was represented by silver hake. Recreational catches are an important component of the yield for certain species and are included in the landings; recreational catches exceeded commercial catches for winter flounder, summer flounder, black sea bass, and bluefish (McBride and Brown 1980). Since 1979 total landings have averaged about 470,000 t or only 40% of the peak landings of 1.2 × 10^6 t during the period 1970–1974, and only about 50% of the MSY level (NOAA 1985). Since 1981, mackerel biomass increased significantly, but herring remained low, and the principal demersal stocks traditionally fished by the U.S. New England fleet generally declined. Excluding sand lance (for which there is no fishery), recovery of fishery yields to MSY levels would appear dependent on continuing increases in the mackerel stock and significant improvement in abundance of stocks of herring and traditional groundfish species and squids (NOAA 1985).

Production in Different Trophic Levels of Northwest Atlantic Shelf Ecosystems

A comprehensive series of primary and secondary production estimates have recently become available for Georges Bank and adjacent waters (O'Reilly and

Busch 1984; Cohen and Grosslein 1987; O'Reilly et al. 1987; Steimle 1987). The various production estimates for Georges Bank are summarized in Table 7-1 together with similar data for three other adjacent Northwest Atlantic shelf regions. Comparisons among the areas suggest there may be significant differences in energy flow and trophic structure between the Georges Bank–Middle Atlantic region and the other areas.

Primary and Secondary Production

In the Gulf of Maine and on the Nova Scotian shelf, two areas just north and east of Georges Bank, primary production estimates are lower than for Georges Bank, whereas on the Middle Atlantic shelf of the United States just southwest of Georges Bank, primary production estimates are nearly as high as those for Georges Bank (Table 7-1).

Secondary production estimates are similar in all areas except the Middle Atlantic (Table 7-1). In the Middle Atlantic, macrobenthos production appears higher than all other areas but macrozooplankton production appears the lowest. Cohen and Grosslein (1987) reviewed sources of error for these various estimates and concluded that none could be discounted on the basis of estimation method. In view of the possible sources of error in estimating secondary production levels, especially benthic production, it is perhaps remarkable that they are so close. In any case, if one assumes that the data are approximately correct, then they imply a somewhat lower transfer efficiency from primary to total secondary production on Georges Bank and the Middle Atlantic shelf.

Fish Production

Fish production estimates for Georges Bank were calculated using an energy balance equation relating consumption, growth, and metabolism. Basic methods are described by Grosslein et al. (1980) for exploitable-size fish and Sissenwine et al. (1984a) for pre-exploitable fish. Production estimates for other Northwest Atlantic areas were derived by applying to them the P:B ratios and the ratio of pre-exploitable to exploitable production characteristic of Georges Bank (Cohen and Grosslein 1987).

Calculated production estimates of fish and squid are two to three times greater on Georges Bank than in the other areas, despite comparable levels of secondary production in all areas. Furthermore, Cohen and Grosslein (1987) concluded that exploitable fish production estimates on Georges Bank probably are conservative. This poses an enigma—higher fish production on Georges Bank with no obvious mechanism to support it.

Table 7-1. Production (kJ $m^{-2}yr^{-1}$) of phytoplankton, secondary producers, and fish in four shelf areas of the Northwest Atlantic; and average landings and MSY estimates (t km^{-2} yr^{-1}) for total finfish and squid [Data derived from Cohen and Grosslein (1987) and Sissenwine et al. (1984a)]

Trophic level	Nova Scotian Shelf	Gulf of Maine	Georges Bank		Middle Atlantic Shelf
Primary production	9,500	10,692	13,926		12,930
Secondary production	1,154	1,271	1,229		1,321
Macrozooplankton	812	862	842		571
Macrobenthos	342	409	387		750
			1964–1966	1973–1975	
Fish and squid	142	133	358	217	133[a]
Pelagic exploitable size			32	30	
Demersal exploitable size			58	37	
Pre-exploitable			268	150	
Fish and squid MSY (t km^{-2} yr^{-1})	3.2	2.8	7,6		3.4(7.6)[b]
Average annual landings[c]	3.1	1.8	6.7		2.9(5.8)[d]

[a]Excludes production of Atlantic menhaden.
[b]Value between parentheses includes sustainable yield of 400,000 t Atlantic menhaden, derived from yield-per-recruit model (Schaaf 1979).
[c]Landings for Northwest Atlantic (excluding menhaden and large pelagics) taken from ICNAF (now NAFO) statistical bulletins for the period 1965-1980 except for Middle Atlantic (1968–1982).
[d]Value between parentheses includes menhaden landings for the period 1973–1977 from Schaaf (1979).

Other measures of fish productivity, namely average landings and MSY estimates, show that, except for the Middle Atlantic region, Georges Bank provides higher fish yields per unit area than the other areas (Table 7-1). In the case of the Middle Atlantic, menhaden were not included in calculated fish production but they are included in the MSY estimate and average landings. With the inclusion of menhaden, which feed directly on phytoplankton, the Middle Atlantic productivity comes up to the level of Georges Bank, which seems more consistent with the relatively high level of primary production in the Middle Atlantic.

It is perhaps too early to conclude that there are fundamental differences in the energy flow and structure of these ecosystems. The data, however, suggest the possibility of such differences.

References

Backus, R. H. (ed.). 1987. Georges Bank. Cambridge, MA: MIT Press.

Banse, K., and Mosher, S. 1980. Adult body mass and annual production/biomass relationships of field populations. Ecol. Monogr. 50:355–379.

Beardsley, R. C., Boicourt, W. C., and Hansen, D. V. 1976. Physical oceanography of the Middle Atlantic Bight. In M. Grant Gross (ed.), Middle Atlantic Continental Shelf and the New York Bight. J. Limnol. Oceanogr. Special Symposia 2:20–34.

Beers, J. R., and Stewart, G. L. 1969. Microzooplankton and its abundance relative to the larger zooplankton and other seston components. J. Mar. Biol. 4:182–189.

Bigelow, H. B. 1926. Plankton of the offshore waters of the Gulf of Maine. Bull. U.S. Bur. Fish. 40(Part II):1–509.

Bigelow, H. B. 1927. Physical oceanography of the Gulf of Maine. Bull. U.S. Bur. Fish. 40(Part II):511–1027.

Bigelow, H. B. 1933. Studies of the waters of the continental shelf, Cape Cod to Chesapeake Bay. I: The cycle of temperatures. Pap. Phys. Oceanogr. 2(4):1–135.

Bigelow, H. B., and Schroeder, W. C. 1953. Fishes of the Gulf of Maine. Bull. U.S. Bur. Fish. 53(74):1–577.

Bigelow, H. B., and Sears, M. 1935. Studies of the waters of the continental shelf, Cape Cod to Chesapeake Bay. II: Salinity. Pap. Phys. Oceanogr. Meteorol. 4(1):1–94.

Bigelow, H. B., and Sears, M. 1939. Studies of the waters on the continental shelf, Cape Cod to Chesapeake Bay. III: A volumetric study of the zooplankton. Mem. Mus. Comp. Zool., Harvard University, Cambridge, MA. 54:179–378.

Bigelow, H. B., Lillick, L. C., and Sears, M. 1940. Phytoplankton and planktonic protozoa of the offshore waters of the Gulf of Maine. I. Numerical distribution. Trans. Am. Philos. Soc., N.S. 31(Part III):149–191.

Briggs, J. C. 1974. Marine Zoogeography. New York: McGraw-Hill.

Brown, B. E., Brennan, J. A., Heyerdahl, E. G., et al. 1976. The effects of fishing on the marine finfish biomass in the Northwest Atlantic from the eastern edge of the Gulf of Maine to Cape Hatteras. Int. Comm. Northwest Atlantic Fish. Res. Bull. 12:49–68.

Bumpus, D. F. 1969. Reversal in the surface drift in the Middle Atlantic Bight area. Deep-Sea Res. 16(Suppl.):17–23.

Bumpus, D. F. 1973. A description of the circulation on the continental shelf on the east coast of the United States. Progr. Oceanogr. 6:111–157.

Bumpus, D. F., and Lauzier, L. 1965. Surface circulation on the continental shelf off eastern North America between Newfoundland and Florida. Serial Atlas of the Marine Environment, Folio F. New York: American Geographical Society.

Bush, K. 1981. Middle Atlantic Bight transports determined from a salinity-heat box model using historical hydrographic and meteorological data. Doctoral dissertation, University of Delaware, Newark.

Butman, B., Beardsley, R. C., Magnell, B., et al. 1982. Recent observations of the mean circulation on Georges Bank. J. Phys. Oceanogr. 12(6):569–591.

Clark, S. H., and Brown, B. E. 1977. Changes of biomass of finfishes and squids from the Gulf of Maine to Cape Hatteras, 1963–74, as determined from research vessel survey data. Bull. U.S. Bur. Fish. 75:1–21.

Clark, S. H., and Brown, B. E. 1979. Trends in biomass of finfish and squids in ICNAF Sub-area 5 and Statistical Area 6, 1964–1977, as determined from research vessel survey data. Invest. Pesq. 43:107–122.

Clarke, G. L. 1940. Comparative richness of zooplankton in coastal and offshore areas of the Atlantic. Biol. Bull. 78:226–255.

Clarke, G. L. 1946. Dynamics of production in a marine area. Ecol. Monogr. 16:323–337.

Clarke, G. L., and Zinn, D. J. 1937. Seasonal production of zooplankton off Woods Hole with special reference to *Calanus finmarchicus*. Biol. Bull. Woods Hole 76:371–383.

Clarke, G. L., Pierce, E. L., and Bumpus, D. F. 1943. The distribution and reproduction of *Sagitta elegans* on Georges Bank in relation to hydrographical conditions. Biol. Bull. Woods Hole 85:201–226.

Cohen, E. B. 1976. An overview of the plankton communities of the Gulf of Maine. Int. Comm. Northwest Atlantic Fish., Res. Bull. 1:89–105.

Cohen, E. B., and Grosslein, M. D. 1987. Production of Georges Bank Compared with Other Shelf Ecosystems. In R. H. Backus (ed.), Georges Bank. Cambridge, MA: MIT Press, pp. 383–391.

Cohen, E. B., and Wright, W. R. 1978. Changes in the plankton on Georges Bank in relation to the physical and chemical environment during 1975–76. Int. Counc. Explor. Sea, C.M. 1978/L:27.

Cohen, E., Grosslein, M., Sissenwine, M., and Steimle, F. 1980. A comparison of energy flow on Georges Bank and in the North Sea. Int. Counc. Explor. Sea, C.M. 1980/L:64.

Cohen, E. B., Grosslein, M., Sissenwine, M., et al. 1982. An Energy Budget of Georges Bank. In M. C. Mercer (ed.), Multispecies Approaches to Fisheries Management Advice. Can. Spec. Publ. J. Fish. Aquat. Sci. 59. pp. 95–107.

Colton, J. B., Jr. 1964. A History of Oceanography in the Gulf of Maine. Spec. Sci. Rept. Fish. 496. Washington, DC: U.S. Fish and Wildlife Service.

Colton, J. B., Jr., Marak, R. R., Nickerson, S., and Stoddard, R. R. 1968. Physical, Chemical and Biological Observations on the Continental Shelf, Nova Scotia to Long Island 1964–66. Data Rept. 23. Washington, DC: U.S. Fish and Wildlife Service.

Cresswell, G. M. 1967. Quasi-synoptic Monthly Hydrography of the Transition Region Between Coastal and Slope Water South of Cape Cod, Massachusetts. WHOI Tech. Rept. 67-35. Woods Hole, MA: Woods Hole Oceanographic Institute.

Crisp, D. J. 1975. Secondary Productivity in the Sea. In D. E. Reichle, J. F. Franklin, and D. W. Goodall (eds.), Productivity of World Ecosystems. Washington, DC: National Academy of Science, pp. 71–90.

Dagg, M. J., and Turner, J. T. 1982. The impact of copepod grazing on the phytoplankton of Georges Bank and the New York Bight. Can. J. Fish. Aquat. Sci. 39:979–990.

Deevey, G. B. 1952. Quantity and composition of the zooplankton of Block Island Sound, 1949. Bull. Bingham Oceanogr. Collect., Yale Univ. 13:120–164.

Deevey, G. B. 1956. Oceanography of Long Island Sound, 1952–1954. V. Zooplankton. Bull. Bingham Oceanogr. Collect., Yale Univ. 15:113–155.

Deevey, G. B. 1960. Relative effects of temperature and food on seasonal variations in length of marine copepods in some Eastern American and Western European waters. Bull. Bingham Oceanogr. Collect., Yale Univ. 17:54–86.

Emery, K. O., and Uchupi, E. 1972. Western North Atlantic Ocean topography, rocks, structure, water life and sediments. Am. Assoc. Pet. Geol. Mem. 17:1–532.

Eppley, R. W. 1980. Estimated Phytoplankton Growth Rates in the Central Oligotrophic Oceans. In P. Falkowski (ed.), Primary Productivity in the Sea. New York: Plenum, pp. 231–242.

Fager, E. W., and McGowan, J. A. 1963. Zooplankton species groups in the North Pacific. Science 140:453–460.

Falkowski, P. G., Hopkins, T. S., and Walsh, J. J. 1980. An analysis of factors affecting oxygen depletion in the New York Bight. J. Mar. Res. 38:479–506.

Fish, C. J. 1936a. The biology of *Calanus finmarchicus* in the Gulf of Maine and Bay of Fundy. Biol. Bull. 70:118–141.

Fish, C. J. 1936b. The biology of *Pseudocalanus minutus* in the Gulf of Maine and Bay of Fundy. Biol. Bull. 70:193–216.

Fish, C. J. 1936c. The biology of *Oithona similis* in the Gulf of Maine and Bay of Fundy. Biol. Bull. 70:168–187.

Ford, W. L., Leonard, J. R., and Banks, R. E. 1952. On the nature, occurrence and origin of cold low salinity water along the edge of the Gulf Stream. J. Mar. Res. 11(3): 281–293.

Green, J., Colton, J. B., Jr., and Bearse, D. T. 1977. Preliminary estimates of secondary production on Georges Bank. Int. Counc. Explor. Sea, C.M. 1977/L:30.

Grice, G. D., and Hart, A. D. 1962. The abundance, seasonal occurrence and distribution of the epizooplankton between New York and Bermuda. Ecol. Monogr. 32:287–308.

Grosslein, M. D., and Azarovitz, T. R. 1982. Fish Distribution. Monogr. 15, MESA New York Bight Atlas Series. Albany, NY: New York Sea Grant Institute.

Grosslein, M. D., Langton, R. W., and Sissenwine, M. P. 1980. Recent fluctuations in pelagic fish stocks of the Northwest Atlantic Georges Bank region, in relationship to species interactions. Rapp. P.-V. Réun. Cons. Int. Explor. Mer 177:374–404.

Hazel, J. E. 1970. Atlantic continental shelf and slope of the United States—Ostracod zoogeography in the southern Nova Scotia and northern Virginia faunal provinces. Geol. Surv. Prof. Pap. 529-E. Washington, DC: U.S. Department of the Interior.

Heinle, D. R. 1966. Production of a calanoid copepod, *Acartia tonsa*, in the Patuxent River estuary. Chesapeake Sci. 7(2):59–74.

Hopkins, T. S., and Garfield, N., III. 1979. Gulf of Maine intermediate water. J. Mar. Res. 37:103–139.

Ingham, M. C., Armstrong, R. S., Chamberlin, J. L., et al. 1982. Summary of the Physical Oceanographic Processes and Features Pertinent to Pollution Distribution in the Coastal and Offshore Waters of the Northeastern United States, Virginia to Maine. NEMP Oceanography Summary Rept. NEMP-IV-82-C-004. Washington, DC: U.S. Department of Commerce.

Judkins, D. C., Wirick, C. D., and Esaias, W. E. 1980. Composition, abundance, and distribution of zooplankton in the New York Bight, September 1974–September 1975. Fish. Bull. U.S. 77:669–683.

Ketchum, B. H., and Keen, D. J. 1955. The accumulation of river water over the continental shelf between Cape Cod and Chesapeake Bay. Deep-Sea Res. 3(Suppl.):346–357.

Malone, T. C. 1977. Plankton systematics and distribution. Monogr. 13, MESA New York Bight Atlas Series. Albany, NY: New York Sea Grant Institute.

Maurer, D., and Toner, R. C. 1983. The seasonal relationship between meroplankton and megabenthos on Georges Bank, Gulf of Maine, Northwest Atlantic. Int. Rev. Ges. Hydrobiol. 68(4):513–538.

McBride, M. M., and Brown, B. E. 1980. The status of the marine fishery resources of the northeastern United States. NOAA Tech. Mem. NMFS-F/NEC-5. Washington, DC: U.S. Department of Commerce.

Morse, W. 1982. Spawning stock biomass estimates of sand lance, *Ammodytes* sp., off northeastern United States, determined from MARMAP plankton surveys, 1974–1980. Int. Counc. Explor. Sea, C.M. 1982/G:59.

Mullin, M. 1969. Production of Zooplankton in the Ocean: The Present Status and Problems. In N. Barnes (ed.), Oceanography and Marine Biology Annual Review. London: Allen and Unwin.

National Oceanic and Atmospheric Administration (NOAA). 1983. Status of the Fishery

Resources Off the Northeastern United States for 1982. NOAA Tech. Mem. NMFS-F/NEFC-22. Washington, DC: U.S. Department of Commerce.

National Oceanic and Atmospheric Administration (NOAA). 1985. Status of the Fishery Resources Off the Northeastern United States for 1985. NOAA Tech. Mem. NMFS-F/NEFC-42. Washington, DC: U.S. Department of Commerce.

Northeast Fisheries Science Center. 1995. Status of the Fishery Resources Off the Northeastern United States for 1994. NOAA Technical Memorandum NMFS-NE-108. Woods Hole, MA: Northeast Fisheries Science Center.

O'Reilly, J. E., and Busch, D. E. 1984. Phytoplankton primary production on the Northwest Atlantic shelf. Rapp. P.-V. Réun. Cons. Int. Explor. Mer 183:255–268.

O'Reilly, J. E., Evans-Zetlin, C., and Busch, D. A. 1987. Primary Production. In R. H. Backus (ed.), Georges Bank. Cambridge, MA: MIT Press, pp. 220–233.

Parr, A. E. 1933. A geographic-ecological analysis of the seasonal changes in temperature conditions in shallow water along the Atlantic coast of the United States. Bull. Bingham Oceanogr. Collect., Yale Univ. IV (3):1–90.

Pastuszak, M., Wright, W. R., and Despres-Patanjo, L. 1981. One year of nutrient distribution in the Georges Bank region in relation to hydrography, 1975–1976. Int. Counc. Explor. Sea, C.M. 1981/C:7.

Ramp, S. R., and Vermersch, J. A., Jr. 1978. Measurements of the deep currents in the Northeast Channel, Gulf of Maine. Int. Counc. Explor. Sea, C.M. 1978/C:40.

Redfield, A. C. 1941. The effects of the circulation of water on the distribution of the calanoid community in the Gulf of Maine. Biol. Bull. 80:86–110.

Redfield, A. C., Smith, H. P., and Ketchum, B. 1937. The cycle of organic phosphorus in the Gulf of Maine. Biol. Bull. 73(3):421–443.

Riley, G. A. 1941. Plankton studies. IV. Georges Bank. Bull. Bingham Oceanogr. Collect., Yale Univ. 7:1–73.

Riley, G. A. 1947. Seasonal fluctuations of the phytoplankton population in New England coastal waters. J. Mar. Res. 6:114–125.

Schaaf, W. E. 1979. An analysis of the dynamic population response of Atlantic menhaden, *Brevoortia tyrannis,* to an intensive fishery. Rapp. P.-V. Réun. Cons. Int. Explor. Mer 177:243–251.

Sherman, K. 1968. Seasonal and Areal Distribution of Zooplankton in Coastal Waters of the Gulf of Maine, 1965 and 1966. Spec. Sci. Rept. Fish. 562. Washington, DC: U.S. Fish and Wildlife Service.

Sherman, K. 1970. Seasonal and Areal Distribution of Zooplankton in Coastal Waters of the Gulf of Maine, 1967 and 1968. Spec. Sci. Rept. Fish. 594. Washington, DC: U.S Fish and Wildlife Service.

Sherman, K. 1976. Seasonal distributions of the copepod food of herring in coastal waters of the Gulf of Maine. Res. Doc. 76/VI/82. Halifax, Nova Scotia: International Commission on Northwest Atlantic Fisheries.

Sherman, K. 1980. MARMAP, A Fisheries Ecosystem Study in the Northwest Atlantic: Fluctuations in Ichthyoplankton-Zooplankton Components and Their Potential for Impact on the System. In F. P. Diemer, F. J. Vernberg, and D. Z. Mirkes (eds.), Advanced Concepts on Ocean Measurements for Marine Biology. Columbia, SC: Belle W. Baruch Institute for Marine Biology and Coastal Research, Univ. of South Carolina Press, pp. 3–37.

Sherman, K., Cohen, E., Sissenwine, M., et al. 1978. Food requirements of fish stocks of the Gulf of Maine, Georges Bank, and adjacent waters. Int. Counc. Explor. Sea, C.M. 1978/Gen: 8 (Symp.).

Sherman, K., Green, J. R., and Ejsymont, L. 1983. Coherence in zooplankton of a large northwest Atlantic ecosystem. Fish. Bull. U.S. 81:855–862.

Sherman, K., Jones, C., Sullivan, L., et al. 1981. Congruent shifts in sand eel abundance in western and eastern North Atlantic ecosystems. Nature 291:486–489.

Sherman, K., Smith, W. G., Green, J. R., et al. 1987. Zooplankton production and the fisheries of the northeast shelf. In R. H. Backus (ed.), Georges Bank. Cambridge, MA: MIT Press, pp. 268–282.

Sherman, K., Smith, W., Morse, W., et al. 1984. Spawning strategies of fishes in relation to circulation, phytoplankton production, and pulses in zooplankton off the northeastern United States. Mar. Ecol. Progr. Ser. 18:1–19.

Sigaev, I. K. 1978. Intra-year variability of geostrophic circulation on the continental shelf off New England and Nova Scotia. Int. Comm. Northwest Atlantic Fish., Selected Pap. 3:97–107.

Sissenwine, M. P., Cohen, E. B., and Grosslein, M. D. 1984. Structure of the Georges Bank ecosystem. Rapp. P.-V. Réun. Cons. Int. Explor. Mer 183:243–254.

State of California: The Resources Agency. 1995. California's Ocean Resources: An Agenda for the State of California. ES/1-6/2 pp. + apps. Sacramento: State of California Printing Office.

Steimle, F. W. 1987. Production by the Benthic Fauna. In R. H. Backus (ed.), Georges Bank. Cambridge, MA: MIT Press, pp. 310–314.

Tremblay, M. J., and Roff, J. C. 1983. Production estimates for Scotian Shelf copepods based on mass specific P/B ratios. Can. J. Fish. Aquat. Sci. 40:749–753.

Waldhauer, R., Matte, A., Draxler, A. F. K., and O'Reilly, J. E. 1981. Seasonal ammonium-nitrogen distribution across the New York Bight shelf. In L. L. Ciaccio and A. C. Christini (eds.), Water Conference, Ramapo College of New Jersey, May 1–2, 1980. Mahwah, NJ: Ramapo College, pp. 274–286.

Walsh, J. J. 1981. Shelf-sea Ecosystems. In A. R. Longhurst (ed.), Analysis of Marine Ecosystems. London: Academic Press, pp. 159–196.

Walsh, J. J. 1983. Death in the sea: Enigmatic phytoplankton losses. Progr. Oceanogr. 12:1–86.

Walsh, J. J., Whitledge, T. E., Barvenik, F. W., et al. 1978. Wind events and food chain dynamics within the New York Bight. Limnol. Oceanogr. 23:659–683.

Wright, W. R. 1976. The limits of shelf water south of Cape Cod 1941–1972. J. Mar. Res. 34:1–14.

III

Population Assessments

8

Zooplankton Dynamics in the Northeast Shelf Ecosystem

Edward G. Durbin and Ann G. Durbin

Zooplankton play a pivotal role in marine food webs, both as a link between the primary producers and higher trophic levels and as a critical food resource for larval fish. Zooplankton consumption rates largely determine the proportion of primary production that enters the pelagic food web; the remainder is consumed in situ by microheterotrophs or settles to the bottom to fuel benthic communities (Parsons and Lalli 1988; Townsend and Cammen 1988). Processes controlling zooplankton population dynamics, or the rates of recruitment, feeding, growth, and mortality, determine secondary production rate and role in the food web. The study of zooplankton dynamics should consider the complete life cycle in relation to local hydrography and seasonal patterns. For example, the timing of the zooplankton production cycle may be an important factor controlling larval fish survival, the degree of coupling with the phytoplankton, and the channeling of primary production between the pelagic and benthic food webs (Cushing 1973; Parsons and Kessler 1987; Townsend and Cammen 1988; Durbin and Durbin 1992).

The present review of zooplankton dynamics on the Northeast Shelf ecosystem is restricted to copepods because they represent the principal route to higher trophic levels in marine systems. Gelatinous zooplankton and microzooplankton (ciliates, tintinnids, and heterotrophic flagellates) are important components of the zooplankton in the Northeast Shelf, however, and at times play a key role in the food web (Kremer 1979; Deason 1982).

Factors Affecting Zooplankton Dynamics

Zooplankton are regulated principally by temperature, food, and predation. We briefly review each of these factors, giving examples from the Northeast Shelf.

Temperature

The annual temperature range on the Northeast Shelf is very large and generally exceeds that elsewhere along the North American coast. In Narragansett Bay, an estuarine embayment, seasonal temperatures range between <0°C and >24°C. The annual temperature range decreases on the shelf, for example, 3–20°C in Gulf of Maine surface water (Bigelow 1927) and 6–24°C in the New York Bight (Judkins et al. 1980). On Georges Bank, the strong vertical mixing further reduces the annual temperature range (Bigelow 1927; Flagg 1987).

The large seasonal change in temperature strongly regulates zooplankton production processes. Narragansett Bay, with its extreme temperatures, illustrates these effects. Narragansett Bay is dominated by two copepod species, *Acartia hudsonica* (formerly *A. clausi*) during winter and spring, and *A. tonsa* during summer and fall (Martin 1965; Hulsizer 1976; Durbin and Durbin 1981). Thermally regulated development rates for *A. hudsonica* are slow at cold winter temperatures but quite rapid for *A. tonsa* during summer (Figure 8-1), when biomass turnover times of less than 1 day are possible (Durbin and Durbin 1981). Potential production of *A. hudsonica* and *A. tonsa* in Narragansett Bay was estimated from temperature-dependent development rates and field measurements of abundance and biomass (Durbin and Durbin 1981). The production estimate assumed that growth was not food-limited, and was not corrected for mortality. Biomass of *A. hudsonica* during spring was about 0.5-fold greater than that of *A. tonsa* during summer (Figure 8-2). Nevertheless, *A. tonsa* potential production exceeded that of *A. hudsonica* by about 2.4-fold because of higher growth rates under warm summer temperatures (Figure 8-3). Mean production/biomass ratio of *A. tonsa* during summer (approximately 80) was also significantly greater than that of *A. hudsonica* during spring (approximately 25).

Davis (1984a) modeled the effect of temperature on production by three copepod species (*Pseudocalanus* sp., *Paracalanus parvus*, and *Calanus finmarchicus*) on Georges Bank. Davis also assumed that food was not a limiting factor. *Pseudocalanus* and *P. parvus* were more abundant on Georges Bank than *Calanus finmarchicus*, but because of their smaller size, they contributed less than *C. finmarchicus* to the total copepod biomass. Model simulations showed that secondary production was dominated by *P. parvus*, with a total

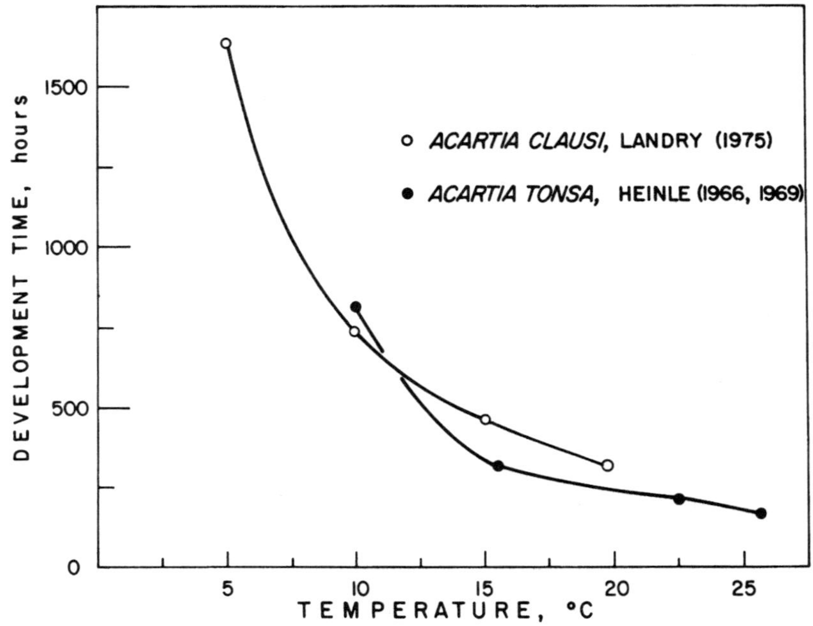

Figure 8-1. Temperature-regulated development times for Acartia clausii [egg to adult (Landry 1975a, b)] and A. tonsa [egg to egg (Heinle 1966, 1969)] (From A. G. Durbin and E. G. Durbin. 1981. Standing stock and estimated production rates of phytoplankton and zooplankton in Narragansett Bay, Rhode Island. Estuaries 4:24–41. Reused by permission. Copyright Estuarine Research Foundation.) Production estimates for A. hudsonica (see Figure 8-3) based on development times for A. clausii (Durbin and Durbin 1981). Bradford (1976) suggested that A. clausii and A. hudsonica may be conspecific.

annual production of 85.9 mg C m^{-3} compared with 52.0 mg C m^{-3} for *Pseudocalanus* sp. and 76.2 mg C m^{-3} for *C. finmarchicus*. The comparatively high production of *P. parvus* reflected its occurrence during the warm, late summer period, whereas *Pseudocalanus* and *C. finmarchicus* were most abundant during spring and early summer when waters were cooler.

The seasonal temperature cycles can have some interesting effects on the timing of the copepod life cycle and on interactions with higher and lower trophic levels. The history of copepods associated with the spring bloom, for example, actually predates the development of the bloom, often by many months (Colebrook 1984; Durbin and Durbin 1992). Over the long term, copepod life cycles may have evolved to take advantage of the spring bloom as a predictable seasonal event (Parsons and Lalli 1988). On shorter, within-season time scales, however, the timing of phytoplankton and zooplankton bio-

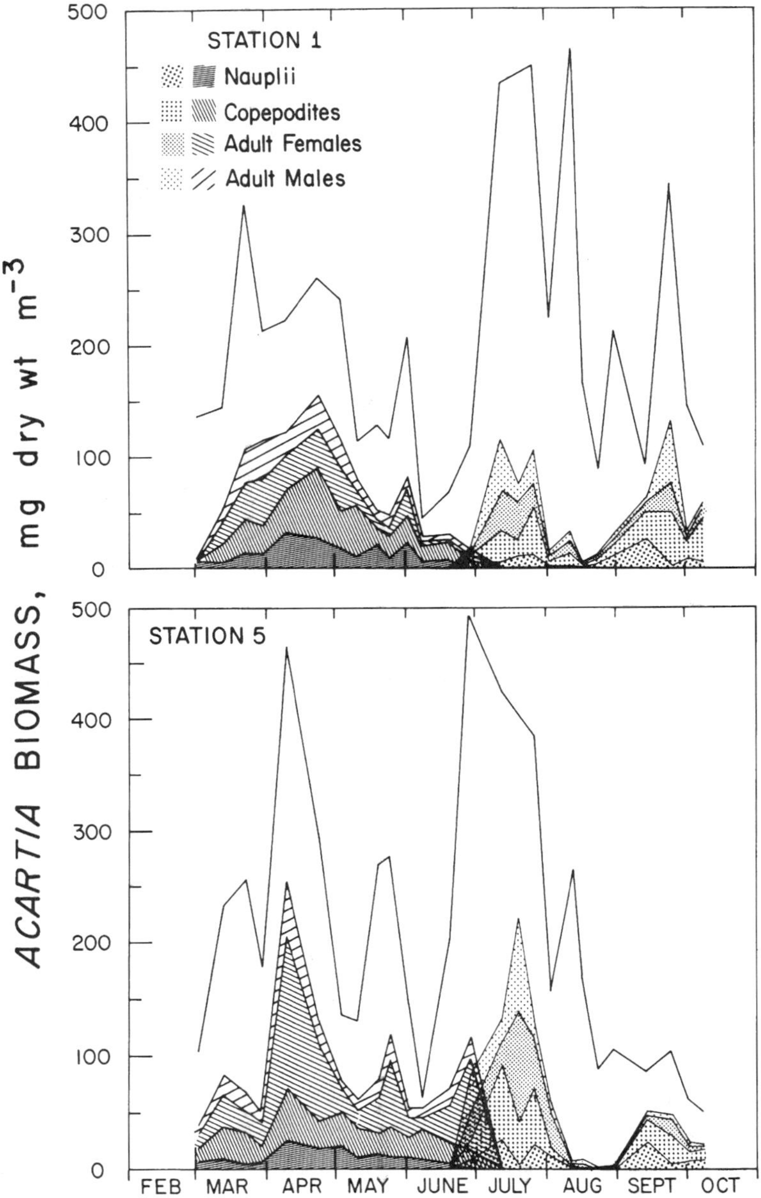

Figure 8-2. Biomass (dry weight) of Acartia hudsonica and A. tonsa nauplii, copepodites, and adults in upper (station 5) and lower (station 1) Narragansett Bay. Solid line represents total zooplankton dry weight. (From A. G. Durbin and E. G. Durbin. 1981. Standing stock and estimated production rates of phytoplankton and zooplankton in Narragansett Bay, Rhode Island. Estuaries 4:24–41. Reused by permission. Copyright Estuarine Research Foundation.)

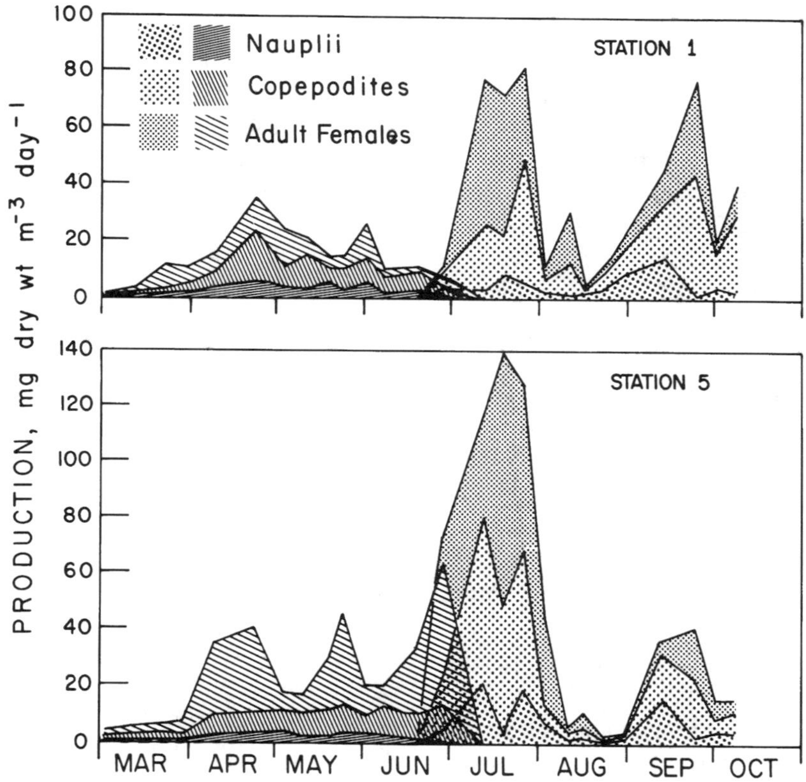

Figure 8-3. Potential production rate of Acartia hudsonica and A. tonsa nauplii, copepodites, and adult females from Narragansett Bay (Durbin and Durbin 1981). (From A. G. Durbin and E. G. Durbin. 1981. Standing stock and estimated production rates of phytoplankton and zooplankton in Narragansett Bay, Rhode Island. Estuaries 4:24–41. Reused by permission. Copyright Estuarine Research Foundation.)

mass maxima may not be closely coupled. For example, in Narragansett Bay the development of *Acartia hudsonica* during winter and spring is governed principally by temperature, and the time at which *A. hudsonica* matures and reaches maximum biomass is largely independent of spring bloom dynamics (Durbin et al. 1992; Durbin and Durbin 1992). The seasonal temperature cycle retards development during autumn and winter but accelerates development during spring, producing a pulse of newly matured adults during mid-spring. The timing of this pulse affects the food web through its influence on copepod reproductive, grazing, and secondary production rates. For example, the production of copepod nauplii affects the food supply for higher trophic levels such as larval fishes (Cushing 1975). An advance of one developmental stage in *A. hudsonica* represents about a 45% increase in biomass and significantly changes population grazing pressure (Durbin and Durbin 1978, 1992). Furthermore, a copepod population maturing early during the winter–spring bloom should consume a larger proportion of the spring bloom than a late-maturing population, channel more of the spring bloom production into the pelagic food web, and leave less to settle to the benthos (Townsend and Cammen 1988).

Where copepods develop over a prolonged period during winter, a small change in temperature has a cumulative effect on development time and can significantly affect the time at which the population reaches maturity (Durbin and Durbin 1992). These effects are illustrated by the specific maturation rate of adults, or the interval over which successive cohorts of copepods mature compared with the interval over which they hatch (Figure 8-4). For *A. hudsonica* in Narragansett Bay, the specific maturation rate reached a maximum during mid-April of 1983–1984. A slight difference of ±1°C in mean temperature strongly affected the predicted pattern of maturation, however, causing a 3-week shift in the timing of peak maturation and a change in the magnitude and shape of the peak. A warm year produced an earlier and prolonged maturation peak, whereas a cooler year produced a delayed and much more intense peak.

Modest interannual variability in temperature can therefore significantly affect zooplankton populations through a cumulative effect on the rate of development. These subtle effects may significantly alter the timing of events and the degree of coupling between different trophic levels, especially during winter and spring. During a warmer year, we would expect a larger population of older copepods to be present during the winter, exerting a greater grazing pressure on the phytoplankton and reducing the size of the bloom and perhaps the amount of phytoplankton settling to the bottom, as suggested by Townsend and Cammen (1988). On the other hand, during a cold year the peak of grazing would occur later and be more intense. Copepod production may become food-limited, and phytoplankton size distribution and species composition may be influenced by heavy grazing pressure.

Food

Copepod feeding rate is strongly affected by food concentration and particle size (Marshall and Orr 1955; Frost 1972; Bartram 1980). Feeding rate increases with food concentration until a critical concentration is reached, above which feeding stabilizes at a maximal rate. Copepods also exhibit a preferred

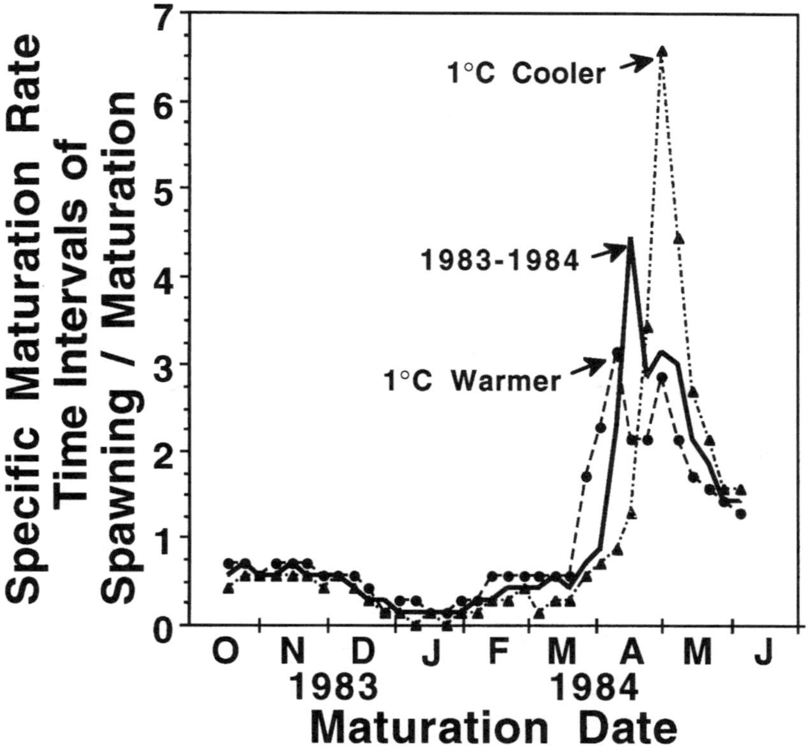

Figure 8-4. Specific maturation rate (SMR = ratio of time interval over which a cohort is spawned to the interval over which it completes the final molt to the adult) for successive cohorts of Acartia hudsonica in lower Narragansett Bay, based on actual temperatures during 1983–1984. The effects on the SMR of a slight warming or cooling (±1°C) from 1983 to 1984 temperatures is also shown.

range of food particle sizes, with feeding efficiency in a number of species declining markedly when particle size is below 10 μm (Frost 1972; Bartram 1980; Berggreen et al. 1988). Thus, simple measures of total phytoplankton biomass, such as chlorophyll *a*, may overestimate the food available to zooplankton in situ. Fractionation of phytoplankton into several size classes (e.g., Malone 1971a, b; Durbin et al. 1975; Peterson and Bellantoni 1987) provides a better measure of available food.

In temperate coastal and shelf waters, phytoplankton size tends to change seasonally, with large diatoms dominant during winter and spring and smaller flagellates during summer and early fall. These seasonal changes in the phytoplankton community alter the quantity and quality of food available to zooplankton, and affect secondary production rates.

Food limitation occurs most frequently during the summer–fall period, when phytoplankton biomass is frequently high but dominated by small algae. In Narragansett Bay, egg production by *Acartia tonsa* was strongly food-limited during the flagellate-dominated summer and fall (Durbin et al. 1983). The mean egg production of *A. tonsa* (25.3 eggs d^{-1}) was only about 17% of the maximum (about 150 eggs d^{-1} at 20°C), indicating that actual production during summer was much less than the potential rates estimated by Durbin and Durbin (1981), assuming non–food-limited growth. The limiting factor appeared to be the small size of the dominant algae rather than low phytoplankton biomass. Similarly, a bloom of a small (1–2 μm) flagellate *Aureococcus anophagefferens* in Narragansett Bay, the Peconic Bays, and Long Island Sound during the summer of 1985 (Cosper et al. 1989; Nuzzi and Waters 1989; Smayda and Villareal 1989) was associated with very low rates of ingestion and egg production by *A. tonsa* (Durbin and Durbin 1989). In Long Island Sound, both *Acartia tonsa* and *Temora longicornis* were shown to be food-limited (Peterson 1985; Beckman and Peterson 1986; Bellantoni and Peterson 1987), especially during summer when the water column was stratified and the <10 μm-size fraction of chlorophyll *a* predominated (Bellantoni and Peterson 1987). In the New York Bight, ingestion rates of *Centropages typicus* that were fed natural particles during the summer and autumn were almost always less than maximal (Dagg and Grill 1980). Natural particle concentrations in the New York Bight were similar to concentrations used in the in situ feeding experiments, indicating that food limitation of *C. typicus* is a common occurrence. In the Chesapeake Bay, an estuary of high primary productivity, Heinle (1966) observed both high and relatively low growth rates of *Acartia tonsa* incubated in situ. Subsequent work indicated that *Eurytemora affinis* in the Chesapeake Bay supplements its phytoplankton diet with detritus to meet energy requirements (Heinle and Flemer 1975; Heinle et al. 1977). Similarly, in the Hudson River plume, in situ grazing studies indicated that detritus was an important food resource for zooplankton (Chervin 1978; Chervin et al. 1981). These studies suggest that detritus and other non-phytoplankton food sources (microzooplankton) are important for zooplankton in estuarine and nearshore waters.

On Georges Bank, there are also indications that zooplankton production is food-limited (Dagg and Turner 1982). Zooplankton consumption was estimated during April, July, and October along transects across the shelf south of Long Island and across Georges Bank. The abundance and size of the dominant copepods and chlorophyll concentration in the field were determined, and ingestion was calculated from equations developed by Mullin and Brooks (1976) relating ingestion of *Calanus pacificus* to body size and temperature. Dagg and Turner concluded that zooplankton did not consume much of the phytoplankton during the spring bloom, but during the fall, grazing was intense, with ingestion rates equal to, or greater than, primary production rates. In contrast, laboratory feeding studies led Davis (1984b) to conclude that the copepods *Paracalanus parvus* and *Pseudocalanus* sp. were generally not food-limited on Georges Bank because of the high concentrations of chlorophyll *a* that are found throughout the year (O'Reilly et al. 1987). As previously noted for Narragansett Bay and Long Island Sound, however, high chlorophyll *a* concentrations on Georges Bank do not necessarily imply non–food-limited zooplankton production, because the total phytoplankton biomass may not be available to the zooplankton as food.

The winter–spring period provides a quite different food environment for zooplankton because of the prevalence of large, chain-forming diatoms. In Narragansett Bay, *A. hudsonica* was not food-limited during the winter–spring bloom except near its end (Durbin et al. 1992). Even then, food limitation was mild, mainly affecting body size rather than egg production rate. Production rates of *A. hudsonica* during this period should therefore approximate the potential production rates estimated by Durbin and Durbin (1981). In contrast, Smith and Lane (1988) found that during spring in the New York Bight, the feeding response of the zooplankton never showed saturation, indicating that ingestion was food-limited even during the spring bloom period. At one shelf station, the proportion of primary production that was ingested increased from 13% to 46% between March and April; at another station it was 100% in both months.

The evidence suggests that over much of the Northeast Shelf, zooplankton production is food-limited, especially during summer and fall. In situ investigations of copepod feeding and production rates, phytoplankton production, and size structure are needed to determine the extent to which potential secondary production is actually achieved even in regions of high primary production such as Georges Bank.

Predation

Predation has been shown to control zooplankton species composition, size structure, and abundance in freshwater lakes [reviewed by Zaret (1980) and Northcote (1988)]. Planktivorous fishes selectively remove larger or more visible prey, whereas invertebrate predators prefer small, easily manipulated prey.

Thus, fishes tend to drive zooplankton community structure toward small-sized or cryptic forms, whereas invertebrate predators have the reverse effect of driving the zooplankton toward large size and elaborate body form (e.g., Brooks and Dodson 1965; Dodson 1974). Zooplankton respond to predators through changes in morphology (Dodson 1989; Stemberger 1990), behavior (Zaret and Suffern 1976; Haney 1988; Gliwicz 1986), and life history (Strickler and Twombly 1975; Hairston et al. 1983), demonstrating the dynamic role of predation in shaping freshwater communities.

The importance of predation in the sea is more difficult to determine. The greater diversity of predators in the sea, the larger size and complexity of most marine ecosystems, and their interconnection with other ecosystems makes it more difficult to quantify predator–prey interactions than in the comparatively smaller, enclosed, and simplified communities of freshwater lakes (Lehman 1988; Nixon 1988). Nevertheless, direct effects of marine planktivorous fishes have been detected in situ (Oviatt et al. 1972; Koslow 1981), and the estimated prey consumption rates of fish (Lasker 1970, 1988; Landry 1978; Mann 1984) and invertebrate predators (Greve 1981; Davis 1984a; Ohman 1986; Schneider 1989) is often large compared with estimates of zooplankton secondary production. Marine zooplankton, in common with their freshwater counterparts, also exhibit diel vertical migration and diel feeding rhythms that appear to be predator-avoidance behaviors (Ohman 1988, 1990; Dagg et al. 1989; Bollens and Frost 1989; Durbin et al. 1990). These observations suggest that, as in freshwater, predation is a dominant factor structuring marine zooplankton communities.

Historically, marine communities have been viewed as driven from the "bottom up," with nutrient, light, and temperature dynamics controlling the primary producers, whose size structure and productivity in turn regulate the food web (e.g., Ryther 1969). The literature on freshwater systems indicates that bottom-up and top-down forces act in concert to regulate ecosystem function: total ecosystem production is determined by bottom-up resource limitation, but predation controls food web structure and the pathways of energy flow (Carpenter et al. 1985; McQueen et al. 1986; Mills and Forney 1988). This paradigm may apply equally well in the sea, and should be explored in future work.

Predator control of marine zooplankton has been most clearly demonstrated during blooms of gelatinous zooplankton. In Narragansett Bay, the ctenophore *Mneniopsis leidyi*, a voracious zooplankton predator, is abundant during the summer and early fall (Kremer and Nixon 1976; Kremer 1979; Deason 1982). Ctenophore population size was related to zooplankton biomass (Deason and Smayda 1982). The initiation of the summer ctenophore pulse was accompanied by a rapid decline in zooplankton abundance and a bloom of large diatoms, an apparent response to reduced grazing pressure (Durbin and Durbin 1981; Deason and Smayda 1982). In some years, termination of the phytoplankton bloom coincided with a decrease in ctenophore abundance and a corresponding increase in zooplankton (Deason and Smayda 1982). These relationships indicate that *M. leidyi* exerts top-down control of summer zooplankton abundance and grazing pressure, indirectly affecting phytoplankton size structure and abundance in a classic "trophic cascade" comparable with those originally described from freshwater lakes (Carpenter et al. 1985; Mills et al. 1987). Control of zooplankton biomass by ctenophores has also been reported off southwestern Nova Scotia (Suthers and Frank 1990), where the ctenophore *Pleurobrachia pileus* appeared to be responsible for a persistent cross-shelf gradient in zooplankton. On Georges Bank, Davis (1984a) concluded from simulation modeling that predation by chaetognaths, ctenophores, and a predatory copepod (*Centropages* sp.) was sufficient to control populations of the dominant herbivorous copepods.

The large populations of plankton-feeding fishes on the Northeast Shelf, including menhaden, bay anchovy, sand lance, herring, and mackerel, suggest that fishes might have a significant impact on zooplankton populations. Sherman et al. (1981) attributed an explosive increase in the abundance of zooplanktivorous sand lance in the Gulf of Maine during the mid-1970s to the depletion of herring and mackerel stocks by overfishing. Sherman et al. did not observe a significant change in zooplankton with the increase in sand lance abundance, but Payne et al. (1990) later demonstrated an inverse correlation between the abundance of sand lance and its preferred prey, the large copepod *Calanus finmarchicus*, during 1982–1988. The results implied that during years of peak abundance, sand lance were capable of depleting their preferred prey. Such a regional impact of fishes on plankton populations has not previously been demonstrated in the ocean, but has been described in large freshwater lakes [e.g., Lake Michigan (Wells 1970; Scavia and Fahnenstiel 1988; Evans 1990)]. Energy budget calculations for Georges Bank by Sissenwine et al. (1984) indicated that zooplankton and fish production were tightly coupled during the period 1973–1975, and that little if any more fish production could be supported within the system. The high levels of predation shown for fishes by Sissenwine et al. (1984) and for carnivorous zooplankton by Davis (1984a) demonstrate that predator control of marine plankton communities is likely to be a widespread phenomenon, even in highly productive environments such as Georges Bank.

Secondary Production on the Northeast Shelf

Secondary production estimates are available for several regions within the Northeast Shelf, including Narragansett Bay, the New York Bight, Georges Bank, and the Gulf of Maine (Table 8-1). Annual secondary production estimates in the different environments are similar and range between 266 and 736 kcal m^2 yr^{-1}. Several different methods have been used for these calculations. Durbin and Durbin (1981) and Davis (1984a) used a form of population dynamics in which the abundance of individual species, the size of individual stages for each species, and estimates of their growth rates are used to calculate production rates. Davis (1984a) estimated mortality from a simulation model and included this in his production estimates, whereas Durbin and Durbin (1981) calculated "potential" production, which did not include mortality. Excluding mortality in this type of analysis underestimates the true production but may not greatly change the results (Kimmerer 1987). In both studies, growth was assumed to be maximal. Subsequent studies in Narragansett Bay have shown that *A. tonsa* is chronically food-limited during the summer (Durbin et al. 1983) and thus, actual production will be lower than the calculated rates. We presently have no direct measures of in situ growth on Georges Bank.

Sissenwine et al. (1984) and Sherman et al. (1987) calculated annual production from estimates of mean annual zooplankton biomass, multiplied by a constant production/biomass ratio to give annual production. Sissenwine et al. (1984) used estimates of mean zooplankton biomass from Riley and Bumpus (1946), and applied a P:B ratio of 7, as suggested by Steele (1974) for North Sea zooplankton. Sherman et al. (1987) estimated mean zooplankton biomass from MARMAP data collected over Georges Bank during 1977–1981. Samples were collected with 333-µm mesh nets and are termed macrozooplankton in Table 8-1. Production of this size class was estimated using Steele's P:B ratio of 7. Microzooplankton, which includes small copepods such as *Paracalanus parvus* and *Pseudocalanus* sp., were estimated to be 43% of the macrozooplankton biomass (Beers and Stewart 1969), and production was calculated using a P:B ratio of 25 (Banse and Mosher 1980; Tremblay and Roff 1983) (see Table 8-1).

Sherman et al. (1987) also calculated production from estimates of zooplankton grazing rate and trophic transfer efficiency. They assumed that 50% of the primary production was grazed by zooplankton (Dagg and Turner 1982; Walsh 1983) and that transfer efficiency was 32% (Steele 1974). Phytoplankton less than 20 µm were consumed only by microzooplankton, and the fraction greater than 20 µm by macrozooplankton. Primary produc-

Table 8-1. Zooplankton secondary production on the northeast continental shelf, estimated from population dynamics (PD), trophic transfer efficiency (TTE), or production/biomass ratio (P:B). Carbon production converted to kilocalories assuming 1 g C = 10 kcal.

			Zooplankton production (kcal m^{-2} y^{-1})		
Region	Reference	Method	Micro	Macro	Total
Narragansett Bay					
Lower bay (St 1)	Durbin and Durbin 1981	PD			475[a]
Upper bay (St 5)	Durbin and Durbin 1981	PD			646[a]
New York Bight	Walsh 1988	TTE			329[b]
Georges Bank	Sissenwine et al. 1984	P:B			329
	Davis 1984	PD	115[c]	33[d]	
	Sherman et al. 1987	P:B	265	173	438
	Sherman et al. 1987	TTE	305	230	535
	Walsh 1988	TTE			266[b]
Gulf of Maine	Sherman et al. 1987	P:B	446	290	736
	Sherman et al. 1987	TTE	287	123	410

[a]Total annual production extrapolated from data for Acartia spp. during the period 1 March–8 October 1976, assuming that Acartia production is about one-half of total zooplankton production. Annual Acartia production = 1.2 × production during 1 March–8 October, and total zooplankton production = 2× annual Acartia production.
[b]Assumes a trophic transfer efficiency of 32% (Steele 1974).
[c]Only Paracalanus parvus and Pseudocalanus spp.
[d]Only Calanus finmarchicus.

tion was estimated from O'Reilly and Busch (1984) and O'Reilly et al. (1987). Approximately 57% of the primary production on Georges Bank and 70% in the Gulf of Maine was the result of the less than 20-μm size fraction (O'Reilly et al. 1987). Zooplankton production estimated by Sherman et al. (1987) from the trophic transfer efficiency and the P:B ratio methods were similar for Georges Bank, but differed by a factor of almost 2 in the Gulf of Maine, where the P:B ratio yielded the higher value (see Table 8-1). Walsh (1988) similarly estimated the role of zooplankton in recycling nitrogen from ingestion rates and trophic efficiencies. His consumption rate estimates and an assumed trophic transfer efficiency of 32% yield secondary production rates of 266 kcal m^2 yr^{-1} and 329 kcal m^2 yr^{-1} for total zooplankton in Georges Bank and the New York Bight, respectively (see Table 8-1).

These production rate estimates all have limitations. The estimates of Durbin and Durbin (1981) and Davis (1984a) assume that production was not food-limited, an assumption that has been demonstrated to be incorrect for *A. tonsa* in Narragansett Bay (Durbin et al. 1983), and that may well also be incorrect for Georges Bank. In this case, the estimates by these authors in Table 8-1 are too high. Whereas the biomass estimate of Sherman et al. (1987) is based on the extensive MARMAP data set, production estimates are based on a P:B ratio for the North Sea, which may not be appropriate in the Gulf of Maine or Georges Bank. Assumptions that a constant proportion of primary production is consumed and that trophic transfer efficiency is constant (Sherman et al. 1987; Walsh 1988) are at best only approximations. Important biological events such as phytoplankton blooms and predator–prey interactions are episodic rather than continuous processes, and there is a need to quantify their variance as well as their seasonal or annual mean rates. More realistic estimates of zooplankton production in ecosystem models will require field measurements of growth and production on time and space scales ranging from the individual event to the seasonal and annual average.

Responses of Zooplankton to Perturbations

Human environmental perturbations that may affect zooplankton include those caused by eutrophication, longer-term climatic changes, changes in predation pressure resulting from fishing pressure, and release of toxic chemicals. In this section we will focus on the first three factors.

Eutrophication

Adverse effects of artificial eutrophication will be strongest in estuaries, embayments, and other nearshore environments. Eutrophication may initially cause a general increase in primary production, but may eventually lead to species shifts and increases in noxious phytoplankton blooms. A series of long-term experiments in the Marine Ecosystem Research Laboratory (MERL) mesocosms at the University of Rhode Island has investigated the effects of different levels of artificial eutrophication on coastal ecosystems (Nixon et al. 1984), and the results are quite instructive. Zooplankton production in the control mesocosms was food-limited (Sullivan and Ritacco 1985a). At low to intermediate levels of enrichment, egg production rates increased, whereas at the higher levels of enrichment, egg production declined to levels typical of the controls despite the higher phytoplankton levels. At these highest levels of enrichment (8× and 32× mean inputs to Narragansett Bay), copepod abundance was reduced and mortality rates were high. Thus, zooplankton secondary production was enhanced by a low to moderate level of enrichment, but declined under high levels of eutrophication. The reduced secondary production under eutrophic conditions may have reflected toxicity of relatively high concentrations of ammonia (Sullivan and Ritacco 1985b) and a shift in phytoplankton composition toward species of lower food quality (Frithsen et al. 1985).

In the natural environment, a change in food quality brought about by eutrophication may significantly affect zooplankton populations. For example, the flagellate *Olisthodiscus luteus* (= *Heterosigma akashiwo*) is abundant in Narragansett Bay during summer and fall (Tomas 1980), forming extremely dense blooms in the upper, more eutrophic, part of the bay (Durbin and Durbin 1981). Laboratory experiments have shown that *O. luteus*, although of suitable size, is a poor food for the copepod *Acartia tonsa* (Tomas and Deason 1981). The "brown tide" bloom of *Aureococcus anophagefferens* similarly illustrates how certain algal blooms may reduce zooplankton secondary production (Durbin and Durbin 1989). The brown tide, although of uncertain origin, was undoubtedly sustained by the high levels of nutrients in coastal waters of the Northeast Shelf. These studies indicate that whereas eutrophication may enhance primary production, the energy is not efficiently transferred to higher trophic levels, and may actually have the reverse effect of decreasing production.

Climate Change

Predictions of global warming associated with the greenhouse effect make it prudent to consider how climatic change may affect zooplankton in the Northeast Shelf. A temperature rise may cause secondary production rates to increase and the geographic distribution of certain species to change. Temperature-induced hydrographic changes may also affect phytoplankton communities and coupling with zooplankton.

Slight changes in mean water temperature affect zooplankton population dynamics because of the strong temperature dependence of growth rate. Global warming is predicted to affect winter temperatures more strongly than summer temperatures. Figure 8-4 suggests that zooplankton developing under cold temperatures will also be the most sensitive to small temperature changes. A 2°C temperature difference scarcely affected specific maturation rate in *Acartia hudsonica* under the relatively warm temperatures of spring and fall, but imposed a large cumulative effect during winter because of the prolonged development at cold temperatures. Consequently, zooplankton species that complete a major part of their life cycle during winter will likely be the first affected by global warming.

Global warming might produce tighter coupling between zooplankton and phytoplankton on the Northeast Shelf, with less phytoplankton settling out to the benthic food web, as suggested by Townsend and Cammen (1988) and Durbin and Durbin (1992). On the other hand, warmer winter temperatures may stimulate benthic filter feeders that either compete with (Daborn

1986) or prey on (Sullivan and Banzon 1990) water column zooplankton. The outcome of these complex interactions is, as yet, difficult to predict. There is evidence, however, that fishes tend to amplify subtle changes within the ecosystem, and that variation in fish population size is a sensitive indicator of environmental change (Jeffries and Terceiro 1985; Jeffries et al. 1989).

Global warming may produce hydrographic changes that alter the timing of phytoplankton events and disrupt the linkage with zooplankton. Wright et al. (1986) reviewed the probable effects of climate change on the physical oceanography of the New England shelf. They predicted an earlier and perhaps more intense development of stratification, leading to an earlier termination of the spring bloom. Potential effects on *Calanus finmarchicus* in the Gulf of Maine can be used to illustrate the impact of these physical oceanographic changes. In the Gulf of Maine, *Calanus* survives unfavorable periods as resting stages (copepodite stage C5 and possibly some C4s) in the deep basins, and then resumes activity in late winter to produce one major spring cohort (Bigelow 1926; Fish 1936). Several smaller generations may be formed later in the year. The production of the spring cohort coincides with the spring bloom. The spring bloom terminates as the water stratifies; phytoplankton numbers decline to a low level and are dominated by small flagellates that are not efficiently consumed by *Calanus*. An earlier termination of the spring bloom may reduce the likelihood that the spring cohort of *Calanus* will complete its development to the overwintering stage. *Calanus* might then be excluded from the Gulf of Maine, Georges Bank, and the New York Bight, where it is presently a biomass dominant and a primary food resource for fish and planktivorous whale populations. The replacement of *Calanus* by smaller copepod species would have a serious impact on the fisheries and marine mammal populations in the region.

Change in Predation Caused by Fishing

Fishing can alter energy flow in an ecosystem through a reduction in the biomass and food consumption of targeted species (Clark and Brown 1977; Sissenwine et al. 1984; Sissenwine 1986), a reduction in fish population age structure toward younger animals with higher feeding rates per unit biomass (Sissenwine et al. 1984; Andersen and Ursin 1977), and shifts in species dominance (Sherman et al. 1981; Skud 1982; Mayo et al. 1992).

The effects on zooplankton of such ecosystem-wide changes in marine fish populations have been examined in only a few cases, however, and the results have been inconclusive. Smith and Eppley (1982) found no significant relationship between northern anchovy (*Engraulis mordax*) and zooplankton biomass in the southern California bight between 1951 and 1966. Koslow (1983) found that, contrary to expectation, zooplankton populations in the North Sea declined during a period when herring stocks were depleted from overfishing, and concluded that zooplankton were regulated by large-scale oceanographic processes and food limitation rather than predation. Mullin and Conversi (1988) found that the initiation of a fishery on Pacific whiting (*Merluccius productus*) had no demonstrable effect on euphausiid abundance, and like Koslow, and Smith and Eppley, concluded that zooplankton abundance was correlated with large-scale oceanographic events, such as El Niño, rather than predation.

In these studies, the finding of a strong link between zooplankton abundance and changes in oceanographic conditions was convincing. The authors experienced some difficulty in the analysis, however, because of limitations in the historical data available to them. Smith and Eppley noted that their data on total zooplankton biomass would not reveal possible effects of selective feeding by anchovies on certain species or size groups of zooplankton. Koslow was unable to determine whether total predation pressure in the ecosystem actually declined during the years of low herring abundance. Andersen and Ursin (1977) suggested that predation from large herring may have been more than compensated for by increased numbers of benthic fishes whose juveniles eat zooplankton, by gelatinous zooplankton that were not well sampled by the continuous plankton recorder (CPR) used in Koslow's analysis, or by a shift toward a younger age structure with higher rates of food consumption. Similarly, the study period of Mullin and Conversi, the mid-1960s, was a period of peak abundance of northern anchovy (Smith 1978), and it was not clear that zooplankton predation pressure declined after fishing reduced the abundance of whiting. These results illustrate the difficulty of establishing cause and effect in predator–prey relationships in complex marine ecosystems using a single-species approach, and the value of long time series of observations, the importance of considering large-scale oceanic processes such as El Niño, and the need to quantify the entire predator field (including not only fishes but predatory zooplankton) and the species composition, size structure, and biomass of the zooplankton prey.

Long-term observations by the National Marine Fisheries Service on fish and plankton communities on the Northeast Shelf documented the replacement of herring and mackerel by sand lance in the Gulf of Maine. Skud (1982) demonstrated that when species dominance changes in response to a natural or human-induced perturbation, predator–prey and competitive interactions may create a new equilibrium that resists a return to the original species dominance. Time-series observations also led to the discovery by Payne et al.

(1990) of an inverse relationship between the abundance of sand lance and *Calanus* during the 1980s. Although requiring independent confirmation, for example, through quantification of sand lance prey consumption, this relationship may prove to be one of the first documented examples of predator control of zooplankton by a marine fish. Thus, whereas the role of predation in governing secondary production is more difficult to establish in the sea than in freshwater, evidence from the Northeast Shelf and elsewhere demonstrates that the problem is tractable when addressed at the level of the ecosystem. Future work should consider top-down as well as bottom-up regulation of marine communities and, through appropriate sampling, measurements of in situ rate processes, experimentation, and modeling, will yield new insight into ecosystem structure, species abundance, and production in the sea.

References

Andersen, K. P., and Ursin, E. 1977. A multispecies extension to the Beverton and Holt theory of fishing, with accounts of phosphorus circulation and primary production. Meddr. Danm. Fisk. Havunders., N.S. 7:319–435.

Banse, K., and Mosher, S. 1980. Adult body mass and annual production/biomass relationships of field populations. Ecol. Monogr. 50:355–379.

Bartram, W. C. 1980. Experimental development of a model for the feeding of neritic copepods on phytoplankton. J. Plankton Res. 3:25–51.

Beckman, B. R., and Peterson, W. T. 1986. Egg production by *Acartia tonsa* in Long Island Sound. J. Plankton Res. 8(5):917–925.

Beers, J. R., and Stewart, G. L. 1969. Microzooplankton and its abundance relative to the larger zooplankton and other seston components. Mar. Biol. 4:182–189.

Bellantoni, D. C., and Peterson, W. T. 1987. Temporal variability in egg production rates of *Acartia tonsa* Dana in Long Island Sound. J. Exp. Mar. Biol. Ecol. 107(3):199–208.

Berggreen, U., Hansen, B., and Kiørboe, T. 1988. Food size spectra, ingestion and growth of the copepod *Acartia tonsa* during development: Implications for determination of copepod production. Mar. Biol. 99: 341–352.

Bigelow, H. B. 1926. Plankton of the offshore waters of the Gulf of Maine. Bull. Bur. Fish., U.S. 40:1–509.

Bigelow, H. B. 1927. Physical oceanography of the Gulf of Maine. Bull. Bur. Fish., U.S. 40:511–1027.

Bollens, S. M., and Frost, B. W. 1989. Zooplanktivorous fish and variable diel vertical migration in the marine planktonic copepod *Calanus pacificus*. Limnol. Oceanogr. 34:1072–1083.

Bradford, J. M. 1976. Partial revision of the *Acartia* subgenus *Acartiura* (Copepoda: Calanoida: Acartiidae). N.Z. J. Mar. Freshw. Res. 10(1):159–202.

Brooks, J. L., and Dodson, S. I. 1965. Predation, body size, and composition of plankton. Science 150:28–35.

Carpenter, S. R., Kitchell, J. F., and Hodgson, J. R. 1985. Cascading trophic interactions and lake productivity. BioScience 35:634–639.

Chervin, M. 1978. Assimilation of particulate organic carbon by estuarine and coastal copepods. Mar. Biol. 49:265–275.

Chervin, M. B., Malone, T. C., and Neale, P. J. 1981. Interactions between suspended organic matter and copepod grazing in the plume of the Hudson River. Estuar. Coast. Shelf Sci. 13:169–184.

Clark, S. H., and Brown, B. E. 1977. Changes in biomass of finfishes and squids from the Gulf of Maine to Cape Hatteras, 1963–74, as determined from research vessel survey data. Fish. Bull., U.S. 75:1–21.

Colebrook, J. M. 1984. Continuous plankton records: Relationships between species of phytoplankton and zooplankton in the seasonal cycle. Mar. Biol. 83:313–323.

Cosper, E. M., Carpenter, E. J., and Cottrell, M. 1989. Primary Productivity and Growth Dynamics of the "Brown Tide" in Long Island Embayments. In E. M. Cosper, V. M. Bricelj, and E. J. Carpenter (eds.), Novel Phytoplankton Blooms. Berlin: Springer-Verlag, pp. 139–158.

Cushing, D. H. 1973. Recruitment and Parent Stock in Fishes. Washington Sea Grant Publ. WSG 73-1. Seattle: University of Washington, Division of Marine Resources.

Cushing, D. H. 1975. Marine Ecology and Fisheries. Cambridge, England: Cambridge University Press.

Daborn, G. R. 1986. Effects of Tidal Mixing on the Plankton and Benthos of Estuarine Regions of the Bay of Fundy. In M. J. Bowman, C. M. Yentsch, and W. T. Peterson (eds.), Tidal Mixing and Plankton Dynamics. Lecture Notes on Coastal and Estuarine Studies, Vol. 17. Berlin: Springer-Verlag, pp. 390–413.

Dagg, M. J., Frost, B. W., and Walser, W. E., Jr. 1989. Copepod diel migration, feeding, and the vertical flux of phaeopigments. Limnol. Oceanogr. 34:1062–1071.

Dagg, M. J., and Grill, D. W. 1980. Natural feeding rates of *Centropages typicus* females in the New York Bight. Limnol. Oceanogr. 25:597–609.

Dagg, M. J., and Turner, J. T. 1982. The impact of copepod grazing on the phytoplankton of Georges Bank and the New York Bight. Can. J. Fish. Aquat. Sci. 39:979–990.

Davis, C. S. 1984a. Predatory control of copepod seasonal cycles on Georges Bank. Mar. Biol. 82:31–40.

Davis, C. S. 1984b. Food concentrations on Georges Bank: Non-limiting effect on development and survival of laboratory reared *Pseudocalanus* sp. and *Paracalanus parvus* (Copepoda: Calanoida). Mar. Biol. 82:41–46.

Deason, E. E. 1982. *Mnemiopsis leidyi* (Ctenophora) in Narragansett Bay, 1975–79: Abundance, size, composition and estimation of grazing. Estuar. Coast. Shelf Sci. 15:121–134.

Deason, E. E., and Smayda, T. J. 1982. Ctenophore-zooplankton-phytoplankton interactions in Narragansett Bay, Rhode Island, USA, during 1972–1977. J. Plankton Res. 4:203–217.

Dodson, S. I. 1974. Adaptive change in plankton morphology in response to size-selective predation: A new hypothesis of cyclomorphosis. Limnol. Oceanogr. 19:721–729.

Dodson, S. I. 1989. Predator-induced reaction norms. BioScience 39:447–452.

Durbin, A. G., and Durbin, E. G. 1981. Standing stock and estimated production rates of phytoplankton and zooplankton in Narragansett Bay, Rhode Island. Estuaries 4:24–41.

Durbin, A. G., and Durbin, E. G. 1989. Effect of the "Brown Tide" on Feeding, Size and Egg Laying Rate of Adult Female *Acartia tonsa*. In E. M. Cosper, V. M. Bricelj, and E. J. Carpenter (eds.), Novel Phytoplankton Blooms. Berlin: Springer-Verlag, pp. 625–646.

Durbin, A. G., and Durbin, E. G. 1992. Seasonal change in size frequency distribution and estimated age in the marine copepod *Acartia hudsonica* during a winter-spring diatom bloom in Narragansett Bay. Limnol. Oceanogr. 37:379–392.

Durbin, A. G., Durbin, E. G., and Wlodarczyk, E. 1990. Diel feeding behavior in the marine copepod *Acartia tonsa* in relation to food availability. Mar. Ecol. Prog. Ser. 68:23–45.

Durbin, E. G., and Durbin, A. G. 1978. Length and weight relationships of *Acartia clausi* from Narragansett Bay, R.I. Limnol. Oceanogr. 23:958–969.

Durbin, E. G., Durbin, A. G., and Campbell, R. G. 1992. Seasonal changes in egg laying rate in the marine copepod *Acartia hudsonica* during a winter-spring diatom bloom in Narragansett Bay. Limnol. Oceanogr. 37:342–360.

Durbin, E. G., Durbin, A. G., Smayda, T. J., and Verity, P. G. 1983. Food limitation of production by adult *Acartia tonsa* in Narragansett Bay, Rhode Island. Limnol. Oceanogr. 28:1199–1213.

Durbin, E. G., Krawiec, R. W., and Smayda, T. J. 1975. Seasonal studies on the relative importance of different size fractions of phytoplankton in Narragansett Bay (USA). Mar. Biol. 32:271–287.

Evans, M. S. 1990. Large-lake responses to declines in the abundance of a major fish planktivore—the Lake Michigan example. Can. J. Fish. Aquat. Sci. 47:1738–1754.

Fish, C. J. 1936. The biology of *Calanus finmarchicus* in the Gulf of Maine and Bay of Fundy. Biol. Bull. 70:118–141.

Flagg, C. N. 1987. Hydrographic Structure and Variability. In R. H. Backus (ed.), Georges Bank. Cambridge, MA: MIT Press, pp. 108–124.

Frithsen, J. B., Lane, P. A., Keller, A. A., and Pilson, M. E. Q. 1985. Effects of Inorganic Nutrient Additions in Coastal Areas: A Mesocosm Experiment Data Report. Volume 2. MERL Series, Rept. No. 4. Kingston: University of Rhode Island.

Frost, B. W. 1972. Effects of size and concentration of food particles on the feeding behavior of the marine planktonic copepod *Calanus pacificus*. Limnol. Oceanogr. 17:805–815.

Gliwicz, Z. M. 1986. Predation and the evolution of vertical migration in zooplankton. Nature 320:746–748.

Greve, W. 1981. Invertebrate predator control in a coastal marine ecosystem: The significance of *Beroe gracilis* (Ctenophora). Kieler Meeresforsch. 5:211–217.

Hairston, N. G., Jr., Walton, W. E., and Li, K. T. 1983. The causes and consequences of sex-specific mortality in a freshwater copepod. Limnol. Oceanogr. 28:935–947.

Haney, J. F. 1988. Diel patterns of zooplankton behavior. Bull. Mar. Sci. 43:583–603.

Heinle, D. R. 1966. Production of a calanoid copepod, *Acartia tonsa*, in the Patuxent River estuary. Chesapeake Sci. 7(2):59–74.

Heinle, D. R. 1969. Temperature and zooplankton. Chesapeake Sci. 10:186–209.

Heinle, D. R., and Flemer, D. A. 1975. Carbon requirements of a population of the estuarine copepod *Eurytemora affinis*. Mar. Biol. 31:235–247.

Heinle, D. R., Harris, R. P., Ustach, J. F., and Flemer, D. A. 1977. Detritus as food for estuarine copepods. Mar. Biol. 40:341–353.

Hulsizer, E. E. 1976. Zooplankton of lower Narragansett Bay, 1972–1973. Chesapeake Sci. 17(4):260–270.

Jeffries, H. P., and Terceiro, M. 1985. Cycle of changing abundances in the fishes of the Narragansett Bay area. Mar. Ecol. Prog. Ser. 25:239–244.

Jeffries, P., Keller, A., and Hale, S. 1989. Predicting winter flounder (*Pseudopleuronectes americanus*) catches by time series analysis. Can. J. Fish. Aquat. Sci. 46:650–659.

Judkins, D. C., Wirick, C. D., and Esaias, W. E. 1980. Composition, abundance, and distribution of zooplankton in the New York Bight, September 1974–September 1975. Fish. Bull., U.S. 77:669–683.

Kimmerer, W. J. 1987. The theory of secondary production calculations for continuously reproducing populations. Limnol. Oceanogr. 32:1–13.

Koslow, J. A. 1981. Feeding selectivity of schools of northern anchovy, *Engraulis mordax*, in the Southern California Bight. Fish. Bull., U.S. 79:131–142.

Koslow, J. A. 1983. Zooplankton community structure in the North Sea and Northeast Atlantic: Development and test of a biological model. Can. J. Fish. Aquat. Sci. 40:1912–1924.

Kremer, P. 1979. Predation by the ctenophore *Mnemiopsis leidyi* in Narragansett Bay, Rhode Island. Estuaries 2:97–105.

Kremer, P., and Nixon, S. 1976. Distribution and abundance of the ctenophore *Mnemiopsis leidyi* in Narragansett Bay. Estuar. Coast. Mar. Sci. 4:627–639.

Landry, M. R. 1975a. Seasonal temperature effects and predicting development rates of marine copepod eggs. Limnol. Oceanogr. 20:434–440.

Landry, M. R. 1975b. The relationship between temperature and the development of life stages in the marine copepod *Acartia clausi* Giesbr. Limnol. Oceanogr. 20:854–857.

Landry, M. R. 1978. Population dynamics and production of a planktonic marine copepod, *Acartia clausii*, in a small temperate lagoon on San Juan Island, Washington. Int. Rev. Ges. Hydrobiol. 63:77–119.

Lasker, R. 1970. Utilization of Zooplankton Energy by a Pacific Sardine Population in the California Current. In J. H. Steele (Ed.), Marine Food Chains. Berkeley: Univ. of California Press, pp. 265–284.

Lasker, R. 1988. Food Chains and Fisheries: An Assessment After 20 Years. In B. J. Rothschild (ed.), Toward a Theory on Biological-Physical Interactions in the World Ocean. Boston: Kluwer Academic, pp. 173–182.

Lehman, J. T. 1988. Ecological principles affecting community structure and secondary production by zooplankton in marine and freshwater environments. Limnol. Oceanogr. 33(4, Pt. 2):931–945.

Malone, T. C. 1971a. The relative importance of nanoplankton and netplankton as primary producers in tropical oceanic and neritic phytoplankton communities. Limnol. Oceanogr. 16:633–639.

Malone, T. C. 1971b. The relative importance of nanoplankton and netplankton as primary producers in the California Current system. Fish. Bull., U.S. 69:799–820.

Mann, K. H. 1984. Fish Production in Open Ocean Ecosystems. In M. J. R. Fasham (ed.), Flows of Energy and Materials in Marine Ecosystems: Theory and Practice. New York: Plenum, pp. 435–458. Marshall, S. M., and Orr, A. P. 1955. The Biology of a Marine Copepod. Edinburgh: Oliver and Boyd.

Martin, J. H. 1965. Phytoplankton-zooplankton relationships in Narragansett Bay. Limnol. Oceanogr. 10:185–191.

Mayo, R. K., Fogarty, M. J., and Serchuk, F. M. 1992. Aggregate fish biomass and yield on Georges Bank, 1960–1987. J. Northwest Atlantic Fish. Sci. 14:59–78.

McQueen, D. J., Post, J. R., and Mills, E. L. 1986. Trophic relationships in freshwater pelagic ecosystems. Can. J. Fish. Aquat. Sci. 43:1571–1581.

Mills, E. L., and Forney, J. L. 1988. Trophic Dynamics and Development of Freshwater Pelagic Food Webs. In S. R. Carpenter (ed.), Complex Interactions in Lake Communities. Berlin: Springer-Verlag, pp. 11–30.

Mills, E. L., Forney, J. L., and Wagner, K. J. 1987. Fish Predation and Its Cascading Effect on the Oneida Lake Food Chain. In W. C. Kerfoot and A. Sih (eds.), Predation: Direct and Indirect Impacts on Aquatic Communities. Hanover, NH: University Press of New England, pp. 118–131.

Mullin, M. M., and Brooks, E. R. 1976. Some consequences of distributional heterogeneity of phytoplankton and zooplankton. Limnol. Oceanogr. 21:784–796.

Mullin, M. M., and Conversi, A. 1988. Biomasses of euphausiids and smaller zooplankton in the California Current: Geographic and interannual comparisons relative to the Pacific whiting, *Merluccius productus*, fishery. Fish. Bull., U.S. 87:633–644.

Nixon, S. W. 1988. Physical energy inputs and the comparative ecology of lake and marine ecosystems. Limnol. Oceanogr. 33:1005–1025.

Nixon, S. W., Pilson, M. E. Q., Oviatt, C. A., et al. 1984. Eutrophication of a Coastal Marine Ecosystem—An Experimental Study Using the MERL Microcosms. In M. J. R. Fasham (ed.), Flows of Energy and Materials in Marine Ecosystems: Theory and Practice. New York: Plenum, pp. 105–135.

Northcote, T. G. 1988. Fish in the structure and function of freshwater ecosystems: A "top-down" view. Can. J. Fish. Aquat. Sci. 45:361–379.

Nuzzi, R., and Waters, R. M. 1989. The Spatial and Temporal Distribution of "Brown Tide" in Eastern Long Island. In E. M. Cosper, V. M. Bricelj, and E. J. Carpenter (eds.), Novel Phytoplankton Blooms. Berlin: Springer-Verlag, pp. 117–137.

O'Reilly, J. E., and Busch, D. A. 1984. Phytoplankton primary production for the Northwestern Atlantic shelf. Rapp. P.-V. Réun. Cons. Int. Explor. Mer 183:255–268.

O'Reilly, J. E., Evans-Zetlin, C., and Busch, D. A. 1987. Primary production. In R. H. Backus (ed.), Georges Bank. Cambridge, MA: MIT Press, pp. 220–233.

Ohman, M. D. 1986. Predator-limited population growth of the copepod *Pseudocalanus* sp. J. Plankton Res. 8:673–713.

Ohman, M. D. 1988. Behavioral responses of zooplankton to predation. Bull. Mar. Sci. 43:530–550.

Ohman, M. D. 1990. The demographic benefits of diel vertical migration by zooplankton. Ecol. Monogr. 60:257–281.

Oviatt, C. A., Gall, A. L., and Nixon, S. W. 1972. Environmental effects of Atlantic menhaden on surrounding waters. Chesapeake Sci. 13(4):321–323.

Parsons, T. R., and Kessler, T. A. 1987. An ecosystem model for the assessment of plankton production in relation to the survival of young fish. J. Plankton Res. 9:125–137.

Parsons, T. R., and Lalli, C. M. 1988. Comparative oceanic ecology of the plankton communities of the subarctic Atlantic and Pacific oceans. Oceanogr. Mar. Biol. Annu. Rev. 26:317–359.

Payne, P. M., Wiley, D. N., Young, S. B., et al. 1990. Recent fluctuations in the abundance of baleen whales in the southern Gulf of Maine in relation to changes in selected prey. Fish. Bull., U.S. 88:687–696.

Peterson, W. T. 1985. Abundance, age structure and in-situ egg production rates of the copepod *Temora longicornis* in Long Island Sound, New York. Bull. Mar. Sci. 37(2):726–738.

Peterson, W. T., and Bellantoni, D. C. 1987. Relationships between water-column stratification, phytoplankton cell size and copepod fecundity in Long Island Sound and off central Chile. S. Afr. J. Mar. Sci. 5:411–421.

Riley, G. A., and Bumpus, D. F. 1946. Phytoplankton-zooplankton relationships on Georges Bank. J. Mar. Res. 6:33–46.

Ryther, J. 1969. Photosynthesis and fish production in the sea. Science 166:72–76.

Scavia, D., and Fahnenstiel, G. L. 1988. From Picoplankton to Fish: Complex Interactions in the Great Lakes. In S. R. Carpenter (ed.), Complex Interactions in Lake Communities. Berlin: Springer-Verlag, pp. 85–100.

Schneider, G. 1989. The common jellyfish *Aurelia aurita*: Standing stock, excretion and nutrient regeneration in the Kiel Bight, Western Baltic. Mar. Biol. 100:507–514.

Sherman, K., Jones, C., Sullivan, L., et al. 1981. Congruent shifts in sand eel abundance in western and eastern North Atlantic ecosystems. Nature 291:486–489.

Sherman, K., Smith, W. G., Green, J. R., et al. 1987. Zooplankton Production and the Fisheries of the Northeastern Shelf. In R. H. Backus (ed.), Georges Bank. Cambridge, MA: MIT Press, pp. 268–282.

Sissenwine, M. P. 1986. Perturbation of a Predator-controlled Continental Shelf Ecosystem. In K. Sherman and L. M. Alexander (eds.), Variability and Management of Large Marine Ecosystems. AAAS Selected Symposium 99. Boulder, CO: Westview Press, pp. 55–85.

Sissenwine, M. P., Cohen, E. B., and Grosslein, M. D. 1984. Structure of the Georges Bank ecosystem. Rapp. P.-V. Réun. Cons. Int. Explor. Mer 183:243–254.

Skud, B. E. 1982. Dominance in fishes: The relation between environment and abundance. Science 216:144–149.

Smayda, T. J., and Villareal, T. A. 1989. The 1985 "Brown-tide" and the Open Phytoplankton Niche in Narragansett Bay During Summer. In E. M. Cosper, V. M. Bricelj, and E. J. Carpenter (eds.), Novel Phytoplankton Blooms. Berlin: Springer-Verlag, pp. 159–187.

Smith, L. S., and Lane, P. V. Z. 1988. Grazing of the spring diatom bloom in the New York Bight by the calanoid copepods *Calanus finmarchicus, Metridia lucens* and *Centropages typicus*. Cont. Shelf Res. 8(5–7):485–509.

Smith, P. E. 1978. Precision of sonar mapping for pelagic fish assessment in the California Current. J. Cons. Int. Explor. Mer 38:33–40.

Smith, P. E., and Eppley, R. W. 1982. Primary production and the anchovy population in the Southern California Bight: Comparison of time series. Limnol. Oceanogr. 27:1–18.

Steele, J. H. 1974. The Structures of Marine Ecosystems. Cambridge, MA: Harvard University Press.

Stemberger, R. S. 1990. Food limitation, spination, and reproduction in *Brachionus calyciflorus*. Limnol. Oceanogr. 35:33–44.

Strickler, J. R., and Twombly, S. 1975. Reynolds number, diapause, and predatory copepods. Int. Ver. Theor. Angew. Limnol. Verh. 19:2943–2950.

Sullivan, B. K., and Banzon, P. V. 1990. Food limitation and benthic regulation of populations of the copepod *Acartia hudsonica* Pinhey in nutrient-limited and nutrient-enriched systems. Limnol. Oceanogr. 35:1618–1631.

Sullivan, B. K., and Ritacco, P. J. 1985a. The response of dominant copepod species to food limitation in a coastal marine ecosystem. Arch. Hydrobiol. Beih. 31:407–418.

Sullivan, B. K., and Ritacco, P. J. 1985b. Ammonia toxicity to larval copepods in eutrophic marine ecosystems: A comparison of results from bioassays and enclosed experimental ecosystems. Aquat. Toxicol. 7:205–217.

Suthers, I. M., and Frank, K. T. 1990. Zooplankton biomass gradient off southwestern Nova Scotia: Nearshore ctenophore predation or hydrographic separation? J. Plankton Res. 12:831–850.

Tomas, C. 1980. *Olisthodiscus luteus*, Chrysophyceae. V. Its occurrence, abundance and dynamics in Narragansett Bay, Rhode Island. J. Phycol. 16:156–166.

Tomas, C. R., and Deason, E. E. 1981. The influence of grazing by two *Acartia* species on *Olisthodiscus luteus* Carter. Mar. Ecol. 2:215–223.

Townsend, D. W., and Cammen, L. M. 1988. Potential importance of the timing of spring plankton blooms to benthic-pelagic coupling and recruitment of juvenile demersal fishes. Biol. Oceanogr. 5:215–229.

Tremblay, M. J., and Roff, J. C. 1983. Production estimates for Scotian Shelf copepods based on mass specific P/B ratios. Can. J. Fish. Aquat. Sci. 40:249–253.

Walsh, J. J. 1983. Death in the sea: Enigmatic phytoplankton losses. Prog. Oceanogr. 12:1–86.

Walsh, J. J. 1988. On the Nature of Continental Shelves. New York: Academic Press.

Wells, L. 1970. Effects of alewife predation on zooplankton populations in Lake Michigan. Limnol. Oceanogr. 15:556–565.

Wright, D. G., Hendry, R. M., Loder, J. W., and Dodson, F. W. 1986. Oceanic changes associated with global increases in atmospheric carbon dioxide: A preliminary report for the Atlantic coast of Canada. Can. Tech. Rep. Fish. Aquat. Sci. 1426.

Zaret, T. M. 1980. Predation and Freshwater Communities. New Haven, CT: Yale University Press.

Zaret, T. M., and Suffern, J. S. 1976. Vertical migration in zooplankton as a predator avoidance mechanism. Limnol. Oceanogr. 21:804–813.

9

The State of Groundfish Resources Off the Northeastern United States*

Vaughn C. Anthony

Abstract

From the early 1960s to 1977, fishing vessels from 13 nations fished heavily off the northeastern United States. Landings of principal groundfish rose to 780,000 tons in 1965 and abundance declined rapidly. Most species of groundfish were being underfished by the United States. With the passage of the Magnuson Fishery Conservation and Management Act in 1976, fishing by foreign nationals stopped and fishing effort by the United States increased. The number of days fished by otter trawlers steadily increased into the mid-1980s before tapering off. Fishing technology improved and the effective fishing effort continued to go up even though the number of days at sea did not. Landings rose with effort until 1983 and then fell dramatically. Groundfish abundance declined steadily after 1978 for the next decade. By 1990–1992, fishing mortality rates for the three major groundfish species (cod, haddock, yellowtail flounder) were twice as great as the management targets. Overall groundfish landings were one-third the maximum sustainable yield (MSY). Landings for haddock and yellowtail flounder were one-tenth the MSY. If the abundance of groundfish were rebuilt to provide the MSY, the catch could increase by two to three times with one-half of the present effort.

Introduction

Finfish and squid resources off the northeastern coast of the United States have experienced significant changes in abundance because of fishing during the past 30 years. Whereas some stocks, such as mackerel, butterfish,

*This article appeared in Fisheries 1993;18(3):12–17.

skates, dogfish, and squids, are underfished, groundfish abundance has fallen to all-time record lows. Before 1960, the fishery resources in the Northeast were fished almost exclusively by the United States, except for scallops, which were also fished by Canada. In the 1960s, however, distant-water fleets expanded their efforts into the New England and mid-Atlantic areas. As many as 13 nations fished these resources until 1977, increasing the catch of most species to record levels. Figure 9-1 indicates the landings of principal groundfish species off the New England and mid-Atlantic coasts from 1960 to 1991 by United States fishermen and by fishermen from all countries. These species are Atlantic cod, haddock, yellowtail flounder, redfish, pollock, silver and red hake, winter flounder, American plaice, witch flounder, windowpane flounder, and summer flounder. In 1965, about 780,000 tons of groundfish were harvested, including 323,000 tons of silver hake, 155,000 tons of haddock, and 96,000 tons of red hake. The total of 574,000 tons compares with a U.S. catch today of only 24,000 tons for the same species. As the abundance of fish declined during the late 1960s, the foreign fleets moved from one stock to another, overfishing haddock, silver hake, red hake, herring, and mackerel. The International Commission for Northwest Atlantic Fisheries (ICNAF), which had been the international management body for such fisheries since the mid-1950s, attempted to control fishing effort with mesh sizes and closed areas. These indirect management controls, however, were no match for the heavy fishing by the large foreign trawlers.

Catch Quotas

Restrictions on total allowable groundfish catches were introduced by ICNAF in 1973 along with national allocations of species catch quotas. Twenty-three stocks were controlled by catch quotas in 1973, and 54 catch quotas were put into place by ICNAF in 1974. Total landings rapidly declined as the catch quotas were reduced and monitored (Figure 9-1). The Magnuson Fishery Conservation and Management Act (MFCMA) was implemented in 1977 and restricted fishing by foreign nationals except for Canada. The use of catch quotas continued under the New England Fishery Management Council until 1982. Landings by the United States after 1973 increased to about 200,000 tons in 1980–1983 (see Figure 9-1). Landings steadily declined after 1983 as the abundance of groundfish stocks also declined. The landings of principal groundfish species by the United States in 1991 were less than 100,000 tons.

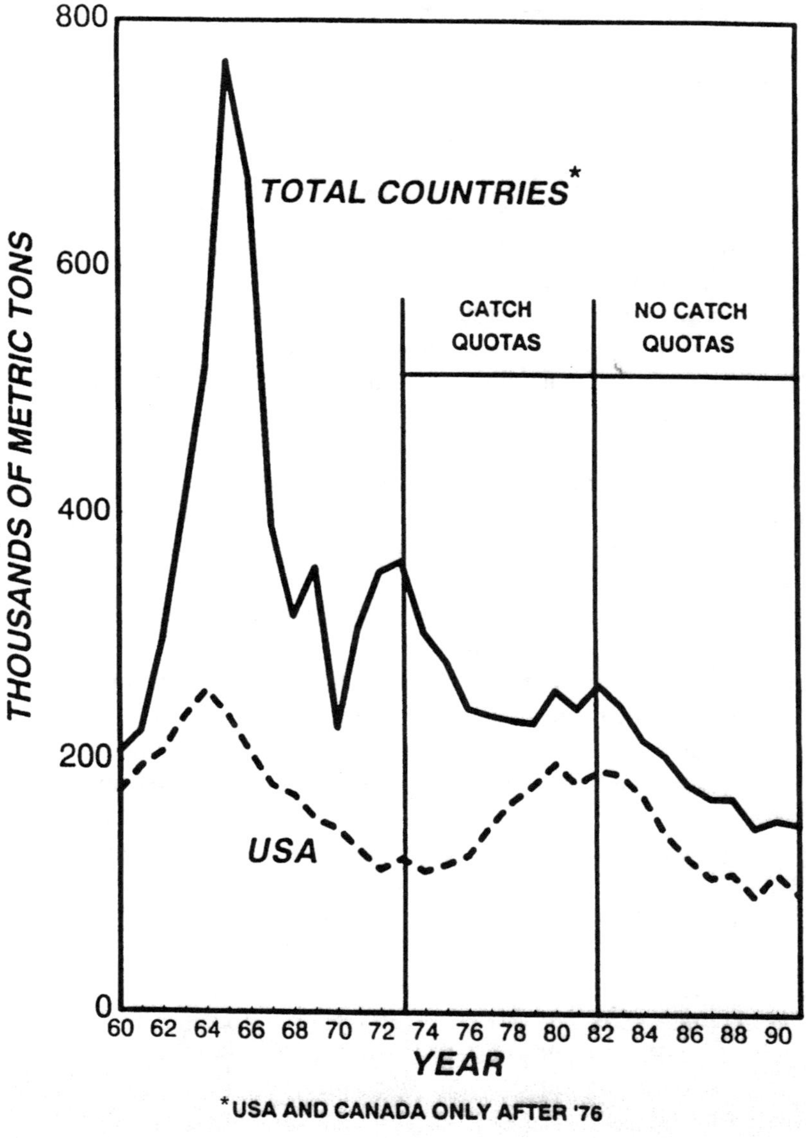

Figure 9-1. Landings of principal groundfish off the northeastern United States by the United States and by all countries, 1960–1991.

Fishing Effort and Abundance

Fishing effort increased markedly on the traditional groundfish species (cod, haddock, flounder) with the passage of the Magnuson Fishery Conservation and Management Act. Figure 9-2 indicates the change in days fished by the otter trawl fleet from 1976 to 1991. Fishing effort increased by 80–100% from the mid-1970s to the mid-1980s. In addition, new vessels replaced old vessels, larger vessels replaced smaller vessels, and electronics, nets, and engines improved. This meant that a day at sea in the late 1980s was much more effective for catching fish than ever before. Under this increase in effort, landings rose by 66% from 1976 to 1983. Catches doubled, or more than doubled, for American plaice, witch flounder, winter flounder, white hake, cusk, Atlantic wolffish, and haddock. Large numbers of small fish were also discarded.

Fishing mortality increased so greatly during this period that 50–60% of the fishable populations of some groundfish were landed annually. The abundance of many groundfish species declined sharply. As a result, catches started to decrease in the early 1980s and they continue to decline today for many species.

Figure 9-3 shows two indices of abundance for the 12 principal groundfish species. These indices were derived from commercial catch-per-day information from the otter trawl fishery and from the autumn bottom trawl research surveys conducted by the Northeast Fisheries Science Center. The research surveys began in 1963. Indices for the time series indicate a great reduction in abundance over a 25-year period. The high levels in the early 1960s were a result of underfishing and an accumulation of old age, slow-growing stocks of fish such as redfish, hakes, and two very good year-classes of haddock. These high catches could not be sustained, and the stocks declined very rapidly until 1974. At this point, restrictions on fishing under ICNAF and good year-classes of cod and haddock produced a sharp increase in abundance. The abundance declined during the 1980s but increased again in 1989 and 1990 because of good year-classes of cod and a good year-class of yellowtail flounder. However, these year-classes were quickly fished out.

Status of Cod, Yellowtail Flounder, and Haddock

Examples of the status of groundfish resources are shown in Figures 9-4 through 9-9 for Georges Bank cod, southern New England yellowtail flounder, and Georges Bank haddock. Figure 9-4 shows indices of abundance for Georges Bank cod and the number of age-2 recruits to the population, as indi-

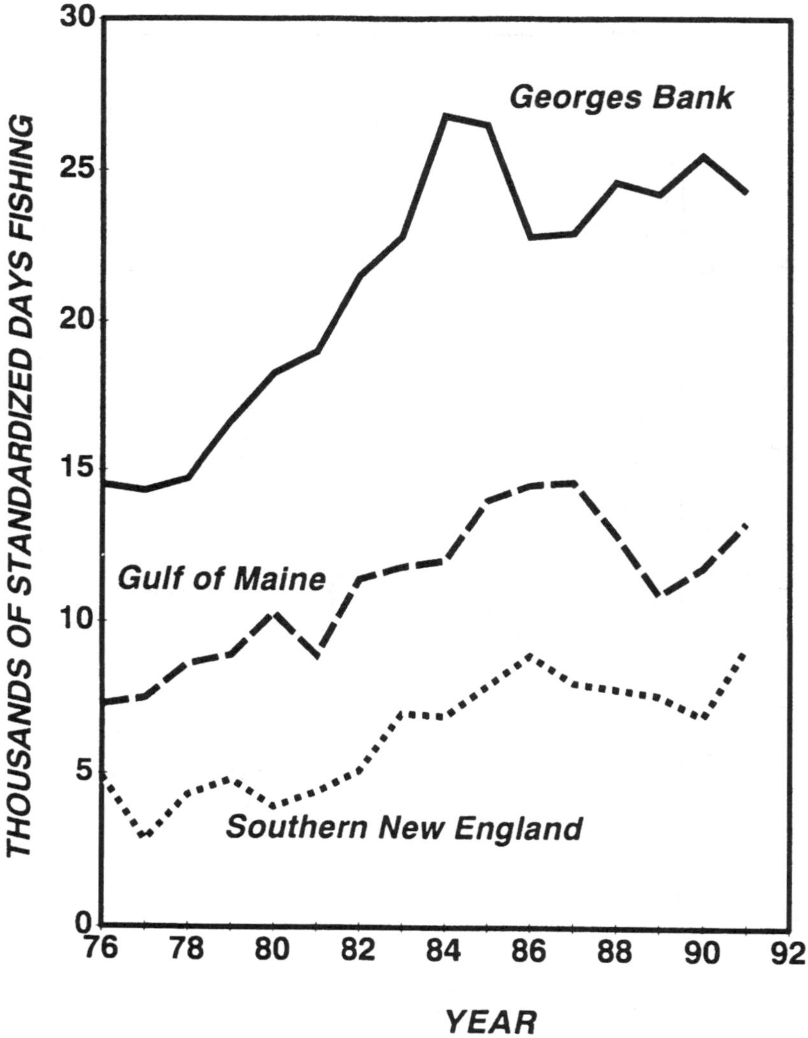

Figure 9-2. Standardized fishing effort by otter trawlers on Georges Bank, the Gulf of Maine, and in southern New England, 1976–1991.

cated by the autumn bottom trawl research survey. The height of each bar indicates the relative number of age-2 cod that are entering the fishery in the year indicated. In 1973 and 1977, for example, recruitment was very good. Recruitment in 1987 and 1990 was slightly better than average. The abundance

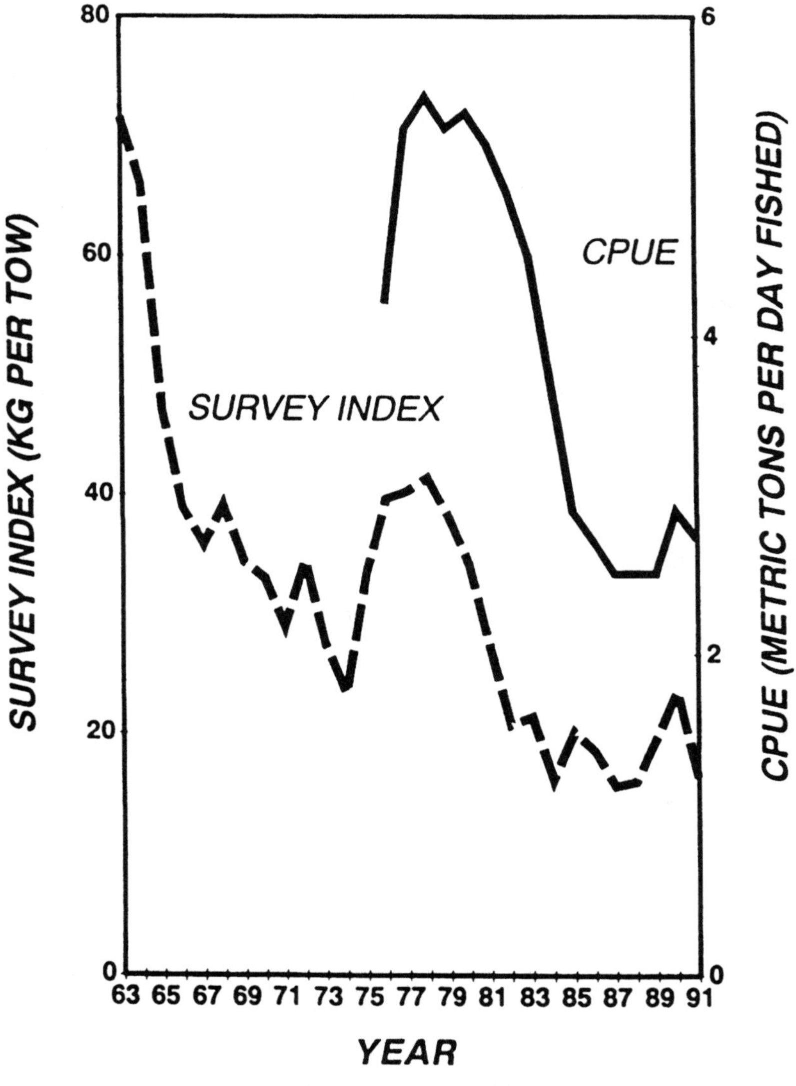

Figure 9-3. Abundance indices for principal northeast groundfish resources from 1963–1991.

index indicates the overall population abundance, including these recruits. During the 1960s, when fishing by foreign nationals was very heavy, the abundance index was quite low. It increased sharply during the 1970s because of the good recruitment in 1973 and 1977. Abundance declined during the 1980s because of heavy fishing and the lack of very strong recruitment for the stock.

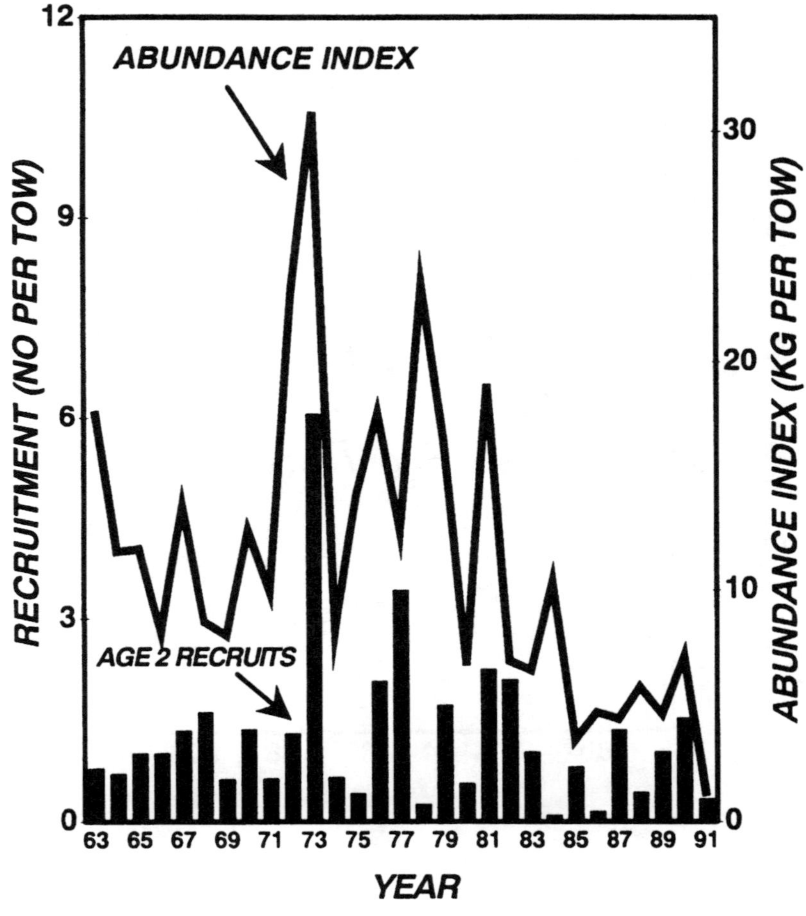

Figure 9-4. Indices of abundance and recruitment for the Georges Bank stock of Atlantic cod from the autumn bottom trawl research survey from 1963–1991.

The index of abundance for 1991 is at a record low level. The analytical assessment for this stock indicates that spawning stock abundance dropped to a record low during 1991, and continued to decline throughout 1992.

Figure 9-5 indicates the reason for the decline in abundance of Georges Bank cod. The fishing mortality rate increased from 0.30 in 1978 to nearly 1.1 by 1991. This is a level of fishing that is three times as high as the management target set by the New England Fishery Management Council and is removing 60% of the exploitable biomass annually. The maximum sustainable yield

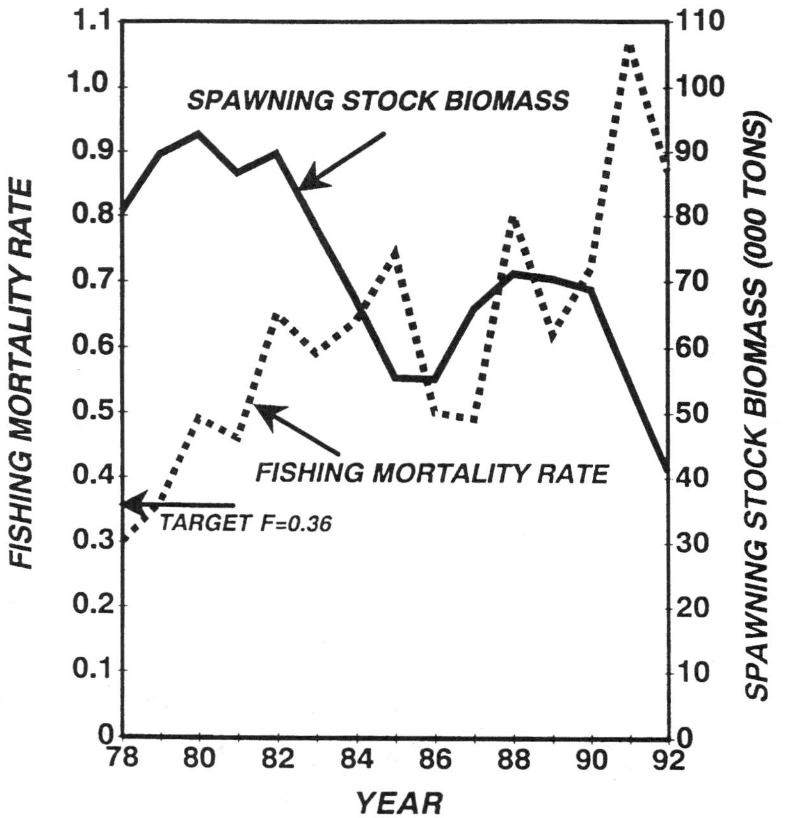

Figure 9-5. Instantaneous fishing mortality rate and spawning stock biomass for the Georges Bank stock of Atlantic cod, 1978–1992.

(MSY) from this resource is achieved when the spawning stock biomass reaches about 105,000 tons. Fishing at a mortality rate of 0.36 (or a removal rate of 28% per year) would produce an annual catch of about 35,000 tons. In 1992, the spawning stock biomass was only slightly above 40,000 tons and declining. The recent catch levels for Georges Bank cod have exceeded 40,000 tons.

Southern New England Yellowtail Flounder

Figure 9-6 shows the change in abundance and recruitment for southern New England yellowtail flounder. During the 1960s, the abundance of yellowtail

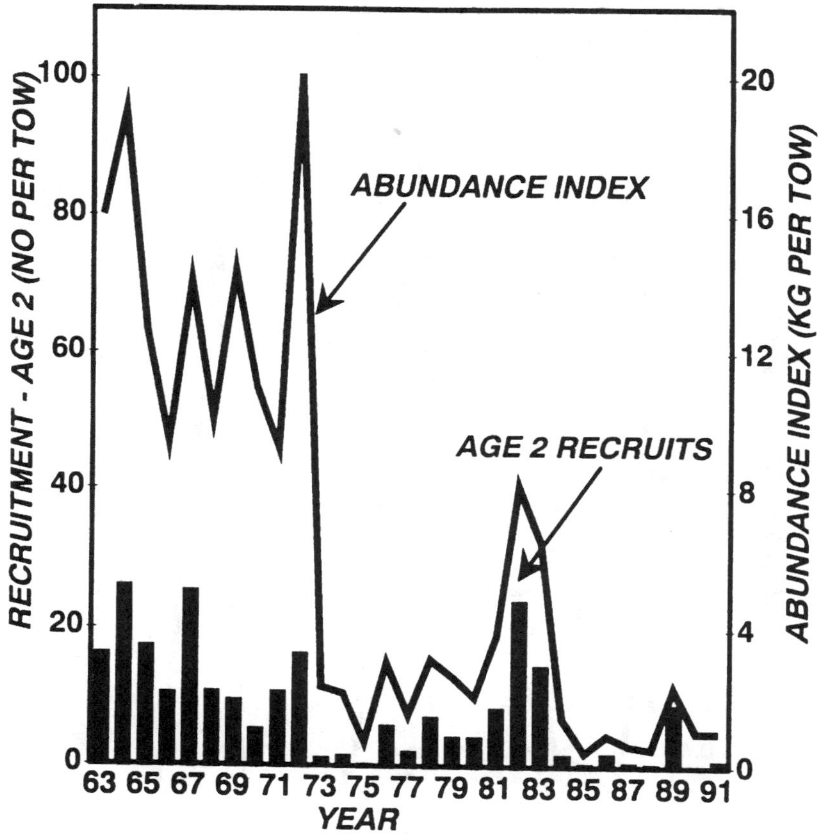

Figure 9-6. Indices of abundance and recruitment for the southern New England stock of yellowtail flounder from the autumn bottom trawl research survey, 1963–1991.

flounder was very high and supported annual catches of about 28,000 tons. The MSY for this stock is about 23,000 tons, which suggests that abundance during the 1960s was at a very high level indeed. The stock dramatically declined during the 1970s and 1980s, with a brief increase in abundance in 1982 and 1983, and then again in 1989. Fish of the 1980 and 1981 year-classes were responsible for the increase in abundance during 1982 and 1983, and the good year-class of 1987 produced significant numbers of age-2 recruits during 1989. These recruits were all rapidly fished out, however.

Fishing mortality (Figure 9-7) for the stock was very high during the 1980s, and in 1992 stood at a level of 1.6. This is 3.3 times the target level

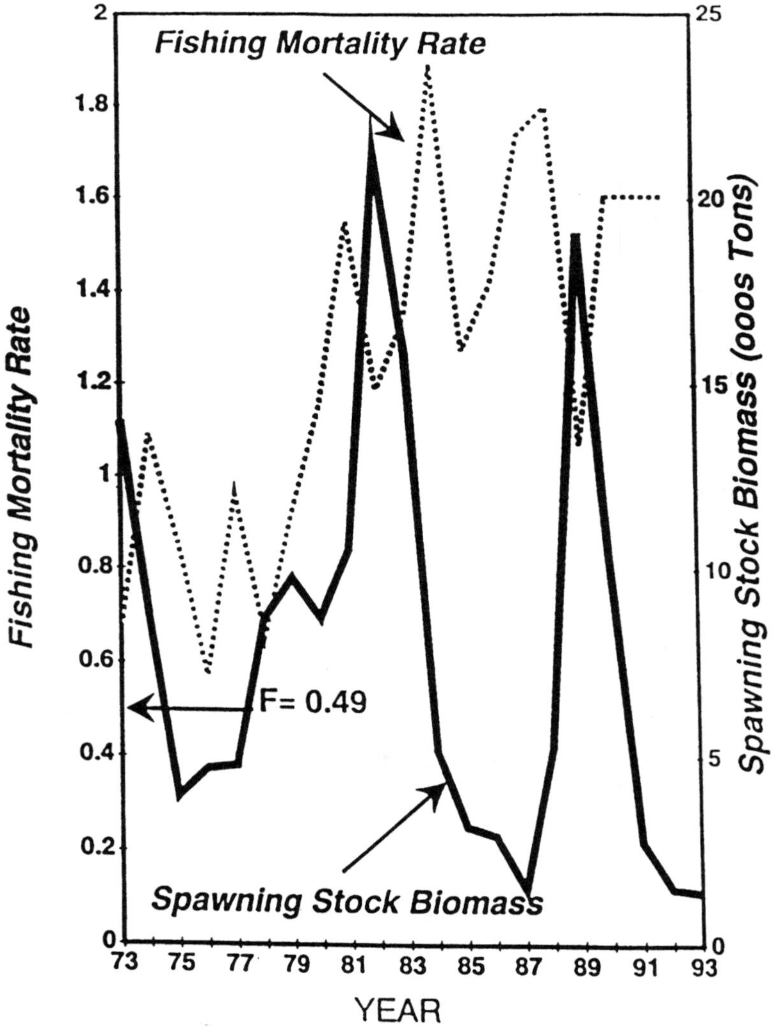

Figure 9-7. Instantaneous fishing mortality rate and spawning stock biomass for the southern New England stock of yellowtail flounder, 1973–1992.

of 0.49 set by the New England Fishery Management Council. The spawning stock biomass of southern New England yellowtail flounder is currently only at about a level of 3,000 tons, and the maximum catch for this stock is only achieved when the spawning stock biomass reaches around 48,000 tons. To complicate matters, this stock of fish is also at the southern end of its range in the north Atlantic and tends to do poorly under warm-water conditions. Productivity and success of this stock therefore is not only a function of managing the fishing mortality and the size of the spawning stock, but also depends in part on nature providing favorable environmental conditions.

Georges Bank Haddock

Figure 9-8 indicates the indices of abundance for Georges Bank haddock along with age-2 recruits. The very strong recruitment levels in 1964 and 1965 are obvious in this figure. This recruitment attracted the former Soviet Union to fish this stock, and in 1964 they caught 80,000 tons. Along with the 58,000 tons taken by the United States and the 18,000 tons taken by Canada, the total catch in that year was 156,000 tons. In the following year, these three nations harvested a total of 122,000 tons of Georges Bank haddock, and the stock quickly declined. This stock had been steadily producing annual catches at its maximum level (MSY) of about 47,000 tons since 1930. After the period of heavy fishing in 1964 and 1965, the next good year-class to come along was the 1975 year-class. Fishing mortality was very low at this time and the year-class spawned in significant numbers in 1978 (Figure 9-9). This spawning produced the second-best recruitment over the period of 1966 to 1992. Fishing mortality increased and these two year-classes were quickly fished out.

The opportunity for rebuilding the haddock resource was lost, and the stock declined to very low levels. After the 1978 year-class, it was all downhill. The current level of fishing mortality is about 0.52, slightly above the target (0.4) set by the New England Fishery Management Council. At the beginning of 1992, only 13,000 tons of spawning stock remained, whereas a spawning stock level of 130,000 tons is required to provide the MSY of just under 50,000 tons annually. The spawning stock biomass therefore would have to increase by ten times to achieve the annual catches that were routinely taken during the 1930s, 1940s, and 1950s.

Fishing Mortality Rates

Table 9-1 indicates the fishing mortality rates in 1992 for Georges Bank and Gulf of Maine cod and in 1990 for Georges Bank and southern New England yellowtail flounder and for Georges Bank haddock. These rates are indicative

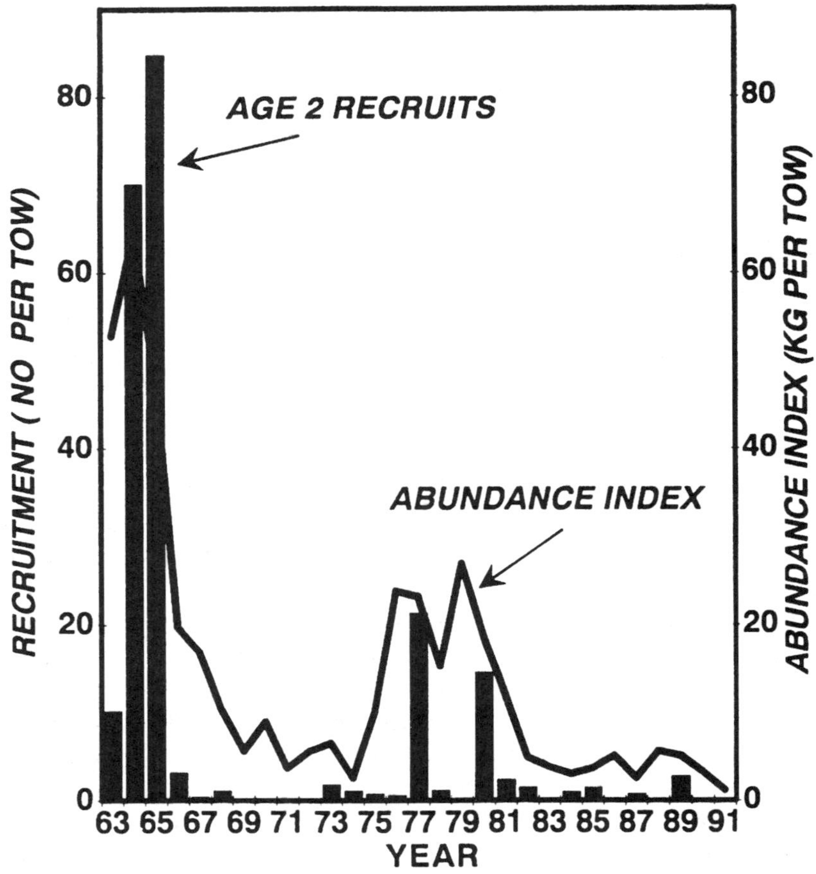

Figure 9-8. Indices of abundance and recruitment for the Georges Bank stock of haddock from the autumn bottom trawl research survey, 1963–1991.

of the groundfish overfishing problem. The table lists the Council's target fishing mortality compared with the most recent mortality estimates. This comparison clearly indicates that considerable reductions will have to be made across the board in order to reach these targets. For cod, a 60–65% reduction is required. A 29% reduction is required for Georges Bank yellowtail flounder, a 70% reduction is required for southern New England yellowtail flounder, and a 23% reduction is required for Georges Bank haddock. On average, about a 50% reduction in fishing mortality rate is needed to reach the mean overall target level.

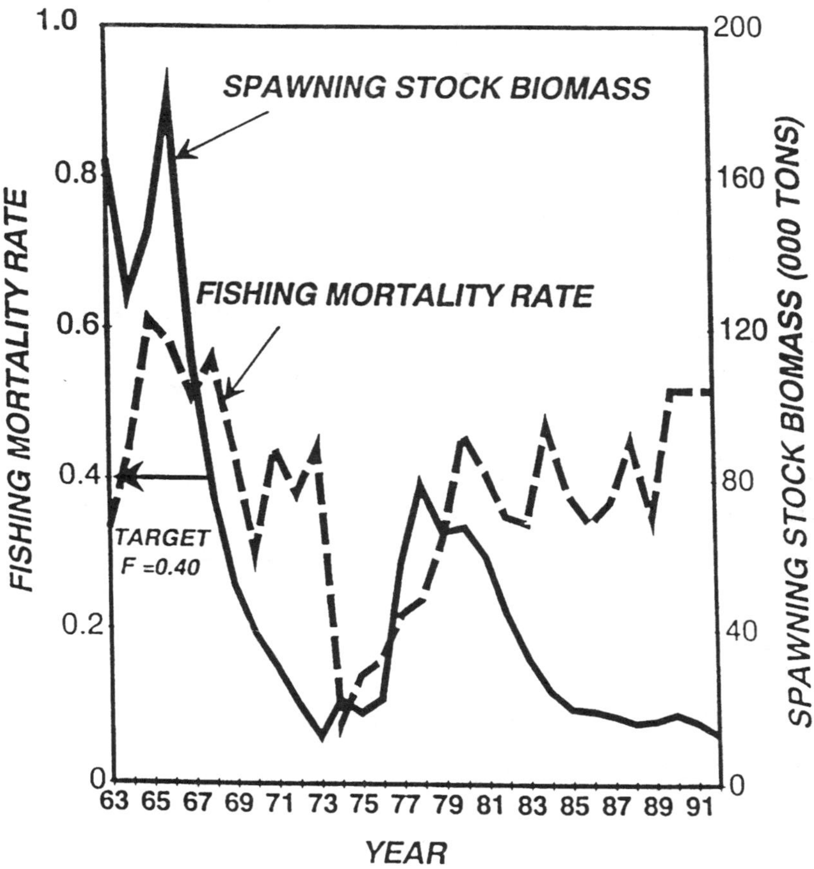

Figure 9-9. Instantaneous fishing mortality rate and spawning stock biomass for the Georges Bank stock of haddock, 1963–1991.

Stock Rebuilding

There is no guarantee that, as the fishing mortality rates are reduced to the target level, stocks will rebuild in the near future. The low stock levels for Georges Bank haddock, in particular, suggest that additional reductions in fishing mortality below the target level might be required to rebuild the stock. At the same time, there is no reason to believe that the fishery management councils would decide to rebuild the stocks to levels that would provide the maximum sustainable yield. The groundfish fishery is a mixed-species fishery and the rates of rebuilding for each species are not independent.

Table 9-1. Fishing mortality rates

Stock	Most recent estimate of fishing mortality	Council target
Georges Bank cod	0.89	0.36
Gulf of Maine cod	1.14	0.40
Georges Bank yellowtail flounder	0.82	0.58
Southern New England yellowtail flounder	1.61	0.49
Georges Bank haddock	0.52	0.40

Table 9-2. Catch in 1992, maximum sustainable yield (MSY), and 1992 spawning stock biomass for principal groundfish stocks

			Spawning stock biomass	
Stock	Catch	MSY	1992	Required for MSY
Gulf of Maine cod	13.0	10	14	30
Georges Bank cod	29.4	35	41	105
Gulf of Maine haddock	0.3	5	2	15
Georges Bank haddock	5.6	47	13	130
Pollock	42.0	54	100	120
Redfish	0.9	14	25	100
Georges Bank yellowtail flounder	1.2	16	3	25
Southern New England yellowtail flounder	2.7	23	2	48
Others	48.0	191	169	381
Total	143.1	395	369	954

Note: Catch data are for the United States and Canada combined and for both commercial and recreational fishing. Values are thousands of metric tons.

Table 9-2 indicates the 1992 catches and the maximum sustainable yields (or catches) that could be obtained from the 13 species currently being regulated by the Multispecies Fishery Management Plan. Along with the catch information in Table 9-2 are the calculated 1992 levels of spawning stock biomass and the levels of spawning stock biomass that would be required to provide the MSY. The 1992 catch data are for the United States and Canada combined and for both commercial and recreational fishing. Table 9-2 indicates the potential for harvesting as much as 400,000 tons of groundfish on a sustainable basis off the northeastern United States if the stocks were rebuilt. This area is one of the most productive fishing areas in the world. The huge amounts of groundfish caught off the northeastern United States as recently as the 1960s clearly support this conclusion. Canada has traditionally caught 25% of these resources, and probably would continue to catch substantial amounts of cod, haddock, and pollock if the stocks were rebuilt. The United

States could, however, increase its recent catch level of less than 100,000 tons by two to three times without overfishing the stocks once they are rebuilt.

Much of this catch (140,000 tons) would have to come from silver hake (whiting) and red hake. The catch of haddock could be dramatically increased (from 5,900 to 52,000 tons) as well as the catch for yellowtail flounder (from 3,900 to 39,000 tons). Catches of cod at the MSY level would actually be less than recent catches or slightly more than the 1992 catch because good year-classes of cod have just passed through the fishery, where they were harvested very heavily. These cod landings are declining rapidly, but 45,000 tons could be taken every year if the stocks were rebuilt. In 1991 and 1992, cod accounted for 43% and 30%, respectively, of the total catch of the regulated groundfish species. The present condition of the spawning stocks, however, suggests that recruitment and therefore catches will not increase in the near future unless fishing mortality dramatically decreases and the spawning stocks are rebuilt. The total spawning stock biomass in 1992 for the 13 regulated groundfish species would have to increase by 2.6 times to reach the level required to produce the MSY for all species.

The benefits of stock rebuilding are obvious. At the completely rebuilt level, catching one-third of the stock each year (instead of 55%) with one-half of the current effort will increase the catch by two to three times. The catch-per-effort would be four times greater and the catch would be sustainable. The alternative of letting the fishery continue as it is suggests that abundance and catches will continue to decline to even lower levels. One consequence of such a high fishing mortality rate is that good recruitment is quickly reduced, and substantial numbers of sub–legal-sized fish are discarded. Inter-annual catches are extremely variable, and substantial economic benefits of good recruitment are foregone because of growth overfishing and discarding. The more a spawning stock is reduced, the longer it takes for stock recovery, and the more infrequent good recruitment becomes.

10

Shifts in Northeast Shelf Cetacean Distributions Relative to Trends in Gulf of Maine/Georges Bank Finfish Abundance

Robert D. Kenney, P. Michael Payne, Dennis W. Heinemann, and Howard E. Winn

Abstract

A significant decrease in the standing stocks of Georges Bank/Gulf of Maine herring and mackerel during the 1960s precipitated a series of ecological responses by the finfish and cetacean fauna of the Northeast Shelf. A population explosion of sand lance followed the decline in the commercial pelagic fish stocks in these regions. Humpback whale distributions shifted from the northern Gulf of Maine to the southwestern Gulf in the late 1970s and early 1980s, reflecting a shift in diet from herring to sand lance. After a crash in Massachusetts Bay sand lance populations in 1986, piscivorous humpback and fin whales were largely replaced by planktivorous right and sei whales. The abundance of white-sided dolphins increased in the Gulf of Maine during the mid-1970s concurrent with a decline in the abundance of white-beaked dolphins. We hypothesize that white-beaked dolphin densities decreased after the decline of herring stocks, despite the increased abundance of sand lance, and that the white-sided dolphin replaced the white-beaked dolphin after the increase in sand lance densities. During the 1980s, the distribution of harbor porpoise also shifted away from Georges Bank waters. We suggest that harbor porpoise shifted their distribution inshore after the complete collapse of the Georges Bank herring stock by 1980. This resulted in increased harbor porpoise density in coastal and nearshore Gulf of Maine waters, with a subsequent increase in incidental take in the Gulf of Maine bottom-gillnet fishery.

Significant changes have occurred in the abundance and biomass of Gulf of Maine/Georges Bank herring (*Clupea harengus*) and Atlantic mackerel (*Scomber scombrus*) stocks during the past several decades (Anthony and Waring 1980; Grosslein et al. 1980; Sissenwine et al. 1982; SAW 1989, 1990). Herring were taken by a foreign distant-water fleet (DWF) purse seine fishery beginning in the 1960s and continuing through 1976 (SAW 1990). Between 1967 and 1976, the herring biomass on Georges Bank was reduced by 75% (Sissenwine et al. 1984), and in 1977, the commercial herring fishery failed completely. There was virtually no evidence of a Georges Bank herring population between 1977 and 1983 (SAW 1989), despite hundreds of days of research trawl and ichthyoplankton sampling (Sissenwine et al. 1984). The standing stock of mackerel also declined during the late 1970s, after DWF landings averaging 310,000 metric tons (mt) annually from 1970 to 1976. Since 1982 herring abundance has increased, being evident primarily on Nantucket Shoals (SAW 1989). In addition, a succession of moderate to good year-classes of Atlantic mackerel during the mid-1980s promoted the recovery of the stock to levels observed in the early 1970s (Overholtz et al. 1991).

After the decline of commercially important herring and mackerel stocks, sand lance (*Ammodytes dubius*) abundance increased significantly in shelf waters from the Gulf of Maine to Cape Hatteras (Meyer et al. 1979; Sherman et al. 1981; Nelson and Ross 1991). Sherman et al. (1981) suggested that sand lance ecologically replaced the other planktivorous finfish in many areas of the Northwest Atlantic. Since 1975, sand lance have become an important link (serving as both predator and competitor) between zooplankton and the larger marine fishes in shelf waters of the northeastern United States (Langton and Bowman 1980; Bowman et al. 1984; Nelson and Ross 1991). In addition, sand lance constitute an extremely important forage species for other marine predators, including seabirds (Powers and Backus 1987), pinnipeds (Payne and Selzer 1989), and cetaceans.

The Northeast Shelf large marine ecosystem (LME) contains a rich cetacean fauna. Kenney and Winn (1986) demonstrated that the western margin of the Gulf of Maine was the most intensively used cetacean habitat on the Northeast Shelf, and concluded that sand lance distribution and abundance patterns constituted the most important underlying mechanism. The significant alterations in the relative abundances of pelagic fish stocks, including herring, mackerel, and sand lance, apparently co-occurred with a number of noticeable shifts in the distribution of at least four species of large baleen whales: the humpback whale (*Megaptera novaeangliae*), fin whale (*Balaenoptera physalus*), sei whale (*B. borealis*), and right whale (*Eubalaena *glacialis*), and three species of smaller odontocete (toothed) cetaceans: the white-beaked dolphin (*Lagenorhynchus albirostris*), Atlantic white-sided dolphin (*L. acutus*), and harbor porpoise (*Phocoena phocoena*). In this chapter, we review previously published studies of distributional shifts in the baleen whales of the Northeast Shelf LME and present new information regarding the smaller odontocetes. We also examine the ecological interrelationships between the large-scale decline of herring and mackerel on Georges Bank, the rise of sand lance as an ecological replacement, and the apparent changes in the distribution and abundance of cetaceans in the region.

Methods

Since 1978, two large-scale cetacean survey programs have provided comprehensive coverage throughout shelf, shelf-edge, and adjacent slope waters of the northeastern United States: (1) the Cetacean and Turtle Assessment Program (CETAP), sponsored by the U.S. Department of the Interior's Bureau of Land Management, and (2) the Cetacean and Seabird Assessment Program, sponsored by the National Marine Fisheries Service/Northeast Fisheries Science Center (NMFS/NEFSC) of the U.S. Department of Commerce's National Oceanic and Atmospheric Administration (NOAA). Despite differences in platforms and objectives, these programs provided parallel information on the distribution of cetaceans in shelf waters of the northeastern United States, resulting in what is probably the most complete record of cetacean occurrence and distribution in existence for any region of the world's ocean.

Study Area

The study areas for both survey programs corresponded closely to the Northeast Shelf LME (Figure 10-1). The NMFS/NEFSC study area (212,686 km^2) consisted of continental shelf, shelf-edge, and adjacent slope waters less than 910 m (500 fathoms [fm] deep) in depth between Cape Hatteras, North Carolina, and Nova Scotia, Canada, although most of the NMFS/NEFSC surveys were conducted in shelf waters less than 183 m (100 fm). The CETAP surveys (CETAP 1982) encompassed the entire U.S. outer continental shelf, from Cape Hatteras north to the Northeast Channel, and seaward to approximately the 2,000-m isobath (278,350 km^2). Four major oceanographic regions (Gulf of Maine, Georges Bank, Southern New England, and Mid-Atlantic) occur within the shelf boundary of the study area. In this chapter, we focus primarily on the southern Gulf of Maine and Georges Bank.

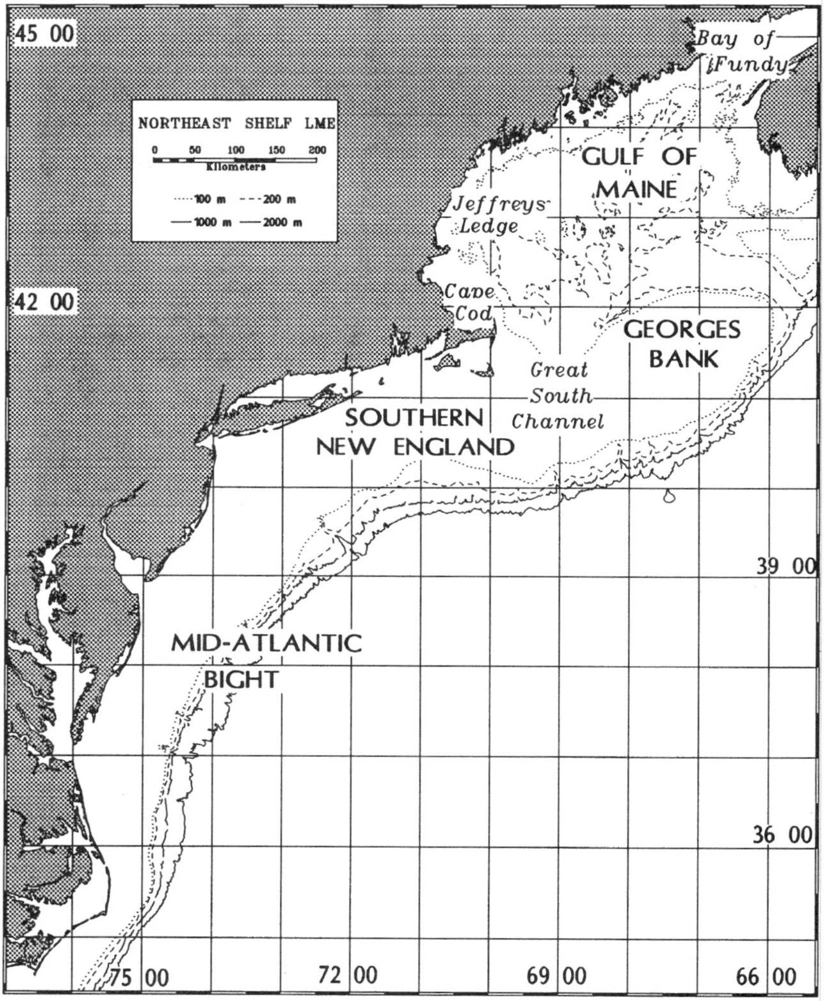

Figure 10-1. The Northeast Shelf large marine ecosystem, identifying the four principal marine regions and major geographical features referred to in the text. This chapter focuses on the Gulf of Maine and Georges Bank regions.

Collection of Survey Data

CETAP AERIAL SURVEYS. The major objectives of CETAP were to provide a spatial and temporal overview of distribution patterns (Hain et al. 1981, 1985; CETAP 1982; Kenney and Winn 1986, 1987b) and to estimate the absolute abundance (Scott et al. 1981; CETAP 1982; Scott and Gilbert 1982) of cetaceans and sea turtles in the study area. These data were used by federal agencies making decisions regarding potential development of offshore oil and gas resources (Edel et al. 1981); therefore, special attention was paid to potential lease sale areas (Sorensen et al. 1984).

The CETAP aerial surveys were conducted from November 1978 to January 1982. Throughout the three-year study, a number of sampling schemes and time intervals for data collection and analyses were used, with some refinements and changes incorporated into the program each year (Edel et al. 1981; Scott et al. 1981; Scott and Gilbert 1982; CETAP 1982). During 1979, dedicated, random-transect aerial surveys were conducted during each of eight time periods throughout the year (Edel et al. 1981). In 1980 and 1981, four dedicated, depth-stratified surveys at twice the sampling density of 1979 were flown each year, usually one per season (CETAP 1982). In 1981, the surveys were limited to areas of high interest to industry (Georges Bank/Nantucket Shoals and Baltimore Canyon), and two slope-water blocks were added. During each year of the CETAP study, special aerial surveys were also conducted to emphasize endangered species not adequately sampled by the dedicated surveys (Winn et al. 1981) or to characterize species occurrence and abundance in specified lease sale areas.

The CETAP aerial surveys provided the first absolute estimates of cetacean abundance in the shelf waters of the northeastern United States from line-transect surveys (Scott et al. 1981; CETAP 1982; Kenney et al. 1985). Only the data collected during the dedicated aerial surveys were used in estimating abundance.

CETAP PLATFORMS OF OPPORTUNITY PROGRAM (POP) SURVEYS. The CETAP program placed trained observers on aircraft and vessels of opportunity operating within the study area (CETAP 1982). The U.S. Coast Guard and NMFS/NEFSC provided the majority of vessel platforms, supplemented by several institutional and foreign research vessels. The aerial portion of this effort was primarily carried out aboard aerial radiothermography (ART) and offshore law enforcement patrol (OLP) surveys conducted by the U.S. Coast Guard, Cape Cod, Massachusetts. Aerial POP surveys continued through 1981, but the shipboard component of this effort was discontinued at the end of 1980.

OTHER CETAP DATA. The CETAP database included two additional types of sighting records. Opportunistic sightings were those reported by researchers, mariners, pilots, or others from the study area during the program. Historical sightings were those records collected by several individuals before the program began and submitted for inclusion in the CETAP database. The historical data are very sparse before the mid-1970s, when the Gulf of Maine Whale Sighting Network was initiated by Dr. Steve Katona at the College of the Atlantic in Bar Harbor, Maine, to collect, archive, and characterize any available cetacean sightings in the Gulf region (e.g., Katona 1976; Ramsdell 1978).

NMFS/NEFSC SHIPBOARD SURVEYS. A long-term series of bottom trawl surveys of groundfish abundance was begun by NMFS/NEFSC in 1963 (Grosslein 1969; Azarovitz 1981). Between 1976 and 1987, hydrographic and plankton surveys were also conducted. The NMFS/NEFSC also initiated the Marine Resources Monitoring Assessment and Prediction (MARMAP) program to provide annual and seasonal monitoring of the physical environment and the distribution, biomass, and population structure of the principal biota on a broad geographic scale (Sherman 1980). In May 1980, NMFS/NEFSC extended the MARMAP strategy by placing a dedicated observer aboard each NMFS/NEFSC groundfish and plankton survey to monitor the distribution and abundance of cetaceans, seabirds, and, to a lesser extent, sea turtles. This effort was designed to provide an extensive assessment of cetacean and seabird populations in shelf and shelf-edge waters (generally <200 m) off the northeastern United States. These standardized shipboard surveys were conducted from May 1980 through December 1988, overlapping with the 1980 (in part) and 1981 CETAP efforts. Sighting data collected during these surveys have been used to monitor small-scale temporal changes in the distribution and relative abundance of these taxa (Powers 1983; Powers and Payne 1983; Powers and Brown 1987; Payne et al. 1984, 1990a, b; Payne and Heinemann 1993); to determine hydrographic factors influencing cetacean distributions (Selzer and Payne 1988); and to describe trophic interactions between cetaceans and seabirds and potential prey (Payne et al. 1986, 1990b; Powers and Backus 1987; Smith et al. 1990; Overholtz et al. 1991).

Data collection procedures have been described previously (Payne et al. 1984, 1986; MBO 1988; Selzer and Payne 1988; Smith et al. 1990). Generally, cetacean observations were recorded continuously by the observer along a predetermined trackline between randomly placed sampling stations as long as the vessel moved on a straight course and at a uniform speed. The sampling stations were randomly placed within each stratum, roughly in proportion to stratum size (Grosslein 1969); therefore, the distribution and intensity of the cetacean sighting effort of the NMFS/NEFSC surveys were dependent entirely on the design of the NMFS/NEFSC fisheries/oceanographic surveys.

Although the collection of the NMFS/NEFSC shipboard sighting data also followed line-transect procedures (Seber 1973; Burnham et al. 1980), the data have thus far been used only to describe relative densities or abundances of cetaceans rather than absolute densities, as were generated by CETAP. These data do not, at present, provide a basis for the estimation of absolute cetacean abundance.

ADDITIONAL DATA FROM THE UNIVERSITY OF RHODE ISLAND. Since the end of CETAP in 1982, a number of additional studies conducted by the University of Rhode Island have generated cetacean sighting records that were added to the archived CETAP sighting data. Most of the surveys were targeted at right whales and were conducted during the spring in the Great South Channel region. A significant component of the data for 1986–1988 in the URI database resulted from a cooperative right whale research program conducted with NMFS/NEFSC, the New England Aquarium (NEA), and the Center for Coastal Studies (CCS). The data therefore included sightings recorded during right whale surveys by CCS in Cape Cod Bay and Massachusetts Bay, by NEA in the lower Bay of Fundy, and by NEA and URI on the Scotian Shelf off the southern tip of Nova Scotia.

Treatment of Survey Sighting Data

For the analyses reported in this chapter, the sighting data for white-sided and white-beaked dolphins and harbor porpoise collected during the CETAP, NMFS/NEFSC, and more recent URI surveys were combined into a common database. Sighting data were classified by number of individuals observed (i.e., 1–10, 11–30, 31–100, or >100 individuals per sighting for dolphins; 1, 2–5, 6–10, and >10 for porpoises), then plotted onto a base map of the study area [following Payne et al. (1990a) and Payne and Heinemann (1993)]. All plots were produced using the PC-based mapping program CAMRIS (Computer Aided Mapping and Resource Inventory System; Ford 1989).

Sighting plots for pertinent time periods were qualitatively compared to look for apparent shifts in distribution between periods. Distributions were also compared with patterns in the available historical sighting information or reported in the literature. Relative densities (sighting frequencies corrected to some standardized measure of effort) could not be used to quantitatively compare distribution patterns in the combined database because of major differences between survey protocols and data collection formats. Neither are there available any estimates of absolute abundance from the more recent NMFS/NEFSC or URI surveys, or before 1979, which could be comparable with the CETAP estimates. Nevertheless, this approach provides a qualitative account of changes in the spatial distribution of cetaceans that occurred throughout the study area between 1978 and 1988, and allows some comparisons with the limited sighting data available for the years before 1978.

Results and Discussion

Baleen Whales

Several species of sympatric baleen whales are known or assumed to feed extensively on sand lance in the western North Atlantic, including humpback, fin, and minke (*Balaenoptera acutorostrata*) whales (Overholtz and Nicolas 1979; Kenney et al. 1981, 1985; Hain et al. 1982, 1992; Mayo 1982; Payne et al. 1986, 1990a, b). The distribution patterns of these individual whale species are the dominant influences in the inshore, western Gulf of Maine habitat use patterns of the total cetacean community described by Kenney and Winn (1986). Both Kenney et al. (1981) and Payne et al. (1986) had hypothesized that the regional and local patterns of spatial distribution of the humpback on the Northeast Shelf were correlated with the occurrence of dense patches of sand lance. Perhaps to a somewhat lesser extent, fin whale distribution patterns are also spatially correlated with sand lance distributions (CETAP 1982; Payne et al. 1990b; Hain et al. 1992).

Payne et al. (1986) were the first to show apparent shifts in Northeast Shelf cetacean distributions correlated with changes in the abundance of fish stocks. They examined humpback whale sighting frequencies from Mt. Desert Rock in the northern Gulf of Maine from about 1972 to 1982, reported in Mullane and Rivers (1982), and observed a sharp decline from 1977 on. This decrease corresponded with an increase in humpback sightings in the southwestern Gulf. They hypothesized that the humpbacks had largely abandoned the more northern areas, where they had been feeding on krill and declining herring stocks, to concentrate in the southwestern Gulf of Maine and feed on superabundant sand lance populations.

A subsequent alteration in sand lance abundance in the southwestern Gulf of Maine provided additional evidence to support the hypothesis that changes in the abundance of fish stocks can cause significant shifts in cetacean distributions. Sand lance catches in the NMFS/NEFSC spring groundfish survey in the stratum encompassing Stellwagen Bank declined from 1982 on, and fell to zero in 1986 (Bowman et al. 1984; Nelson and Ross 1991). Using data from commercial whale-watching vessels operating on southern Stellwagen, Payne et al. (1990b) showed a decline in both fin and humpback whales in the area, with the mean number of humpbacks seen per whale-watching trip dropping from 8.99 in 1985 to 0.23 in 1986. At the same time, two species of whales rarely observed by Stellwagen Bank whale-watchers occurred regularly. The changes in the mean number of right whales and sei whales seen per trip in 1985–1986 were 0.00–0.95 and 0.00–0.26, respectively.

The distributions of both right and sei whales were negatively correlated to that of sand lance (Payne et al. 1990b). Both species are planktivores, preying primarily on dense swarms of copepods, notably *Calanus finmarchicus*, in the Northwest Atlantic (Mitchell and Chapman 1977; Jonsgard and Darling 1977; Kenney et al. 1985, 1986a; Winn et al. 1986; Wishner et al. 1988; Hamilton and Mayo 1990; Mayo and Marx 1990). Sand lance similarly prey on copepods (Reay 1970; Meyer et al. 1979). Mitchell (1975) suggested that right and sei whales compete for prey resources, and they may, in fact, compete with sand lance (Kenney et al. 1986a; Payne et al. 1990b). Furthermore, although blue whales (*Balaenoptera musculus*) are very seldom reported in the Gulf of Maine (Katona et al. 1983), they were observed in the study area during the summers of 1986 and 1987 (Wenzel et al. 1988; Payne et al. 1991). Although blue whales feed almost exclusively on euphausiids (Jonsgard 1955, 1966; Nemoto 1957, 1959; Mitchell 1975; Yochem and Leatherwood 1985), they have also been known to take copepods of the genus *Calanus* (Nemoto 1970; Kawamura 1980). Spring *Calanus* densities in the Stellwagen area in 1986 were an order of magnitude higher than in any other year examined (Payne et al. 1990b). It seems very possible that the crash in sand lance stocks in the area led to a sharp increase in copepod populations, as well as replacement of the normally resident piscivorous whales by planktivorous whale species.

White-Sided and White-Beaked Dolphins

Atlantic white-sided dolphins are the most abundant cetacean in the Gulf of Maine, and one of the two or three most abundant in the Northeast Shelf LME, with an estimated population size in 1979–1981 of 36,000–40,000 animals (CETAP 1982; Kenney et al. 1985). In the combined CETAP–NMFS/NEFSC–URI data, their sighting distribution is principally in the Gulf of Maine and Georges Bank regions, with a concentration north of 40°30' N from the Great South Channel and western Georges Bank, northwest along the 100-m contour east of Cape Cod, and throughout the western half of the Gulf of Maine (CETAP 1982; Selzer and Payne 1988; Payne et al. 1991; Figures 10-2 and 10-3). South and west of Cape Cod, sightings were sparse and limited largely to southern New England waters. There is some indication that Hudson Canyon may be a local area of concentration (Payne et al. 1991). White-sided dolphins are abundant in Gulf of Maine/Georges Bank waters over three seasons of the year, with relatively fewer sightings during the winter (CETAP 1982; Selzer and Payne 1988). The fact that white-sided dolphins are largely absent from the region during the winter and that no southern area has a significant number of winter sightings suggests a seasonal inshore/offshore migration (CETAP 1982).

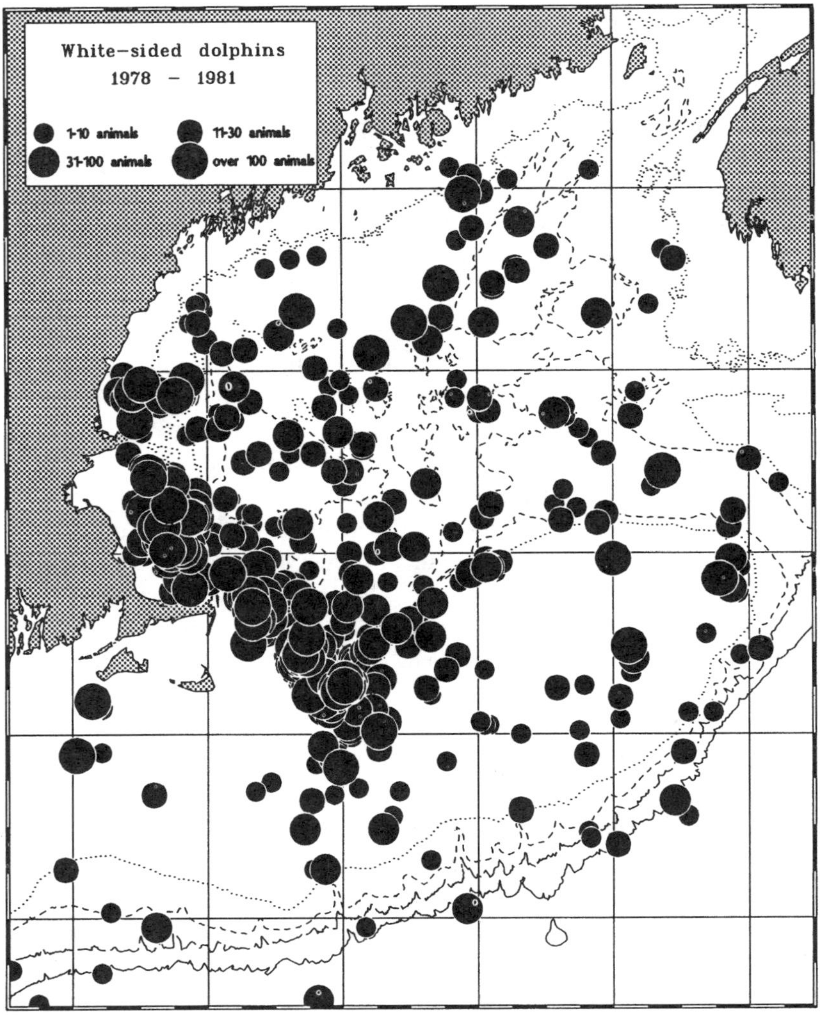

Figure 10-2. The distribution of white-sided dolphin (Lagenorhynchus acutus) sightings throughout the study area, 1978–1981.

This spatial distribution of white-sided dolphin sightings closely parallels that of the sand lance and the baleen whales, which prey on sand lance, as discussed earlier. In fact, in Gulf of Maine and Georges Bank waters, white-sided dolphins have been frequently observed in close association with humpback and fin whales in multispecies feeding aggregations (Hain et al. 1981; Mayo 1982). White-sided dolphins have also been directly

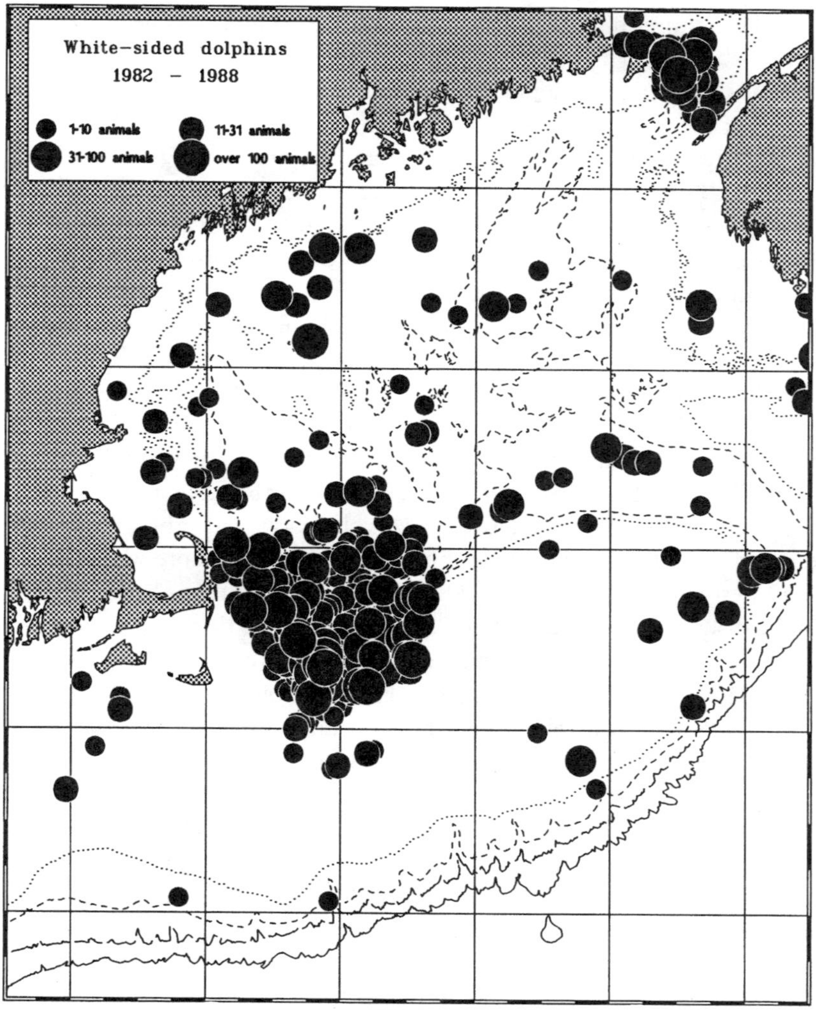

Figure 10-3. The distribution of white-sided dolphin (Lagenorhynchus acutus) sightings throughout the study area, 1982–1988.

observed feeding on sand lance in these mixed-species associations. These patterns strongly suggest that white-sided dolphin distributions are largely controlled by sand lance distribution and abundance patterns (Selzer and Payne 1988). This close predator–prey relationship between white-sided dolphins and sand lance in the study area can be further demonstrated, although indirectly.

The sighting distribution of white-sided dolphins between 1978 and 1981 extended in a patchy continuum from the southwestern Gulf of Maine/Great South Channel north into the western and north-central waters of the Gulf of Maine (see Figure 10-2). During this period, sightings were concentrated in the Great South Channel, from Stellwagen Bank north to Jeffreys Ledge, and from Cashes Ledge and Basin north to the central Gulf of Maine south of Mount Desert Island (CETAP 1982; Payne et al. 1991). However, this sighting distribution shifted southeastward from Stellwagen Bank into the Great South Channel and northwestern Georges Bank in 1982–1988 (see Figure 10-3). (Note that sighting effort patterns strongly influence observed distributions. For example, the exceptionally dense concentrations of sightings seen in Figure 10-3 in the area east of Cape Cod and north of the Great South Channel, as well as in the lower Bay of Fundy, result at least in part from concentration of effort by right whale surveys from URI and NEA, respectively. Thus, conclusions based on these distributions must be tempered somewhat.) This shift occurred concurrently with the decline in the abundance of sand lance in the nearshore waters of the southwest Gulf of Maine from 1982 to 1986–1987 and simultaneous shifts in the distributions of Stellwagen Bank baleen whales discussed earlier. The increased densities of white-sided dolphins on Georges Bank observed during 1986 NMFS/NEFSC surveys were the largest reported since 1978–1979 (CETAP 1982). In corroboration, the maximum estimated regional abundances of white-sided dolphins in the area north of the Great South Channel to date have been 9,500–10,000 animals on 30 May 1985 and 7 July 1987 (Kenney et al. 1986b; Kenney and Winn 1987a).

On the other hand, white-sided dolphins seem to have been relatively rare in Gulf of Maine waters before the mid-1970s. Schevill (1956) considered a small group of about 12 individuals in the Great South Channel area in September 1954 worthy of publication, "since [the species] has not often been reported from our waters." White-sided dolphin sighting frequencies in coastal and nearshore waters have increased significantly since then, although part of the increase is certainly because of the large increase in sighting effort in the study area since the mid-1970s. Sergeant et al. (1980) reported that, in the 1970s, the number of white-sided dolphin sightings increased in coastal and offshore waters from Newfoundland to Cape Cod over sighting frequencies in the 1950s and 1960s, apparently reflecting a "real increase of the western North Atlantic population." The increase in white-sided dolphin sightings also co-occurred with a significant increase in the number of strandings from the Gulf of St. Lawrence to the Gulf of Maine (Katona et al. 1976; Mead 1979; Sergeant et al. 1980; Mitchell 1981; Sergeant 1982), all strongly suggesting that

Table 10-1. Sightings per year of Atlantic white-sided dolphins (WSDO) and white-beaked dolphins (WBDO) in Gulf of Maine/Georges Bank waters, 1974–1981

Species	1974	1975	1976	1977	1978	1979	1980	1981
WSDO	1	11	21	73	65	216	257	138
WBDO	1	11	24	5	2	25	7	1

the absolute abundance of white-sided dolphins has increased in shelf waters north of Cape Cod in recent decades.

The abundance and distribution of the white-sided dolphin and the other Gulf of Maine *Lagenorhynchus* species, the white-beaked dolphin, have apparently undergone considerable change during the past several decades, especially during the years just before the beginning of intensive and extensive survey effort in 1979 and 1980. The white-beaked dolphin was the *Lagenorhynchus* species most often reported in the Gulf of Maine as recently as the early 1970s (Katona 1976). In addition, white-beaked dolphins were apparently much more abundant around Cape Cod in the 1950s and 1960s than at present (Katona et al. 1983). Unfortunately, however, the regular, systematic collection of cetacean sighting records for the Northeast Shelf appears to have commenced after the decline in white-beaked dolphin populations in the Gulf of Maine was well under way. The combined white-beaked and white-sided dolphin sighting data from 1974 to 1981 do show a trend from equal sighting frequencies in 1974 and 1975 to strong dominance by white-sided dolphins (Table 10-1). One can also see a marked decline in white-beaked dolphin numbers in 1979–1981, the years of most intensive survey effort. Since 1981, we have recorded only two sightings of white-beaked dolphins in the Northeast Shelf LME, both in 1983. White-beaked dolphins should now be considered extremely rare in Gulf of Maine and Georges Bank waters, whereas the white-sided dolphin is the most abundant dolphin in the region (CETAP 1982; MBO 1988).

Katona et al. (1983) suggested that the two *Lagenorhynchus* species segregate seasonally and geographically by water temperature, and ecologically by prey preferences. Sigurjonsson et al. (1989) also suggested that white-beaked dolphins displace the smaller white-sided dolphins in cooler, nearshore Icelandic waters. Since there is evidence for a long-term warming trend in Northeast Shelf waters (Hertzman, this volume), we cannot simply reject the hypothesis that changes over the last few decades in white-sided and white-beaked dolphin distributions and abundance are temperature-related, either directly or through effects on prey. On the other hand, an alternative hypothesis based only on demonstrated changes in the stocks of various prey species can be formulated, which may be a reasonable explanation for the concurrent changes in both dolphin species.

The preferred prey species of white-beaked dolphins in Canada and Iceland are schooling fish and squid, the former being composed mainly of Atlantic herring and capelin (*Mallotus villosus*) (Sergeant and Fisher 1957). The apparent decline in abundance of white-beaked dolphins in the Gulf of Maine/Georges Bank regions was concurrent with the significant reduction of herring and mackerel standing stocks in the Gulf of Maine and Georges Bank regions between 1967 and 1976. This was despite subsequent increases in populations of sand lance, a species that is ecologically similar to herring. White-sided dolphins are observed to feed extensively on sand lance in the Gulf of Maine. Therefore, white-beaked dolphins may be more stenophagous than white-sided dolphins, with a diet more restricted to larger prey items, resulting in the replacement of white-beaked dolphins by white-sided dolphins in this region.

This apparent replacement of one dolphin species by another could have been caused by a major increase in populations of white-sided dolphin after the nearly simultaneous departure of a major competitor (white-beaked dolphins) and explosion in prey (sand lance) abundance. We feel a more likely alternative is that, in addition to some increase in white-sided dolphin numbers, the collapse of Georges Bank herring stocks, coupled with increased sand lance abundance in nearshore waters, precipitated an offshore–inshore shift in their distribution. Since white-sided dolphins were considered uncommon before and during the 1960s, and so little is known regarding their distribution before the mid-1970s, it is possible that their earlier distribution was different, possibly mid-shelf to shelf-edge/slope, similar to that of the common dolphin, *Delphinus delphis* (CETAP 1982; Selzer and Payne 1988), rather than the more nearshore distribution that recent sighting data indicate.

Several earlier authors have characterized the white-sided dolphin as primarily an inhabitant of slope waters (Sergeant and Fisher 1957; Waters and Rivard 1962; Sergeant et al. 1970; Geraci et al. 1978). This species has also been reported from Newfoundland over deep waters (Mercer 1973), from slope waters (>500 fm) south of Iceland (Sigurjonsson et al. 1989), and from the shelf-edge of southwest Norway and northwest Europe (Jonsgard and Nordli 1952). Sightings were also reported from Georges Bank slope waters during the mid-1970s (Katona 1976; Ramsdell 1978; Testaverde and Mead 1980). A plot of the available historical sighting data through 1977 shows concentrations near Cape Cod, near Jeffreys Ledge, and in the northern Gulf of Maine (Figure 10-4). These are very likely a result of effort bias, because of commercial whale-watching near the Cape, and the Gulf of Maine Whale Sighting

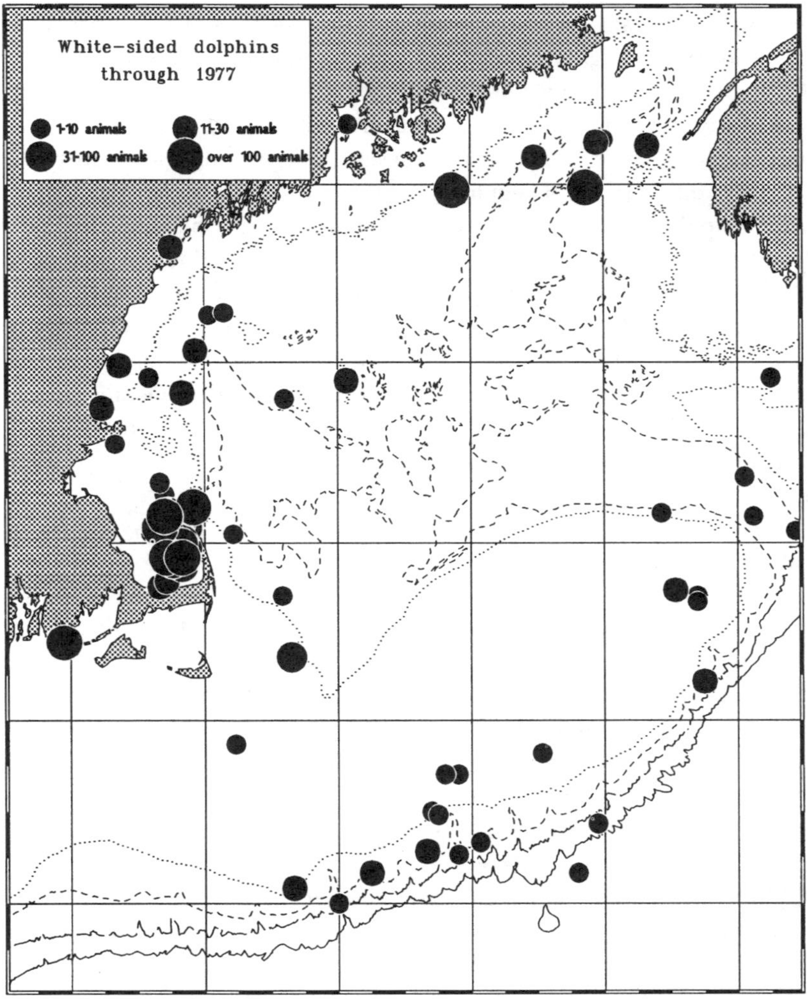

Figure 10-4. The distribution of white-sided dolphin (Lagenorhynchus acutus) sightings throughout the study area before 1978.

Network in the other two locations. However, there were also a number of sightings along the outer part of Georges Bank (see Figure 10-4)—in fact, a number roughly similar to those during 1978–1981 (see Figure 10-3), despite the very much greater survey effort during the later period. Such a recent distributional shift into inshore areas of the Gulf of Maine (in response to increased prey densities) would explain the lack of confirmed white-sided dolphin sightings near Cape Cod during the 1950s and 1960s, as well as the subsequent increase in sightings after the sand lance population explosion in the 1970s. If, in fact, the population does migrate off-shelf on a seasonal basis as the lack of winter sightings suggests, an offshore–inshore shift in the main feeding areas might be viewed simply as an extension of normal migratory patterns.

One additional bit of supporting evidence for the hypothesis that white-sided dolphins may have shifted away from a primarily shelf-edge/slope distribution to a more nearshore one comes from the distribution patterns of other *Lagenorhynchus* species. There are four additional species in the genus that occur outside of the North Atlantic: Pacific white-sided dolphins (*L. obliquidens*) in the North Pacific, and dusky (*L. obscurus*), hourglass (*L. cruciger*), and blackchin (*L. australis*) dolphins in the Southern Ocean. Three of the four species are characterized as primarily pelagic, with only the blackchin dolphin having a mainly coastal distribution (Watson 1981; Leatherwood and Reeves 1986; Corkeron 1988).

Harbor Porpoise

The other small odontocete species whose apparent shift in distribution patterns may be the result of changes in prey stocks is the harbor porpoise, although the data in this case are less clear. During the combined 1978–1981 CETAP and NMFS/NEFSC surveys, harbor porpoise sightings were distributed into two spatially and seasonally disjunct clusters north of 40°00'N (CETAP 1982; Figure 10-5). One cluster of sightings occurred primarily during the spring in southwestern Gulf of Maine/Great South Channel and Georges Bank waters between approximately 42°00'N and 42°30'N. The second main area of harbor porpoise occurrence, with much higher regional abundances, was in the upper Gulf of Maine north of about 42°30'N during summer through early autumn. Porpoise sightings were relatively rare during late fall and winter, again suggesting offshore migration during those seasons (Prescott and Fiorelli 1980). Although sightings have been recorded as far south as Cape Hatteras, the largest concentrations of harbor porpoise during any season have occurred in the Gulf of Maine (CETAP 1982).

The distribution of harbor porpoise sightings in 1982–1988 was quite different (Figure 10-6). Sightings were much reduced everywhere, probably at least in part because of reduced survey effort. The dense concentration of sightings in the Bay of Fundy is another example of effort bias, resulting from the NEA right whale surveys. However, the frequency and distribution of sightings in the southern portion of the study area seem to be reduced more

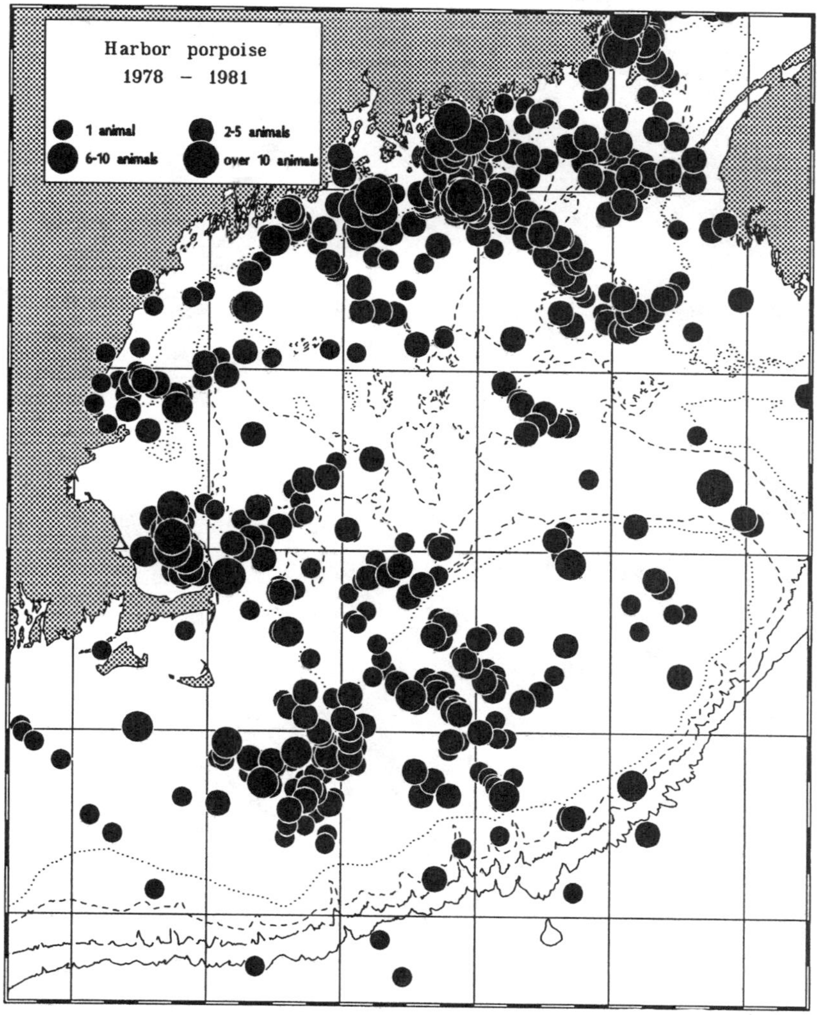

Figure 10-5. The distribution of harbor porpoise (Phocoena phocoena) sightings throughout the study area, 1978–1981.

than one might expect from effort differences alone. In 1978–1981, there were many sightings in the area of the western half of Georges Bank, the Great South Channel, the periphery of Nantucket Shoals, and immediately east of Cape Cod (CETAP 1982; Payne et al. 1991; see Figure 10-5). Most of these occurred in 1979 and 1980 during CETAP surveys, and most were during May and June, as the porpoise population presumably migrated northward from

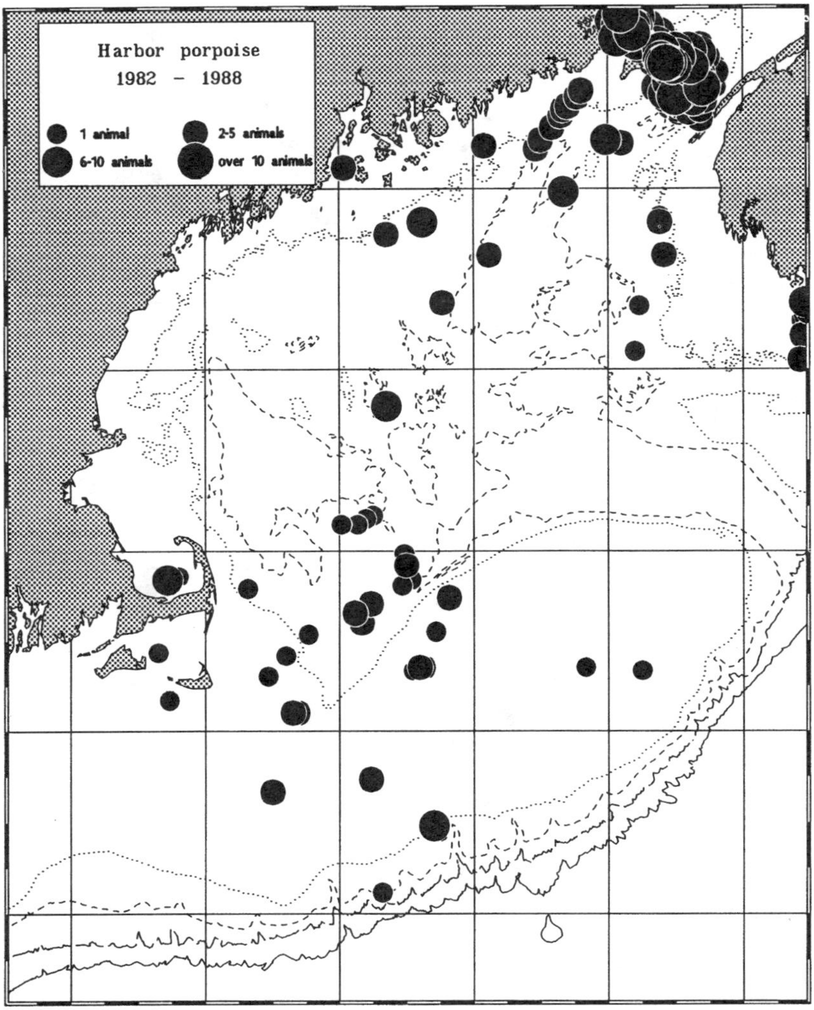

Figure 10-6. The distribution of harbor porpoise (Phocoena phocoena) sightings throughout the study area, 1982–1988.

probable wintering areas. The number of harbor porpoises sighted on Georges Bank and in surrounding areas between 1982 and 1988 was greatly reduced (see Figure 10-6), despite the intensive springtime URI survey effort east of Cape Cod, as evidenced in Figure 10-3.

One hypothesis that might account for the reduced number of sightings on Georges Bank during the latter part of the study period is that at least part of the Gulf of Maine harbor porpoise subpopulation (Gaskin 1984) shifted from Georges Bank toward nearshore Gulf of Maine waters in response to the complete collapse of the Georges Bank herring stock by 1980. Several sources of information provide support for this hypothesis. The seasonal and shorter-term movements of harbor porpoise in the lower Bay of Fundy are correlated quite closely with those of herring and mackerel, their main prey in the area (Smith and Gaskin 1974; Gaskin et al. 1975; Smith et al. 1983; Gaskin and Watson 1985; Read and Gaskin 1985; Recchia and Read 1989). Few sightings in the central area of Nantucket Shoals were reported by CETAP (1982) during the aerial surveys. However, in only two days of shipboard surveys on Nantucket Shoals during February 1991, more harbor porpoise sightings were reported than were recorded in the combined 1978–1988 data from this area during January and February (D. Wiley, International Wildlife Coalition, Falmouth, Massachusetts, personal communication, 1991; and A. Read, Woods Hole Oceanographic Institute, Woods Hole, Massachusetts, personal communication, 1991). The increased herring abundance reported by NMFS between 1982 and 1989 was evident primarily on Nantucket Shoals (SAW 1989). Additionally, anecdotal information suggests that sightings of harbor porpoise by nearshore Gulf of Maine fishermen have increased in recent years. This recent harbor porpoise "awareness" by fishermen has probably been the direct result of a recent increase in harbor porpoise incidental take in the Gulf of Maine/Bay of Fundy bottom-gillnet fishery (Gilbert and Wynne 1988; Read and Gaskin 1988, 1990; Read 1990).

As with the dolphin distributional changes, it is not possible to rule out potential effects of long-term temperature trends on the distribution and seasonal occurrence of harbor porpoise in Northeast Shelf waters. Porpoise distributions might have begun changing well before the pulse of DWF fishing effort on Georges Bank. For example, Connor (1971) and Ulmer (1977) reported small numbers of harbor porpoise in summer in lower New York Bay during the 1940s. It is difficult to determine what "small number" implied; however, harbor porpoise are now rarely observed outside the upper Gulf of Maine and Bay of Fundy during the summer (CETAP 1982). In addition, harbor porpoise are probably the most difficult to sight of all cetaceans of the region. They are small, occur singly or in very small groups, tend to be "shy," and overall are quite inconspicuous. Slight changes in sighting conditions (e.g., sea state, sun glare, platform speed) can strongly affect sightability. Data from harbor porpoise surveys might therefore be more expected than the data for other cetaceans to show differences because of varying survey platforms or sampling protocols (Kraus et al. 1983).

Conclusions and Recommendations

An increasing number of studies are concluding that cetacean distributions in general are largely controlled by the distribution and abundance of prey species (CETAP 1982; Gaskin 1982; Kenney and Winn 1986; Katona and Whitehead 1988). We have reviewed several studies that have demonstrated clear correlation between changes in the distributions of Northeast Shelf baleen whales and ecological shifts in prey patterns after a period of intense fishing effort. We have also presented new evidence suggesting that similar changes in distributions of smaller odontocetes are also related to changing prey patterns. Comparable shifts in the distribution of cetaceans following and correlated with changes in the distribution and abundance of prey have been reported previously (Lien et al. 1979; Perkins and Beamish 1979; Bryant et al. 1981; Whitehead and Lien 1982; Whitehead 1983; Lynch and Whitehead 1984; Whitehead and Carscadden 1985). For example, after the collapse of heavily exploited offshore capelin stocks, the distribution of humpback whales shifted from offshore feeding grounds on the Grand Banks to nearshore Newfoundland waters, resulting in increased whale entanglements and mortality in inshore fixed fishing gear (Lien et al. 1979; Whitehead and Carscadden 1985). This situation is analogous to the depletion of offshore Georges Bank herring stocks, followed by a shift in the distribution of harbor porpoise to nearshore Gulf of Maine waters and subsequent increased gillnet mortality, as we have suggested in this chapter.

Payne et al. (1990b) suggested that shifts in the spatial distribution and abundance of baleen whales in the Gulf of Maine can be characterized as a series of ecological responses to human-induced changes (through commercial fishing) in the abundance of planktivorous prey fishes. Although more difficult to demonstrate, we also suggest that commercial fishing practices during the 1960s and 1970s on Georges Bank have indirectly resulted in the redistribution of at least three species of smaller odontocete cetaceans during the same time interval. The collapse of offshore finfish stocks may have also indirectly resulted in the observed increase in harbor porpoise entanglement and mortality in inshore gillnets during the 1980s.

There remains the difficulty of partitioning the observed variability in whale and dolphin distribution and abundance patterns into natural and anthropogenic components. For example, long-term alterations in ocean temperatures (which themselves could have natural or anthropogenic causes) might affect cetacean populations directly, or indirectly through effects on trophic webs and prey populations. Only long-term monitoring of both the cetacean populations and their habitats can effectively provide the data necessary to understand the cause-and-effect mechanisms involved. Intensive studies such as CETAP are invaluable for baseline characterization of populations, but far too expensive for long-term monitoring. Programs such as the NMFS/NEFSC surveys, which are "piggy-backed" on other monitoring efforts, constitute an efficient sampling method at relatively low expense, and are therefore a cost-effective means for long-term monitoring. Cetacean populations range over extremely wide areas, so monitoring studies must be conducted at broad spatial scales to be able to effectively detect population-level changes. We strongly recommend that a monitoring regime such as the NMFS/NEFSC survey program be reinstated as soon as possible. We also recommend that the data be collected in a format that enables easier quantification of survey effort, which will make future attempts to demonstrate shifts in distribution and abundance between time intervals more rigorous and less speculative. The lack of systematic cetacean sighting data before the mid- to late-1970s has forced us in this chapter to speculate based on sparse data and sketchy reports in the literature for the period when DWF fishing caused dramatic shifts in pelagic fish populations and the most drastic changes in odontocete distributions may have occurred. It is our hope that future researchers, with more complete data to address these types of questions, might not have to stick their necks out quite so far.

References

Anthony, V. A., and Waring, G. 1980. The assessment and management of the Georges Bank herring fishery. Rapp. P.-V. Réun. Cons. Int. Explor. Mer 177:72–111.

Azarovitz, T. R. 1981. A Brief Historical Review of the Woods Hole Laboratory Trawl Survey Time Series. In W. G. Doubleday and D. Rivard (eds.), Bottom Trawl Surveys. Can. Spec. Publ. Fish. Aquat. Sci. Vol. 58. Ottawa: Department of Fisheries and Oceans, pp. 62–67.

Bowman, R., Warzocha, J., and Morris, T. 1984. Trophic relationships between Atlantic mackerel and American sand lance. Int. Counc. Explor. Sea (ICES). Doc. C.M. 1984/H:27. Copenhagen: International Council for the Exploration of the Sea.

Bryant, P. J., Nichols, G., Bryant, T. B., and Miller, K. 1981. Krill availability and the distribution of humpback whales in southeastern Alaska. J. Mamm. 62:427–430.

Burnham, K. P., Anderson, D. R., and Laake, J. F. 1980. Estimation of density from line transect sampling of biological populations. Wildl. Monogr. 72:1–202.

Cetacean and Turtle Assessment Program (CETAP). 1982. A Characterization of Marine Mammals and Turtles in the Mid- and North-Atlantic Areas of the U.S. Outer Continental Shelf, Final Report. Contract No. AA551-CT8-48. Washington, DC: U.S. Dept. of the Interior, Bureau of Land Management.

Connor, P. F. 1971. The mammals of Long Island, New York. New York State Mus. Sci. Serv. Bull. 416:1–78.

Corkeron, P. 1988. Distribution and Ecology. In R. Harrison and M. M. Bryden (eds.), Whales, Dolphins and Porpoises. New York: Facts on File, pp. 84–98.

Edel, R. K., Hyman, M. A. M., and Tyrell, M. F. 1981. Introduction to 1979 Annual Report. In Cetacean and Turtle Assessment Program. A Characterization of Marine Mammals and Turtles in the Mid- and North-Atlantic Areas of the U.S. Outer Continental Shelf, Annual Report for 1979. Contract No. AA551-CT8-48. Washington, DC: U.S. Dept. of the Interior, Bureau of Land Management, pp. I.1–I.66.

Ford, R. G. 1989. CAMRIS—Computer Aided Mapping and Resource Inventory System. Portland, OR: Ecological Consulting, Inc.

Gaskin, D. E. 1982. The Ecology of Whales and Dolphins. London: Heinemann.

Gaskin, D. E. 1984. The harbour porpoise *Phocoena phocoena* (L.): Regional populations, status, and information on direct and indirect catches. Rept. Int. Whaling Comm. 34:569–586.

Gaskin, D. E., Smith, G. J. D., and Watson, A. P. 1975. Preliminary study of harbor porpoises (*Phocoena phocoena*) in the Bay of Fundy using radiotelemetry. Can. J. Zool. 53:1466–1471.

Gaskin, D. E., and Watson, A. P. 1985. The harbor porpoise, *Phocoena phocoena*, in Fish Harbour, New Brunswick, Canada: Occupancy, distribution, and movements. Fish. Bull., U.S. 83:427–442.

Geraci, J. R., Testaverde, S. A., St. Aubin, D. J., and Loop, T. H. 1978. A Mass Stranding of the Atlantic White-sided Dolphin, *Lagenorhynchus acutus*: A Study into Pathobiology and Life History. Report No. MM5-AC-008. Washington, DC: Marine Mammal Commission.

Gilbert, J. R., and Wynne, K. 1988. Marine Mammal Interactions with New England Gillnet Fisheries. Final Rept., Contract No. NA-84-EA-C-00070. Woods Hole, MA: National Marine Fisheries Service, Northeast Fisheries Center.

Grosslein, M. D. 1969. Groundfish survey program of BCF Woods Hole. Commer. Fish. Rev. 31 (8–9):22–35.

Grosslein, M. D., Langton, R. W., and Sissenwine, M. P. 1980. Recent fluctuations in pelagic fish stocks of the northwest Atlantic, Georges Bank region, in relation to species interactions. Rapp. P.-V. Réun. Cons. Int. Explor. Mer 177:374–404.

Hain, J. H. W., Carter, G. R., Kraus, S. D., et al. 1982. Feeding behavior of the humpback whale, *Megaptera novaeangliae*, in the western North Atlantic. Fish. Bull., U.S. 80:259–268.

Hain, J. H. W., Edel, R. K., Hays, H. E., et al. 1981. General distribution of cetaceans in the continental shelf waters of the northeastern United States. In Cetacean and Turtle Assessment Program. A Characterization of Marine Mammals and Turtles in the Mid- and North-Atlantic Areas of the U.S. Outer Continental Shelf, Annual Report for 1979. Contract No. AA551-CT8-48. Washington, DC: U.S. Dept. of the Interior, Bureau of Land Management, pp. II.1–II.345.

Hain, J. H. W., Hyman, M. A. M., Kenney, R. D., and Winn, H. E. 1985. The role of cetaceans in the shelf edge region of the western North Atlantic. Mar. Fish. Rev. 47:13–17.

Hain, J. H. W., Ratnaswamy, M. J., Kenney, R. D., and Winn, H. E. 1992. The fin whale, *Balaenoptera physalus*, in waters of the northeastern United States continental shelf. Rept. Int. Whaling Comm. 42:653–669.

Hamilton, P. K., and Mayo, C. A. 1990. Population characteristics of right whales (*Eubalaena glacialis*) observed in Cape Cod and Massachusetts Bays, 1978–1986. Rept. Int. Whaling Comm., Spec. Issue 12:203–208.

Jonsgard, A. 1955. The stocks of blue whales (*Balaenoptera musculus*) in the North Atlantic Ocean and adjacent Arctic waters. Norsk Hvalfangst-Tid. 44:505–519.

Jonsgard, A. 1966. The distribution of Balaenopteridae in the North Atlantic Ocean. In K. S. Norris (ed.), Whales, Dolphins, and Porpoises. Berkeley: University of California Press, pp. 114–124.

Jonsgard, A., and Darling, K. 1977. On the biology of the eastern North Atlantic sei whale, *Balaenoptera borealis* Lesson. Rept. Int. Whaling Comm., Spec. Issue 1:124–129.

Jonsgard, A., and Nordli, O. 1952. Concerning a catch of white-sided dolphins (*Lagenorhynchus acutus*) on the west coast of Norway, winter 1952. Norsk Hvalfangst-Tid. 5:229–232.

Katona, S. K. 1976. Whales in the Gulf of Maine: 1976. Rept. No. MM6AC-018. Washington, DC: Marine Mammal Commission.

Katona, S. K., Rough, V. A., and Richardson, D. T. 1983. A Field Guide to the Whales, Porpoises and Seals of the Gulf of Maine and Eastern Canada: Cape Cod to Newfoundland (3rd ed.). New York: Scribner's.

Katona, S. K., Testaverde, S. A., and Barr, B. 1976. Observations on a white-sided dolphin, *Lagenorhynchus acutus*, probably killed in gill nets in the Gulf of Maine. Fish. Bull., U.S. 76:475–476.

Katona, S. K., and Whitehead, H. 1988. Are Cetacea ecologically important? Oceanogr. Mar. Biol. Ann. Rev. 26:553–568.

Kawamura, A. 1980. A review of food of balaenopterid whales. Sci. Rept. Whales Res. Inst. 32:155–197.

Kenney, R. D., Goodale, D. R., Scott, G. P., and Winn, H. E. 1981. Spatial and temporal distribution of humpback whales in the CETAP study area. In Cetacean and Turtle Assessment Program. A Characterization of Marine Mammals and Turtles in the Mid- and North-Atlantic Areas of the U.S. Outer Continental Shelf, Annual Report for 1979. Contract No. AA551-CT8-48. Washington, DC: U.S. Dept. Interior, Bureau of Land Management, pp. V.1–V.32.

Kenney, R. D., Hyman, M. A. M., Owen, R. E., et al. 1986a. Estimation of prey densities required by western North Atlantic right whales. Mar. Mamm. Sci. 2:1–13.

Kenney, R. D., Hyman, M. A. M., and Winn, H. E. 1985. Calculation of Standing Stocks and Energetic Requirements of the Cetaceans of the Northeast United States Outer Continental Shelf. NOAA Tech. Memo. NMFS-F/NEC-41. Woods Hole, MA: National Marine Fisheries Service, Northeast Fisheries Center.

Kenney, R. D., and Winn, H. E. 1986. Cetacean high-use habitats of the northeast U. S. continental shelf. Fish. Bull., U.S. 84:345–357.

Kenney, R. D., and Winn, H. E. 1987a. Aerial Surveys of Right Whales in the Great South Channel, Spring 1987. Final Rept., Order No. 14624. Washington, DC: Minerals Management Service.

Kenney, R. D., and Winn, H. E. 1987b. Cetacean biomass densities near submarine canyons compared to adjacent shelf/slope areas. Cont. Shelf Res. 7:107–114.

Kenney, R. D., Winn, H. E., and Brown, C. W. 1986b. Aerial Surveys for Right Whales in the Great South Channel, Spring 1985. Final Rept., Contract No. NA-84-EA-C-00079. Woods Hole, MA: National Marine Fisheries Service, Northeast Fisheries Center.

Kraus, S. D., Gilbert, J. R., and Prescott, J. H. 1983. A comparison of aerial, shipboard, and land-based survey methodology for the harbor porpoise, *Phocoena phocoena*. Fish. Bull., U.S. 81:910–913.

Langton, R. W., and Bowman, R. 1980. Food of Fifteen Northwest Atlantic Gadiform Fishes. NOAA Tech. Rept. NMFS SSRF 740. Seattle: National Marine Fisheries Service.

Leatherwood, S., and Reeves, R. R. 1986. Porpoises and Dolphins. In D. Haley (ed.), Marine Mammals (2nd ed.). Seattle: Pacific Search Press, pp. 110–131.

Lien, J., Johnson, S., and Merdsoy, B. 1979. Whale distribution in Newfoundland during 1979. Osprey 11(2):21–32.

Lynch, K., and Whitehead, H. 1984. Changes in the Abundance of Large Whales Off Newfoundland and Labrador, 1976–1983, With Special Reference to the Finback Whale. Doc. SC/36/02. Cambridge, UK: International Whaling Commission.

Manomet Bird Observatory (MBO). 1988. Cetacean and Seabird Assessment Program–Effort Evaluation. Final Rept., Contract No. 50-EANF-6-00028. Woods Hole, MA: National Marine Fisheries Service, Northeast Fisheries Center.

Mayo, C. A. 1982. Observations of Cetaceans: Cape Cod Bay and southern Stellwagen Bank, Massachusetts, 1975–1979. Rept. No. MMC-80/07. Washington, DC: U.S. Marine Mammal Commission.

Mayo, C. A., and Marx, M. K. 1990. Feeding behaviour of northern right whales, *Eubalaena glacialis*, in Cape Cod Bay, and associated zooplankton characteristics. Can. J. Zool. 68:2214–2220.

Mead, J. G. 1979. An Analysis of Cetacean Strandings Along the Eastern Coast of the United States. In J. R. Geraci and D. J. St. Aubin (eds.), Biology of Marine Mammals: Insights Through Strandings. Rept. No. MMC-77/13. Washington, DC: U.S. Marine Mammal Commission, pp. 54–68.

Mercer, M. C. 1973. Observations on distribution and intraspecific variation in pigmentation patterns of odontocete Cetacea in the western North Atlantic. J. Fish. Res. Board Can. 30:1111–1130.

Meyer, T. L., Cooper, R. A., and Langton, R. W. 1979. Relative abundance, behavior, and food habits of the American sand lance, *Ammodytes americanus*, from the Gulf of Maine. Fish. Bull., U.S. 77:243–253.

Mitchell, E. D. 1975. Trophic relationships and competition for food in northwest Atlantic whales. In Proceedings of the Canadian Society of Zoology Annual Meeting, June 1974, Fredericton, New Brunswick, pp. 123–133.

Mitchell, E. D. 1981. Canada progress report on cetacean research, June 1979–May 1980. Rept. Int. Whaling Comm. 31:174–175.

Mitchell, E., and Chapman, D. G. 1977. Preliminary assessment of stocks of Northwest Atlantic sei whales (*Balaenoptera borealis*). Rept. Int. Whaling Comm., Spec. Issue 1:117–120.

Mullane, S. J., and Rivers, A. 1982. Mt. Desert Rock, Maine—Annual Report. Bar Harbor, ME: Allied Whale, College of the Atlantic.

Nelson, G. A., and Ross, M. R. 1991. Biology and population changes of northern sand lance (*Ammodytes dubius*) from the Gulf of Maine to the Middle Atlantic Bight. J. Northwest Atlantic Fish. Sci. 11:11–27.

Nemoto, T. 1957. Foods of baleen whales in the North Pacific. Sci. Rept. Whales Res. Inst. 12:33–89.

Nemoto, T. 1959. Food of baleen whales with reference to whale movements. Sci. Rept. Whales Res. Inst. 14:149–290.

Nemoto, T. 1970. Feeding Patterns of Baleen Whales in the Ocean. In J. H. Steele (ed.), Marine Food Chains. Berkeley: University of California Press, pp. 241–252.

Overholtz, W. J., Murawski, S. A., and Foster, K. L. 1991. Impact of predatory fish, marine mammals, and seabirds on the pelagic fish ecosystem of the northeastern USA. ICES Mar. Sci. Symp. 193:198–208.

Overholtz, W. J., and Nicolas, J. R. 1979. Apparent feeding by the fin whale, *Balaenoptera physalus*, and humpback whale, *Megaptera novaeangliae*, on the American sand lance, *Ammodytes americanus*, in the northwest Atlantic. Fish. Bull., U.S. 77:285–287.

Payne, P. M., and Heinemann, D. W. 1993. Distribution of pilot whales (*Globicephala* spp.) in shelf/shelf-edge and slope waters of the northeastern United States, 1978–1988. Rept. Int. Whaling Comm., Spec. Issue 14:51–68.

Payne, P. M., Heinemann, D. W., and Selzer, L. A. 1991. A Distributional Assessment of Cetaceans in Shelf/Shelf-edge and Adjacent Slope Waters of the Northeastern United States Based on Aerial and Shipboard Surveys, 1978–1988. Final Report. Woods Hole, MA: National Marine Fisheries Service, Northeast Fisheries Center.

Payne, P. M., Heinemann, D. W., and Smith, T. D. 1990a. Seasonal Distribution of Minke Whales in the Shelf and Shelf-edge Waters of the Northeastern U.S. Doc. SC/42/NHMi32. Cambridge, UK: International Whaling Commission.

Payne, P. M., Nicolas, J. R., O'Brien, L., and Powers, K. D. 1986. The distribution of the humpback whale *Megaptera novaeangliae* on Georges Bank and in the Gulf of Maine in relation to densities of the sand eel *Ammodytes americanus*. Fish. Bull., U.S. 84:271–277.

Payne, P. M., and Selzer, L. A. 1989. The distribution, abundance and selected prey of the harbor seal, *Phoca vitulina concolor*, in southern New England. Mar. Mamm. Sci. 5(2):173–192.

Payne, P. M., Selzer, L. A., and Knowlton, A. R. 1984. Distribution and Density of Cetaceans, Marine Turtles and Seabirds in the Shelf Waters of the Northeastern United States, June 1980–December 1983, from Shipboard Observations. Final Rept., Contract No. NA-81-FA-C-00023. Woods Hole, MA: National Marine Fisheries Service, Northeast Fisheries Center.

Payne, P. M., Wiley, D. N., Young, S. B., et al. 1990b. Recent fluctuations in the abundance of baleen whales in the southern Gulf of Maine in relation to changes in selected prey. Fish. Bull., U.S. 88:687–696.

Perkins, J. S., and Beamish, P. C. 1979. Net entanglements of baleen whales in the inshore fishery of Newfoundland. J. Fish. Res. Board Can. 36:512–528.

Powers, K. D. 1983. Pelagic Distributions of Marine Birds Off the Northeastern United States. NOAA Tech. Mem. NMFS-F/NEC-27. Woods Hole, MA: National Marine Fisheries Service, Northeast Fisheries Center.

Powers, K. D., and Backus, E. H. 1987. Energy Transfer to Seabirds. In R. H. Backus (ed.), Georges Bank. Cambridge, MA: MIT Press, pp. 372–374.

Powers, K. D., and Brown, R. G. B. 1987. Seabirds. In R. H. Backus (ed.), Georges Bank. Cambridge, MA: MIT Press, pp. 359–371.

Powers, K. D., and Payne, P. M. 1983. Distribution of Marine Mammals, Birds and Turtles Near the Deepwater Dumpsite 106 (DWD-106) and the Philadelphia Dumpsite (PDS). In J. B. Pearce, D. C. Miller, and C. Berman (eds.), Northeast Monitoring Program, 106-Mile Characterization Update. NOAA Tech. Memo. NMFS-F/NEC-26. Woods Hole, MA: National Marine Fisheries Service, Northeast Fisheries Center.

Prescott, J. H., and Fiorelli, P. M. 1980. Review of the Harbor Porpoise (*Phocoena phocoena*) in the U.S. Northwest Atlantic. Rept. No. MMC-78/08. Washington, DC: Marine Mammal Commission.

Ramsdell, K. 1978. Whales in the Gulf of Maine 1977. Annual Report of the Gulf of Maine Whale Sighting Network. Bar Harbor, ME: College of the Atlantic.

Read, A. J. 1990. Gillnet and Trap Fisheries in the Northwest Atlantic. Doc. SC/090/G6. Cambridge, UK: International Whaling Commission.

Read, A. J., and Gaskin, D. E. 1985. Radio-tracking the movements and activities of harbor porpoises, *Phocoena phocoena* (L.), in the Bay of Fundy, Canada. Fish. Bull., U.S. 83:543–552.

Read, A. J., and Gaskin, D. E. 1988. Incidental catch of harbor porpoises by gill nets. J. Wildl. Manage. 52:517–523.

Read, A. J., and Gaskin, D. E. 1990. The Effects of Incidental Mortality on Harbour Porpoises (*Phocoena phocoena*) in the Bay of Fundy and Gulf of Maine. Doc. SC/42/SM21. Cambridge, UK: International Whaling Commission.

Reay, P. J. 1970. Synopsis of Biological Data on North Atlantic Sand Eels of the Genus *Ammodytes*: *A. tobianus*, *A. dubius*, *A. americanus*, and *A. marinus*. FAO Fisheries Synopsis No. 82. Rome: Food and Agriculture Organization of the United Nations.

Recchia, C. A., and Read, A. J. 1989. Stomach contents of harbour porpoises, *Phocoena phocoena* (L.), from the Bay of Fundy. Can. J. Zool. 67:2140–2146.

Schevill, W. E. 1956. *Lagenorhynchus acutus* off Cape Cod. J. Mamm. 37:128–129.

Scott, G. P., and Gilbert, J. R. 1982. Problems and progress in the U.S. BLM-sponsored CETAP surveys. Rept. Int. Whaling Comm. 32:587–600.

Scott, G. P., Kenney, R. D., Gilbert, J. R., and Edel, R. K. 1981. Estimates of Cetacean and Turtle Abundance in the CETAP Study Area with an Analysis of Factors Affecting Them. In Cetacean and Turtle Assessment Program. A Characterization of Marine Mammals and Turtles in the Mid- and North-Atlantic Areas of the U.S. Outer Continental Shelf, Annual Report for 1979. Contract No. AA552-CT8-48. Washington, DC: U.S. Dept. of the Interior, Bureau of Land Management, pp. IV.1–IV.70.

Seber, G.A.F. 1973. The Estimation of Animal Abundance. New York: Hafner.

Selzer, L. A., and Payne, P. M. 1988. The distribution of white-sided (*Lagenorhynchus acutus*) and common dolphins (*Delphinus delphis*) versus environmental features of the continental shelf of the northeastern United States. Mar. Mamm. Sci. 4:141–153.

Sergeant, D. E. 1982. Mass strandings of toothed whales (Odontoceti) as a population phenomenon. Sci. Rept. Whales Res. Inst. 34:1–47.

Sergeant, D. E., and Fisher, H. D. 1957. The smaller Cetacea of eastern Canadian waters. J. Fish. Res. Board Can. 14:83–115.

Sergeant, D. E., Mansfield, A. W., and Beck, B. 1970. Inshore records of Cetacea for eastern Canada. J. Fish. Res. Board Can. 27:1903–1915.

Sergeant, D. E., St. Aubin, D. J., and Geraci, J. R. 1980. Life history and northwest Atlantic status of the Atlantic white-sided dolphin, *Lagenorhynchus acutus*. Cetology 37:1–12.

Sherman, K. 1980. MARMAP, A Fisheries Ecosystem Study in the Northwest Atlantic: Fluctuations in Ichthyoplankton–Zooplankton Components and Their Potential for Impact on the System. In P. Diemer, F. J. Vernberg, and D. Z. Mirkes (eds.), Advanced Concepts in Ocean Measurements for Marine Biology. Columbia, SC: Belle W. Baruch Institute for Marine Biology and Coastal Research, University of South Carolina Press, pp. 9–37.

Sherman, K., Jones, C., Sullivan, L., et al. 1981. Congruent shifts in sand eel abundance in western and eastern North Atlantic ecosystems. Nature 291:486–489.

Sigurjonsson, J., Gunlaugsson, T., and Payne, P. M. 1989. NASS-87: Shipboard sighting surveys in Icelandic and adjacent waters, June–July 1987. Rept. Int. Whaling Comm. 39:395–409.

Sissenwine, M. P., Brown, B. E., Palmer, J. E., and Essig, R. J. 1982. Empirical Examination of Population Interactions for the Fishery Resources Off the Northeastern USA, pp. 82–94. In M. C. Mercer (ed.), Multispecies Approaches to Fisheries Management Advice. Can. Spec. Publ. Fish. Aquat. Sci. Vol. 59. Ottawa: Department of Fisheries and Oceans, pp. 82–94.

Sissenwine, M. P., Overholtz, W. J., and Clark, S. H. 1984. In Search of Density Dependence. In B. R. Melteff (ed.), Proceedings of the Workshop on Marine Mammal–Fishery Interactions of the East Bering Sea, 18–21 October 1983, Anchorage, Alaska. Fairbanks: University of Alaska Sea Grant, pp. 119–138.

Smith, G. J. D., and Gaskin, D. E. 1974. The diet of harbour porpoises (*Phocoena phocoena* (L.)) in coastal waters of eastern Canada, with special reference to the Bay of Fundy. Can. J. Zool. 52:777–782.

Smith, G. J. D., Read, A. J., and Gaskin, D. E. 1983. Incidental catch of harbor porpoise, *Phocoena phocoena* (L.), in herring weirs in Charlotte County, New Brunswick, Canada. Fish. Bull., U.S. 81:660–662.

Smith, T., Payne, P. M., Heinemann, D. W., et al. 1990. Simultaneous fishery resource and seabird and cetacean sighting surveys: Advantages and disadvantages. North Atlantic Studies 2 (1&2):90–103.

Sorensen, P. W., Medved, R. J., Hyman, M. A. M., and Winn, H. E. 1984. Distribution and abundance of cetaceans in the vicinity of human activities along the continental shelf of the northwestern Atlantic. Mar. Environ. Res. 12:69–81.

Stock Assessment Workshop (SAW). 1989. Status of Georges Bank Herring. SAW 10, Working Paper 13. Woods Hole, MA: National Marine Fisheries Service, Northeast Fisheries Center.

Stock Assessment Workshop (SAW). 1990. Report of the Eleventh NEFC Stock Assessment Workshop, Fall 1990. Woods Hole, MA: National Marine Fisheries Service, Northeast Fisheries Center.

Testaverde, S. A., and Mead, J. G. 1980. Southern distribution of the Atlantic whitesided dolphin *Lagenorhynchus acutus* in the western North Atlantic. Fish. Bull., U.S. 78:167–169.

Ulmer, F. A. 1977. New Jersey's Dolphins and Porpoises. Occ. Papers New Jersey Audubon Soc. 137:1–11.

Waters, J. H., and Rivard, C. J.-J. 1962. Terrestrial and Marine Mammals of Massachusetts and other New England States. Brockton, MA: Standard Modern Printing.

Watkins, W. A., and Schevill, W. E. 1982. Observations of right whales, *Eubalaena glacialis*, in Cape Cod waters. Fish. Bull., U.S. 80:875–880.

Watson, L. 1981. Sea Guide to Whales of the World. New York: E. P. Dutton.

Wenzel, F., Mattila, D. K., and Clapham, P. J. 1988. *Balaenoptera musculus* in the Gulf of Maine. Mar. Mamm. Sci. 4:172–175.

Whitehead, H. P. 1983. Structure and stability of humpback whale groups off Newfoundland. Can. J. Zool. 61:1391–1397.

Whitehead, H. P., and Carscadden, J. 1985. Predicting inshore whale abundance—whales and capelin off the Newfoundland coast. Can. J. Fish. Aquat. Sci. 42:976–981.

Whitehead, H., and Lien, J. 1982. Changes in the Abundance of Whales and Whale Damage, Along the Newfoundland Coast 1973–1981. Doc. SC/34/01. Cambridge, UK: International Whaling Commission.

Winn, H. E., Goodale, D. R., Hyman, M. A. M., et al. 1981. Right Whale Sightings and the Right Whale Minimum Count. In Cetacean and Turtle Assessment Program. A Characterization of Marine Mammals and Turtles in the Mid- and North-Atlantic Areas of the U.S. Outer Continental Shelf, Annual Report for 1979. Contract No. AA551-CT8-48. Washington, DC: U.S. Dept. of the Interior, Bureau of Land Management, pp. VI.1–VI.37.

Winn, H. E., Price, C. A., and Sorensen, P. W. 1986. The distributional biology of the right whale (*Eubalaena glacialis*) in the western North Atlantic. Rept. Int. Whaling Comm., Special Issue 10:129–138.

Wishner, K., Durbin, E., Durbin, A., et al. 1988. Copepod patches and right whales in the Great South Channel off New England. Bull. Mar. Sci. 43:825–844.

Yochem, P. K., and Leatherwood, S. 1985. Blue whale *Balaenoptera musculus*. In S. H. Ridgway and R. Harrison (eds.), Handbook of Marine Mammals. Vol. 3: The Sirenians and Baleen Whales. London: Academic Press, pp. 193–240.

11

The State of Marine Bird Populations from Cape Hatteras to the Gulf of Maine

David C. Schneider and Dennis W. Heinemann

Abstract

The coast and shelf environments from the Gulf of Maine to Cape Hatteras are used by coastally breeding birds (terns, gulls, skimmers) and by a far larger number of nonbreeding residents—shorebirds from the Arctic and seabirds from elsewhere in the Atlantic. These groups experience different environmental stresses. For coastally breeding birds, the primary stresses are habitat loss or modification, disturbance and predation at colonies, and bioaccumulation of fat-soluble toxins. For migrant shorebirds, the primary stresses are habitat loss and disturbance of high-tide roosts. For the offshore seabirds, the potential stresses are chronic or catastrophic oil spills, although heavy metal intoxication during periods of weather-induced starvation might also be important. Changes in abundance of seabirds are typically recorded at colonies, a method that for most species cannot be used to monitor the response of seabirds to ecosystem change in the northeastern United States. We examined annual changes in seabird populations as recorded in opportunistic counts from research and commercial fishing vessels over the period 1978–1988. The numbers of five species of ship-following birds decreased during this period, despite an increase in fishing days; the most salient hypothesis is a reduction in rate of discarding, as catch per unit effort decreased. The populations of five other species oscillated at periods on the order of several years; the most salient hypothesis for these is a change in the production of juveniles. For another seven species, the most salient hypothesis is the occurrence of distributional shifts in response to changes in food abundance. All of these hypotheses were developed from exploratory analysis; they remain to be tested rigorously. One of the major problems in understanding the effects of anthropogenic stresses on large marine ecosystems is the problem of scale-up from measurements that are necessarily limited in time and space. A combination of formal models and experimental verification will be needed to evaluate the effects of large-scale stresses on marine ecosystems, including the avian components of these systems.

The coast from the Gulf of Maine to Cape Hatteras includes one of the largest conurbations in the world, tens of millions of people living in cities from Boston to Washington, D.C. The stresses that are placed on the environment by this highly industrialized conurbation include release of organic and toxic materials, selective removal of species targeted by an industrial fishery, and alterations of the coastal hydroregime and land use. The effects of these stresses on coastal environments have long been evident, but it is now becoming increasingly clear that these stresses have effects offshore. These include stock collapses and replacements (Sherman et al. 1988) brought on by the interaction of heavy fishing with natural fluctuations in recruitment.

Marine birds are an abundant, highly visible, and valued component of this ecosystem. As such, they lend themselves to the role of indicator species in both coastal and offshore habitats. The purpose of this chapter is to review briefly the community structure of birds in this ecosystem, list the known and potentially significant stresses on the avian community, and examine long-term trends.

The marine birds in this ecosystem can be divided into three groups: breeding birds, nonbreeding shorebirds, and nonbreeding pelagic seabirds. The breeding birds, primarily gulls, terns, and herons (Buckley and Buckley 1984), are, however, far outnumbered by the nonbreeding residents belonging to the other two groups. The shorebirds (plovers and sandpipers) breed in the Arctic, shift to coastal habitats to build fat reserves over several weeks, and then migrate to wintering grounds in the tropics. The other group of nonbreeding residents, pelagic seabirds, forage in shelf waters beyond sight of land, breeding in colonies as distant as Europe, Greenland, and the south Atlantic. Despite the absence of large colonies, the numbers and biomasses of seabirds are similar to those of other mid-latitude continental shelf ecosystems that have substantial breeding populations (Schneider et al. 1987).

In the past, coastal marine birds were under substantial stress from market hunting, either for plumes (breeders), or for food (plovers and sandpipers). This stress radically altered the structure of marine bird communities. Terns, herons, sandpipers, and plovers were nearly exterminated from the ecosystem. Since then, terns and herons have recovered in numbers, although they will never reach historical levels because of the loss of breeding habitat. Sandpipers and plovers have also recovered, perhaps to levels observed before market hunting. Offshore seabirds were not substantially affected by this stress.

All marine foraging birds in this ecosystem are at risk from oil spills, either chronic or catastrophic. Two well-documented examples are the 1969 West Falmouth spill, which inundated a coastal marsh, and the 1976 *Argo Merchant* spill on Nantucket Shoals (Powers and Rumage 1978). The potential for mortality is substantial; the best estimate of mortality for the *Exxon Valdez* spill was 105 murres killed (Piatt and Lensink 1990). Another risk to marine birds in this ecosystem is the bioaccumulation of toxins in fat-soluble form. The most famous case is that of DDT, but PCBs and toxic metals (especially mercury) are also bioaccumulated through food chains. This can result in substantial body burdens and intakes by apex predators such as marine birds. The toxic effects are often delayed rather than acute—reduced reproductive success resulting from eggshell thinning (Bourne 1976) or intoxication after mobilization of fat reserves during periods of weather-induced starvation. Bioaccumulation and release of toxins is potentially important in this highly industrialized region, with known pollutant hotspots such as New Bedford Harbor.

Seabirds, shorebirds, and coastal breeders each face slightly different stresses. The offshore seabirds experience fishery-induced changes in prey species. The two coastal groups face stresses resulting from changes in the physical habitat and consequent changes in prey availability or abundance. The coastal groups also face stress from continued disturbance at breeding colonies or at spatially restricted roosting sites, in the case of shorebirds.

A knowledge of long-term trends in abundance is important for two reasons. First, detection of changes in abundance relative to a particular environmental stress requires some knowledge of natural or background variation in numbers and of the sources of these changes. This is a matter of statistical power—how much survey effort is required to detect a given change (e.g., 20% decrease) given the variability that one can normally expect from year to year? This is also a matter of working out causes—what factors normally result in sudden decreases in numbers, in the absence of anthropogenic stress?

The second reason for examining long-term change is, of course, to isolate the role of anthropogenic stresses. If trends occur over prolonged periods, correlative analysis can point to possible causes, which can then be tested by more direct means. A good example of an association in need of direct analysis is the continued reproductive failure in arctic terns associated with the development of an industrial fishery on sand lance in the Shetland Islands (Monaghan et al. 1989).

Long-term trends in marine bird populations are usually monitored at breeding colonies for a number of reasons. Most marine birds nest at a small number of isolated colonies, and hence, the reproductively active part of the population is confined to a limited area. This makes it possible to execute accurate surveys and to examine reproductive success. With some investment of effort over time, it becomes possible to measure recruitment (to reproductive status) via ringing schemes.

Monitoring at colonies is unfortunately inapplicable to most of the marine bird species in this ecosystem because they are nonbreeding residents. Shorebirds must be surveyed either at high-tide roosts or over the intertidal areas where they feed. Nonbreeding seabird populations, like many commercially important fish species, must be surveyed on the spatial scale of the entire ecosystem. Avian feeding activity shifts rapidly enough that surveys within restricted areas cannot be relied on to give an index of trends in abundance at larger spatial scales.

At-sea surveys are seldom used for measuring long-term changes in abundance of seabirds, in part because of the spatial scale required and the fact that information about recruitment is usually obtainable at colonies. However, seabirds that occupy the Northeast Shelf ecosystem generally breed elsewhere. From 1976 through 1988, the Manomet Observatory conducted a low-cost opportunistic survey. Data from 1978 through 1980 were used to map distributional patterns (Powers 1983; Powers and Brown 1987). The same data were used to estimate spatial variation in the flux of prey to seabirds (Powers and Backus 1987). Repeatable seasonal patterns were also evident for common species (Powers 1983). In this chapter, we use the survey data to examine change over the period from 1978 to 1988. During this period, there were substantial changes in pelagic fish populations, which are preyed on by many seabirds.

Methods

Community structure was described from available publications. Stresses on marine birds were compiled from the published literature in two categories: either known for the study area, or known from elsewhere (e.g., Ferns 1993; Brown and Nettleship 1984) and judged likely based on the authors' experience with marine birds in this ecosystem. Long-term changes in seabird abundance were measured from ships of opportunity. In addition to commercial fishing vessels, counts were made by an observer aboard MARMAP research cruises, which systematically covered the entire shelf area each year. Counts were made in a strip 300 m wide, from one side of the vessel, while steaming at a constant speed. Counts were standardized to the area surveyed in 10-minute periods. Further details can be found in Powers (1983). The standardized counts must be considered relative rather than absolute density estimates because of uncontrolled biases and sources of error. Average counts by year (numbers per km^2) were computed for the shelf area within ICNAF divisions 6C, 6B, 6A, 5Z, 5Y, and parts of 4X, corresponding to the four geographic subareas of Clark and Brown (1977): Middle Atlantic, Southern New England, Georges Bank, and Gulf of Maine. The annual averages were then displayed as deviations from the 10-year mean in units of standard deviations. Thus, a value of –1 on a graph indicates that the mean for that year was one standard deviation below the 10-year mean. Linear trends were assessed by computing the regression against time, then using randomization tests (Manly 1991) to evaluate statistical significance of trends. A cutoff of α = 10% was used to distinguish species with trends from those showing no trend. Based on our combined experience with marine birds in this ecosystem, we list the most salient explanation for the observed patterns of change in common offshore species. To identify the most salient hypothesis, we discarded those inconsistent with available evidence, and then used knowledge of the population biology of each species to decide on the most relevant among those remaining. Because these hypotheses are based on substantial research experience, we consider tests of them to be far more informative than tests of null hypotheses of no change.

Results

Coastal Breeders: Community Structure

A total of 20 species of marine birds breed along this coast (Table 11-1). Gulls and terns are the most common. Leach's storm petrels are also important within this area, although total numbers are small relative to colonies in Newfoundland, which has colonies numbering in the millions. Avian community structure changes from north to south, particularly at the transition from the rocky coastlines north of Cape Cod to the sand and mud south of the Cape. Several species (Leach's storm petrel, great cormorant, razorbill, and Atlantic puffin) are restricted to the Gulf of Maine, where they are at the southern limit of their range. The several species of tern nest on beaches, sandspits, and dredge spoils, for the most part south of Cape Cod.

Coastal Breeders: Stresses

Known stresses on breeding species in this ecosystem include disturbance (CWB 1981), changes in land use and hydroregime (Buckley and Buckley 1981, 1983), toxic pollutants (Nisbet 1980), and increases in predators associated with human activity—pets, rats, and herring gulls (Buckley and Buckley 1984). The most important of these is change in land use at colony sites.

Table 11-1. Probable populations (pairs) of breeding seabirds, Cape Hatteras to Gulf of Maine, for the period 1977–1983

Common name	Latin name	Number of pairs
Herring gull	Larus argentatus smithsonianus	92,000
Laughing gull	Larus atricilla	65,000
Common tern	Sterna h. hirundo	29,000
Leach's storm petrel	Oceanodroma l. leucorhoa	19,000
Double-crested cormorant	Phalacrocorax a. auritus	17,000
Great black-backed gull	Larus marinus	17,500
Least tern	Sterna antillarum	7,500
Black skimmer	Rynchops n. niger	4,250
Royal tern	Sterna m. maxima	4,000
Roseate tern	Sterna d. dougalli	3,300
Black guillemot	Cepphus grylle atlantis	2,700
Forster's tern	Sterna forsteri	2,300
Arctic tern	Sterna paradisaea	1,325
Gull-billed tern	Sterna nilotica aranea	1,000
Sandwich tern	Sterna sandvicensis acuflavidus	140
Atlantic puffin	Fratercula a. arctica	125
Great cormorant	Phalacrocorax c. carbo	30
Razorbill	Alca t. torda	25
Manx shearwater	Puffinus p. puffinus	1–10
Caspian tern	Sterna caspia	2–4

Source: Reprinted with permission from PA Buckley, FG Buckley. Seabirds of the North and Middle Atlantic Coast of the United States: Their Status and Conservation. In JP Croxall, PGH Evans, RW Schreiber (eds), Status and Conservation of the World's Seabirds. ICBP Technical Publ. 2, 1984;101–133. Cambridge, UK: International Council for Bird Preservation.

Coastal Breeders: Changes in Numbers

Egg-collecting and feather hunting before 1900 largely eradicated gulls, terns (Drury 1973–1974), and other breeding populations (Buckley and Buckley 1984). Population estimates for coastal breeders for the entire coast were made in 1977 (Erwin 1979). Changes since then have not been reported at the spatial scale of the ecosystem. Census data from 1974 to 1983 on Long Island indicate that numbers of great black-backed gulls, common terns, and least terns have increased steadily, whereas numbers of herring hulls, roseate terns, and black skimmers fluctuated with no clear trend (Buckley and Buckley 1984). On the ecosystem scale, the number of roseate terns has fluctuated at around 3,000 pairs in 18 colonies (Table 11-2).

Migratory Shorebirds: Community Structure

Four species of migratory shorebird breed in the coastal zone between Cape Hatteras and Nova Scotia: American oystercatcher *(Haematopus palliatus)*,

Table 11-2. Colony number and size of breeding population of roseate terns in the Northwest Atlantic

Year	Number of pairs	Number of colonies
1977	2,333	21
1984	3,016	26
1985	2,853	19
1986	2,850	14
1987	2,899	21
1988	3,263	21
1989	3,188	18
1990	3,302	18
1991	3,611	18

Table 11-3. Commonly occurring shorebirds, Cape Hatteras to Gulf of Maine

Common name	Latin name
Black-bellied plover	Pluvialis squatarola
Ruddy turnstone	Arenaria interpres
Semipalmated plover	Charadrius semipalmatus
Short-billed dowitcher	Limnodromus griseus
Red knot	Calidris canutus
Whimbrel	Numenius phaeopus
Greater yellowlegs	Tringa melanoleuca
Sanderling	Calidris alba
Dunlin	Calidris alpina
Purple sandpiper	Calidris maritima
Least sandpiper	Calidris minutilla
Semipalmated sandpiper	Calidris pusilla

piping plover (*Charadrius melodus*), Wilson's plover (*Charadrius wilsonia*), and willet (*Catoptrophorus semipalmatus*). None are common and one, the piping plover, is endangered. A little more than ten species of migratory shorebirds are common in intertidal habitats in this ecosystem (Table 11-3). All of these species nest in the Arctic. These migrants become abundant for periods of 4–10 weeks (depending on species) at stopovers during southward migration (July through September). Overwintering is restricted by ice cover (north of Cape Cod) and by subfreezing temperatures. Many of the same species are abundant for brief periods during the spring migration period. Population sizes are not known. Large numbers occur in a limited number of places, of which the Bay of Fundy is the most notable in eastern North America. Substantial numbers occur at locations in Massachusetts (Newburyport and Scituate estuaries, Plymouth and Barnstable lagoons, Monomoy Island), in

New Jersey (Brigantine and Cape May), and the barrier islands running from Assateague southward to Cape Hatteras.

Migratory Shorebirds: Stresses

Limited breeding habitat as a result of human population pressure is a known stress for breeding shorebirds, particularly piping plovers. A known stress on arctic breeding shorebirds in this ecosystem is increased flight time to intertidal feeding areas from high-tide roosts as a result of recreational activity on beaches. Potential stresses in this ecosystem are ingestion of fat-soluble toxic compounds and reduction of intertidal feeding habitat. Present numbers are near the carrying capacity of the environment; prey depletion for common species is 50–90% over a six-week period (Schneider and Harrington 1982).

Migratory Shorebirds: Long-Term Changes

Market hunting eliminated virtually all shorebirds from coastal habitats early in the century. One species, the Eskimo curlew (*Numenius borealis*), nearly went extinct. Stress caused by hunting ceased when legislation was enacted early in the century. Numbers recovered in coastal habitats from 1920 onward.

Seabirds: Community Structure

The offshore seabird community consists of 46 regularly occurring species (Powers 1983), of which half are terns, gulls, and jaegers. All species are highly seasonal in abundance, occupying the region either for an extended period (e.g., fulmars from December through June) or for brief periods (e.g., gannets in March and in December). Three of the common species breed in the southern hemisphere; these become abundant in June–July (sooty shearwaters), in May through August (Wilson's storm petrel), and in both June–July and October–November (greater shearwater). A number of species from breeding colonies in Greenland and eastern Canada drift southward, residing to the north of Long Island through the winter (Iceland and glaucous gulls, kittiwakes, Leach's storm petrel, dovekies, razorbills, murres, and puffins). A few species (laughing gull, royal terns) rarely occur at sea north of Cape Cod. During winter and spring, seabird densities in southern New England and the Middle Atlantic Bight are similar to those on Georges Bank and the Gulf of Maine. During the summer and fall, densities rise in the northern areas, large-

Table 11-4. Annual occupancy and feeding mode of commonly occurring seabirds, Cape Hatteras to Gulf of Maine

Common name	Latin name	Feeding method	Occupancy
Greater shearwater	Puffinus gravis	SS, PP	253
Herring gull	Larus argentatus	SS, dip, scavenging	243
Northern fulmar	Fulmarus glacialis	SS, scavenging	189
Great black-backed gull	Larus marinus	SS, dip, scavenging	177
Wilson's stormpetrel	Oceanites oceanicus	SS, pattering	172
Black-legged kittiwake	Rissa tridactyla	SS, dip, scavenging	135
Phalaropes	Phalaropus spp.	SS	80
Northern gannet	Sula bassanus	Plunge-diving	40
Unidentified large gulls	Larus spp.	SS, dip, scavenging	33
Cory's shearwater	Calonectris diomedea	SS, PP	29
Sooty shearwater	Puffinus griseus	SS, PP	24
Atlantic puffin	Fratercula arctica	Pursuit diving	8
Laughing gull	Larus atricilla	SS, dip, scavenging	5
Dovekie	Alle alle	Pursuit diving	4
Razorbill	Alca torda	Pursuit diving	3
Unidentified large auk	Uria spp. or A. torda	Pursuit diving	3
Murres	Uria spp.	Pursuit diving	2
Leach's stormpetrel	Oceanodroma leucorhoa	SS, pattering	2
Manx shearwater	Puffinus puffinus	SS, PP	0.6
Audubon's shearwater	Puffinus lherminieri	SS, PP	0.2

SS = surface feeding; PP = pursuit and plunging.

Note: Occupancy (density · area · duration) calculated from Appendix 5 in Powers (1983) is expressed in units of mega bird-days per year.

ly because of the influx of shearwaters. At the same time, densities drop in the southern areas, possibly because of reduced availability of prey to surface foragers as stratification intensifies (Powers 1983).

Population sizes cannot be estimated because the turnover rate as species move through the area is unknown. However, it is possible to calculate occupancy, which is the product of three quantities: the estimated density, the area in which the estimate was made, and the period for which the estimate was made (Schneider and Hunt 1982). Marine birds require a minimum daily food intake to maintain their high metabolic rates, so occupancy is a measure of each species' trophic role in the ecosystem. Table 11-4 shows occupancy for shelf waters from 35° N to 44° N latitude, as calculated from Appendix 5 in Powers (1983). Occupancy is high for southern hemisphere breeders (greater shearwaters and Wilson's storm petrel) in the summer and fall; it is high in the winter and spring for northern hemisphere breeders (northern fulmar, herring gulls, great black-backed gulls, and black-legged kittiwakes).

Table 11-5. Number of counts and area surveyed, Gulf of Maine to Cape Hatteras

Year	Number of counts	Area surveyed (km^2)
1978	2,860	2,579
1979	3,548	3,266
1980	1,453	1,733
1981	2,114	2,564
1982	3,098	3,681
1983	3,591	4,491
1984	4,056	5,537
1985	3,365	4,505
1986	2,267	2,971
1987	3,103	4,047
1988	1,286	1,652

Seabirds: Stresses

Oil spills are known to cause mortality in diving species (murres and sea ducks) in this ecosystem (Powers and Rumage 1978). The mortality rate from either catastrophic or chronic spillage is not known. Changes in pelagic fish stocks are potentially stressing to piscivores such as shearwaters, gannets, kittiwakes, terns, gulls, murres, razorbills, and puffins. The most likely responses are switching to alternative prey (e.g., squid), or shifting to feeding grounds further south (e.g., gannets) or further north (e.g., alcids). Ingestion of plastic is a potential stress (Rothstein 1973), but effects on recruitment and mortality are unknown.

Seabirds: Long-Term Trends

Distributional studies related to offshore oil development have been conspicuously absent in this region, unlike other areas of the United States (NAS 1990). The only surveys have been those done by Manomet Observatory from 1976 to 1988, sponsored by the National Marine Fisheries Service. This program was unusual in that it allowed comparison of avian numbers at sea over a period of 11 years. Table 11-5 shows the area surveyed in each year; Table 11-6 shows the 11-year means for each species, the standard deviation used to scale the magnitude of change, the rate of annual change (birds yr^{-1}), and the type-I error rates on assuming these rates are significant. Species were grouped for graphical display according to similarity in patterns of change. Northern fulmars, great black-backed gulls, and herring gulls showed substantial and significant declines (see Table 11-6) from 1978 to 1980 (Figure 11-1A). Greater and sooty shearwaters also show substantial and significant

Table 11-6. Eleven-year means and standard deviations in relative density (birds km^{-2}) of marine birds, Gulf of Maine to Cape Hatteras

Common name	Mean	Standard deviation	$\hat{\beta}$	p
Greater shearwater	8.13	19.81	–3.66	0.03
Herring gull	3.94	4.572	–1.02	0.0
Northern fulmar	2.52	4.227	–0.97	0.005
Great black-backed gull	3.08	4.216	–0.93	0.018
Wilson's storm petrel	2.30	1.591	–0.38	0.009
Black-legged kittiwake	1.47	1.204	–0.37	<0.0001
Red phalarope	2.68	2.783	0.24	0.41
Northern gannet	1.07	0.5837	–0.007	0.91
Cory's shearwater	0.286	0.2957	–0.05	0.08
Sooty shearwater	1.19	3.321	–0.60	0.09
Atlantic puffin	0.015	0.0082	0.0005	0.58
Laughing gull	0.115	0.0679	–0.0056	0.55
Dovekie	0.108	0.1238	–0.0014	0.93
Leach's storm petrel	0.073	0.0499	–0.0006	0.92
Ring-billed gull	0.024	0.0116	0.00005	0.97
Pomarine jaeger	0.025	0.0140	–0.0035	0.006
Common loon	0.043	0.0306	–0.0034	0.29
Common tern	0.073	0.0506	0.005	0.35

Note: Rate of annual change (β = birds km^{-2}yr^{-1}); p value is for H_0: $\beta = 0$.

declines (see Table 11-6) a year earlier than the fulmars and large gulls (Figure 11-1B). All five of these species rely on offal from ships. Trawler effort (fishing days per year) nearly doubled from 1978 to 1984 on Georges Bank, with smaller increases in southern New England and the Gulf of Maine (Anthony 1993). Consequently, the sharp drop in numbers cannot be attributed to changes in activity by trawlers. The most salient hypothesis for these five species is that the declines reflect a reduction in rate of discard in the early 1980s, with shearwaters responding earlier than gulls or fulmars.

Neither ring-billed gulls nor common terns showed any long-term trend (see Table 11-6). Instead, relative densities oscillated at periods on the order of a few years (Figure 11-1C). These two species are present during different seasons and do not rely on the same prey (Powers 1983; Powers and Backus 1987). Both species breed in adjacent coastal areas. The most salient hypothesis for these two species is that the observed patterns reflect annual changes in production of juveniles.

Northern gannets and Atlantic puffins (Figure 11-1D) showed somewhat more frequent oscillations than the previous two species, again with no long-term trends (see Table 11-6). These two species are present during different seasons and rely on different prey; they breed at a limited number of colonies in

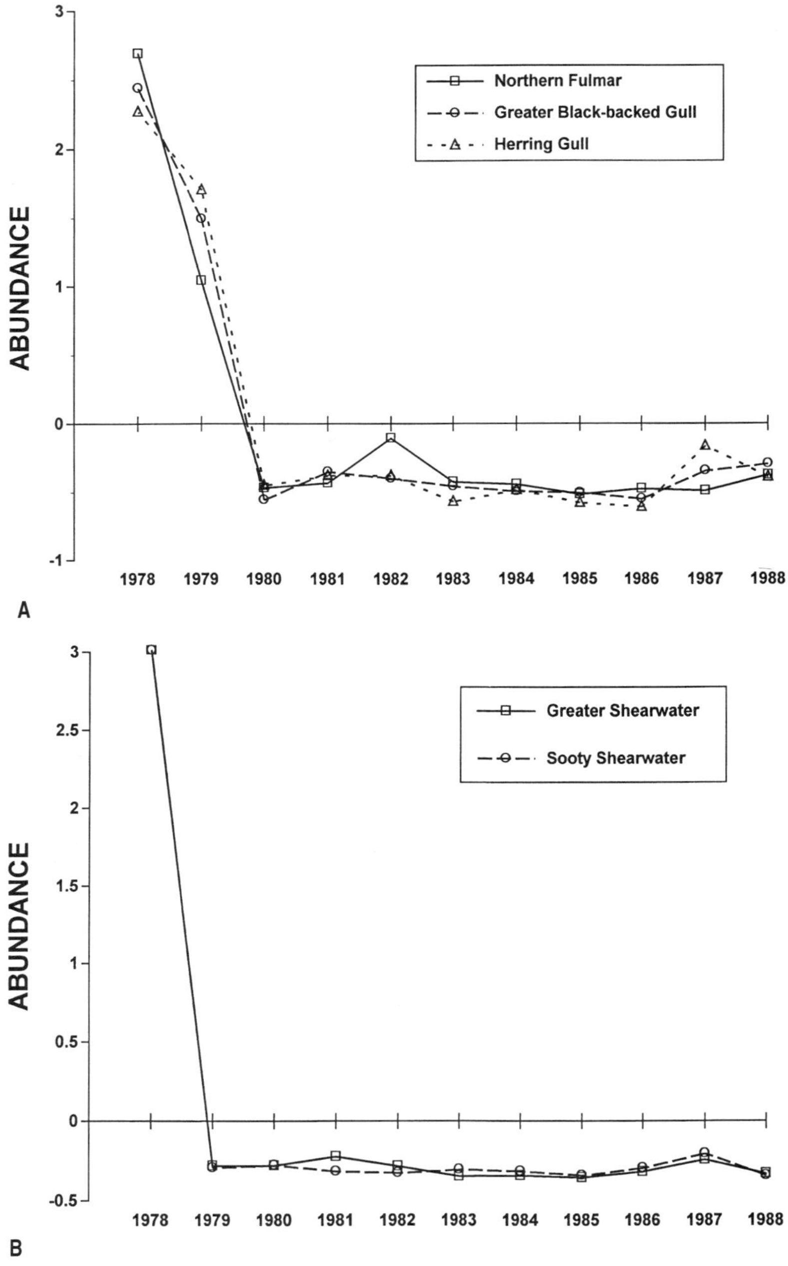

Figure 11-1. Trends in the abundance of seabirds, Gulf of Maine to Cape Hatteras. Units are standard deviations from the 11-year mean (see Tables 11-5, 11-6, and text). A. Northern fulmar, greater black-backed gull, and herring gull. B. Greater shearwater and sooty shearwater.

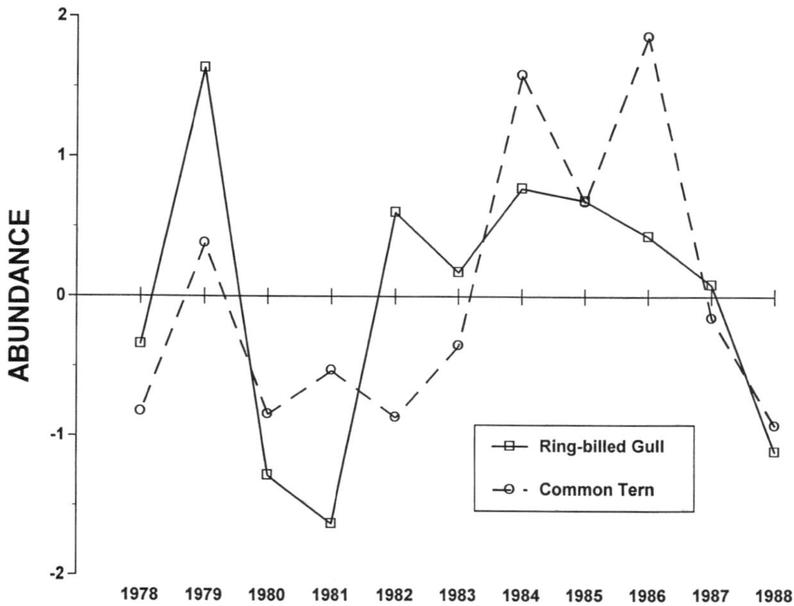

C

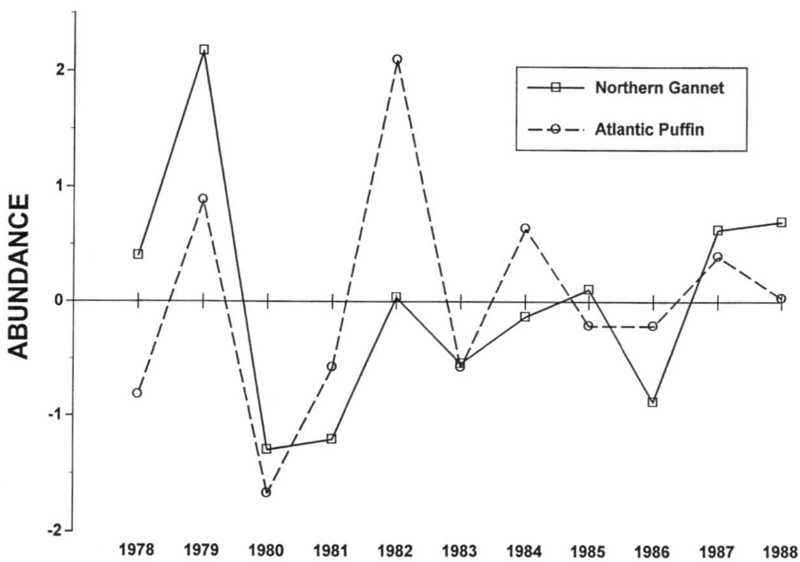

D

Figure 11-1 (continued). C. Ring-billed gull and common tern. D. Northern gannet and Atlantic puffin.

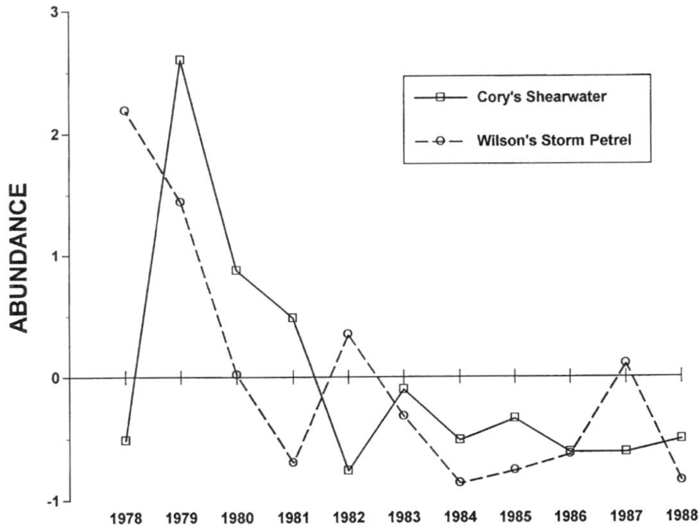

Figure 11-1 (continued). E. Cory's shearwater and Wilson's storm petrel.

eastern Canada (Brown and Nettleship 1984). The most salient hypothesis for puffins is that of annual change in production of juveniles, consistent with the recruitment failure reported for 1981 in the only major puffin colony in the eastern Atlantic (Brown and Nettleship 1982).

Cory's shearwater (which breeds in the North Atlantic) and Wilson's storm petrel (which breeds in the Antarctic) showed similar patterns (Figure 11-1E) of significant decline (see Table 11-6). Both species feed in large numbers in the same area; the most salient hypothesis is that abundance reflects change in prey availability at the sea surface—euphausiids, small fish, and perhaps sand lance *(Ammodytes* spp.) moving toward the sea surface at night.

Black-legged kittiwakes and pomarine jaegers showed similar patterns (Figure 11-1F) of substantial and significant decline (see Table 11-6). This similarity cannot be attributed to the kleptoparasitic activity of the jaeger, which does not occur in exactly the same places as the kittiwake. The most salient hypothesis for kittiwakes is decline because of a series of poor recruitment years.

Common loons and laughing gulls showed peaks in the early part of the time period (Figure 11-1G) with no significant trend (see Table 11-6). Foraging methods and diet of these species differ. The most salient hypothesis is fluctuation because of annual variability in the production of juveniles.

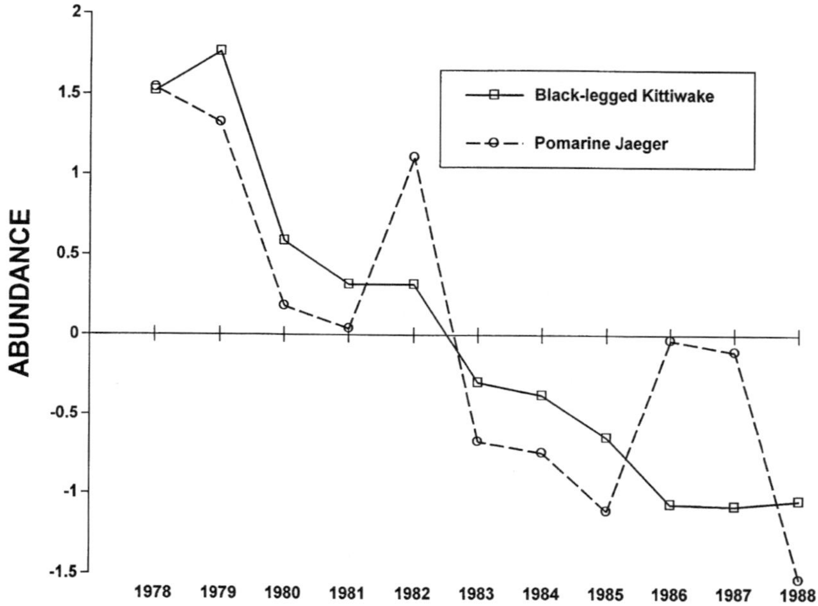

F

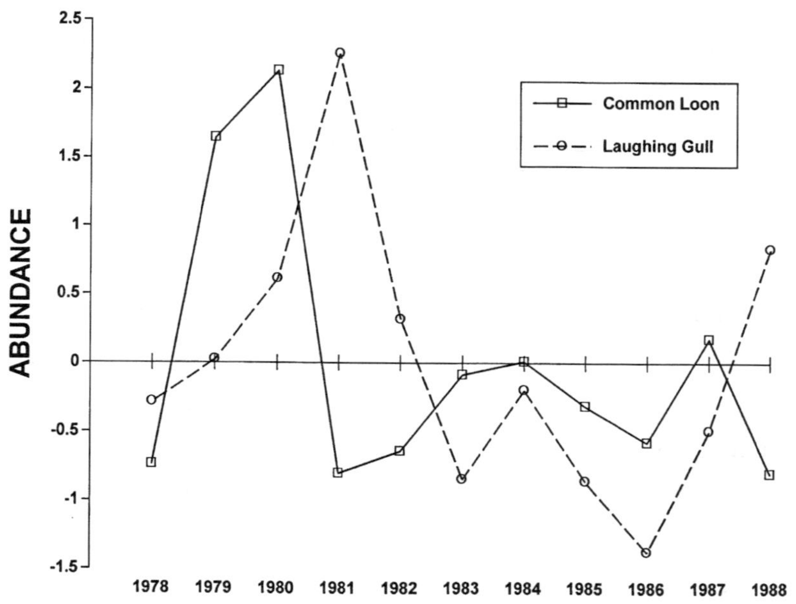

G

Figure 11-1 (continued). F. Black-legged kittiwake and pomarine jaeger. G. Common loon and laughing gull.

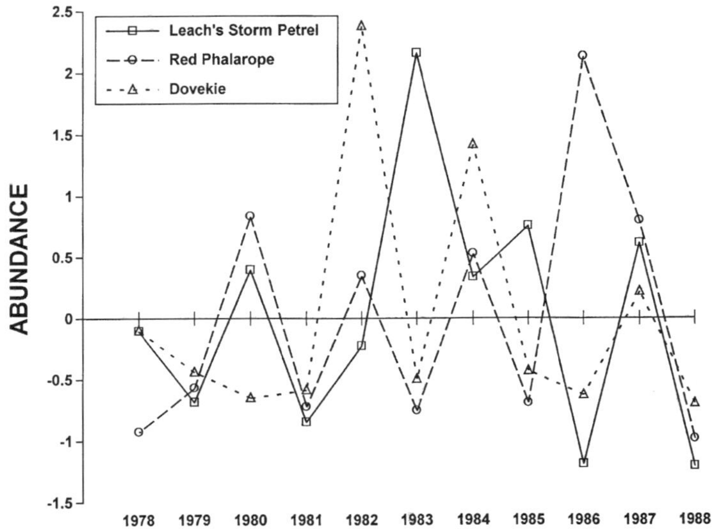

Figure 11-1 (continued). H. Leach's storm petrel, red phalarope, and dovekie.

The numbers of Leach's storm petrel, red phalarope, and dovekie showed marked oscillation from year to year (Figure 11-1H), with no significant longer-term trend (see Table 11-6). All three species rely on zooplankton, either at the surface (petrels, phalaropes) or in swarms beneath the surface (dovekies). The most salient hypothesis is that the observed trends reflect shifts in distribution brought about by annual fluctuations in zooplankton abundance.

Discussion

Three different avian groups have been reviewed: coastally breeding populations, migrant shorebirds, and offshore residents that are mostly nonbreeding. All three groups rely on marine food webs: coastal fish and invertebrates for the breeding birds, intertidal invertebrates for the shorebirds, and small fish, squid, or large zooplankton for the offshore seabirds. The major stresses from ecosystem change differ substantially among the three groups. For the coastally breeding birds, the stresses from ecosystem change are potential: loss of breeding habitat and bioaccumulation of fat-soluble toxins. For the shorebirds, the stresses are also potential: loss of feeding habitat and bioaccumulation of toxins. For the offshore birds, the potential stresses are catastrophic oil spills and ingestion of plastic debris. The numbers of ship-following or ship-attracted species have experienced declines during the decade. The most salient hypothesis for this decline is reduction in the rate of discards and offal released at sea during a period of increasing fishing effort. Substantial changes in marine food webs occurred from 1978 to 1988, but there is little evidence that this altered the abundance of marine birds. The feeding habits (see Table 11-4) of marine birds are highly stereotyped, but the prey taken is diverse. Dietary composition can change substantially from year to year. In light of their dietary flexibility, we hypothesize that marine birds in this ecosystem are more at risk from a drop in total production or availability of edible size classes than from even radical changes in a single prey species.

In this chapter, we have listed what we believe to be the most salient of several competing explanations for observed patterns of change in abundance of pelagic seabirds in this system. These hypotheses are not necessarily correct. They need to be tested. We believe that testing of salient hypotheses, based on substantial research experience with all three avian groups, will be more informative than testing null hypotheses of no change.

All three avian groups are characterized by life histories with low mortality (on the order of 10% yr^{-1} or less) after reaching reproductive age. Recruitment varies substantially from year to year. Because of these life history characteristics, the major stresses on marine bird populations will be those that fluctuate at the scale of decades, rather than those that fluctuate on the order of less than a decade. In the shelf and coastal ecosystem from Cape Hatteras to the Gulf of Maine, direct mortality resulting from drowning in drift nets or hunting is not substantial, as it is in some parts of the world. In this ecosystem, decadal scale increases in the frequency of oil spills, reductions in feeding or nesting habitat, or occurrences of bioaccumulating toxins are likely to be stressful. Decadal scale changes in prey could be important, particularly if the total production or availability were to change.

Based on the life history characteristics of marine birds and on the lack of sources of acute mortality in reproductive-aged birds in this ecosystem, we recommend monitoring at the frequency of decades. In particular, we recommend that all three groups be censused once a decade, with attention to modern techniques of increasing reliability, such as stratified survey designs. The success of decadal monitoring will, of course, depend on the spatial coverage achieved. Furthermore, we recommend that levels of toxins in tissues be measured at the same time. Finally, we recommend that any increase in the frequency of oil spills in any area within this region should immediately trigger a more frequent monitoring program of offshore birds in that area. A decadal survey is, of course, too infrequent to identify potential causes of change in abundance. It is instead a minimum frequency to detect anthropogenic change at time-scales of natural rates of change, which in marine birds are determined by generation times on the order of ten years. By reporting standard deviations (see Table 11-5), we provide the necessary information to test whether estimates from any future surveys deviate significantly from the baseline established in the 1980s.

The need for multispecies management of marine resources has long been recognized because of interactions such as cannibalism, reliance on the same set of prey species, and the incompatibility of single-species regulation. A number of international conservation agreements have adopted ecosystem goals, notably in the Antarctic (Sherman and Ryan 1988) and in the northwest Atlantic, with the adoption of a "second tier" quota (less than the sum of single species quotas) by ICNAF (Anthony and Murawski 1986). Seabirds prey on early life-stages of marine fish and squid. Consequently, a better understanding of mortality and recruitment variability during these early life history stages can be obtained by computing the removal rates for these species by birds. Removal can be calculated from occupancy, daily metabolic requirements, and dietary composition (as percentage of volume or percentage of mass). Seabirds are now being integrated into multispecies management of fisheries in the North Sea through calculation of removals based on diet, occupancy, and energy requirements. Marine birds need to be incorporated into multispecies management in a similar fashion for the western North Atlantic (ICES 1994) and the Bering Sea (Schneider and Shuntov 1994).

Marine birds are valued components of ecosystems, and hence, monitoring of these groups is important. Most of the species in this ecosystem are highly mobile, and many cannot be monitored when they are tied to breeding colonies. As a result, surveys of marine birds in this system, notably the shorebirds and the offshore seabirds, must be extensive. The resources available for direct measurement of avian abundance are necessarily limited. The result is that any survey of marine birds will require substantial scale-up of local measurements. This problem of scale-up is one of the major impediments to understanding the effects of anthropogenic stresses on marine birds, as it is with any widely distributed population. A combination of formal models and experimental verification will be needed to evaluate the effects of large-scale stresses on marine ecosystems, including its avian components.

ACKNOWLEDGMENTS

We thank the Manomet Observatory for permission to use their pelagic survey data. The survey was funded by a series of contracts from the U.S. National Marine Fisheries Service, supplemented by grants and donations to the Observatory. We also thank George Hunt for comments on an earlier draft.

References

Anthony, V. C. 1993. The state of groundfish resources off the northeastern United States. Fisheries 18:12–17.

Anthony V. C., and Murawski, S. A. 1986. Managing Multispecies Fisheries with Catch Quota Regulations—The ICNAF Experience. Cooperative Research Report 139. Copenhagen: International Council for the Exploration of the Sea, Annex 7: pp. 42–57.

Bourne, W. R. P. 1976. Seabirds and Pollution. In R. Johnston (ed.), Marine Pollution (Vol. 6). London: Academic Press, pp. 403–502.

Brown, R. G. B., and Nettleship, D. N. 1982. Capelin and seabirds in the northwest Atlantic. In D. N. Nettleship, G. A. Sanger, and P. F. Springer (eds.), Marine Birds—Feeding Ecology and Commercial Fisheries. Ottawa: Canadian Wildlife Service, pp. 184–194.

Brown, R. G. B., and Nettleship, D. N. 1984. The Seabirds of Northeastern North America: Their Present Status and Conservation Requirements. In J. P. Croxall, P. G. H. Evans, R. W. Schreiber (eds.), Status and Conservation of the World's Seabirds. ICBP Technical Publication 2. pp. 85–100.

Buckley, P. A., and Buckley, F. G. 1981. The endangered status of North American Roseate Terns. Colonial Waterbirds 4:166–173.

Buckley, P. A., and Buckley, F. G. 1983. Conservation of colonial waterbirds. Oceanus 26:55–61.

Buckley, P. A., and Buckley, F. G. 1984. Seabirds of the North and Middle Atlantic Coast of the United States: Their Status and Conservation. In J. P. Croxall, P. G. H. Evans, R. W. Schreiber (eds.), Status and Conservation of the World's Seabirds. Cambridge, UK: International Council for Bird Preservation Technical Publication 2. pp. 101–133.

Clark, S. H., and Brown, B. E. 1977. Changes of biomass of finfishes and squids from the Gulf of Maine to Cape Hatteras, 1963–74, as determined from research vessel survey data. Bull. U.S. Bur. Fish. 75:1–21.

Colonial Waterbirds. 1981. Symposium on Disturbance. Colonial Waterbirds 4:1–70.

Drury, W. 1973–1974. Population changes in New England seabirds. Bird-banding 44:267–313; 45:1–15.

Erwin, R. M. 1979. Coastal Waterbird Colonies: Cape Elizabeth, Maine to Virginia. Contract Report FWS/OBS 79/10. Washington, DC: U.S. Fish and Wildlife Service.

Ferns, P. N. 1993. Bird Life of Coasts and Estuaries. Cambridge, UK: Cambridge University Press.

International Council for Exploration of the Sea (ICES). 1994. Report of the Study Group on Seabird/Fish Interactions. ICES C.M.1994/L:3. Copenhagen: International Council for the Exploration of the Sea.

Manly, B. J. F. 1991. Monte Carlo Methods in Biology. London: Chapman Hall.

Monaghan, P., Uttley, J. D., and Okill, J. D. 1989. Terns and sandeels: Seabirds as indicators of changes in marine fish populations. J. Fish. Biol. 35:339–340.

Murawski, S. A. 1991. Can we manage our multispecies fisheries? Fisheries 16:5–13.

National Academy of Science (NAS). 1990. Assessment of the U.S. Outer Continental Shelf Environmental Studies Program. II. Ecology. Washington, DC: National Academy of Science Press.

Nisbet, I. C. T. 1980. Effects of toxic pollutants on productivity in colonial waterbirds. Trans. Linn. Soc. N.Y. 9:103–113.

Piatt, J. F., and Lensink, C. J. 1990. Exxon Valdez bird toll. Nature 342:865–866.

Powers, K. D. 1983. Pelagic Distribution of Marine Birds Off the Northeastern United States. NOAA Tech. Memo. NMFS-F/NEC-27. Washington, DC: U.S. Department of Commerce.

Powers, K. D., and Backus, E. H. 1987. Energy Transfer to Birds. In R. H. Backus and D. W. Bourne (eds.), Georges Bank. Cambridge, MA: MIT Press, pp. 372–375.

Powers, K. D., and Brown, R. G. B. 1987. Seabirds. In R. H. Backus and D. W. Bourne (eds.), Georges Bank. Cambridge, MA: MIT Press, pp. 359–371.

Powers, K. D., and Rumage, W. T. 1978. Effect of the 'Argo Merchant' oil spill on bird populations off the New England coast, 15 December 1976–January 1977. In: In the Wake of the 'Argo Merchant.' Proceedings of a Conference and Workshop, January 1978. Kingston, RI: Center for Oceanographic Management Studies, University of Rhode Island.

Rothstein, S. 1973. Plastic particle pollution of the surface of the Atlantic Ocean: Evidence from a seabird. Condor 75:344–366.

Schneider, D. C., and Harrington, B. A. 1982. Timing of shorebird migration in relation to food depletion. Auk 98:801–811.

Schneider, D. C., and Hunt, G. L. 1982. Carbon flux to seabirds in waters with different mixing regimes in the southeastern Bering Sea. Mar. Biol. 67:337–344.

Schneider, D. C., Hunt, G. L., and Powers, K. D. 1987. Energy flux to pelagic birds: A comparison of Bristol Bay (Bering Sea) and Georges Bank (northwest Atlantic). In J. P. Croxall (ed.), Seabirds—Feeding Biology and Role in Marine Ecosystems. Cambridge, UK: Cambridge University Press, pp. 259–277.

Schneider, D. C., and Shuntov, V. P. 1994. The trophic organization of the marine bird community in the Bering Sea. Rev. Fish. Sci. 1:311–335.

Sherman, K., Grosslein, M., Mountain, D., et al. 1988. The continental shelf ecosystem off the northest coast of the United States. In H. Postma and J. J. Ziljstra (eds.), Ecosystems of the World 27: Continental Shelves. Amsterdam: Elsevier, pp. 279–337.

Sherman, K., and Ryan, A. F. 1988. Antarctic marine living resources. Oceanus 31:59–62.

12

Zooplankton Prey Field Variability During Collapse and Recovery of Pelagic Fish in the Northeast Shelf Ecosystem

Kenneth Sherman, John Green, Andrew Solow, Steven A. Murawski, Joseph Kane, Jack Jossi, and Wallace Smith

Abstract

Since the mid-1970s, three pelagic fish species—herring (Clupea harengus), mackerel (Scomber scombrus), and sand eel (Ammodytes spp.)—have undergone significant changes in biomass. The herring and mackerel stocks that declined from dominant to subordinate positions in the 1970s recovered to a present combined biomass exceeding 5 million metric tons. Sand eels, whose population levels exploded in the late 1970s and early 1980s, are now in decline. The effects of changes in biomass of several million metric tons among these three species of zooplanktivores are examined in relation to possible density-dependent variability in zooplankton and pelagic fish within the Northeast Shelf ecosystem.

Changes in Fish Abundance Within the Ecosystem

In an effort to reduce scientific uncertainties in fisheries management, over the past few years increasing attention has been focused on synthesizing available information on factors influencing the natural productivity of marine fish biomass from an ecosystems perspective (Sherman et al. 1993). One of the results of this effort has been a series of reports on the large-scale changes in fishery biomass yields of the Northeast Shelf ecosystem during the past 30 years (Sissenwine 1986; Murawski 1991; Anthony 1993).

From 1963 to the present, highly valued groundfish species (gadoids, flounders) have declined to unprecedented low levels. Other demersal species with low market value [spiny dogfish (*Squalus acanthias*) and skates*] have increased in abundance (Figure 12-1). Other species, the principal pelagics [Atlantic herring (*Clupea harengus*) and Atlantic mackerel (*Scomber scombrus*)], have undergone precipitous declines for the decade 1968 through 1978, followed by population recoveries from the 1980s to the present time (see Figure 12-1). Their recovery is attributed to reduced fishing effort (Murawski 1991). The observed biomass "flip" of the principal pelagics from high levels in the 1960s to historically low levels in the 1970s and then to the present high biomass has been the dominant factor in the population explosion of sand lance (*Ammodytes* spp.) during the latter half of the 1970s (Sherman et al. 1981), and their precipitous four-year decline in abundance from 1982 to 1985. Based on predator–prey studies, it was concluded that predation on sand lance in their early, planktonic developmental stages by the principal pelagics depressed the population of sand lance, after a period of approximately six years (1976–1982) when the predation pressure was significantly lower (Fogarty et al. 1991) (Figure 12-2).

Long-Term Trends in Herring and Mackerel Abundance

The principal pelagics and sand lance are zooplanktivores (Bowman et al. 1984; Maurer 1976). The long-term inverse relationship in abundance of the principal pelagics as discussed by Skud (1982) appears to have shifted during the past 30 years from the trends in a 90-year time-series of landings data examined for 1870 through 1960. Based on an analysis of herring and mackerel catches in relation to temperature trends, Skud (1982) concluded that when one of the species was dominant, the other was subordinate in response to a combination of temperature- and density-dependent competition between the two species. He argued that density (abundance) of the two species is "governed by the carrying capacity of the ecosystem and [that] the species' apparent response to climate change is dependent on its position in

*Species of skates include little skate (*Raja erinacea*), winter skate (*R. ocellata*), barndoor skate (*R. laevis*), thorny skate (*R. radiata*), briar skate (*R. eglanteria*), leopard skate (*R. garmani*), and smooth-tailed skate (*R. senta*).

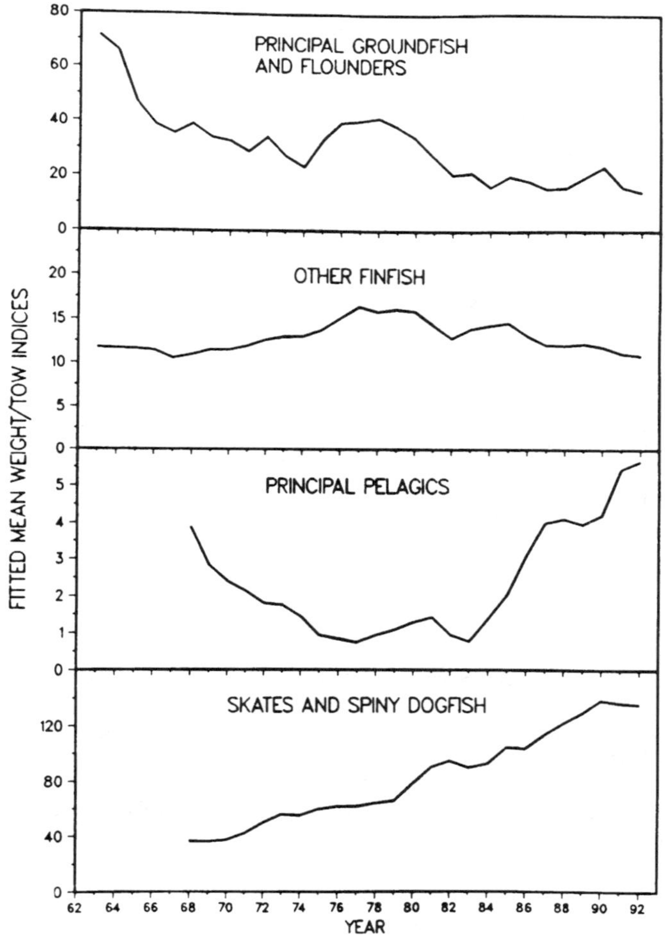

Figure 12-1. Trends in indices of aggregate abundance (catch in weight per survey trawl haul) for four species groups, reflecting the major changes in fishery resources, 1962–1992. (Reprinted from Northeast Fisheries Science Center. 1993. Status of the Fishery Resources Off the Northeastern United States for 1993. NOAA Tech. Mem. NMFS-F/NEC-101. Washington, DC: National Oceanic and Atmospheric Administration.)

the dominance hierarchy and its population size relative to its equilibrium density." Since 1960, this apparent relationship has been modified by the high levels of fishing mortality from the late 1960s through the mid-1970s for Atlantic herring and mackerel (Murawski 1991). Since 1982, the depressed mackerel stock has been undergoing significant rebuilding from an estimated stock biomass level of approximately 650,000 metric tons (mt) in 1982 to 2.8

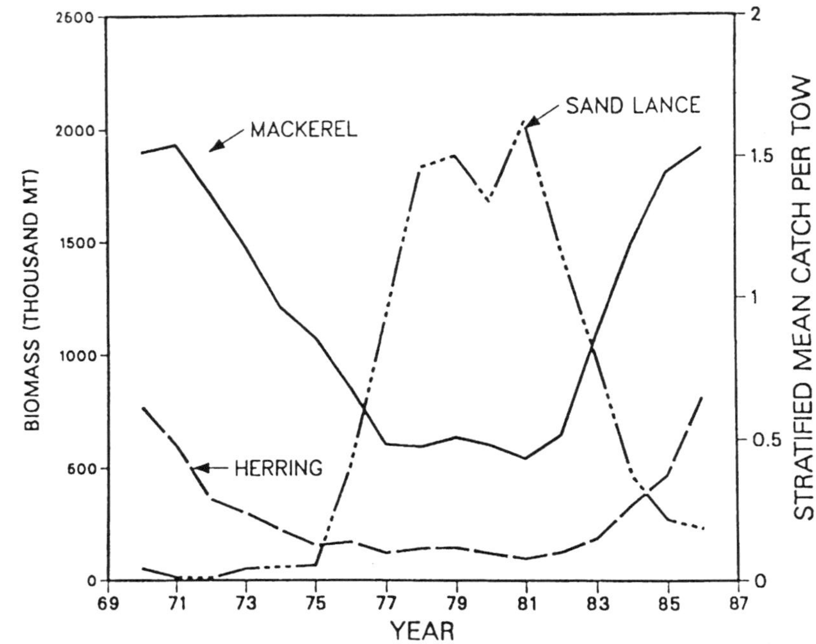

Figure 12-2. Trends in biomass of mackerel (age 1+) and herring (age 3+) derived from virtual population analysis, and trends in relative abundance (stratified mean catch in kilograms per tow) of sand lance (age 2+) based on research vessel surveys. (Reprinted from M. Fogarty, E. Cohen, W. Michaels, and W. Morse. 1991. Interactions Among Herring, Mackerel, and Sand Lance in the Northwest Atlantic. ICES Mar. Sci. Symp. Multispecies Models Relevant to Management of Living Resources. Vol. 193. Copenhagen: International Council for the Exploration of the Sea, pp. 120–124.)

million metric tons (mmt) in 1992. The rebuilding of the herring stock during the same decade has progressed from less than 100,000 mt to 2.7 mmt. Neither species can be considered subordinate. Both have undergone recovery and are simultaneously at high abundance levels (Figures 12-3 and 12-4) with, as yet, no apparent adverse impact on the "carrying capacity" of the ecosystem, as suggested by Skud (1982).

Zooplankton Composition and Abundance

The zooplankton of the Northeast Shelf ecosystem has been examined for broad-area spatial and temporal trends in abundance and species composition since the mid-1960s (Sherman et al. 1983, 1988; Kane 1993; Jossi and Goulet

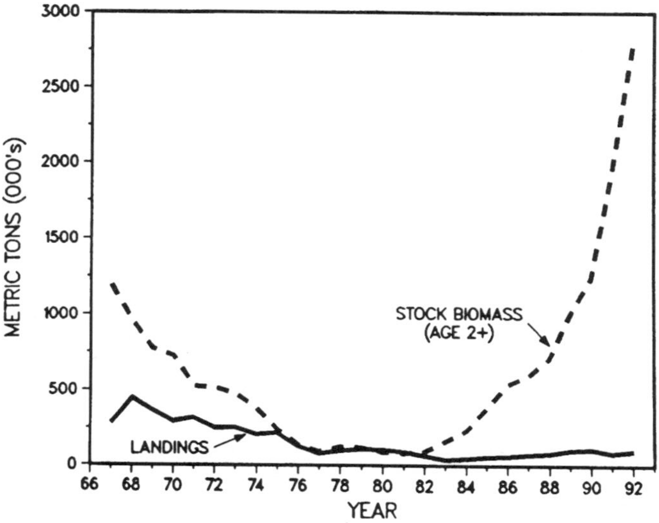

Figure 12-3. Atlantic herring commercial landings and stock biomass, 1967 through 1992 (thousand metric tons). (Reprinted from Northeast Fisheries Science Center. 1993. Status of the Fishery Resources Off the Northeastern United States for 1993. NOAA Tech. Mem. NMFS-F/NEC-101. Washington, DC: National Oceanic and Atmospheric Administration.)

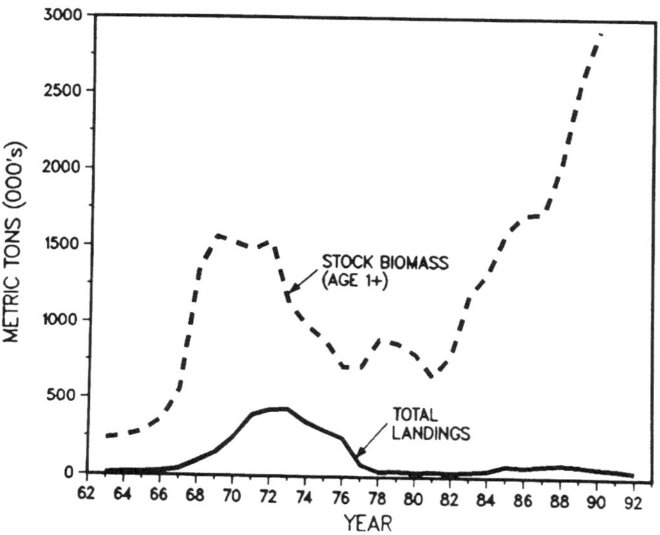

Figure 12-4. Atlantic mackerel commercial landings and stock biomass, 1963 through 1992 (thousand metric tons). (Reprinted from Northeast Fisheries Science Center. 1993. Status of the Fishery Resources Off the Northeastern United States for 1993. NOAA Tech. Mem. NMFS-F/NEC-101. Washington, DC: National Oceanic and Atmospheric Administration.)

1993), and during four earlier periods. The first was the classic measurements of zooplankton volumes and species made by Bigelow (1926) during the first two decades of the century (1912 through 1920), and the second, the volume measurements made by Bigelow and Sears (1934) in the late 1920s through the 1940s. The third covered the late 1930s to the 1960s, with the biomass and species demographic studies of Fish (1936a,b), Clarke and Zinn (1937), Clarke (1940), Redfield (1941), Clarke et al. (1943), Riley and Bumpus (1946), Deevey (1952, 1956, 1960a,b), and Grice and Hart (1962). The fourth includes the more contemporary measurements of Judkins et al. (1980), Dagg and Turner (1982), Davis (1984a,b), Townsend and Cammen (1988), and the recent study of Durbin and Durbin (see Chapter 8), who examined the regulatory processes of the Northeast Shelf zooplankton (e.g., temperature, food, and predation).

During the decade 1977 to 1986, samples were collected using 0.333-mm mesh bongo nets towed obliquely through the water column to a maximum depth of 200 m. Tows of 15-minutes duration at 1.5 knots were made at 200 sampling stations in a grid network at 35-km spacing within the entire ecosystem, six or more times per year (Sibunka and Silverman 1984, 1989) (Figure 12-5). In addition to point-samplings with nets, the Northeast Fisheries Science Center (NEFSC) staff has completed monthly continuous plankton recorder (CPR) transects across the Gulf of Maine since 1961. This transect series has been augmented by a second transect from New York to Bermuda, across the Middle Atlantic Bight of the Northeast Shelf ecosystem, since 1976 (Jossi and Goulet 1993).

The total zooplankton measured as displacement volumes during the Marine Resource Monitoring Assessment and Prediction Program (MARMAP) surveys of the shelf from 1977 to 1981 did not differ significantly from the earlier estimates of zooplankton standing stock of the shelf ecosystem from 1912 through the 1960s (Sherman et al. 1988). The zooplankton composition included 394 taxa, with 50 dominant in at least one location in one or more seasons, including copepods, chaetognaths, barnacle larvae, cladocerans, appendicularia, doliolids, brachyuran larvae, echinoderm larvae, and thaliaceans (Sherman et al. 1988).

Ecosystem Level Response to Pelagic Perturbations

The more recent observations of zooplankton volumes on the shelf, reported by Kane (1993) for the Georges Bank area of the Shelf ecosystem and Jossi and Goulet (1993) for the Gulf of Maine and the Middle Atlantic Bight, reveal

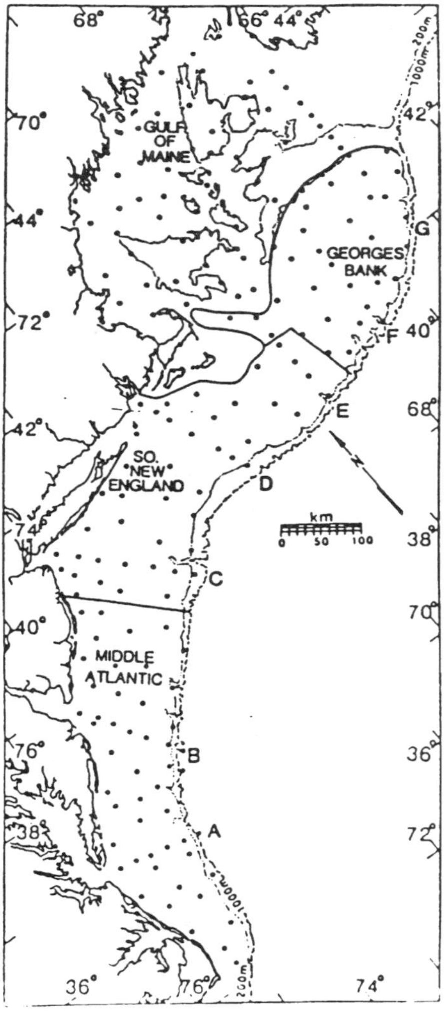

Figure 12-5. Zooplankton sampling locations on MARMAP surveys of the Northeast Continental Shelf ecosystem.

interannual variability in zooplankton abundance. Examination of the CPR data for the transect across the Middle Atlantic Bight from 1976 through 1990 revealed a declining trend for the MARMAP decade of 1977–1986 ($p = 0.026$). However, the interannual variability observed from net tows for the entire time-series was not statistically significant as a declining trend ($p = 0.30$) (Figure 12-6). The numbers of zooplankton on the northern CPR transect across the Gulf of Maine generally increased in abundance from 1961 through 1990 ($p = 0.054$), interrupted by a downward trend during the MARMAP

Figure 12-6. Continuous plankton recorder trend data for the transect across the Middle Atlantic Bight, 1976–1990. Rank correlation test for trend; p = 0.026, 1977–1986; p = 0.30, 1976–1986. The ordinate values represent $\log_e$ transformed numbers of total zooplankton/100 m^3 of water strained/year.

decade ($p = 0.06$) (Figure 12-7). Based on MARMAP net sampling for Georges Bank, Kane reported volumes higher than the 10-year median biomass for 1977, 1978, and 1979, and lower for 1982, 1983, and 1984. According to Kane, the higher volumes may have contributed to the increases in sand lance (*Ammodytes* spp.) during the late 1970s. In contrast, Payne et al. (1990) reported an inverse relationship between sand lance and the copepod, *Calanus finmarchicus*, during the population explosion of sand lance, suggesting density-dependent grazing control of the standing stock of zooplankton. This observation, however, was limited to a relatively restricted area of the Northeast Shelf ecosystem known as the Stellwagen Bank area (Payne et al. 1990). On Georges Bank, where sand lance, herring, and mackerel are present in relatively high numbers in spring, zooplankton biomass in spring (measured as displacement volume per 100 m^3) was abundant during the population explosion of sand lance, 1971–1981, with median annual volumes equal to or exceeding the 22-year (1971 through 1992) long-term median value of 43 cc/100 m^3 (Figure 12-8).

Examination of the temperature information for the Northeast Shelf ecosystem during the last two decades indicates that the temperatures during 1978 to 1982, a period of population depression for the principal pelagics, was

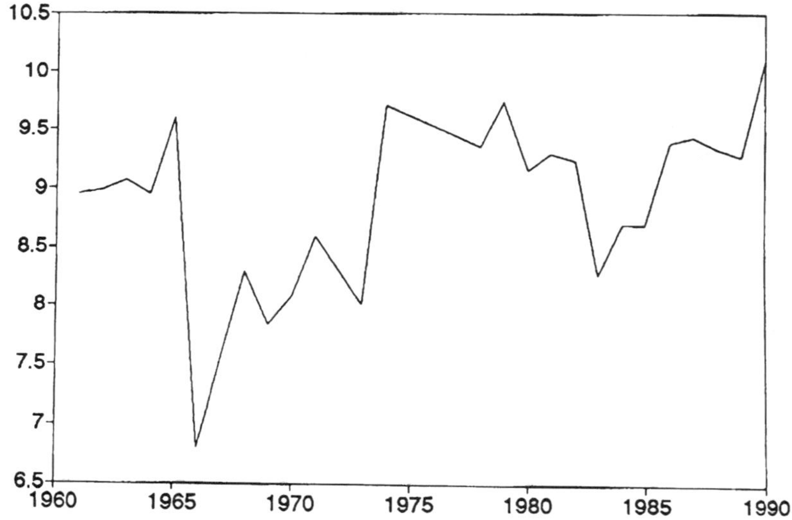

Figure 12-7. Continuous plankton recorder trend data for the transect across the Gulf of Maine, 1961–1990. Rank correlation test for trend; $p = 0.054$, 1961–1990; $p = 0.062$, 1977–1986. The ordinate values represent $\log_e$ transformed numbers of total zooplankton/100 m^3 of water strained/year.

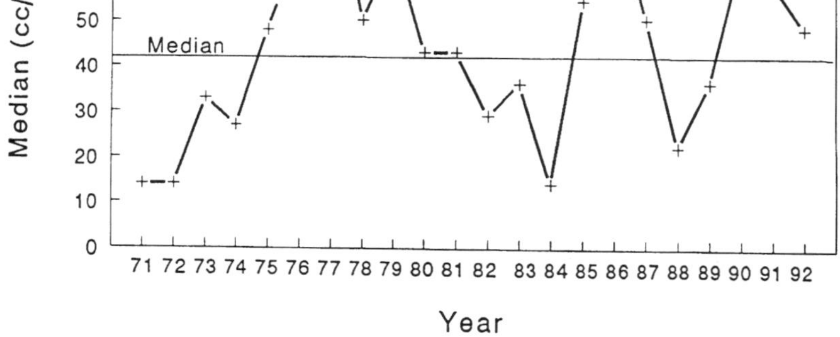

Figure 12-8. Annual zooplankton median volumes (cc/100 m^3) for Georges Bank during early spring, 1971–1992, with the 22-year median for the total time series; N = 733. Data for 1976 are not available.

intermediate between a slightly warmer period of the mid-1980s (1983–1986) and moderately cooler waters of the mid-1960s (Holzwarth and Mountain 1992). The influence of temperature on the survival of early developmental stages of sand lance, mackerel, and herring is not well understood and requires further study. The positive and negative correlations between temperatures and herring and mackerel dominance reported by Skud (1982) are not supported by the present study, wherein both species show population declines and increases attributed to excessive fishing effort in the 1970s and included sharp declines in the economically important demersal species during the same period (Anthony 1993; Murawski 1991). Whereas excessive fishing mortality has persisted for the demersals, it has been significantly reduced for the principal pelagic species and is considered the principal factor in their recovery (Murawski 1991). The lowest period of larval herring abundance from 1971 through 1992 occurred from 1976 to 1984, spanning the intermediate and warm temperature periods. This is in contrast to the warming and cooling correlation and analysis of changes in herring distribution reported by Skud (1982), and lends support to the predominant influence of excessive fishing mortality on the decline of both herring and mackerel abundance (Figure 12-9). The influence of temperature on the interannual variability of a dominant zooplankter, the copepod *Calanus finmarchicus*, on Georges Bank is suggestive of a relationship (Meise-Munns et al. 1990). Further studies are presently under way to examine more closely the relationship between seasonal and interannual environmental variability and changes in zooplankton production (GLOBEC 1991, 1992).

Earlier observations indicate that the zooplankton component of the Northeast Shelf ecosystem has been relatively coherent in resilience, abundance, and biodiversity from the first decade of the century to the 1980s (Sherman et al. 1988). Based on the present analysis of nearly 10,000 zooplankton volume measurements (9,942) for the MARMAP decade (1977 through 1986), we find that the zooplankton was in a declining trend during the 1980s. The declining trend is highly significant for the whole ecosystem (Figure 12-10A), and for three of the four subareas, the trend is significant at the level of $p < 0.05$ (Figure 12-10B, C, and D). The declining trend for Georges Bank was more variable than for the other three subareas ($p = 0.10$) (Figure 12-10E). The persistent downward trend coincided with the increase in abundance level of the principal pelagics and the increase in biomass of another pelagic zooplanktivore, the butterfish (*Peprilus triacanthus*) (Figure 12-11). During the years 1977, 1978, and 1979, the highest numbers of zooplankters, based on CPR sampling (see Figure 12-6), coincided with the first three years of the seven-year depressed state of the herring and mackerel bio-

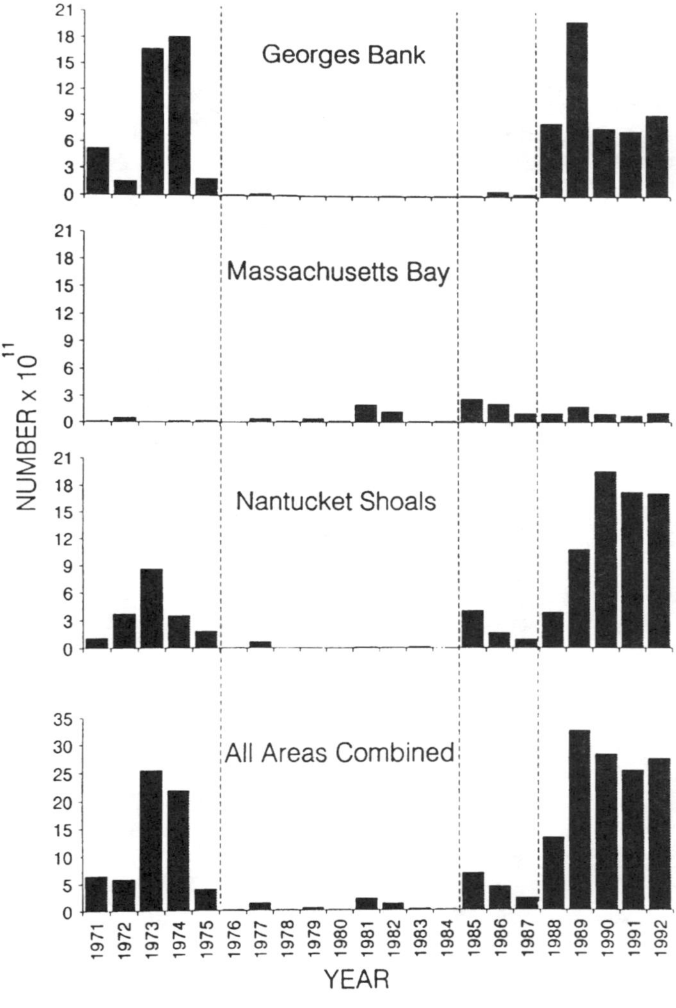

Figure 12-9. Changes in abundance of Atlantic herring Clupea harengus larvae in the Georges Bank area between 1971 and 1990. Intervals represent multiyear periods that reflect the most apparent changes in spawning patterns.

mass (see Figures 12-3 and 12-4). From a top-down examination of the ecosystem, the zooplankton component has been sufficiently robust, even through the period of declining trend, to sustain the recovery of both herring and mackerel to a level of 5.5 mmt. In addition to this biomass, the zooplankton prey field is needed to support the recovery of the depressed gadoid and flounder stocks during their larval stages, posing another significant biomass to be considered

Figure 12-10. Mean annual volumes in cc/100 m^3 for the Northeast Continental Shelf ecosystem, 1977–1986. Data were combined for the four subareas—Gulf of Maine, Southern New England, Middle Atlantic Bight, and Georges Bank—based on a rank correlation test for trend for the area weighted total (A), p = 0.001. The results of the test are given separately for each of the four subareas: (B) Gulf of Maine, p = 0.006; (C) Southern New England, p = 0.006; (D) Middle Atlantic Bight, p = 0.040; and (E) Georges Bank, p = 0.130.

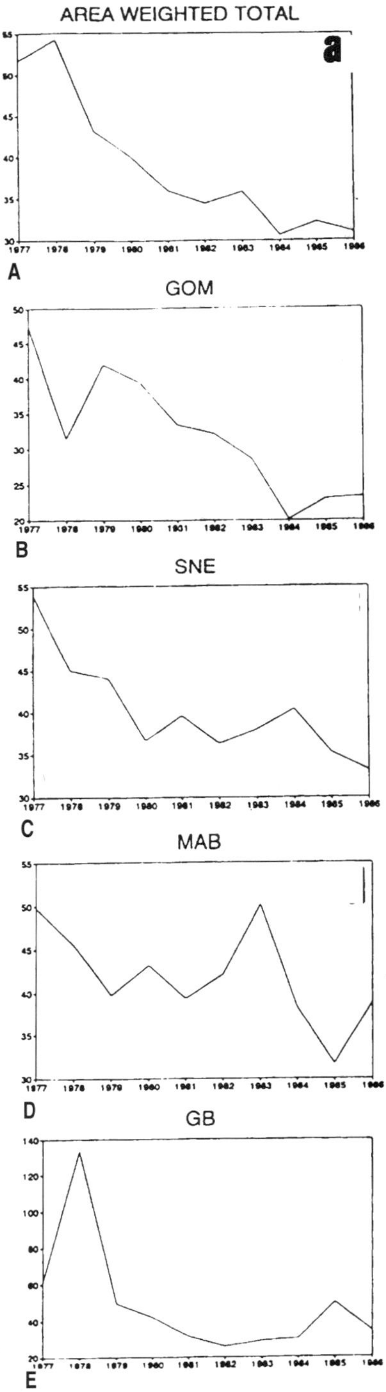

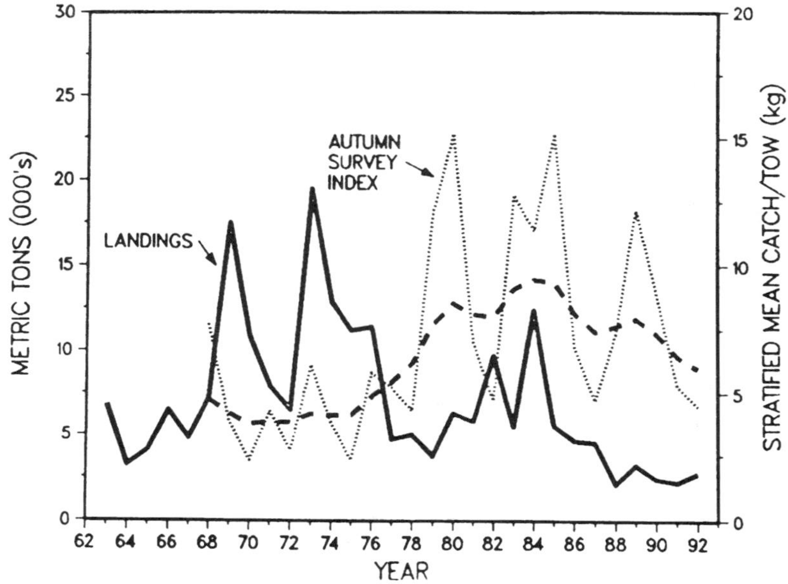

Figure 12-11. Commercial catches of butterfish for the Northeast Continental Shelf ecosystem (1,000 metric tons) and stratified mean catch per tow (kilograms) based on the autumn bottom trawl survey index, Northeast Fisheries Science Center, 1963–1992. (Reprinted from Northeast Fisheries Science Center. 1993. Status of the Fishery Resources Off the Northeastern United States for 1993. NOAA Tech. Mem. NMFS-F/NEC-101. Washington, DC: National Oceanic and Atmospheric Administration.)

in estimating the carrying capacity of the ecosystem. Whether the decline reflects a response to increasing predation by the growing biomass of the pelagic fish species, or a biofeedback response to an environmental signal remains an important unanswered question. From the spring and autumn volume data, based on bongo tows over Georges Bank from an extended data set encompassing 1971 through 1992, it appears that zooplankton standing stock may again be increasing (Figure 12-12A and B).

For insights to the possibility of an environmentally induced biofeedback, the zooplankton time series was examined for shifts in biodiversity using the annual time-series of log abundance of five zooplankton species—*Calanus finmarchicus*, *Pseudocalanus* spp., *Centropages typicus*, *Metridia lucens*, and *Centropages hamatus*—for four regions: Georges Bank, Gulf of Maine, Middle Atlantic Bight, and Southern New England, for the period 1977–1987. Annual time-series were constructed by averaging all samples within a year and also by taking the maximum sample within a year. In terms of temporal behavior,

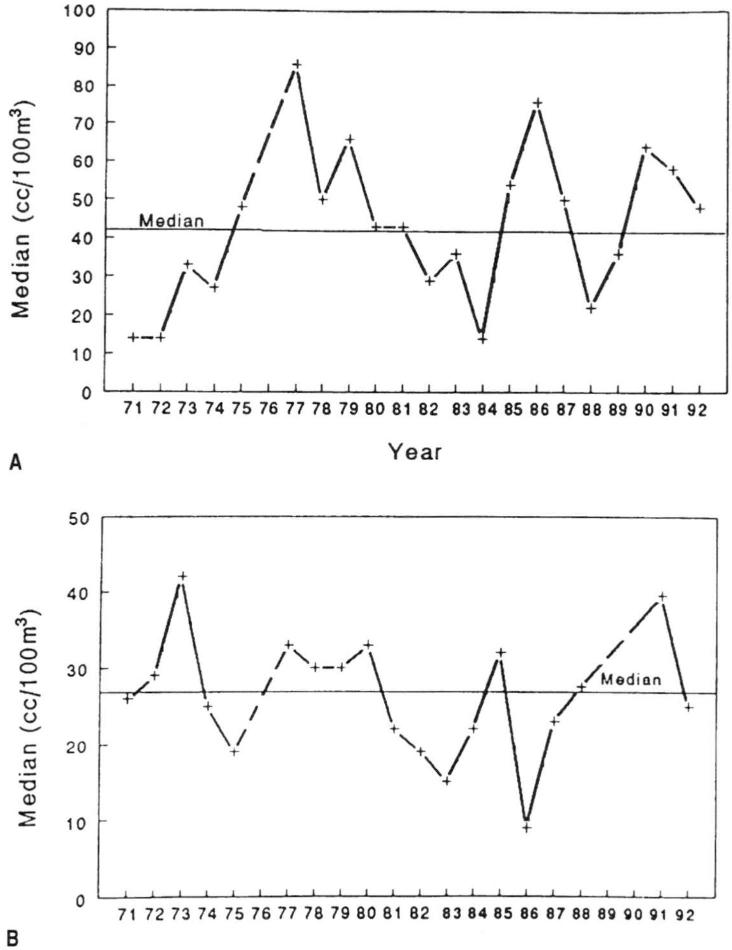

Figure 12-12. The annual median zooplankton standing stock (displacement volume/100 m^3) for (A) early spring, 1971 to 1992, and (B) autumn, 1971 to 1992, based on bongo net tows made during spring and fall bottom trawl surveys of the Northeast Continental Shelf ecosystem. Data missing for 1976, 1989, and 1990.

there were no real differences between these two approaches, so the average values were analyzed. The methods described by Solow (1994) were used to extract an overall trend from the five species for each of the four regions. There was no significant trend for the Middle Atlantic Bight ($p = 0.785$) or Southern New England ($p = 0.128$). There were significant trends for Georges Bank ($p = 0.031$) and the Gulf of Maine ($p = 0.014$). The trends, shown in Figure 12-13,

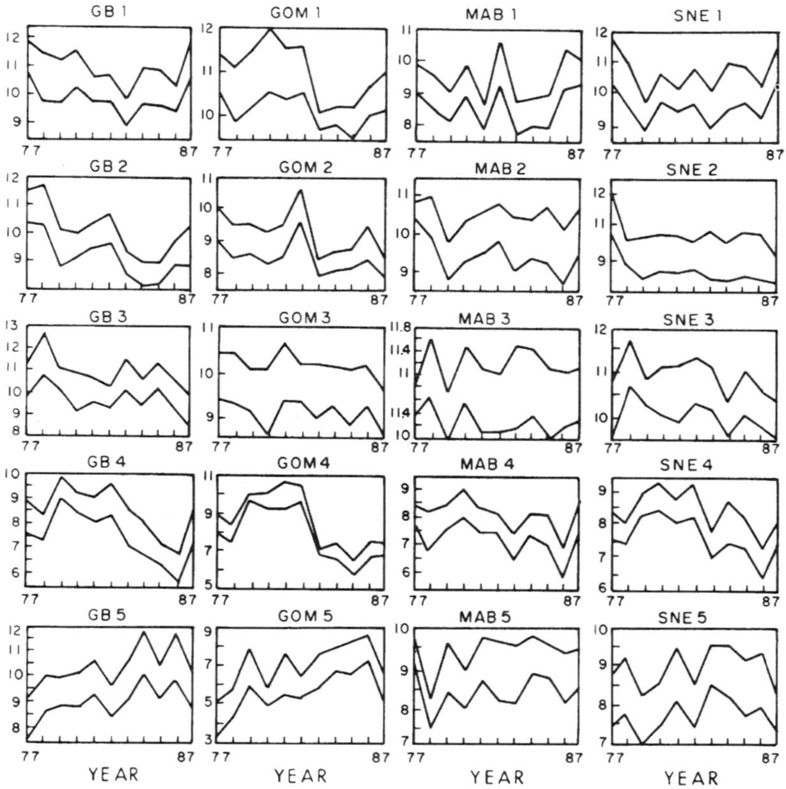

Figure 12-13. Abundance trends for five species of dominant zooplankton species—Calanus finmarchicus (1), Pseudocalanus spp. (2), Centropages typicus (3), Metridia lucens (4), and Centropages hamatus (5)—within four regions of the Northeast Shelf ecosystem—Georges Bank (GB), Gulf of Maine (GOM), Middle Atlantic Bight (MAB), Southern New England (SNE)—for the period 1977 through 1987. Annual time-series are based on averaging all samples within one year. Plots show relationship between $\log_e$ transformed sample means (lower trend line) and $\log_e$ transformed maximum sample abundances (upper trend line) for each year within the time-series. Number designations for each species are given in parentheses.

are highly correlated. The correlations between these trends and the five species are given in Table 12-1. From an examination of biodiversity, it appears that the dominant copepod community is undergoing a shift from *Calanus finmarchicus, Pseudocalanus* spp., *Metridia lucens*, and *Centropages typicus* toward an increasing abundance of *Centropages hamatus* on Georges Bank and in the Gulf of Maine (Figure 12-14). This apparent shift in species abundance may be environmentally induced, as it is unlikely that the principal pelagics would selec-

Table 12-1. Correlations between the five copepod species and the time trends found by the minimum/maximum autocorrelation factor analysis for Georges Bank and the Gulf of Maine subareas of the Northeast Shelf ecosystem

Species	Georges Bank	Gulf of Maine
Calanus finmarchicus	–0.32	–0.64
Pseudocalanus spp.	–0.82	–0.54
Centropages typicus	–0.44	–0.35
Metridia lucens	–0.69	–0.77
Centropages hamatus	0.70	0.76

Source: A. R. Solow. 1994. Detecting change in the composition of a multispecies community. Biometrics 50:556–565.

tively avoid preying on *C. hamatus*. The ecosystem monitoring program presently under way by the Northeast Fisheries Science Center is designed to provide more definitive information on density-dependent predation between the principal pelagics and their prey field, and density-independent environmental signals that may be the cause of apparent changes in the dominance structure of the zooplankton community. Predator–prey studies being conducted as part of the National Oceanic and Atmospheric Administration's Coastal Ocean Program are examining the fine-scale temporal and spatial interrelation between the principal pelagics and their prey-field.

The present state of the ecosystem remains stressed. The shift since the 1960s from a demersal assemblage of fish dominated by gadoids to the current one, dominated by spiny dogfish and skates and coinciding with the increased abundance of the pelagic fish species (see Figure 12-1), raises an ecological question. How will fishing effort reduction measures initiated in 1994 (New England Fishery Management Council 1994) affect the process of recovery from the depressed state of highly valued demersal species (e.g., cod, haddock, flounders)? The consequence of reduced biomass of gadoids and flounders has been an increase in fisheries targeting abundant elasmobranch species. Nevertheless, these species are at historically high levels. Implications of the removal of an increasing fraction of the elasmobranch biomass on other fish components of the ecosystem will be closely monitored. The ecosystem level consequences associated with this adaptive management strategy have been reviewed by Sissenwine and Cohen (1991), who list several alternative species that could become major fish predators. With regard to the zooplankton component of the system, the changed state of the fish community has introduced uncertainty to any forecast of recovery for the demersal populations of cod, haddock, and flounders that are dependent on the zooplankton prey field during their early developmental stages. Unlike the gadoids and flounders, the present high bio-

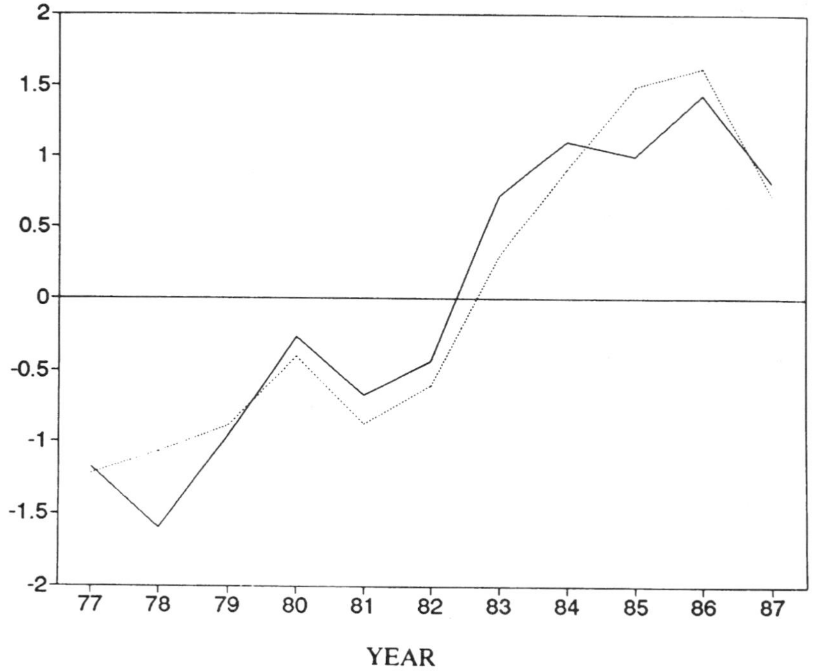

Figure 12-14. Minimum-maximum autocorrelation factor (MAF) plot (standardized to mean 0 and variance 1) showing the monotone trend toward Centropages hamatus for Gulf of Maine (dotted line) and Georges Bank (solid line) from 1977 to 1987. Randomization test probability values calculated following Solow (1994) were p = 0.031 for Georges Bank, and p = 0.014 for the Gulf of Maine. The trends for the Southern New England (p = 0.128) and Middle Atlantic Bight (p = 0.785) subareas were not significant and are not plotted.

mass of elasmobranchs is independent of the zooplankton prey field during early development. The estimated biomass level of herring and mackerel before the 1960s is not well-documented and is likely, according to data presented by Skud (1982), to have been less than the present estimated level of 5.5 mmt. The questions posed by Skud with regard to the carrying capacity of the ecosystem will require further study as the Northeast Shelf undergoes significant alteration to an ecological state within which greater numbers of predators will be dependent on a zooplankton prey field that may be undergoing a shift in biodiversity. The recent simulations of the pelagic components of the ecosystem by Overholtz et al. (1991) will be extended to include the zooplankton component in an effort to examine the capacity of the ecosystem to support recovering populations of pelagic and demersal fish and baleen whale species.

References

Anthony, V. C. 1993. The state of groundfish resources off the northeastern United States. Fisheries 18(3):12–17.

Bigelow, H. B. 1926. Plankton of the Offshore Waters of the Gulf of Maine. Bull. Bur. Fish. XL Part 2. Washington, DC: Government Printing Office.

Bigelow, H. B., and Sears, M. 1939. Studies of the waters on the continental shelf, Cape Cod to Chesapeake Bay. III. A volumetric study of the zooplankton. Mem. Mus. Comp. Zool., Harvard Univ. 54:179–378.

Bowman, R., Warzocha, J., and Morris, T. 1984. Trophic Relationships Between Atlantic Mackerel and American Sand Lance. ICES C.M. 1984/H:27. Cophenhagen: International Council for Exploration of the Sea.

Clarke, G. L. 1940. Comparative richness of zooplankton in coastal and offshore areas of the Atlantic. Biol. Bull., Woods Hole 78:226–255.

Clarke, G. L., Pierce, E. L., and Bumpus, D. F. 1943. The distribution and reproduction of *Sagitta elegans* on Georges Bank in relation to hydrographical conditions. Biol. Bull., Woods Hole 85(3):201–226.

Clarke, G. L., and Zinn, D. J. 1937. Seasonal production of zooplankton off Woods Hole with special reference to *Calanus finmarchicus*. Biol. Bull., Woods Hole 73:464–487.

Dagg, M. J., and Turner, J. T. 1982. The impact of copepod grazing on the phytoplankton of Georges Bank and the New York Bight. Can. J. Fish. Aquat. Sci. 39:979–990.

Davis, C. S. 1984a. Predatory control of copepod seasonal cycles on Georges Bank. Mar. Biol. 82:31–40.

Davis, C. S. 1984b. Food concentrations on Georges Bank: Non-limiting effect on development and survival of laboratory reared *Pseudocalanus* sp. and *Paracalanus parvus* (Copepoda: Calanoida). Mar. Biol. 82:41–46.

Deevey, G. B. 1952. Quantity and composition of the zooplankton of Block Island Sound, 1949. Bull. Bingham Oceanogr. Collect., Yale Univ. 13:120–164.

Deevey, G. B. 1956. Oceanography of Long Island Sound, 1952–1954. V. Zooplankton. Bull. Bingham Oceanogr. Collect., Yale Univ. 15:113–155.

Deevey, G. B. 1960a. The zooplankton of the surface waters of the Delaware Bay region. Bull. Bingham Oceanogr. Collect., Yale Univ. 17(Article 2):5–53.

Deevey, G. B. 1960b. Relative effects of temperature and food on seasonal variations in length of marine copepods in some Eastern American and Western European waters. Bull. Bingham Oceanogr. Collect., Yale Univ. 17(Article 2):54–86.

Fish, C. J. 1936a. The biology of *Calanus finmarchicus* in the Gulf of Maine and Bay of Fundy. Biol. Bull., Woods Hole 70(1):118–141.

Fish, C. J. 1936b. The biology of *Pseudocalanus minutus* in the Gulf of Maine and Bay of Fundy. Biol. Bull., Woods Hole 70(2):193–216.

Fogarty, M., Cohen, E., Michaels, W., and Morse, W. 1991. Interactions among herring, mackerel, and sand lance in the Northwest Atlantic. ICES Mar. Sci. Symp. Multispecies Models Relevant to Management of Living Resources. 193:120–124. Copenhagen: International Council for the Exploration of the Sea.

Global Ocean Ecosystems Dynamics (GLOBEC). 1991. Report Number 1. Initial Science Plan. February 1991. Washington, DC: Joint Oceanographic Institutions Inc.

Global Ocean Ecosystems Dynamics (GLOBEC). 1992. Northwest Atlantic Implementation Plan. June 1992, Report No. 6. Davis, CA: U.S. GLOBEC Scientific Steering Committee Coordinating Office, Division of Environmental Studies, University of California, Davis.

Grice, G. D., and Hart, A. D. 1962. The abundance, seasonal occurrence, and distribution of the epizooplankton between New York and Bermuda. Ecol. Monogr. 32(4):287–308.

Holzwarth, T., and Mountain, D. 1992. Surface and Bottom Temperature Distributions from the Northeast Fisheries Center Spring and Fall Bottom Trawl Survey Program, 1963–1987; with Addendum for 1988–1990. Northeast Fish. Sci. Center Ref. Doc. 90-03. Woods Hole, MA: Northeast Fisheries Science Center.

Jossi, J. W., and Goulet, J. R., Jr. 1993. Zooplankton trends: U.S. northeast shelf ecosystem and adjacent regions differ from northeast Atlantic and North Sea. ICES J. Mar. Sci. 50:303–313.

Judkins, D. C., Wirick, C. D., and Esaias, W. E. 1980. Composition, abundance, and distribution of zooplankton in the New York Bight, September 1974–September 1975. Fish. Bull., U.S. 77:669–683.

Kane, J. 1993. Variability of zooplankton biomass and dominant species abundance on Georges Bank, 1977–1986. Fish. Bull., U.S. 91:464–474.

Maurer, R. 1976. A preliminary analysis of inter-specific trophic relationships between the sea herring, *Clupea harengus* Linnaeus and the Atlantic mackerel, *Scomber scombrus* Linnaeus. ICNAF Res. Doc. 76/VI/121; Serial No. 3967 (D.c.9). Halifax, Nova Scotia: International Commission on Northwest Atlantic Fisheries.

Meise-Munns, C., Green, J. R., Ingham, M. C., and Mountain, D. 1990. Interannual variability in the copepod populations of Georges Bank and the Western Gulf of Maine. Mar. Ecol. Prog. Ser. 65:225–232.

Murawski, S. A. 1991. Can we manage our multispecies fisheries? Fisheries 16(5):5–13.

New England Fishery Management Council. 1994. Amendments 5 and 6 to the New England Groundfish Management Plan. Peabody, MA: New England Fishery Management Council.

Northeast Fisheries Science Center (NEFSC). 1993. Status of the Fishery Resources Off the Northeastern United States for 1993. NOAA Tech. Mem. NMFS-F/NEC-101. Washington, DC: National Oceanic and Atmospheric Administration.

Overholtz, W. J., Murawski, S. A., and Foster, K. L. 1991. Impact of predatory fish, marine mammals, and seabirds on the pelagic fish ecosystem of the northeastern USA. ICES Mar. Sci. Symp. 193:198–208.

Payne, P. M., Wiley, D. N., Young, S. B., et al. 1990. Recent fluctuations in the abundance of baleen whales in the southern Gulf of Maine in relation to changes in selected prey. Fish. Bull., U.S. 88:687–696.

Redfield, A. C. 1941. The effects of the circulation of water on the distribution of the calanoid community in the Gulf of Maine. Biol. Bull., Woods Hole 80(1):86–110.

Riley, G. A., and Bumpus, D. F. 1946. Phytoplankton–zooplankton relationships on Georges Bank. J. Mar. Res. 6:54–73.

Sherman, K., Alexander, L. M., Gold, B. D. (eds.). 1993. Large Marine Ecosystems: Stress, Mitigation, and Sustainability. Washington, DC: AAAS Press.

Sherman, K., Green, J. R., Goulet, J. R., and Ejsymont, L. 1983. Coherence in zooplankton of a large Northwest Atlantic ecosystem. MARMAP Contribution No. MED/NEFC 82-68. Fish. Bull., U.S. 81:855–862.

Sherman, K., Grosslein, M., Mountain, D., et al. 1988. The Continental Shelf Ecosystem Off the Northeast Coast of the United States. pp 279–337. In H. Postma and J. J.

Ziljstra (eds.), Ecosystems of the World 27: Continental Shelves. Amsterdam: Elsevier, pp. 279–337.

Sherman, K., Jones, C., Sullivan, L., et al. 1981. Congruent shifts in sand eel abundance in western and eastern North Atlantic ecosystems. Nature 291(5815):486–489.

Sibunka, J. D., and Silverman, M. J. 1984. MARMAP Surveys of the Continental Shelf from Cape Hatteras, North Carolina, to Cape Sable, Nova Scotia (1977–1983). Atlas No. 1. Summary of Operations. NOAA Tech. Mem. NMFS-F/NEC-33. Washington, DC: National Oceanic and Atmospheric Administration.

Sibunka, J. D., and Silverman, M. J. 1989. MARMAP Surveys of the Continental Shelf from Cape Hatteras, North Carolina, to Cape Sable, Nova Scotia (1984–1987). Atlas No. 3. Summary of Operations. NOAA Tech. Mem. NMFS-F/NEC-68. Washington, DC: National Oceanic and Atmospheric Administration.

Sissenwine, M. P. 1986. Perturbation of a predator-controlled continental shelf ecosystem. pp. 55–85. In K. Sherman and L. M. Alexander (eds.), Variability and Management of Large Marine Ecosystems. AAAS Selected Symposium 99. Boulder, CO: Westview Press, pp. 55–85.

Sissenwine, M. P., and Cohen, E. B. 1991. Resource Productivity and Fisheries Management of the Northeast Shelf Ecosystem. In K. Sherman, L. M. Alexander, and B. D. Gold (eds.), Food Chains, Yields, Models, and Management of Large Marine Ecosystems. Boulder, CO: Westview Press, pp. 107–123.

Skud, B. E. 1982. Dominance in fisheries: A relation between environment and abundance. Science 216:144–149.

Solow, A. R. 1994. Detecting change in the composition of a multispecies community. Biometrics 50:556–565.

Townsend, D. W., and Cammen, L. M. 1988. Potential importance of the timing of spring plankton blooms to benthic–pelagic coupling and recruitment of juvenile demersal fishes. Biol. Oceanogr. 5:215–229.

IV

Environmental Stresses on the Ecosystem

13

Coastal Sediment Contamination in the Northeast Shelf Large Marine Ecosystem

Thomas P. O'Connor

Abstract

The National Oceanic and Atmospheric Administration's National Status and Trends (NS&T) Program has analyzed samples of fine-grained surface sediment collected at 233 sites throughout the coastal and estuarine United States. Although a little less than one-third of the sites are along the U.S. margin (Virginia through Maine) of the Northeast Shelf large marine ecosystem (LME), that group includes more than two-thirds of all sites displaying "high" concentrations for most contaminants. The tendency for coastal contamination to be disproportionately high in the northeast United States is, to some extent, a function of the distribution of human population, as a little less than half the total U.S. coastal population resides in this region. Another factor is the presence of large urban areas discharging contaminants directly into adjacent, poorly flushed marine waters. In the regions of the LME farther away from the coast, sediment contamination is not at high levels. The spatial confinement of contamination limits its influence on living resources of the LME.

In 1984, the National Oceanic and Atmospheric Administration (NOAA) inaugurated its National Status and Trends (NS&T) Program for Marine Environmental Quality. Since 1984, in the Benthic Surveillance Project, benthic fishes have been collected in estuaries or very nearshore, and their livers excised and stored for subsequent chemical analysis. Since 1986, in the Mussel Watch Project, bivalve mollusks have been collected on the shore at intertidal sites or very nearshore. Surface sediment samples have been collected nearshore at all sites in both projects. These data have been used to define the national distribution of contamination (O'Connor 1990; NOAA 1991a) and can be specifically applied to the Northeast Shelf large marine ecosystem (LME).

Sites Sampled and Chemicals Measured

Because the NS&T objective is to quantify contamination on a national scale, sites were selected to represent their general surroundings. Ends of waste discharge pipes, waste disposal sites, and other points dominated by single discharges were avoided. On average, the sites within estuaries and embayments were separated by about 20 km, whereas along open coastlines the distance could be 100 km. The elements and organic compounds measured in NS&T sediment samples and used in this discussion are listed in Table 13-1. In this chapter, sums of individual organic compound concentrations have been aggregated into groups termed total polycyclic aromatic hydrocarbons (tPAH), total DDT (tDDT), and total polychlorinated biphenyls (tPCB).

Field and Analytical Methods

In the Benthic Surveillance Project, sediment samples were obtained with a specially constructed box corer or a standard Smith-MacIntyre bottom grab.* In the Mussel Watch Project, the samples were obtained with a box corer or with a Kynar-coated Van Veen grab sampler. In the Benthic Surveillance Project, subsamples for organic analyses were composites of surface skims from the top 3 cm of each grab or core. A small corer was used to subsample the top 3 cm of each box core or grab sample for trace metal and other analyses. In the Mussel Watch Project, the subsamples for all composites were surface skims from the top 1 cm of each box core or grab. Composites were made from three grabs or cores from each of the three stations at a site. Sediment analyses in any given year for a single site consisted of organic analyses of three composites (one from each station), inorganic analyses of three separate composites, and grain-size and other measurements on a third set of three composites. A more detailed presentation of the sampling protocols is included in Shigenaka and Lauenstein (1988).

*The National Marine Fisheries Service does not approve, recommend, or endorse any proprietary product or proprietary material mentioned in this chapter.

Table 13-1. Chemicals and related parameters measured in sediment samples

DDT and its metabolites
- 2,4'-DDD
- 4,4'-DDD
- 2,4'-DDE
- 4,4'-DDE
- 2,4'-DDT
- 4,4'-DDT

Chlorinated pesticides other than DDT
- Aldrin
- Cis-chlordane
- Trans-nonachlor
- Dieldrin
- Heptachlor
- Heptachlor epoxide
- Hexachlorobenzene
- Lindane (gamma-BHC)
- Mirex

Polychlorinated biphenyls[a]
- Dichlorobiphenyls
- Trichlorobiphenyls
- Tetrachlorobiphenyls
- Pentachlorobiphenyls
- Hexachlorobiphenyls
- Heptachlorobiphenyls
- Octachlorobiphenyls
- Nonachlorobiphenyls

Polycyclic aromatic hydrocarbons[b]
- 2-ring
  - Biphenyl
  - Naphthalene
  - 1-Methylnaphthalene
  - 2-Methylnaphthalene
  - 2,6-Methylnaphthalene
  - Acenaphthylene
- 3-ring
  - Fluorene
  - Phenanthrene
  - 1-Methylphenanthrene
  - Anthracene
- 4-ring
  - Fluoranthene
  - Pyrene
  - Benz(a)anthracene
- 5-ring
  - Chrysene
  - Benzo(a)pyrene
  - Benzo(e)pyrene
  - Perylene
  - Dibenz(a,h)anthracene

Major elements
- Al Aluminum
- Fe Iron
- Mn Manganese
- Si Silicon

Trace elements
- Sb Antimony
- As Arsenic
- Cd Cadmium
- Cr Chromium
- Cu Copper
- Pb Lead
- Hg Mercury
- Ni Nickel
- Se Selenium
- Ag Silver
- Sn Tin
- Zn Zinc

Related parameters
- Grain size
- Total organic carbon

[a]PCBs quantified by level of chlorination in 1984 through 1987. Beginning with samples collected in 1988, 18 individual compounds (congeners) were quantified.
[b]Five PAH compounds added in 1988: acenaphthylene, 1,6,7-trimethyl naphthalene, benzo(b) and benzo(k) fluoranthene, indeno(1,2,3-cd)pyrene, and benzo(ghi)perylene.

The methods used for analysis of organic chemicals in sediments are described in a technical report prepared by NOAA's National Analytical Facility (MacLeod et al. 1985) and in papers on NS&T results in the Gulf of Mexico by Wade et al. (1988) and Sericano et al. (1990). For elemental analysis, sediments were totally digested in concentrated nitric and hydrochloric acids and in hydrofluoric acid for subsequent analysis by atomic absorption or, for some elements, x-ray fluorescence by the Battelle Ocean Sciences Laboratory. Battelle Ocean Sciences (1987) and Texas A&M Geochemical and Environmental Research Group (1988) both provide details of analytical methods used in the Mussel Watch Project.

Accounting for Sand

Before comparing concentrations of contaminants among sites, it was necessary to recognize that contaminants are associated with particle surfaces and that differences in contaminant concentrations among sites can be generated simply by differences in particle sizes of sediments. To compensate for this, sediment data for each sample were adjusted by dividing raw concentrations by the fraction-by-weight of sediment particles less than 63 μm in diameter (i.e., the fine-grained or silt and clay fraction). This is equivalent to assuming that no contaminants are associated with sand-sized particles and that the only effect of sand in a sample is to dilute its level of contamination.

This method of adjustment can yield misleading results for sediments that are primarily sand; if such sediments contain detectable levels of contamination, the adjusted level will appear to be very high because it will have been divided by a low number. The opposite will occur when sandy sediment has contaminant levels that are below detection. Adjusting such values still leaves them below the level of detection, or essentially zero. Such zeroes, primarily resulting from the diluting effect of sand, would be misleading in the context of other samples from a site that was not so sandy. To avoid concluding that a mean sediment concentration is unusually high or low when, in fact, it simply includes data from a sandy sample, no concentration data were used if they were from samples containing less than 20% fine-grained material. Therefore, the adjusting procedure increased a raw concentration by, at most, a factor of five.

Geographic Distribution of Sediment Contamination

NOAA has summarized all sediment data available through 1989 based on analytical results from 1,426 composite samples and 266 sites; at 233 of these, fine-grained samples were collected at least once (NOAA 1991a). The distribution of mean concentrations among sites is highly skewed toward low values; when plotted as logarithms, the data are approximately normally distributed (Figure 13-1 shows this for Cd and tPCB concentrations). This provides a convenient

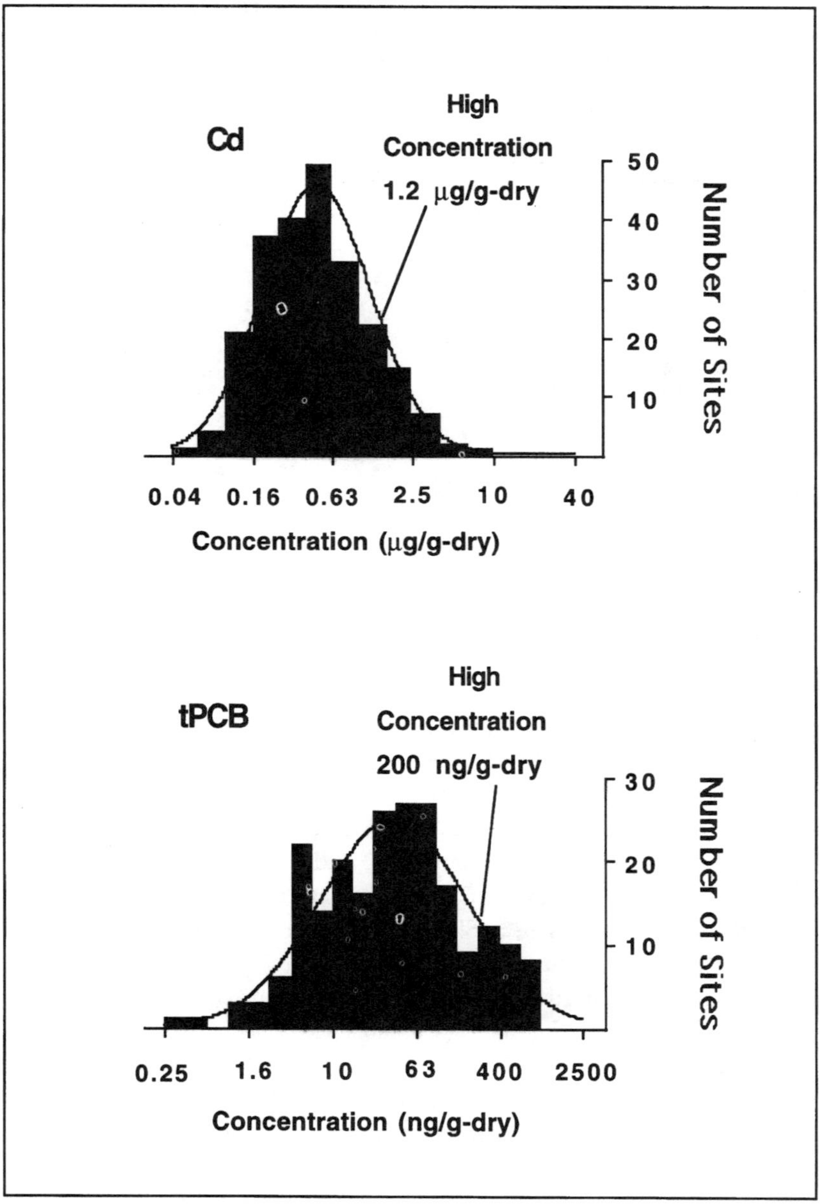

Figure 13-1. Distributions of Cd and total PCB (tPCB) concentrations in sediment on arithmetic and logarithmic scales.

Table 13-2. "High" concentrations (dry weight) of chemicals in the NS&T samples from throughout the coastal United States and numbers of sites with concentrations above those levels in the entire nation and in the Northeast Shelf LME

			Number of sites with higher concentration		
Chemical	"High" concentration		In United States*	In NE Shelf LME	% in NE Shelf LME
Ag	1.2	μg/g	37	26	70
As	24		33	9	27
Cd	1.2		41	24	58
Cr	230		30	8	27
Cu	84		35	23	66
Hg	0.49		34	23	68
Ni	69		29	4	14
Pb	89		39	25	64
Zn	270		40	24	60
tPAH	4,000	ng/g	39	26	67
tDDT	37		33	15	45
tPCB	200		39	25	64

*In a normal population of 233, about 39 (17%) should exceed the mean by more than one standard deviation.

method for defining the "high" concentrations listed in Table 13-2 as the mean plus one standard deviation of the lognormal distribution.

This definition of "high" has been used by NOAA (1991a) and O'Connor (1990) to describe the national distribution of contamination; their reports list all the NS&T sites, with chemicals whose mean concentration exceeds the "high" value indicated for each. Table 13-3 is the portion of that list for the coastline between the mouth of Chesapeake Bay and Canada—i.e., the landward boundary of the Northeast Shelf LME. Whereas it has only 29% (67/233) of the 233 fine-grained sites sampled throughout the coastal United States, it contains about two-thirds of the sites with sediment concentrations above "high" levels for most chemicals (see Table 13-2). Three chemicals not so strongly associated with the Northeast Shelf LME—As, Cr, and Ni—are the three elements whose "high" concentrations have a strong natural component. The fourth such chemical, tDDT, is certainly not natural but is uniquely high on the West Coast.

Because there is a statistical correlation between sedimentary concentrations of most chemicals and numbers of people living within 20 km of each site (NOAA 1991a), the disproportionate number of "high" concentrations in the Northeast is, to some extent, a function of population distribution. Culliton et al. (1990) listed the 1988 population in counties bordering marine waters in the

Table 13-3. NS&T sites in the Northeast Shelf LME, the mean fine fraction in fine-grained samples or an indication "sandy" (meaning that all samples from the site contained >80% sand), and specific chemicals whose mean concentrations were at the "high" end of the overall, national lognormal distribution (i.e., greater than one standard deviation above the geometric mean)

Code	General location	Specific location	State	% Fine	Chemicals with "high" concentrations
MAC	Machias Bay		ME	66	
FRN	Frenchmans Bay		ME	93	
PNB	Penobscot Bay		ME	97	
PBSI	Penobscot Bay	Sears Island	ME	90	
PBPI	Penobscot Bay	Pickering Island	ME	49	tPAH
MSSP	Merriconeag Sound	Stover Point	ME	38	
CSC	Casco Bay		ME	81	
CAKP	Cape Arundel	Kennebunkport	ME	Sandy	
MER	Merrimack River		MA	Sandy	
CAGH	Cape Ann	Gap Head	MA	27	Pb, tPAH
SHFP	Salem Harbor	Folger Point	MA	52	Ag, Cd, Cr, Cu, Hg, Pb, tPAH, tDDT
SAL	Salem Harbor		MA	67	Ag, Cd, Cr, Cu, Hg, Pb, Zn, tPAH, tPCB, tDDT
BHDI	Boston Harbor	Deer Island	MA	78	Ag, Cd, Cr, Cu, Hg, Pb, tPAH, tPCB, tDDT
BHDB	Boston Harbor	Dorchester Bay	MA	85	Ag, Cd, Cr, Cu, Hg, Pb, tPAH, tPCB, tDDT
BHHB	Boston Harbor	Hingham Bay	MA	27	Ag, Cd, Cr, Cu, Hg, Pb, tPCB
BHBI	Boston Harbor	Brewster Island	MA	Sandy	
BOS	Boston Harbor		MA	59	Ag, Cd, Cr, Cu, Hg, Pb, Zn, tPAH, tPCB, tDDT
QUI	Quincy Bay		MA	75	Ag, Cd, Cr, Cu, Hg, Pb, tPAH, tPCB, tDDT
DBCI	Duxbury Bay	Clarks Island	MA	Sandy	
CCNH	Cape Cod	Nauset Harbor	MA	68	
BBCC	Buzzards Bay	Cape Cod Canal	MA	63	Cd
BBRH	Buzzards Bay	Round Hill	MA	65	tPCB
BBAR	Buzzards Bay	Angelica Rock	MA	29	Ag, Pb, tPAH, tPCB
BBGN	Buzzards Bay	Goosebury Neck	MA	37	
BUZ	Buzzards Bay		MA	77	tPCB
NBMH	Narragansett Bay	Mount Hope Bay	RI	90	Ag, Cu, Hg, Pb
NBPI	Narragansett Bay	Patience Island	RI	Sandy	
NBDI	Narragansett Bay	Dyer Island	RI	38	Ag, Pb, tPAH
NBDU	Narragansett Bay	Dutch Island	RI	60	
NAR	Narragansett Bay		RI	66	Ag, Cu, Pb, tPCB
BIBI	Block Island	Block Island	RI	71	
LIGB	Long Island	Gardiners Bay	NY	Sandy	
ELI	East Long Is. Sound		CT	Sandy	

Table 13-3 (continued).

LICR	Long Island Sound	Connecticut River	CT	50	tPCB, tPAH
LINH	Long Island Sound	New Haven	CT	Sandy	
LIHR	Long Island Sound	Housatonic River	CT	Sandy	
WLI	West Long Is. Sound		NY	81	Ag, Cd, Cu, Hg, Pb, Sn, Zn, tPAH, tPCB
LIHU	Long Island Sound	Huntington Hrb.	NY	46	Ag, Cd, Cu, Pb, Zn
LIPJ	Long Island Sound	Port Jefferson	NY	Sandy	
LIMR	Long Island Sound	Mamaroneck	NY	74	Ag, Cd, Cu, Hg, Pb, Zn, tPAH
LIHH	Long Island Sound	Hempstead Hrb.	NY	91	Ag, Cd, Cu, Hg, Pb, Zn, tPAH, tPCB, tDDT
LITN	Long Island Sound	Throgs Neck	NY	71	Ag, Cd, Cu, Hg, Pb, Zn, tPAH, tPCB, tDDT
MBTH	Moriches Bay	Tuthill Point	NY	56	Ag, Hg, Pb, tDDT
HRJB	Hudson-Raritan Est.	Jamaica Bay	NY	64	Ag, As, Cd, Cr, Cu, Hg, Pb, Zn, tPAH, tPCB, tDDT
HRUB	Hudson-Raritan Est.	Upper Bay	NY	67	Ag, As, Cd, Cu, Hg, Pb, Zn, tPAH, tPCB
HRLB	Hudson-Raritan Est.	Lower Bay	NY	71	Ag, As, Cd, Cr, Cu, Hg, Pb, Zn, tPAH, tPCB, tDDT
HRRB	Hudson-Raritan Est.	Raritan Bay	NJ	71	Ag, As, Cd, Cr, Cu, Hg, Pb, Zn, tPAH, tPCB, tDDT
RAR	Raritan Bay		NJ	73	Ag, As, Cd, Cr, Cu, Hg, Pb, Zn, tPAH, tPCB, tDDT
NYSH	New York Bight	Sandy Hook	NJ	64	Ag, As, Cd, Cr, Cu, Hg, Pb, Zn, tPAH, tPCB, tDDT
NYLB	New York Bight	Long Branch	NJ	Sandy	
NYSR	New York Bight	Shark River	NJ	Sandy	
BIBL	Barnegat Inlet	Barnegat Light	NJ	Sandy	
GRB	Great Bay		NJ	77	Hg
AIAC	Absecon Inlet	Atlantic City	NJ	Sandy	
DBCM	Delaware Bay	Cape May	NJ	62	Ag, As, Zn
DEL	Delaware Bay		DE	48	tPCB
DBFE	Delaware Bay	False Egg Is. Pt.	NJ	47	Cd, tDDT
DBBD	Delaware Bay	Ben Davis Shoal	NJ	68	
DBAP	Delaware Bay	Arnolds Pt. Shoal	DE	76	
DBHC	Delaware Bay	Hope Creek	NJ	22	Ni, Zn
DBWB	Delaware Bay	Woodland Beach	DE	31	Zn
DBKI	Delaware Bay	Kelly Island	DE	58	
DBCH	Delaware Bay	Cape Henlopen	MD	77	
BAL	Baltimore Harbor		MD	93	Ag, As, Cd, Cr, Cu, Hg, Ni, Pb, Sn, Zn, tPAH, tPCB
UCB	Upper Ches. Bay		MD	82	Ni, Zn, tPAH, tPCB

CBBO	Chesapeake Bay	Bodkin Point	MD	99	Ni, Sn, Zn, tPAH, tPCB
CBMP	Chesapeake Bay	Mountain Pt. Bar	MD	98	As, Zn, tPAH
CBHP	Chesapeake Bay	Hackett Point Bar	MD	98	Zn, tPAH
CBCP	Chesapeake Bay	Choptank River	MD	93	
CBHG	Chesapeake Bay	Hog Point	MD	Sandy	
MCB	Middle Ches. Bay		VA	59	As, Zn
PRRP	Potomac River	Ragged Point	VA	96	
PRSP	Potomac River	Swan Point	MD	38	Cd, Zn
CBIB	Chesapeake Bay	Ingram Bay	VA	78	
RRRR	Rappahannock R.	Ross Rock	VA	94	
CBCC	Chesapeake Bay	Cape Charles	VA	71	
CBDP	Chesapeake Bay	Dandy Point	VA	46	
CBJR	Chesapeake Bay	James River	VA	72	
LCB	Lower Ches. Bay		VA	51	
ELZ	Elizabeth River		VA	66	Cd, Cu, Hg, Pb, Zn, tPAH, tPCB
CBCI	Chincoteague Bay	Chinco. Inlet	VA	Sandy	
QIUB	Quinby Inlet	Upshur Bay	VA	38	

Northeast as 39 million people; the corresponding populations along the Southeast, Gulf of Mexico, and West Coasts were 9 million, 14 million, and 29 million, respectively. Thus, 43% of the U.S. population living along marine coasts is located along the coastal edge of the Northeast Shelf LME.

Seaward Extent of Contamination

Combining data from C. E. Harrington (personal communication, NOAA, Coast and Geodetic Survey 1991) on areas seaward of coastal headlands with data from NOAA (1990) on areas within estuaries gives a total area for the United States' portion of the Northeast Shelf LME (i.e., the area in estuaries between Maine through Virginia and seaward to the 200-nautical-mile [nm] extent of the Exclusive Economic Zone) of 395,000 km^2. From the viewpoint of productive fisheries, the important portion of the Northeast Shelf LME is the 230,000 km^2 in estuaries and over the continental shelf (the area with depths less than 200 m). The shallow portion of that, the area within 3 nm of the coast, is about 35,000 km^2 and, of that, 27,000 km^2 is in estuaries (40% of the estuarine area being Long Island Sound and Chesapeake Bay). A review of available data will show that "high" levels of sediment contamination in the Northeast Shelf LME are restricted almost exclusively to this coastal portion, or 15% of the productive part of the LME. This is the expected distribution, because population centers along the edge are the source of chemical contamination and, regardless of whether contaminants are directly discharged into water or result from atmospheric fallout, concentrations would drop off exponentially with distance from the coast as a simple consequence of dilution.

However, although the distribution appears unremarkable, comparisons with offshore data must exclude data from coarse-grained sediments, and here lies a major difference between NS&T sites and sites on the continental shelf. The continental shelf has a sandy bottom (Shepard 1963). For the most part, fine-grained material escaping the mouths of rivers and estuaries bypasses the shelf and is not deposited until it reaches the deeper ocean (Schubel and Okubo 1972). There are basins on the shelf, such as the Christiaensen Basin in the apex of the New York Bight, where fine-grained material can be found. The largest of these, the "Mud Patch," a depression in the continental shelf west of Georges Bank and south of the Nantucket Shoals where modern fine-grained sediments are accumulating (Bothner et al. 1981), is over 12,000 km^2 in area and larger than Chesapeake Bay.

Data on chemical concentrations in sediments of the offshore Northeast Shelf LME are available from studies centering on the Middle Atlantic Bight, Georges Bank, and the Gulf of Maine, and, on slightly smaller scales, the New York Bight within the Middle Atlantic Bight, the New York Bight apex within the New York Bight, and Massachusetts Bay within the Gulf of Maine.

Bothner (1977) analyzed Cr, Cu, Ni, Pb, and Zn in sediments collected at 52 sites along transects crossing the shelf in the Middle Atlantic Bight. The bulk of these samples were sand and had low trace metal concentrations. Seven of the samples were taken beyond the edge of the shelf at depths in excess of 200 m, and contained slightly more than 20% fine-grained sediment. The elemental concentrations in these sediments, after adjusting for grain size, would not have exceeded the "high" concentrations in Table 13-2.

Bothner et al. (1987) collected and analyzed sediments at depths between 1,500 and 2,100 m in the mid-Atlantic. In these cases, more than 95% of the sediment samples were fine-grained, and the concentrations of Cd, Cr, Cu, Hg, Ni, Pb, and Zn could be compared directly with the NS&T data. None of the sediment samples collected at 14 sites had concentrations in the "high" range. Similarly, Bothner et al. (1987) collected sediments at 14 sites off the North Atlantic slope and rise at depths between 2,050 and 2,200 m. In no case did the concentrations of Cd, Cr, Cu, Hg, Ni, Pb, or Zn exceed the "high" value.

Sampling on smaller scales in the Middle Atlantic Bight, however, has yielded sediments with chemical concentrations in the "high" range. Invariably, these samples are collected in the apex of the New York Bight and are samples contaminated with sewage sludge dumped at the (no longer used) 12-mile dumpsite, or they are collected very nearshore. In 1980, the National

Marine Fisheries Service Northeast Monitoring Program occupied 54 sites in the Middle Atlantic Bight (Reid et al. 1982). Seventeen of those sites had sediments with more than 20% fine-grained material: all 10 of the Long Island Sound sites and 7 of the 24 sites in the apex of the New York Bight. After adjusting for grain size, Cd, Pb, and Zn concentrations were "high" at all seven fine-grained sites in the apex. Copper was "high" at five, Hg at three, and Cr at one. In conformance to the NS&T data (NOAA 1991a), the three sites at the western end of Long Island Sound had "high" concentrations for most elements. Boehm (1980) analyzed sediments from 40 of these same 54 sites for PCBs, including nine fine-grained sites, and four fine-grained sites for PAHs. In two of the seven fine-grained samples collected near waste disposal sites, PCB concentrations were greater than 100 ng/g but less than the "high" value in Table 13-2 of 200 ng/g. The two fine-grained sites in the apex analyzed for PAHs had concentrations of tPAH greater than 4,000 ng/g (for approximately the same set of 18 compounds listed in Table 13-1), as did a site in the western end of Long Island Sound.

Over a wider scale than the Middle Atlantic Bight, the NOAA National Marine Fisheries Service Northeast Monitoring Program sampled 28 stations on the continental shelf from Chesapeake Bay to the Gulf of Maine (NOAA 1983). Ten of those sites contained more than 20% fine-grained material (Boehm 1983). Sediments from eight sites, including five with fine-grained material, were analyzed for PCBs and PAHs. After adjusting for grain size, only one site, once again in the apex of the New York Bight, had more than 100 ng/g tPCB. Similarly, that site was the sole example of a tPAH concentration in excess of 4,000 ng/g. Sites where neither tPCB nor tPAH were at "high" concentrations were in Massachusetts Bay, Buzzards Bay, the New York Bight (seaward of the apex), and the "Mud Patch."

General information on concentrations of Ag, Cd, Cr, Cu, Hg, Ni, Pb, and Zn for those shelf-wide 28 sites were summarized by NOAA (1983). Whereas concentrations were higher in fine-grained sediments, the only such sediments that may have had concentrations greater than those listed in Table 13-2 were taken from waste disposal sites in the New York Bight apex. It was stated that sediments from the "Mud Patch," despite being fine-grained, had tenfold lower concentrations of trace metals than did those from the New York Bight apex.

From 1981 through 1983, Bothner et al. (1985) made seasonal collections of sediments at 19 sites on and near Georges Bank, two of which were in the "Mud Patch." All samples were analyzed for Cd, Cr, Cu, Hg, Ni, Pb, and Zn. At 15 of the 19 sites, the sediments were more than 95% sand and gravel, so comparisons with NS&T data are inappropriate. At four sites where the sediments consisted of at least 20% fine-grained particles, the concentrations for those elements (after adjustment) were below the "high" values listed in Table 13-2. Bothner et al. separated the <60-μm particles from the predominant sand-sized and gravel-sized sediment in samples from 11 of the coarse-grained sites. The results of analyzing that fraction are directly comparable with the NS&T results. With a few minor exceptions for individual samples but not replicates or blends of replicates, the concentrations of Cd, Cr, Cu, Hg, Ni, Pb, or Zn in the fine fractions of these sediments were below the "highs" of Table 13-2.

Boehm and Farrington (1984) analyzed PAH compounds in sediments from 22 stations on Georges Bank and its environs, including two in the "Mud Patch" and one in Murray Basin, a depositional site in the Gulf of Maine, east of Cape Cod. Sediments at these latter three and one other site had the highest tPAH concentrations, and were the only four with more than a 10% clay content. However, no concentrations would have been in the "high" category.

Boehm et al. (1984) analyzed PCB and PAH concentrations at 14 sites in Massachusetts Bay (as well as at six within Boston Harbor). Although quantitative data on grain size were not provided, sediments at 13 of the sites were described as mud. Sediments from Boston Harbor had concentrations of tPAH and tPCB in the "high" range, but only two of the Massachusetts Bay samples were "high" for tPAH and none for tPCB. Trace element data for the 14 Massachusetts Bay sites (Shea et al. 1991) showed no concentrations of Cd, Cr, Cu, Ni, or Zn in the "high" range, but Pb was "high" at one site (a site with high PAH) and Ag was high at another.

Analytical data from a more recent series of sediment collections in Massachusetts Bay by Battelle Ocean Sciences have been included in a summary document by Shea et al. (1991). Fifteen sites at the shoreward side of Massachusetts Bay but outside Boston Harbor were sampled in triplicate on three separate cruises. At seven sites, samples contained more than 20% fine-grained material. In no cases, after adjusting for grain-size, did concentrations of As, Cd, Cr, Cu, Ni, Pb, or Zn exceed "high" concentrations. However, Hg concentrations averaged very near the 0.49 μg/g "high" level at six of the seven fine-grained sites, and tPAH exceeded the 4,000 ng/g "high" level at five sites. It was pointed out by Shea et al. that the concentrations of all chemicals at these Massachusetts Bay sites were less than those in Boston Harbor. Nevertheless, concentrations for two chemicals were "high" when viewed from a national perspective.

Two groups have reported tPAH concentrations in Gulf of Maine sediments. Windsor and Hites (1979) sampled a ten-site transect extending from the inner part of Boston Harbor, to Massachusetts Bay, and into the basins of the Gulf of Maine. Larson et al. (1986) analyzed PAH compounds at 17 stations in the Gulf of Maine. Windsor and Hites (1979) included some alkylated PAH compounds that are not part of the NS&T tPAH; nevertheless, the only sites with more than 4,000 ng/g tPAH were in Boston Harbor. One site in Massachusetts Bay with a tPAH of 3,400 might have exceeded the "high" level if data on grain size had been provided and concentrations could have been adjusted. Larson et al. (1986) described 11 of their 19 sites as clay or silt, but since no site had tPAH concentrations in excess of 400 ng/g, none would have been in the "high" category.

Distribution of Living Resources

This review of data on chemical concentrations in sediments of the Northeast Shelf LME has demonstrated that, with the exceptions of waste disposal sites and a part of Massachusetts Bay, "high" levels of contamination are limited to near-coastal sites. Nevertheless, although this is a small portion of the Northeast Shelf LME, it is the source of about half its living marine resources. Table 13-4 lists the 22 species whose landings each constituted at least 1% of the total value of the U.S. fisheries from the LME in 1989 (NOAA 1991b). Together, they account for 87% of the total landings and for 94% of the total value. At least 10% of the commercial harvest for 9 of these 22 species occurs in estuaries or within 3 nm of the coast. In terms of dollars, these nine species account for 47% of the value of landings from the Northeast Shelf LME. State-specific landing data for 1989 from NMFS show that 88%, 77%, and 76% of blue crabs, oysters, and menhaden, respectively, were taken within Chesapeake Bay, so that their exposure to contamination is defined by the contaminant distribution within the Bay rather than the entire Northeast Shelf LME. In general, the important Northeast Shelf LME species exposed to coastal contamination are lobsters; hard, surf, and soft clams; and summer and winter flounder. For the two flounder species, there is coastal and recreational fishery that in the Northeast Shelf LME equaled 15% and 43% of the commercial landings for summer and winter flounder, respectively, in 1989 (NOAA 1991b).

Distribution of Biological Responses to Contamination

The "high" concentrations of contaminants found in coastal sediments and listed in Table 13-2 are not necessarily of biological significance. Long and Morgan (1990) searched the technical literature for reports where both sedimentary

Table 13-4. Fishery statistics for the Northeast Shelf LME

Species	Landings (1,000 mt)	Value ($ millions)	1989 Landings (% of total)	Value (% of total)	% Coastal or estuarine
Lobster	24.0	148.9	3.71	19.68	82
Hard clam	4.2	50.6	0.65	6.69	100
Blue crab	45.6	42.0	7.05	5.55	100
Menhaden	287.8	31.3	44.51	4.14	100
Surf clam	30.4	30.7	4.70	4.06	26
Oyster meats	2.4	22.3	0.37	2.95	100
Summer flounder	6.2	22.2	0.96	2.93	37
Winter flounder	6.6	19.6	1.02	2.59	10
Soft clam	2.9	19.4	0.45	2.56	100
Subtotal	410.1	387.0	63.42	51.15	
Sea scallops	14.4	126.6	2.23	16.73	4
Cod	35.6	47.8	5.51	6.32	2
Loligo squid	23.0	21.8	3.56	2.88	7
Bluefin tuna	1.1	19.7	0.17	2.60	1
Swordfish	2.7	17.2	0.42	2.27	7
Ocean quahog	23.1	16.4	3.57	2.17	1
Yellowtail flounder	5.6	13.9	0.87	1.84	1
Goosefish	11.3	12.6	1.75	1.67	
Pollock	10.5	9.9	1.62	1.31	1
Silver hake	17.8	9.4	2.75	1.24	
Witch flounder	2.4	9.0	0.37	1.19	
American plaice	3.5	8.8	0.54	1.16	
Northern shrimp	3.6	7.8	0.56	1.03	1
Total*	646.6	756.5	100.00	100.00	

*Total includes landings and value of species not accounting for more than 0.6% of total landings or 1.0% of total value.
Source: Landings and values for 1989 from NOAA (1991b). Holliday and O'Bannon (1991) supply 1990 landing data divided between that taken within 3 nm of the coast and that taken further offshore; they did not provide data for goosefish, silver hake, witch flounder, or American plaice, but synopses for the latter three species show that they are targets of offshore fisheries (NOAA 1991b).

chemical concentrations and biological responses were measured. The concentrations that were found to be generally (although not always) associated with biological effects are all greater than the "high" concentrations in Table 13-2. However, because the literature they searched involved so many different types of biological responses and because sediments always contain more than a single chemical, the listing of concentrations found to be associated with effects will always be subject to question. In 1988, Scott (SAIC, Narragansett RI, personal communication) collected sediment at NS&T sites with "high" levels of contamination in western Long Island Sound and in the Hudson/Raritan estuary. In a standard bioassay based on 10-day survival of amphipods, these sediments proved nontoxic. This emphasizes that the error involved in extrapolating from sediment contamination to biological effects probably overestimates effects.

As discussed by O'Connor (1990) and O'Connor and Ehler (1991), there are numerous reports of extreme levels of chemical contamination in "hot spots" of limited spatial extent. Such sites are not representative of their surroundings and are avoided in the NS&T Program. It should be pointed out, however, that these are also locations where sediments can be found that are toxic in bioassay tests. The issue here is not to compare the NS&T assessment of coastal contamination with what can be found by seeking out extreme contamination on small scales, but to compare it with the much larger scale of the Northeast Shelf LME.

Generally, surveys for actual field evidence of biological responses to contamination have been conducted in the coastal or estuarine portions of the Northeast Shelf LME. An exception is Ziskowski et al. (1987), who collected data on external evidence of disease of finfishes from throughout the Northeast Shelf LME. Although certainly not all diseases of marine organisms are attributable to chemical contamination, diseases that tend to co-occur with contaminated sediments are often considered to be caused, or at least exacerbated, by the contamination. Even in those cases, the cause may be over-enrichment of the sea floor with organic matter rather than the concentrations of trace elements or specific organic compounds. Ziskowski et al. (1987) found a 0.70% incidence of fin rot in 10,192 winter flounder collected in inshore waters compared with 0.49% among 6,527 winter flounder from offshore areas. These are low incidences but statistically different ($p < 0.1$), indicating a tendency for the response to be more common inshore.

At four sites in Massachusetts Bay, Stegeman et al. (1987) measured the activity of mixed function oxidase (MFO) induced in the livers of winter flounder through their metabolism of xenobiotic organic compounds. Induction was found at all sites and was higher in Boston Harbor and Plymouth Bay than in Buzzards Bay or Nantucket Shoals. This result means that fish from all those areas are exposed to contamination. Although there are no reports of MFO induction having been sought in winter flounder from farther offshore, it is most likely that it would be found. Induction has been found even in deep-sea fish from off the continental shelf, near Hudson Canyon and Carson Canyon, and off Newfoundland (Stegeman et al. 1986).

In response to reports of shell disease in lobsters and crabs taken from offshore shelf and canyon areas, Sindermann et al. (1989) reviewed the locations and prevalences of external diseases in crustaceans. They concluded that although shell disease is a natural stage in the relationship between crustaceans and chitin-degrading bacteria, prevalences greater than 15% are found only in contaminated areas such as the New York Bight apex. There was no basis for claiming contaminant-induced shell disease offshore.

Among mollusks, hematopoietic neoplasia in soft clams is found throughout the coastal Northeast Shelf LME (Leavitt et al. 1990; Farley et al. 1991), and oysters in Chesapeake Bay suffer the diseases MSX and dermo because of infections by *Haplosporidium nelsoni* and *Perinsus marinus*, respectively (Haven 1987). There is no clear causal link between these diseases and chemical contamination. Because the host mollusks are limited to coastal waters, so are the diseases.

Last, although the primary concern is public health, it should also be recognized that fisheries that are closed or restricted because of chemical contamination are a lost resource. Such closures are limited to coastal waters. The Hudson River stock of striped bass carries a high body burden of PCBs, forcing a ban on their harvest in New York waters (Faber 1986). Commercial fishing is banned in New Bedford Harbor, Massachusetts, because of extreme PCB contamination (Weaver 1984). There are advisories on consumption of fish from Boston Harbor and consumption of lobster hepatopancreas from all waters off Massachusetts (Schwartz et al. 1991).

Whereas outright bans on harvest are rare, mollusk harvesting in 1990 was restricted on 20% of the 3.25 million hectares of molluscan shellfish beds in the coastal and estuarine Northeast Shelf LME (NOAA 1991c) because of contamination with coliform bacteria. In all likelihood, cases in which chemical contaminant levels in mollusks warrant measures to protect human health are already cases where shellfishing is restricted for bacteriologic reasons.

Conclusions

Chemical analyses of fine-grained coastal and estuarine sediments by the NOAA NS&T Program have yielded a definition of "high" concentrations for several trace elements and groups of organic compounds. The coastal and estuarine area of the Northeast Shelf LME has more than its proportional share of sites with "high" levels. Fine-grained sediments in the 85% of the Northeast Shelf LME that is beyond 3 nm from the coast are relatively rare, but those collected from basins or separated from sandy sediments do not show "high" levels except near waste disposal sites in the apex of the New York Bight and in a portion of Massachusetts Bay. In some parts of the estuarine and coastal Northeast Shelf LME, chemical contamination has restricted harvesting of fish and shellfish, and in limited locations fish exhibit adverse responses to contamination. The offshore portion of the Northeast Shelf LME, however, is not an area threatened by chemical contaminants.

References

Battelle Ocean Sciences. 1987. Phase 2 Work/Quality Assurance Project Plan for Contract No. 50-DGNC-5-0263: Collection of Bivalve Molluscs and Surficial Sediments and Performance of Analyses for Organic Chemicals and Toxic Trace Elements. Report to National Oceanic and Atmospheric Administration (NOAA) from Battelle Ocean Sciences, Duxbury, MA.

Boehm, P. D. 1980. New York Bight Benthic Sampling Survey: Coprostanol, Polychlorinated Biphenyl and Polycyclic Aromatic Hydrocarbon Measurements in Sediments. Final Report, NOAA Contract NA-80-FAC-00038. Highlands, NJ: NOAA/NMFS/NEFC, Sandy Hook Laboratory.

Boehm, P. D. 1983. Chemical Contaminants in Northeast United States Marine Sediments. NOAA Technical Report NOS 99. Rockville, MD: NOAA National Ocean Service.

Boehm, P. D., and Farrington, J. W. 1984. Aspects of the polycyclic aromatic hydrocarbon geochemistry of recent sediments in the Georges Bank region. Environ. Sci. Technol. 18:840–845.

Boehm, P. D., Steinhauer, W., and Brown, J. 1984. Organic Pollutant Biogeochemistry Studies, Northeast U.S. Marine Environment. Part 1. The State of Organic Pollutant (PCB, PAH, coprostanol) Contamination of the Boston Harbor–Massachusetts Bay–Cape Cod Bay System: Sediments and Biota. Part 2. Organic Geochemical Studies in the Hudson Canyon and Gulf of Maine Areas. Final Report from Battelle Ocean Sciences to NOAA National Marine Fisheries Service. Woods Hole, MA.

Bothner, M. B., Campbell, E. Y., DiLisio, G. P., et al. 1987. Analysis of Trace Metals in Bottom Sediments in Support of Deepwater Biological Process Studies on the U.S. Mid-Atlantic Continental Slope and Rise. Final Report Prepared for U.S. Department of the Interior. Minerals Management Service, Washington, DC.

Bothner, M. B., Rendigs, R. R., Campbell, E. Y., et al. 1985. The Georges Bank Monitoring Program 1985: Analysis of Trace Metals in Bottom Sediments During the Third Year of Monitoring. U.S. Geol. Survey Circ. 988. Department of the Interior, U.S. Geological Survey, Washington, DC.

Bothner, M. H. 1977. Trace Metals in Surface Sediments of the Middle Atlantic Continental Shelf. In R. Miller, P. Cousins, W. Dillon, et al. (eds.), Middle Atlantic Outer Continental Shelf Environmental Studies. Vol. III: Geological Studies. Report for the U.S. Department of the Interior, Bureau of Land Management. Gloucester Point, VA: Virginia Institute of Marine Science, pp. 9-1–9-18.

Bothner, M. H., Spiker, E. C., Johnson, P. P., et al. 1981. Geochemical evidence for modern sediment accumulation on the continental shelf off Southern New England. J. Sediment. Petrol. 51:281–292.

Culliton, T. J., Warren, M. A., Goodspeed, T. R., et al. 1990. 50 Years of Population Change Along the Nation's Coasts 1960–2010. The Second Report of a Coastal Trend Serties. Rockville, MD: NOAA Office of Oceanography and Marine Assessment.

Faber, H. 1986. State to Ban All Commercial Striped Bass Fishing. New York Times, April 22. p. A1.

Farley, C. A., Plutschak, D. L., and Scott, R. F. 1991. Epizootiology and distribution of transmissible sarcoma in Maryland softshell clams, *Mya arenaria*, 1984–1988. Environ. Health Persp. 90:35–41.

Haven, D. 1987. The American Oster (*Crassostria virginica*) in Chesapeake Bay. In S. K. Majumdar, L. W. Hall, Jr., and H. M. Austin (eds.), Contaminant Problems and Management of Living Chesapeake Bay Resources. Easton, PA: Pennsylvania Academy of Science, pp. 165–176.

Holliday, M. C., and O'Bannon, B. K. 1991. Fisheries of the United States, 1990. Current Fishery Statistics No. 9000, Silver Spring, MD: NOAA, National Marine Fisheries Service.

Larson, P. F., Gadbois, D. F., and Johnson, A. C. 1986. Polycyclic aromatic hydrocarbons in Gulf of Maine sediments: Distributions and mode of transport. Mar. Environ. Res. 18:231–244.

Leavitt, D. F., McDowell-Capuzzo, J., Smolowitz, R. M., et al. 1990. Hematopoietic neoplasia in *Mya arenaria*: Prevalence and indices of physiological condition. Mar. Biol. 105:313–321.

Long, E. R., and Morgan, L. G. 1990. The Potential for Biological Effects of Sediment-sorbed Contaminants Tested in the National Status and Trends Program. NOAA Technical Memorandum NOS OMA 52. Seattle: NOAA Office of Oceanography and Marine Assessment, Ocean Assessments Division.

MacLeod, W. D., Jr., Brown, D. W., Friedman, A. S., et al. 1985. Standard Analytical Procedures of the NOAA National Analytical Facility, 1985–1986: Extractable Toxic Organic Compounds (2nd ed.). NOAA Tech. Mem. NMFS F/NWC-92. Seattle: NOAA National Marine Fisheries Service.

National Oceanic and Atmospheric Administration (NOAA). 1983. Annual NEMP Report on the Health of the Northeast Coastal Waters of the United States, 1981. NOAA Tech. Mem. NMFS-F/NEC-20. Rockville, MD: NOAA National Marine Fisheries Service.

National Oceanic and Atmospheric Administration (NOAA). 1990. Estuaries of the United States: Vital Statistics of a National Resource Base. A Special NOAA 20th Anniversary Report. Rockville, MD: NOAA Office of Oceanography and Marine Assessment.

National Oceanic and Atmospheric Administration (NOAA). 1991a. National Status and Trends Program—Second Summary of Data on Chemical Contaminations in Sediments from the National Status and Trends Program. NOAA Tech. Mem. NOS OMA 59. Rockville, MD: NOAA Office of Oceanography and Marine Assessment.

National Oceanic and Atmospheric Administration (NOAA). 1991b. Status of the fishery Resources Off the Northeastern United States for 1990. NOAA Tech. Mem. NMFS-F/NEC-81. Rockville, MD: NOAA National Marine Fisheries Service.

National Oceanic and Atmospheric Administration (NOAA). 1991c. The 1990 National Shellfish Register of Classified Estuarine Waters. Rockville, MD: NOAA Office of Oceanography and Marine Assessment, Strategic Assessment Branch.

O'Connor, T. P. 1990. Coastal Environmental Quality in the United States, 1990: Chemical Contamination in Sediments and Tissues. A Special NOAA 20th Anniversary Report. Rockville, MD: NOAA Office of Oceanography and Marine Assessment.

O'Connor, T. P., and Ehler, C. N. 1991. Results from NOAA National Status and Trends Program on distributions and effects of chemical contamination in the coastal and estuarine United States. Environ. Monit. Assess. 17:33–49.

Reid, R. N., O'Reilly, J. E., and Zdanowicz, V. S. (eds.). 1982. Contaminants in New York Bight and Long Island Sound Sediments and Demersal Species, and Contaminant

Effects on Benthos: Summer 1980. Tech. Mem. NMFS-F/NEC-16. Woods Hole, MA: NOAA National Marine Fisheries Service.

Schubel, J. R., and Okubo, A. 1972. Comments on the Dispersal of Suspended Sediments Across the Continental Shelves. In D. J. P. Swift and O. H. Pilkey (eds.), Shelf Sediment Transport: Process and Pattern. Stroudsburg, PA: Dowden, Hutchinson, and Ross, pp. 333–346.

Schwartz, J. P., Duston, N. M., and Batdorf, C. A. 1991. PCBs in Winter Flounder, American Lobster, and Bivalve Molluscs from Boston Harbor, Salem Harbor, and Coastal Massachusetts: 1984–1989. Publ. No. 16,966-63-250-10-91-C.R. Salem, MA: Massachusetts Office of Environmental Affairs, Division of Marine Fisheries, Cat Cove Marine Laboratory.

Sericano, J. L., Atlas, E. L., Wade, T. L., and Brooks, J. M. 1990. NOAA's Status and Trends Mussel Watch Program: Chlorinated pesticides and PCBs in oysters (*Crassostrea virginica*) and sediments from the Gulf of Mexico, 1986–1987. Mar. Environ. Res. 29:161–203.

Shea, D., Lewis, D. A., Buxton, B. E., et al. 1991. The Sedimentary Environment of Massachusetts Bay: Physical, Biological, and Chemical Characteristics. Boston: Massachusetts Water Resources Authority.

Shepard, F. P. 1963. Submarine Geology (2nd ed.). New York: Harper and Row.

Shigenaka, G., and Lauenstein, G. G. 1988. National Status and Trends Program for Marine Environmental Quality: Benthic Surveillance and Mussel Watch Projects Sampling Protocols. NOAA Tech. Mem. NOS OMA 40. Rockville, MD: NOAA Office of Oceanography and Marine Assessment.

Sindermann, C. J., Csulak, F., Saywer, T. K., et al. 1989. Shell diseases of crustaceans in the New York Bight. NOAA Tech. Mem. NMFS-F/NEC-74. Woods Hole, MA: NOAA National Marine Fisheries Service.

Stegeman, J. J., Kloepper-Sams, P. J., and Farrington, J. W. 1986. Monooxygenase induction and chlorobiphenyls in deep-sea fish *Coryphaenoides armatus*. Science 231:1287–1289.

Stegeman, J. J., Teng, F. Y., and Snowberger, E. A. 1987. Induced cytochrome P450 in winter flounder (*Pseudopleuronectes americanus*) from coastal Massachusetts evaluated by catalytic assay and monoclonal antibody probes. Can. J. Fish. Aquat. Sci. 44:1270–1277.

Texas A&M University Geochemical and Environmental Group. 1988. Texas A&M University, Geochemical and Environmental Research Group: Second Annual Report on Analyses of Bivalves and Sediments for Organic Chemicals and Trace Elements. Report to the National Oceanic and Atmospheric Administration. College Station, TX: Texas A&M Research Foundation.

Wade, T. L., Atlas, E. L., Brooks, J. M., et al. 1988. NOAA Gulf of Mexico Status and Trends Program: Trace organic contaminant distribution in sediments and oysters. Estuaries 11:171–179.

Weaver, G. 1984. PCB contamination in and around New Bedford, MA. Environ. Sci. Technol. 18:22A–27A.

Windsor, J. G., and Hites, R. A. 1979. Polycyclic aromatic hydrocarbons in Gulf of Maine sediments and Nova Scotia soils. Geochim. Cosmochim. Acta 43:27–33.

Ziskowski, J. J., Despres-Patanjo, L., Murchelano, R. A., et al. 1987. Disease in commercially valuable fish stocks in the Northwest Atlantic. Mar. Pol. Bull. 18:496–504.

14

Biological Responses to Toxic Contaminants in the Northeast Shelf Large Marine Ecosystem

Harris H. White and Andrew Robertson

Abstract

This paper reviews dozens of published studies and attempts to depict an organized view of the extent of biological effects of contaminants in the northeastern coastal environment. We identified four general strategies that were utilized to relate biological responses to toxic contaminants: field/field strategy (biological response of feral organisms collected from or exposed to natural field conditions); field/lab strategy (biological response of feral organisms exposed to artificially contaminated laboratory regimes); lab/field strategy (biological response of laboratory organisms exposed to natural or field conditions); and lab/lab strategy (biological response of laboratory organisms to laboratory regimes). Papers utilizing the first three strategies have been drawn on for this review.

The literature demonstrates that biological responses to contaminants have been detected in many estuarine areas of the Northeast, particularly those locales near urban and industrial centers. Toxic effects have been noted more often in the inner New York Bight Apex and Long Island Sound than in other Northeast estuaries, but this is mostly a result of more intensive studies in the New York area. Organisms collected from or exposed to certain portions of Boston Harbor and the Elizabeth River (Virginia) have also exhibited toxic responses, but there are fewer studies in these estuaries. Biological responses based on only one or two studies have also been detected in a number of other Northeast estuaries.

We found that a great deal of the work in the Northeast was merely descriptive and did not provide the basis for definitive conclusions about biological responses to contaminants. Attempts to relate biological response measurements to population and community effects were conspicuously few for the Northeast. Even on the individual organism level, the significance of a biological response to contaminants was rarely investigated. We identified the pressing need for more coherent, broadly focused, and long-lived programs to monitor biological responses to contaminants in the Northeast.

Measurements of anthropogenic contaminants in marine and estuarine environments have been made over many decades. A number of regional, national, and even international measurement programs have been developed to determine and track the levels of contaminants in sediments and living tissues over large geographical areas.

In recent years, there has been growing focus on evaluating what toxic effects, if any, anthropogenic compounds may have on organisms inhabiting contaminated environments. However, unlike geographically large-scale assessments of contaminant concentrations, the effort to assess biological effects over wide areas is still in its infancy. In part, the paucity of uniform biological response measurements applied over large areas may be attributed to disagreement within the scientific community over what constitutes an important biological response; it may also be attributed to the more complicated nature of biological response measurements vis-à-vis chemical contamination measurements. Whatever the reason, area-wide measurements of biological responses to toxic contaminants can be very useful in many environmental management decisions, and yet such measurements are often unavailable.

Certainly this is the case for the estuaries and continental shelf regions of the northeast United States. With only a few exceptions, biological response measurements in this region have been highly localized to specific estuaries or coastal waters. In this chapter, we review several dozen decidedly heterogeneous and incommensurable studies from the published literature and attempt to depict as organized a view as possible of the extent of biological effects of contaminants in the Northeast.

In our review, we excluded papers that reported on the response of laboratory organisms to laboratory regimes (e.g., most toxicity bioassays) because these studies are not explicitly linked to the environment of the Northeast. Furthermore, we limited our review to the literature on biological effects of contaminants, excluding studies on bioaccumulation and tissue residues. We view the latter more as a measure of exposure than a measure of effect.

Local Assessments

Maine

Despite the known presence of contaminants in some Maine estuaries, there are few studies on biological effects. Over the period 1967–1977, Yevich and Barszcz (1977) conducted histologic examinations on 14,000 soft-shell clams, as well as on 17 other marine invertebrates, from 11 oil spill sites, mostly in the

Northeast. Many of the sites were also contaminated by heavy metals and other organics. They found that only soft-shell clams, *Mya arenaria,* collected from two sites in Maine (Searsport and Harpswell Neck), consistently exhibited neoplasms. Gonadal neoplasms were noted in about 6% of the clams from Searsport, and hematopoietic neoplasms were noted in about 10% of the animals from Harpswell Neck. There was no relationship, however, between neoplasm incidence and sex, age, or season, nor any indication that the disease is caused by or exacerbated by anthropogenic contaminants.

Oxidative metabolism by cytochrome P-450 monooxygenase is a process determining the biological fate of many xenobiotic organic compounds. Therefore, the finding of cytochrome P-450 induction in feral organisms is often viewed as evidence that the organisms have been exposed recently to organic contaminants. Foureman et al. (1983) noted cytochrome P-450 activity in a large fraction (as much as 80% or higher) of winter flounder (*Pseudopleuronectes americanus*) from the vicinity of Mount Desert Island, Maine, a relatively uncontaminated environment. However, the presence of P-450 activity was highly variable and could not be correlated to liver weight, body weight, gonad weight, feeding status, sex, month, or year. The authors speculated that low levels of PAH or other organics, exposure history, genetics, organic compounds of natural botanic origin, or other factors may explain the presence and high variation of P-450 activity in fish from this fairly clean environment.

Massachusetts: Boston Harbor

Cytochrome P-450 activity was the focus of a study by Stegeman et al. (1987) in winter flounder collected from Boston Harbor, Plymouth Bay, Nantucket Sound, and Buzzards Bay (outside New Bedford Harbor). Interestingly, the induction was strongest in fish from Boston Harbor and Plymouth, which are, respectively, among the most contaminated and least contaminated estuaries in Massachusetts. Furthermore, some of the specimens from New Bedford, the sediments of which contain high levels of many organic contaminants, exhibited virtually no P-450 induction. Unfortunately, tissue residues were not analyzed in any fish, so it was impossible to determine which contaminants might be the primary inducers. Elskus et al. (1989) further explored the enigmatic relationship between P-450 activity and tissue contaminant concentrations, and found that very high levels of PCB contamination in winter flounder could actually suppress P-450 catalytic activity, further confusing the relationship between environmental contaminants and P-450 induction.

Smolowitz et al. (1989), on the other hand, did find among samples from Boston Harbor a general agreement in winter flounder between hepatic pathologic conditions and cytochrome P-450 activity.

Varanasi et al. (1989) analyzed hepatic DNA adducts in winter flounder from Boston Harbor and found relatively high levels of adducts. However, the study was only preliminary in nature, and they did not compare the DNA effect in winter flounder from less contaminated reference areas.

A suite of studies has been conducted on the occurrence of neoplasms, non-neoplastic lesions, and other abnormalities in liver tissue of winter flounder from Boston Harbor. For example, Murchelano and Wolke (1985, 1991) and Bodammer and Murchelano (1990) found adult winter flounder from Boston Harbor often exhibited a variety of neoplastic lesions and non-neoplastic cellular anomalies. The affected fish were always adults, although Moore et al. (1989) showed that cellular alterations preceding neoplasia could be found even in younger fish from Boston Harbor. The percentage of affected fish varied widely, from a few percent to 100%, but was generally higher than that reported in winter flounder populations elsewhere, such as on Georges Bank, in Narragansett Bay, and in uncontaminated Northeast estuaries. Again, no tissue residues were analyzed, so the offending toxicants, if any, could not be correlated with the prevalence of lesions.

Carr et al. (1991) compared a number of biochemical and histopathological conditions in winter flounder collected from Boston Harbor and Plymouth Beach, and demonstrated a difference in response of animals from the two sites. In particular, 50–64% of the fish from Boston Harbor exhibited hepatic lesions, whereas only 3% of the specimens from Plymouth Beach were similarly affected. Pereira et al. (1992) found that Boston Harbor winter flounder bearing liver lesions had significantly reduced capacity (relative to fish without lesions from Boston Harbor) to produce vitellogenin, a precursor protein manufactured by the liver and crucial to egg yolk development in the larvae.

Massachusetts: Buzzards Bay

Yevich and Barszcz (1977) looked for signs of neoplasms in scallops (*Aequipecten irradians*) at West Falmouth, site of a 1970 oil spill, but found no specimens with neoplasms in seven years of sampling after the spill.

Hematopoietic neoplasia (leukemia) occurs in many populations of softshell clams throughout the Northeast. Reinisch et al. (1984) found one of the highest reported prevalences (88.2%) of the disease in the clam population of New Bedford Harbor, an environment severely contaminated by PCBs and many other anthropogenic organic compounds, although Leavitt et al. (1990)

found the prevalence to be seasonally variable and much lower (39.3%, including all stages of the disease). The latter investigators also determined the disease prevalence in clams of Little Buttermilk Bay, a clean site at the northeastern tip of Buzzards Bay, and found it to be less than half that of New Bedford Harbor. They showed that development of the neoplasia is linked to season as well as age, whereas the role of contaminants in compromising the animals' defense mechanisms is only speculative.

Nacci and Jackim (1989) tested a DNA unwinding assay on mussels caged in New Bedford Harbor and noted a significant increase in DNA strand breaks in gill tissues of animals held at the site for as little as three days. However, the work is preliminary and, as the authors point out, the DNA unwinding assay may not reflect true genotoxicity in the whole animal.

Gardner et al. (1989) found that the incidence of winter flounder liver neoplasms was high in New Bedford Harbor (26%), Boston Harbor (Quincy Bay, 23%), and Black Rock Harbor, Connecticut (18%), all areas of severe sediment contamination. However, there is some year-to-year variation, and fish from some contaminated sites, such as Gaspee Point in upper Narragansett Bay, exhibited much lower incidences of neoplasia.

Rhode Island: Narragansett Bay

Murchelano and Wolke (1985) examined liver tissues excised from winter flounder and another resident flatfish, windowpane flounder (*Scophthalmus aquosus*), from upper Narragansett Bay and New Haven Harbor, two relatively contaminated environments surrounded by urbanized areas. Younger fish did not exhibit any liver lesions, but in fish over 25 cm long, about 10% exhibited non-neoplastic liver lesions and 3.4% exhibited neoplastic liver lesions. Although the abundance of lesions in these fish was not as egregious as liver lesions in fish from Boston Harbor, it certainly exceeded the zero incidence of lesions in specimens from Georges Bank and other uncontaminated, nearshore environments.

Black et al. (1988) compared the survival rate and weight of eggs and larvae of winter flounder collected from Gaspee Point, near Providence, and a clean site near Fox Island in the southern part of the Bay. Although there were significant differences between the two stations in hydrocarbon residues in the eggs, no interstation differences were observed in the eggs with regard to percentage of fertilization, embryo or larval survival, or length or weight at hatch. The experiment was repeated in Buzzards Bay, comparing New Bedford Harbor, Apponagansett Bay, and, again, Fox Island. Eggs from New Bedford Harbor were the most contaminated with PCBs (39.6 $\mu g\ g^{-1}$ dry weight), eggs from Fox Island the least contaminated (1.08 μg g^{-1} dry weight), and eggs from Apponagansett moderately contaminated (15.7 μg g^{-1} dry weight). Although New Bedford eggs attained the lowest weight and length at hatch, other measures of reproductive success (percentage of fertilization, percentage of embryo survival, final length of larvae, etc.) were unaffected.

Connecticut and New York: Long Island Sound

The benthic infauna were the focus of a study by Tietjen (1977). He analyzed the population structure of nematodes in sediment samples from 18 stations across the width and length of Long Island Sound. He also determined the concentrations of contaminants in the samples and found no correlation between heavy metal or organic carbon content of the sediment, and population density, species composition, or species diversity of the nematode populations.

Another negative finding was reported by Yevich and Barszcz (1977), who were unable to find neoplasms in soft-shell clams at oil spill sites in New Haven and Bridgeport, Connecticut. These are areas that are generally contaminated by a variety of anthropogenic chemicals, so the failure to find neoplasms may indicate general resistance of this organism to the effects of toxicants.

Stiles et al. (1991) looked at the early reproductive success of the hard clam (*Mercenaria mercenaria*) from five sites near urban areas (Greenwich, Norwalk, Bridgeport, Milford, and New Haven) along the northern shore of Long Island Sound. Mature clams were harvested and spawned in the laboratory. The fertilized eggs were cultured in water collected from just above the sediment in their native sites. A variety of embryonic developmental endpoints were used as indices of reproductive success. The results were mixed, although the embryos from Bridgeport clams, which had the highest tissue contaminant levels, exhibited the worst overall performance. Depending on the endpoint, there was greater impact on embryonic development at different sites. In some instances and some sites, there was no impact on development relative to embryos held in clean reference seawater. Of importance, Stiles et al. analyzed the gonads of the parental stock from which the experimental eggs were spawned and determined that although there were some differences among the five sites, all the gonads had fairly low concentrations of PCBs, copper, and cadmium. Therefore, it is possible that other unmeasured anthropogenic or natural factors were responsible for the variable embryonic development.

Pereira et al. (1992) followed the seasonal cycle of vitellogenin, an egg yolk precursor protein, through monthly sampling of winter flounder at a relatively clean Long Island Sound site (Shoreham, New York) and two degraded areas (Black Rock and New Haven Harbors, Connecticut). Large fish from the degraded areas exhibited lower blood vitellogenin and lowered ovarian lipid levels (suggesting impaired vitellogenin uptake) relative to the specimens collected from the clean site. Regrettably, the authors did not determine the tissue residue levels of any contaminants, which would have established more firmly that toxicants could be contributing to the differences in vitellogenin production.

Nelson et al. (1991) compared the viability and larval development rate of winter flounder larvae from various sites in Long Island Sound (three of them heavily urbanized) and two sites in Boston Harbor. Only New Haven Harbor and Boston Harbor consistently produced low percentages of viable hatch and small larvae. Moreover, at the various sites, there were no significant differences in concentration of PCBs in livers of the spawned adults, in sediments, or in eggs, so the effect at New Haven and Boston could not be attributed to that contaminant.

With regard to biological responses to toxicants, one of the most intensively studied sites in the Northeast is Black Rock Harbor, near Bridgeport, Connecticut, which was chosen by the U.S. Environmental Protection Agency and the U.S. Army Corps of Engineers as the venue for their Field Verification Program. This six-year program in the mid-1980s focused research effort on evaluation of environmental effects of disposal of material dredged from this harbor. Black Rock Harbor was selected for the study, among other reasons, because of the high levels of many anthropogenic contaminants in its sediments. A variety of biological responses to the dredged material was measured for a number of different species and test conditions.

Zaroogian et al. (1985), for instance, attempted to demonstrate the utility of adenylate energy charge (AEC) as an index of biological response to contaminants. They obtained the marine polychaete, *Nephthys incisa*, and the blue mussel, *Mytilus edulis*, from uncontaminated sites. The organisms were first acclimated to laboratory conditions, and then exposed to 100%, 50%, and 0% mixtures of Black Rock Harbor sediment and clean sediment from a reference site. The results were inconclusive: 50% contaminated sediment produced a greater reduction of AEC than 100% contaminated sediment. The authors suggested that animals in the more contaminated sediments minimized their filtering activities, thereby conserving AEC, but this was speculation.

In contrast, Nelson et al. (1985) measured the impact of 100%, 50%, and 0% mixtures of Black Rock Harbor sediment and clean sediment on the scope-for-growth (SFG) index in mussels. They demonstrated that exposures of the animals to higher concentrations of the contaminated sediment reduced the SFG index.

Pesch et al. (1985) utilized the polychaete *N. incisa* collected from a relatively clean site in central Long Island Sound to test the ability of Black Rock Harbor sediments to produce sister chromatid exchanges. For the most part, however, the worm's chromosomes were unresponsive.

Using a bioassay approach, Gentile et al. (1985) used laboratory-cultured mysids (*Mysidopsis bahia*) and laboratory-acclimated amphipods (*Ampelisca abdita*) to test the impact of Black Rock Harbor sediments on mortality, growth, and reproduction. They determined that all of the measured variables responded to the contaminated treatments, with the reproduction measure being the most sensitive endpoint for both animals.

Gardner and Yevich (1988) conducted a large study that was primarily interested in the histopathological effects of Black Rock Harbor sediments. Although they examined a total of three species of fish and six species of invertebrates, their focus was on winter flounder and the oyster *Crassostrea virginica*. Laboratory studies explored the effects of exposure to Black Rock Harbor sediments. Field studies focused on winter flounder collected from Black Rock Harbor and central Long Island Sound as well as from a reference site in Narragansett Bay, and on oysters, which were caged at sites in Black Rock Harbor and central Long Island Sound. Results clearly indicated widespread pathological effects from exposure to Black Rock Harbor sediments, ranging from inflammation to hyperplasia to neoplasms of various tissues. The authors also fed winter flounder with mussels that had been held in Black Rock Harbor sediments and had presumably accumulated contaminants; this increased the frequency of pathological responses, suggesting that the trophic route of contaminant exposure may also be significant.

New York: Hudson River

In an early survey of Atlantic tomcod (*Microgadus tomcod*) in the Hudson River estuary, Smith et al. (1979) found high prevalences (up to 25%) of gross liver abnormalities in the fish population. Cormier (1986) and Cormier et al. (1989) described the histopathology of widespread hepatocellular carcinomas in the tomcod population and compared it to the normal hepatocyte morphology in tomcod from the uncontaminated Pawcatuck River, Rhode Island. Klauda et al. (1981) attempted to determine the cause of the histopathologic effects in tomcods of the Hudson River by comparing the PCB levels of affected fish to PCB levels in fish exhibiting normal gross liver appearance. However, they were unable to demonstrate a relationship between PCB accumulation and liver abnormality, although the sample size was very small.

Mehrle et al. (1982) used vertebral strength to compare the biological response of striped bass (*Morone saxatilis*) taken from the Hudson River, the Potomac and Nanticoke Rivers in Chesapeake Bay, and a fish hatchery in North Carolina. They also analyzed the concentration of PCBs, some chlorinated pesticides, and a number of heavy metal contaminants in the tissues of the fish. As a whole, the Hudson River fish bore the highest levels of contaminants, whereas the hatchery-reared fish had the lowest levels. Vertebrae of the Hudson River striped bass had the least strength and ruptured under the lowest force, whereas the hatchery fish had the strongest vertebrae, indicating a general relationship between vertebral strength and environmental contaminants.

New York and New Jersey: Sandy Hook/Raritan Bay

Environmental response to cleanup efforts was examined in the lower Raritan estuary by Dean and Haskin (1964). Until 1958, the estuary was depopulated by severe pollution. A new sewer system began operation in the lower Raritan Valley in 1958, and the authors followed the recovery of the benthic population for three years. Previously uninhabited stretches regained their normal community assemblages.

While conducting ecological studies on the caridean shrimp, *Crangon septemspinosa*, Gopalan and Young (1975) noticed that a substantial portion of the populations from Sandy Hook Bay were affected by a disease characterized by rotting appendages and black erosions of the exoskeleton. Further collections revealed that between 9% and 15% of the shrimp from Sandy Hook Bay and 7% from western Long Island Sound were infected, whereas diseased specimens were rarely encountered in collections from Woods Hole, Massachusetts, or Beaufort, North Carolina. The authors also kept both diseased and normal-appearing specimens in aquaria for six weeks. Antibiotic was added to some of the aquaria, and the disease was found to be significantly reduced as a result, suggesting a possible infectious etiology. The disease may possibly be exacerbated by organic wastes in contaminated environments.

In routine histologic examinations of oysters from Raritan Bay, Farley (1988) noted extensive acute inflammatory lesions in many of the tissues. When sections of affected animals were stained, the lesions were found to contain large amounts of copper. The amount of staining was more intense in the summer, and much lower in oysters from Great Bay, a relatively uncontaminated New Jersey estuary 100 km to the south. This study presented an interesting (but perhaps uncommon) situation in which the causative agent of biological response could be explicitly identified.

From 1967 to 1971, Mahoney et al. (1973) conducted a field survey of fin erosion in the fishes of the Sandy Hook/Raritan Bay estuary. The principal species affected were the bluefish (*Pomatomus saltatrix*), summer flounder (*Paralichthys dentatus*), winter flounder, and weakfish (*Cynoscion regalis*). Incidence of the disease varied from 0 to 70%, depending on the year, season, and species, and was nearly always much more prevalent in the fish caught in the estuary than in fish caught even a few kilometers outside Lower Bay. The authors believe the disease was bacterial in nature, and they conjectured that the disease may have been exacerbated by the highly contaminated environment.

A similar result was found by Ziskowski and Murchelano (1975) in a 1973–1974 survey of five species of flatfish in Sandy Hook/Raritan Bay. Overall incidence of fin erosion ranged from 3.9% to 16.3%, significantly higher than the 0–3% incidence found in Great Bay. Ziskowski et al. (1980) described an abnormal, bent fin ray condition in winter flounder from Raritan Bay/Sandy Hook, but the incidence was extremely low (about 0.5%), so it may represent an inconsequential biological response.

New York and New Jersey: New York Bight Apex

Tietjen (1980, 1982) examined the distribution of meiofauna in sediments of the New York Bight apex. He demonstrated that the density of meiofauna, particularly nematodes (the dominant taxon), is not correlated with sediment type, organic carbon content, or heavy metal concentrations. However, in contrast to his findings in Long Island Sound (Tietjen 1977), the species composition in the New York Bight apex seemed to be affected by environmental factors associated with the sewage sludge and dredge spoil dumpsites in the apex. Species diversity was inversely correlated with Cr, Cu, Pb, and Zn, and normally occurring nematodes were replaced by pollution-tolerant species such as *Sabatieria pulchra*. Similar results were obtained by Steimle et al. (1982) in an intensive survey of the New York Bight apex during 1973–1974. Their data indicated benthic diversity changes, especially severe reduction of certain sensitive amphipod species in the vicinity of the dredge spoil and sewage sludge dumpsites.

A useful introduction to vertebrate and invertebrate diseases represented in specimens from the New York Bight was presented by Murchelano (1982). He discussed microbial diseases, the etiology of which is generally known, neoplastic diseases, and diseases of unknown etiology, such as fin erosion, skeletal anomalies, and black gill disease. From his review, it is evident that some of these diseases and abnormal conditions are more prevalent in the Bight than in less contaminated areas, whereas others are uncommon or comparisons with the prevalence in uncontaminated areas have not been adequate.

Young and Pearce (1975) collected lobsters (*Homarus americanus*) and rock crabs (*Cancer irroratus*) from the vicinity of the sewage sludge and dredge spoil disposal areas in the New York Bight apex. Approximately half the specimens of both species exhibited areas of skeletal erosion. Animals collected from areas removed from the dumpsites had a low incidence of such lesions. Moreover, lobsters and crabs from these cleaner areas were held for six weeks in aquaria containing sediments collected from the disposal sites, and virtually all the specimens developed skeletal lesions.

Fin erosion is one of the better studied diseases of New York Bight fish. Intensive trawl surveys conducted within the New York Bight and adjacent coastal areas to the northeast and south during 1973 and 1974 revealed that 14.1% of the winter flounder from the Bight apex exhibited fin rot disease (Ziskowski and Murchelano 1975). In comparison, only 1.9% of the winter flounder from Great Bay, New Jersey, showed signs of the disease. The authors suggested that there was a strong relationship between fin erosion and environmental contamination.

This work was continued in 1974–1975 (Murchelano and Ziskowski 1976) and again demonstrated the focus of fin erosion to be in the Bight apex. During this latter survey, 3.9% of the winter flounder from the apex showed fin erosion, compared with 0.7% outside the apex. Even within the apex, fish collected in areas of high carbon deposits (and presumably high concentrations of contaminants) had nearly twice the disease prevalence of fish from areas with lower carbon sediments. Given the mobility of winter flounder, however, the significance of this finding might be questioned.

Murchelano and Ziskowski (1982) summarized winter flounder fin erosion survey data for the entire five-year period from 1973 to 1977. Although there was a steady decline in prevalence at all sites during the study period, the apex had the highest average prevalence of the disease (7.4%). Murchelano and Ziskowski (1976) found that 3.4% of the summer flounder from Raritan Bay/Sandy Hook exhibited fin erosion, whereas only 0.6% from the Bight apex, and 0% in Great Bay, demonstrated the condition.

Murchelano and Ziskowski (1976) also held winter flounder in nylon mesh cages placed on the sediment at the sewage sludge dumpsite and at a less contaminated control site within the apex. After 12–29 days the fish were recovered. Specimens that had been placed at the sewage sludge site invariably exhibited fin erosion and were in substantially worse condition than those at the control site. Murchelano and Ziskowski (1982) stated that the caudal fin lesions, at least, were probably the results of effects of the cages rather than the environment.

In their 1973–1977 study, Murchelano and Ziskowski (1982) found that PCBs and trace metals were not consistently elevated in liver and muscle tissue of winter flounder with fin erosion, suggesting that the role of environmental contaminants is unclear. Sherwood (1982), on the other hand, found that fish with fin erosion had significantly elevated PCB levels in muscle, liver, and brain tissue.

Longwell and Hughes (1982) and Longwell et al. (1992) reported on their study of development rates, viability, and chromosome abnormalities in mackerel (*Scomber scombrus*) embryos collected from surface waters in the New York Bight. During the 1974 spawning season, they found significantly lower egg viability in areas of the Bight that generally were impacted by contaminants from the Hudson–Raritan estuary and the dumpsites in the apex. During the 1977 season, they also determined the concentrations of a suite of heavy metals and organic contaminants in the surface water and zooplankton at each station where mackerel eggs were sampled. Embryonic effects were associated with certain heavy metals in the surface water, as well as aromatic and chlorinated hydrocarbons in the zooplankton.

Chesapeake Bay: York and Elizabeth Rivers

For the most part, the York River has been seen as a relatively uncontaminated estuary, although Thiyagarajah and Bender (1988) found one specimen in a sample of ten toadfish (*Opsanus tau*) from the vicinity of a refinery outfall that had pancreatic and liver lesions. In the absence of samples from other sites, the significance of this finding is difficult to interpret.

The nearby Elizabeth River, on the other hand, is one of the most contaminated estuaries in the Northeast and has been the subject of a number of studies on biological responses to toxicants. Sediments from certain Elizabeth River sites bear very high levels of organic and inorganic contaminants. As a demonstration of the potency of some of these sediments, Hargis et al. (1984) held spot (*Leiostomus xanthurus*) from the clean Ware River in an experimental aquarium with highly contaminated sediments from the Elizabeth River, and other spot from the Ware River in a control aquarium with uncontaminated sediments from the York River. By the end of the 28-day experiment, 56% of the fish in the contaminated sediment tank had died, whereas there was 100% survival in the control tank. Moreover, most of the exposed fish showed evidence of injury, severe fin erosion, and heavy integumental lesions, whereas the control fish remained healthy. Even when spot were held in an aquarium that received the effluent of the contaminated sediment tank, there was some mortality, and nearly half of the specimens developed lesions.

In a similar demonstration of Elizabeth River sediment toxicity, Roberts et al. (1989) exposed laboratory-acclimated juvenile spot to dilutions of Elizabeth River sediments and were able to calculate LC-50 values. They furthermore judged that even chronic effects might be produced when more than 10% Elizabeth River sediment was present.

Although field populations of Elizabeth River fish also exhibit toxicant responses, the incidence is much lower than one might think from the laboratory results mentioned above. For instance, at an Elizabeth River site highly contaminated by petroleum, creosote, and other anthropogenic compounds, Hargis et al. (1989) collected and necropsied mummichogs (*Fundulus heteroclitus*), but found only 2% of them bearing neoplastic lesions. No fish were examined from clean reference sites, so the finding of neoplasms in the Elizabeth River specimens cannot be interpreted definitively.

Hargis and Zwerner (1988) examined 70,000 specimens of eight fish species collected from heavily contaminated Elizabeth River sites. Depending on species and site, the occurrence of lens cataracts, fin erosion, and skin ulcerations ranged from 0.01% to 5.66% of the specimens, with the higher number generally in the area of higher PAH contamination. The incidence of these abnormalities at a reference site, the Nansemond River, was essentially nil. In similar studies, however, Huggett et al. (1987) and Bender et al. (1988) surveyed cataracts, skin lesions, and fin erosion in Elizabeth River fish and showed an incidence as high as 10–30% at the more heavily contaminated sites, much higher values than in the previously cited reference.

Weeks and Warinner (1984) captured spot (*Leiostomus xanthurus*) and hogchoker (*Trinectes maculatus*) from the Elizabeth River and the relatively uncontaminated Ware River in Virginia, and quantified the phagocytic efficiency of macrophages isolated from the kidneys. Cells harvested from the Elizabeth River fish were severely impaired relative to those from the Ware. Interestingly, the authors also captured fish from the Elizabeth River and held them in clean water for several weeks before testing. After this recovery period, macrophage phagocytic efficiency had returned to normal, demonstrating that the effect was related to exposure to the Elizabeth River environment, but was reversible. A nearly identical experiment, with similar results, was conducted by Weeks et al. (1986) using the same two species of fish from the Elizabeth and York Rivers, and quantifying the chemotactic ability of kidney macrophages.

Other Chesapeake Bay Estuaries

Two papers describing preliminary surveys of other estuaries in Chesapeake Bay reported the discovery of neoplasms. May et al. (1987) found neoplasms in about 15% of the white perch (*Morone americana*) livers they examined from fish collected in 15 estuarine tributaries of the Bay. However, the estuaries were not specified. Similarly, Bunton and Baksi (1988) found 2 of 21 specimens of white perch from Chesapeake Bay exhibiting liver cholangiomas.

Continental Shelf and Slope

Stegeman et al. (1986) found cytochrome P-450 induction in the livers of the rattail (*Coryphaenoides armatus*), a deep-sea fish they had collected live on the continental slope near the Hudson Canyon. This indicates that some deep-sea fish have an induced capacity to biotransform xenobiotic contaminants and therefore that biochemical effects of xenobiotics may be occurring even in the deep ocean.

Discussion

We noted a wide array of biological responses described in the references we reviewed for the Northeast Shelf and estuarine ecosystem. We identified four general types of studies that were used to relate biological responses to toxic contaminants.

1. *Field/field.* Biological measurements on feral organisms collected from field sites, feral organisms collected and then caged in field sites, or feral organisms collected and then exposed to laboratory regimes containing media (water or sediment) from field sites.
2. *Field/laboratory.* Biological response of feral organisms exposed to artificially contaminated laboratory regimes. Note that the longer a feral organism is held in and acclimated to laboratory conditions before testing, the closer the *field/laboratory* comes to a *laboratory/laboratory* situation. The line between the two strategies is gray.
3. *Laboratory/field.* Biological response of laboratory-cultured or laboratory-acclimated organisms that are caged in field sites or laboratory organisms exposed to laboratory regimes containing media (water or sediment) from field sites.
4. *Laboratory/laboratory.* Biological response of laboratory-cultured or laboratory-acclimated organisms to artificially contaminated laboratory regimes. We did not review the literature using this strategy, which includes most toxicity bioassays, because these studies are not explicitly linked to the Northeast environment.

In our review, we discovered that most biological responses to contaminants have been measured and evaluated in the Northeast using field/field studies. A great deal of the work is merely descriptive; organisms were collected from known contaminated sites and a variety of anomalies or injuries were measured. This, however, constitutes nothing beyond step one in the scientific method, viz., observation of a system so that coherent and well-defined experiments can be designed. Rarely do such preliminary observations provide the basis for definitive conclusions.

It is not our intent to review the many unique strengths and shortcomings characteristic of field versus laboratory studies. We will simply note that, as a rule, the ability to manipulate treatment conditions and control extraneous factors is enhanced in the laboratory setting over the field setting. This allows one to pinpoint contaminants with toxic effects and often to quantify dose-response relationships. However, laboratory-oriented strategies sacrifice realism, because many field conditions are far too complex and variable to be simulated in the laboratory. Lack of realism in laboratory strategies diminishes their utility in evaluating and predicting field responses to contaminants.

The tension between the control of the laboratory and the realism of the field is not an unbridgeable dichotomy. The cage experiments that were used in a number of the studies we reviewed provide one approach. Although no mesocosm studies were cited, the chief raison d'être for these large enclosures is to bridge the laboratory-field gap. A third approach, used in several papers, is to corroborate field/field measurements with field/laboratory measurements, where field organisms can be subjected to definable treatment conditions and their responses seen to mimic the responses of wild organisms. In some select instances, there is already a body of literature describing such field/laboratory responses. A fourth approach, not often applicable, is to measure changes in field responses before and after a very large change in environmental conditions (e.g., the installation of a sewer system with major reductions in contaminant loading to an estuary), and to assume the likelihood that major changes in field responses are ascribable to the change in environmental conditions. A fifth possible approach to rendering some control in field studies is to apply epidemiologic techniques. This is a strategy that requires very rigorously and statistically designed sampling schemes and an expertise that has not adequately been brought to bear in environmental science generally, and certainly not in studies of the Northeast ecosystem.

Many research efforts have not taken any of the above approaches and have shied away from addressing the starkly different philosophies and limitations of field- and laboratory-oriented studies. Indeed, many of the papers we reviewed did not even demonstrate that field organisms had actually been exposed to toxicants, and at what level of concentration, weakening any conclusions about biological responses to those toxicants. The problem is seen most clearly when the variable being measured responds broadly to a host of natural and anthropogenic factors. To an extent, all biological responses fall into this category, but responses such as scope for growth, adenylate energy charge, and other measurements of normal biological function are the ones most in need of rigorously controlled conditions. When such functions (e.g., reproduction and growth) are tied to seasonal or annual variability, the need for controls is all the stronger.

Paucity of Studies

What strikes one immediately in a review of the literature for the Northeast is the dearth and narrowness of studies of biological responses to toxicants. Although the severe contamination of some estuaries and ocean dumpsites is extensively documented, measurements of effects are sparse and often examine only one response in one species or taxon over a short time period. There has been little opportunity to revisit old sites or reexamine old conclusions. The study by Steimle and Caracciolo-Ward (1989), however, did reassess the benthic macrofauna of the Raritan estuary and showed that it is not as degraded as previously reported and is, in many ways, similar to other relatively uncontaminated estuaries in the Northeast.

There is probably more literature pertaining to biological effects of contaminants on marine organisms in the New York–Long Island Sound area than elsewhere in the Northeast. Much pertinent information up until 1980 is presented or referred to in the volume *Ecological Stress and the New York Bight: Science and Management* (Mayer 1982). Still, even the New York region is far from being properly assessed.

In contrast, information on biological responses to contaminants in Narragansett Bay and the Chesapeake Bay and its estuaries is minimal. Organisms in the Elizabeth River have been studied to a substantial extent, but the other important estuaries of the Chesapeake have received less attention. For instance, May et al. (1987) commented on the "total lack of reference data on the extent of neoplastic disease in fish from the Chesapeake Bay." Similarly, Delaware Bay has been virtually ignored in this regard.

The intensive effort invested by the U.S. Army Corps of Engineers and the U.S. Environmental Protection Agency at Black Rock Harbor in Long Island Sound is perhaps unique. Given the extreme temporal, spatial, and interspecies variability of biological responses, such an approach is essential.

Area-Wide Assessments

There are only a handful of area-wide assessments that allow one to directly compare biological responses for the entire Northeast ecosystem. Ziskowski et al. (1987) provided one such study when they looked at a number of diseases and morphologic anomalies in ten economically important fish species throughout the Northeast ecosystem, including nearshore and offshore collection sites. Only fin erosion was consistently more prevalent in nearshore environments or near urban centers. Similarly, Zdanowicz et al. (1986) reported that some pathologic conditions in winter flounder paralleled the distributions of sediment contaminants throughout the Northeast estuaries, whereas other pathologic conditions did not. Longwell et al. (1992) showed that adult mackerel, windowpane flounder (*Scophthalmus aquosus*), and winter flounder from the New York Bight apex and Long Island Sound had higher frequencies of micronucleated erythrocytes (evidence of mitotic abnormality) than did fish from the Gulf of Maine, Georges Bank, or offshore regions of the Middle Atlantic Bight. Hillman et al. (in review) looked at neoplasia in mussels and oysters collected by the National Status and Trends Mussel Watch, a program of the U.S. National Oceanic and Atmospheric Administration (NOAA). They reported a very low incidence (generally less than 5%) of neoplasia in *Mytilus edulis* in Northeast estuaries and were unable to correlate the neoplasms that did occur to any of a wide suite of anthropogenic contaminants found in the tissues of the populations from which the infected organisms were taken. Johnson et al. (1994) investigated the relationship between exposure to xenobiotic compounds and ovarian development in winter flounder by sampling fish from a number of contaminated and uncontaminated sites in the Northeast, including Boston Harbor, Raritan Bay, and nearby estuaries. In general, they found that indicators of contaminant exposure were elevated and prevalences of lesions were highest in fish from Boston Harbor and Raritan Bay (although some measures of reproductive effect did not correlate well with contaminant exposure). Harshbarger and Clark (1990) summarized historical data on epizootic neoplasms in teleosts throughout North America, and reached the general conclusion that tumor-bearing fish are generally located in areas where chemical contaminants are concentrated. However, the data set is not extensive, and the conclusion is somewhat gratuitous, since only the occurrence of tumors was reported, and not the frequency in the sampled populations. The initial phase of a more thorough assessment has recently been completed for that portion of the Northeast ecosystem between Raritan Bay and Salem Harbor, Massachusetts (Johnson et al. 1992). They examined neoplastic, preneoplastic, and other histopathologic conditions of winter flounder livers and found a clear relationship between disease incidence and contaminant concentrations.

These are among the very few papers that have measured biological responses over the heterogeneous Northeast ecosystem. The results are mixed: depending on the measuring system and biological response, there may or may not be indications of significant biological effects in regions of known contamination. This merely highlights the need for an intensive and better-organized, area-wide assessment for the whole Northeast ecosystem.

Moreover, if the Northeast environment, or portions thereof, is an ecosystem in its own right, one would expect to find interactive effects throughout the ecosystem. This might occur as contaminants entering one estuary of the Northeast ecosystem are transported to and have an impact on the biota of other areas of the system. Alternatively, an interactive effect might occur as organisms move or are transported between contaminated estuaries. To date, there are virtually no studies in the Northeast that examine such true, regional effects, where two or more estuaries behave as a single ecosystem. Indeed, there is often a tendency to *minimize* possible interactions with environments outside the immediate study locale, because such interactions merely introduce confounding influences and weaken the investigator's ability to correlate contaminants and biological responses at the site of interest.

Occasionally there have been attempts to demonstrate interactions between an estuary and the environment outside the estuary, but such relationships have been difficult to identify. The 1976 hypoxia along the New Jersey coast, for instance, was an event caused primarily by natural factors. Contributory effects of contaminant loading from the inner New York Bight apex, only a few kilometers away, were minimal (Segar and Berberian 1976; Atwood et al. 1979).

For the most part, however, the literature on biological responses to contaminants in the Northeast is nothing more than a collection of works on biological responses in individual estuaries. Biological responses that extend outside the estuary of contaminant loading may exist, but they are probably limited.

The Migration Problem

As mentioned before, very little attention has been given in the literature to documenting the actual exposure of organisms to contaminants. The issue is most critical in the assessment of motile organisms—and one must bear in mind that virtually all marine organisms are motile during one or more stages in their lives. Hence, the finding of a biological response in animals collected from one site must always be evaluated in light of the possible migrations and movements of those animals over their exposure history. Disagreement and lack of information on the migration of even well-studied species, such as winter flounder, confuse the assessment of any biological responses to toxicants and leave open to question whether the findings of response or no response can be attributed to migration and exposure patterns (Johnson et al. 1994).

The "So What" Question

Perhaps the most troubling issue, not unique to the Northeast, in the assessment of biological response to contaminants, is the question of the significance of the measured response. Indices such as mixed-function oxygenase induction certainly are not significant in their own right and are more properly viewed as measurements of exposure rather than effect. Moreover, as several papers revealed, cytochrome P-450 activity can be found even in animals from purportedly uncontaminated environments. Other responses, such as DNA adducts, have a variable importance to the welfare of the organism, depending on (among other things) where in the DNA chain the foreign compound is adducted, what compensation the organism can make, and how rapidly the adduct can be removed. Consequently, DNA adducts do not necessarily imply a significant effect.

The significance of even more directly deleterious biological responses must be treated with caution. For instance, the presence of tumors, lethal deformities in eggs and larvae, retarded growth, and other patently injurious effects spell trouble for individual organisms but may have little bearing on recruitment and growth of a population. The impact of climate, hydrography, food supply, fishing pressure, and other factors unrelated to contaminants may totally dominate population dynamics of some species.

Unfortunately, attempts to relate individual biological response measurements to population and community effects are conspicuously few for the Northeast ecosystem. The obstacles to achieving such an assessment in the New York Bight are ably articulated by Wolfe et al. (1982), but their conclusions are applicable elsewhere as well. Boesch (1982) wrote one of the very few papers for the Northeast that deals in depth, and as quantitatively as possible, with ecosystem consequences of biological response. Last, as White (1984) pointed out, every biological response, even on the ecosystem level, must be judged in terms of its significance to humans. Protecting the naturally occurring dermovirus in Chesapeake Bay would be no more egregious than protecting the Bay's oysters, were it not that human interest justifies that distinction.

A Philosophy for the Future

Programs to monitor environmental contaminant concentrations have developed over the last two decades. There now are national and even international efforts that emphasize uniformity, statistical design, and strict quality control. The Northeast ecosystem has been included in this well-planned effort. By contrast, programs to monitor biological responses to contaminants in the Northeast are disjointed, short-lived, narrowly focused on the individual scientist's area of expertise, and severed from the context of the larger ecosystem. The study by Johnson et al. (1992) is one of the first to apply uniform methods and intensive sampling to a sizeable portion of the Northeast coast. The increased power of this approach allowed them to draw conclusions about which contaminants were most toxic and which least, and to identify some of the natural factors that influence the biological responses.

The words of Wolfe et al. (1982) were intended for the New York Bight, but they can be applied equally elsewhere:

> Although varying degrees of ecological change can be documented readily . . . , and although the degradation can be ascribed to pollutant inputs of human origin, the evidence that changes are caused by toxicants . . . is only circumstantial. No ecological or community-level change can be ascribed to any specific toxicant. If such cause-effect relationships are to be established, a balanced program of field and laboratory research must be undertaken. Greater emphasis must be placed on experimental approaches to ecosystem analysis in order to deduce causal relationships and mechanisms.

Conclusions

1. Four general strategies have been used to assess biological responses to toxicants in the Northeast ecosystem. The most prevalent is the *field/field* strategy, which merely describes the temporal or spatial distribution of an effect, often without relating it to a causative agent.
2. Boston Harbor, the inner New York Bight and its estuarine tributaries, and the Elizabeth River appear to be localized areas where biological responses to toxicants have been noted most often. However, the lack of some responses in these locations and the presence of responses in less contaminated environments have also been noted, confusing the picture.

3. Most of the estuaries in the Northeast, including scores of smaller estuaries entering bays or the Atlantic proper, have been very poorly surveyed for biological responses to toxicants, and the Northeast as a whole is not well studied. Conclusions about the extent of biological effects, or long-term changes in effects, are tenuous at best.
4. Except for some measurements in the New York Bight, there appear to be very few biological responses to contaminants in any oceanic waters more than a few kilometers from shore.
5. Virtually no information has been gathered on the impact of detected biological responses on population, community, or ecosystem function.
6. In terms of biological responses to contaminants, there is little basis for claiming that the Northeast environment behaves as a single ecosystem.

References

Atwood, D. K., Whitledge, T. E., and Sharp, J. H. 1979. Chemical Factors. In R. L. Swanson and C. J. Sindermann (eds.), Oxygen Depletion and Associated Benthic Mortalities in New York Bight, 1976. NOAA Professional Paper 11. Washington, DC: U.S. Department of Commerce, pp. 79–123.

Bender, M. E., Hargis, W. J., Jr., Huggett, R. J., and Roberts, M. H., Jr. 1988. Effects of polynuclear aromatic hydrocarbons on fishes and shellfish: An overview of research in Virginia. Mar. Environ. Res. 24:237–241.

Black, D. E., Phelps, D. K., and Lapan, R. L. 1988. The effect of inherited contamination on egg and larval winter flounder, *Pseudopleuronectes americanus*. Mar. Environ. Res. 25:45–62.

Bodammer, J. E., and Murchelano, R. A. 1990. Cytological study of vacuolated cells and other aberrant hepatocytes in winter flounder from Boston Harbor. Cancer Res. 50:6744–6756.

Boesch, D. F. 1982. Ecosystem Consequences of Alterations of Benthic Community Structure and Function in the New York Bight Region. In G. F. Mayer (ed.), Ecological Stress and the New York Bight: Science and Management. Columbia, SC: Estuarine Research Federation, pp. 543–568.

Bunton, T. E., and Baksi, S. M. 1988. Cholangioma in white perch (*Morone americana*) from the Chesapeake Bay. J. Wildl. Dis. 24:137–141.

Carr, R. S., Hillman, R. E., and Neff, J. M. 1991. Field assessment of biomarkers for winter flounder. Mar. Poll. Bull. 22:61–67.

Cormier, S. M. 1986. Fine structure of hepatocytes and hepatocellular carcinoma of the Atlantic tomcod, *Microgadus tomcod* (Walbaum). J. Fish Dis. 9:179–194.

Cormier, S. M., Racine, R. N., Smith, C. E., et al. 1989. Hepatocellular carcinoma and fatty infiltration in the Atlantic tomcod, *Microgadus tomcod* (Walbaum). J. Fish. Dis. 12:105–116.

Dean, D., and Haskin, H. H. 1964. Benthic repopulation of the Raritan River Estuary following pollution abatement. Limnol. Oceanogr. 9:551–563.

Elskus, A. A., Stegeman, J. J., Susani, L. C., et al. 1989. Polychlorinated biphenyls concentration and cytochrome P-450E expression in winter flounder from contaminated environments. Mar. Environ. Res. 28:25–30.

Farley, C. A. 1988. Histochemistry as a tool for examining possible pathologic cause-and-effect relationships between heavy metal and inflammatory lesions in oysters, *Crassostrea virginica*. Mar. Environ. Res. 24:271–275.

Foureman, G. L., White, N. B., Jr., and Bend, J. R. 1983. Biochemical evidence that winter flounder (*Pseudopleuronectes americanus*) have induced hepatic cytochrome P-450-dependent monooxygenase activities. Can. J. Fish. Aquat. Sci. 40:854–865.

Gardner, G. R., Pruell, R. J., and Folmar, L. C. 1989. A comparison of both neoplastic and non-neoplastic disorders in winter flounder (*Pseudopleuronectes americanus*) from eight areas in New England. Mar. Environ. Res. 28:393–397.

Gardner, G. R., and Yevich, P. P. 1988. Comparative histopathological effects of chemically contaminated sediment on marine organisms. Mar. Environ. Res. 24:311–316.

Gentile, J. H., Scott, K. J., Lussier, S., and Redmond, M. 1985. Application of laboratory population responses for evaluating the effects of dredged material. Tech. Rept. D-85-8. Narragansett, RI: U. S. Environmental Protection Agency.

Gopalan, U. K., and Young, J. S. 1975. Incidence of shell disease in shrimp in the New York Bight. Mar. Poll. Bull. 6:149–153.

Hargis, W. J., Jr., Roberts, M. H., Jr., and Zwerner, D. E. 1984. Effects of contaminated sediments and sediment-exposed effluent water on an estuarine fish: Acute toxicity. Mar. Environ. Res. 14:337–354.

Hargis, W. J., and Zwerner, D. E. 1988. Effects of certain contaminants on eyes of several estuarine fishes. Mar. Environ. Res. 24:265–270.

Hargis, W. J., Jr., Zwerner, D. E., Thoney, D. A., et al. 1989. Neoplasms in mummichogs from the Elizabeth River, Virginia. J. Aquat. Animal Health 1:165–172.

Harshbarger, J. C., and Clark, J. B. 1990. Epizootiology of neoplasms in bony fish of North America. Sci. Total Environ. 94:1–32.

Hillman, R. E., Lordo, R. A., Menton, R. G., et al. Relationship of environmental contaminants to occurrence of neoplasia in mussels (*Mytilus edulis*) from east and west coast Mussel Watch sites. (in review).

Huggett, R. J., Bender, M. E., and Ungar, M. A. 1987. Polynuclear Aromatic Hydrocarbons in the Elizabeth River, Virginia. In K. L. Dickson, A. W. Maki, and W. Brungs (eds.), Fate and Effects of Sediment-bound Chemicals in Aquatic Systems. New York: Pergamon Press, pp. 327–341.

Johnson, L. L., Stehr, C. M., Olson, O. P., et al. 1992. National Status and Trends Program, National Benthic Surveillance Project: Northeast Coast, Fish Histopathology and Relationships Between Lesions and Chemical Contaminants (1987–89). NOAA Tech. Mem. NMFS-NWFSC-4. Washington, DC: U. S. Department of Commerce.

Johnson, L. L., Stein, J. E., Collier, T. K., et al. 1994. Indicators of reproductive development in pre-spawning female winter flounder (*Pleuronectes americanus*) from urban and non-urban estuaries on the northeast coast of the United States. Sci. Total Environ. 141:241–260.

Klauda, R. J., Peck, T. H., and Rice, G. K. 1981. Accumulation of polychlorinated biphenyls in Atlantic tomcod (*Microgadus tomcod*) collected from the Hudson River Estuary, New York. Bull. Environ. Contam. Toxicol. 27:829–835.

Leavitt, D. F., Capuzzo, J. M., Smolowitz, R. M., et al. 1990. Hematopoietic neoplasia in *Mya arenaria*: Prevalence and indices of physiological condition. Mar. Biol. 105: 313–321.

Longwell, A. C., Chang, S., Hebert, A., et al. 1992. Pollution and developmental abnormalities of Atlantic fishes. Environ. Biol. Fishes 35:1–21.

Longwell, A. C., and Hughes, J. B. 1982. Cytologic, Cytogenetic, and Embryologic State of Atlantic Mackerel Eggs from Surface Waters of the New York Bight in Relation to Pollution. In G. F. Mayer (ed.), Ecological Stress and the New York Bight: Science and Management. Columbia, SC: Estuarine Research Federation, pp. 381–388.

Mahoney, J. B., Midlige, F. H., and Deuel, D. G. 1973. A fin rot disease of marine and euryhaline fishes in the New York Bight. Trans. Am. Fish. Soc. 102:596–605.

May, E. B., Lukacovic, R., King, H., and Lipsky, M. M. 1987. Hyperplastic and neoplastic alterations in the livers of white perch (*Morone americana*) from the Chesapeake Bay. J. Nat. Cancer Inst. 79:137–143.

Mayer, G. F. (ed.). 1982. Ecological Stress and the New York Bight: Science and Management. Columbia, SC: Estuarine Research Federation.

Mehrle, P. M., Haines, T. A., Hamilton, S., et al. 1982. Relationship between body contaminants and bone development in east-coast striped bass. Trans. Am. Fish. Soc. 111:231–241.

Moore, M. J., Smolowitz, R., and Stegeman, J. J. 1989. Cellular alterations preceding neoplasia in *Pseudopleuronectes americanus* from Boston Harbor. Mar. Environ. Res. 28:425–429.

Murchelano, R. A. 1982. Some Pollution-associated Diseases and Abnormalities of Marine Fishes and Shellfishes: A Perspective for the New York Bight. In G. F. Mayer (ed.), Ecological Stress and the New York Bight: Science and Management. Columbia, SC: Estuarine Research Federation, pp. 327–346.

Murchelano, R. A., and Wolke, R. E. 1985. Epizootic carcinoma in the winter flounder, *Pseudopleuronectes americanus*. Science 228:587–589.

Murchelano, R. A., and Wolke, R. E. 1991. Neoplasms and nonneoplastic liver lesions in winter flounder, *Pseudopleuronectes americanus*, from Boston Harbor, Massachusetts. Environ. Health Persp. 90:17–26.

Murchelano, R. A., and Ziskowski, J. 1976. Fin Rot Disease Studies in the New York Bight. In M. G. Gross (ed.), Proceedings of the Symposium on the Middle Atlantic Continental Shelf and the New York Bight. Lawrence, KS: Allen Press, pp. 329–336.

Murchelano, R. A., and Ziskowski, J. J. 1982. Fin Rot Disease in the New York Bight (1973–1977). In G. F. Mayer (ed.), Ecological Stress and the New York Bight: Science and Management. Columbia, SC: Estuarine Research Federation, pp. 347–358.

Nacci, D., and Jackim, E. 1989. Using the DNA alkaline unwinding assay to detect DNA damage in laboratory and environmentally exposed cells and tissues. Mar. Environ. Res. 28:333–337.

Nelson, D. A., Miller, J. E., Rusanowsky, D., et al. 1991. Comparative reproductive success of winter flounder in Long Island Sound: A three-year study (biology, biochemistry, and chemistry). Estuaries 14:318–331.

Nelson, W. G., Black, D., and Phelps, D. 1985. Utility of the Scope for Growth Index to Assess the Physiological Impact of Black Rock Harbor Suspended Sediment on the Blue Mussel, *Mytilus edulis*: A Laboratory Evaluation. Tech. Rept. D-85-6. Narragansett, RI: U.S. Environmental Protection Agency.

Pereira, J. J., Ziskowski, J., Mercaldo-Allen, R., et al. 1992. Vitellogenin in winter flounder (*Pleuronectes americanus*) from Long Island Sound and Boston Harbor. Estuaries 15:289–297.

Pesch, G. G., Mueller, C. M., Pesch, C. E., et al. 1985. Application of Sister Chromatid Exchange in Marine Polychaetes to Black Rock Harbor Sediment. Tech. Rept. D-85-1. Narragansett, RI: U.S. Environmental Protection Agency.

Reinisch, C. L., Charles, A. M., and Stone, A. M. 1984. Epizootic neoplasia in soft shell clams collected from New Bedford Harbor. Hazardous Waste 1:73–81.

Roberts, M. H., Jr., Hargis, W. J., Jr., Strobel, C. J., and De Lisle, P. F. 1989. Acute toxicity of PAH contaminated sediments to the estuarine fish, *Leiostomus xanthurus*. Bull. Environ. Contam. Toxicol. 42:142–149.

Segar, D. A., and Berberian, G. A. 1976. Oxygen Depletion in the New York Bight: Causes and Consequences. In M. G. Gross (ed.), Proceedings of the Symposium on the Middle Atlantic Continental Shelf and the New York Bight. Lawrence, KS: Allen Press, pp. 220–239.

Sherwood, M. J. 1982. Fin Erosion, Liver Condition, and Trace Contaminant Exposure in Fishes from Three Coastal Regions. In G. F. Mayer (ed.), Ecological Stress and the New York Bight: Science and Management. Columbia, SC: Estuarine Research Federation, pp. 359–377.

Smith, C. E., Peck, T. H., Klauda, R. J., and McLaren, J. B. 1979. Hepatomas in Atlantic tomcod *Microgadus tomcod* (Walbaum) collected in the Hudson River estuary in New York. J. Fish Dis. 2:313–319.

Smolowitz, R. M., Moore, M. J., and Stegeman, J. J. 1989. Cellular distribution of cytochrome P-450E in winter flounder liver with degenerative and neoplastic disease. Mar. Environ. Res. 28:441–446.

Stegeman, J. J., Kloepper-Sams, P. J., and Farrington, J. W. 1986. Monooxygenase induction and chlorobiphenyls in the deep-sea fish *Coryphaenoides armatus*. Science 231:1287–1289.

Stegeman, J. J., Teng, F. Y., and Snowberger, E. A. 1987. Induced cytochrome P450 in winter flounder (*Pseudopleuronectes americanus*) from coastal Massachusetts evaluated by catalytic assay and monoclonal antibody probes. Can. J. Fish. Aquat. Sci. 44:1270–1277.

Steimle, F., Caracciolo, J., and Pearce, J. B. 1982. Impacts of Dumping on New York Bight Apex Benthos. In G. F. Mayer (ed.), Ecological Stress and the New York Bight: Science and Management. Columbia, SC: Estuarine Research Federation, pp. 213–223.

Steimle, F. W., and Caracciolo-Ward, J. 1989. A reassessment of the status of the benthic macrofauna of the Raritan Estuary. Estuaries 12:145–156.

Stiles, S., Choromanski, J., Nelson, D., et al. 1991. Early reproductive success of the hard clam (*Mercenaria mercenaria*) from five sites in Long Island Sound. Estuaries 14:332–342.

Thiyagarajah, A., and Bender, M.E. 1988. Lesions in the pancreas and liver of an oyster toadfish, *Opsanus tau* (L.), collected from the lower York River, Virginia, USA. J. Fish Dis. 11:359–364.

Tietjen, J. H. 1977. Population distribution and structure of free-living nematodes of Long Island Sound. Mar. Biol. 43:123–136.

Tietjen, J. H. 1980. Population structure and species composition of the free-living nematodes inhabiting sands of the New York Bight Apex. Estuar. Coast. Mar. Sci. 10:61–73.

Tietjen, J. H. 1982. Potential Roles of Nematodes in Polluted Ecosystems and the Impact of Pollution of Meiofauna. In G. F. Mayer (ed.), Ecological Stress and the New York Bight: Science and Management. Columbia, SC: Estuarine Research Federation, pp. 225–233.

Varanasi, U., Reichert, W. L., and Stein, J. E. 1989. ^{32}P-postlabeling analysis of DNA adducts in liver of wild English sole (*Parophrys vetulus*) and winter flounder (*Pseudopleuronectes americanus*). Cancer Res. 49:1171–1177.

Weeks, B. A., and Warinner, J. E. 1984. Effects of toxic chemicals on macrophage phagocytosis in two estuarine fishes. Mar. Environ. Res. 14:327–335.

Weeks, B. A., Warinner, J. E., Mason, P. L., and McGinnis, D. S. 1986. Influence of toxic chemicals on the chemotactic response of fish macrophages. J. Fish Biol. 28:653–658.

White, H. H. 1984. Mussel Madness: Use and Misuse of Biological Monitors of Marine Pollution. In H. H. White (ed.), Concept in Marine Pollution Measurements. College Park, MD: Maryland Sea Grant College, pp. 325–337.

Wolfe, D. A., Boesch, D. R., Calabrese, A., et al. 1982. Effects of Toxic Substances on Communities and Ecosystems. In G. F. Mayer (ed.), Ecological Stress and the New York Bight: Science and Management. Columbia, SC: Estuarine Research Federation, pp. 67–86.

Yevich, P. P., and Barszcz, C. A. 1977. Neoplasia in soft-shell clams (*Mya arenaria*) collected from oil-impacted sites. Ann. N.Y. Acad. Sci. 298:409–426.

Young, J. S., and Pearce, J. B. 1975. Shell disease in crabs and lobsters from New York Bight. Mar. Poll. Bull. 6:101–105.

Zaroogian, G. E., Pesch, C., Schauer, P., and Black, D. 1985. Laboratory evaluation of adenylate energy charge as a test for stress in *Mytilus edulis* and *Nephthys incisa* treated with dredged material. Tech. Rept. D-85-3. Narragansett, RI: U.S. Environmental Protection Agency.

Zdanowicz, V. S., Gadbois, D. F., and Newman, M. W. 1986. Levels of Organic and Inorganic Contaminants in Sediments and Fish Tissues and Prevalences of Pathological Disorders in Winter Flounder from Estuaries of the Northeast United States, 1984. In Oceans '86 Conference Record. Washington, DC: Marine Technological Society, pp. 578–585.

Ziskowski, J. J., Anderson, V. T., Jr., and Murchelano, R. A. 1980. A bent fin ray condition in winter flounder, *Pseudopleuronectes americanus*, from Sandy Hook and Raritan Bays, New Jersey, and Lower Bay, New York. Copeia 1980:895–899.

Ziskowski, J. J., Despres-Patanjo, L., Murchelano, R. A., et al. 1987. Disease in commercially valuable fish stocks in the Northwest Atlantic. Mar. Poll. Bull. 18:496–504.

Ziskowski, J., and Murchelano, R. 1975. Fin erosion in winter flounder. Mar. Poll. Bull. 6:26–29.

15

Metal Concentrations in Winter Flounder, American Lobster, and Bivalve Molluscs from Boston Harbor, Salem Harbor, and Coastal Massachusetts: A Summary of Data on Tissues Collected from 1986 to 1991

Jack P. Schwartz, Nina M. Duston, and Carol A. Batdorf

Abstract

Concentrations of trace metals (Cd, Cr, Cu, Pb, Zn, Hg) were measured in winter flounder (Pseudopleuronectes americanus), American lobster (Homarus americanus), soft shell clam (Mya arenaria), blue mussel (Mytilus edulis), ocean quahog (Arctica islandica), and surf clam (Spisula solidissima). Flounder, lobster, and soft shell clam were collected in Boston and Salem harbors. Blue mussel, ocean quahog, surf clam, flounder, and lobster were collected at random locations throughout coastal marine waters of Massachusetts. Cadmium, chromium, copper, lead, and zinc concentrations were determined using flame atomic absorption spectrophotometry (AAS). Mercury was determined by cold vapor AAS. All metal concentrations were determined using edible tissues and reported as parts per million (ppm) wet weight (μg/g wet wt.). The survey was undertaken as part of an ongoing contaminant monitoring program for chemical contaminants in marine fishery resources of Massachusetts, with the objective of assessing spatial, temporal, and seasonal differences in tissue metal concentrations between the two harbors and coastal Massachusetts.

Trace metal concentrations overall were similar to levels previously reported for the same species from Massachusetts and other coastal areas of New England. All seasonal differences in metal concentrations detected among species indicated that tissue concentrations are higher in spring than in fall/winter. Within the survey areas, we found that mercury concentrations in Boston Harbor winter flounder were higher than in Salem Harbor or in coastal flounder. Mercury levels in Boston Harbor flounder decreased over a four-year period. Chromium concentrations in Salem Harbor lobster and soft shell clam were higher than in Boston Harbor samples. Copper concentrations in Boston Harbor soft shell clam were higher than in clams from Salem Harbor. Cadmium concentrations in coastal lobster were higher than levels found in Boston Harbor and Salem Harbor lobster. Cadmium concentrations in coastal lobster increased over a six-year period. The decrease in mercury concentrations in Boston Harbor flounder and lack of an increase over time for any detected metals in Boston Harbor and Salem Harbor suggests that efforts to control and reduce the input of metals to the harbors may be having an effect on reducing tissue concentrations. Further monitoring is necessary to discern between natural and anthropogenic sources of trace metal concentrations in Massachusetts marine species.

For several years, the Contaminant Monitoring Program within the Massachusetts Division of Marine Fisheries (MDMF) has been gathering information on contaminant concentrations in edible tissues of certain marine species of commercial and recreational importance to the Commonwealth. The purpose of collecting the data is to establish baseline levels of contaminants and determine spatial and temporal trends. In many respects, the MDMF monitoring program has the same objectives as the National Status and Trends Program (NOAA 1984, 1987, 1989), which monitors contaminants in mussels throughout coastal United States. However, the MDMF Contaminant Monitoring Program collects contaminant information for several species in Massachusetts' territorial marine waters, including samples from Boston Harbor and Salem Harbor.

This is the fourth report from the Contaminant Monitoring Program and a continuation of a previous report on trace metal concentrations in edible tissues (Duston et al. 1990). The species analyzed were winter flounder (*Pseudopleuronectes americanus*), American lobster (*Homarus americanus*), soft shell clam (*Mya arenaria*), ocean quahog (*Arctica islandica*), surf clam (*Spisula solidissima*), and blue mussel (*Mytilus edulis*). All of these species live either in or on marine sediments where most contaminants accumulate. Multiple bivalve species were selected because no one bivalve is sufficiently abundant for sampling in all marine waters of coastal Massachusetts. The metals for which concentrations were determined were cadmium, chromium, copper, lead, mercury, and zinc. The field collections in this report were taken between 1986 and 1991. Over 750 samples were analyzed for trace metals for this report.

Field Methods

Field collections were obtained according to methods previously described (Duston et al. 1990; Schwartz et al. 1991). In Boston and Salem harbors, soft shell clam was collected from three intertidal stations using a hand-held clam fork, and winter flounder and lobster samples were collected using a Yankee otter trawl towed from the MDMF R/V *FC Wilbour*. Lobster samples designated by "BL" in Boston Harbor and "SL" in Salem Harbor were obtained by MDMF personnel with the assistance of commercial lobstermen using lobster pots.

Winter flounder, lobster, ocean quahog, surf clam, and blue mussel were collected in Massachusetts coastal waters by random sampling during the MDMF Resource Assessment Surveys aboard the NOAA R/V *Gloria Michelle*. Samples were obtained using a three-quarter–size North Atlantic type-2 seam otter trawl towed at 2.5 knots for 20 minutes. Otter trawl configurations were 11.9-m headrope–15.5-m footrope rigged with a 7.6-cm rubber disk sweep; 19.2-m 9.5-mm chain bottom legs; 18.3-m 9.5-mm wire toplegs, and 1.8 × 1.0-m, 147-kg wooden otter trawl doors. Legal-size specimens were used for the survey except that some flounder as small as 25 cm were included.

Samples from the two harbors were either delivered fresh to the laboratory the same day or frozen in polyethylene bags by field personnel until later delivery. Samples from the Resource Assessment Surveys were wrapped in polyethylene bags and frozen at sea. Samples were tagged in the field with site, date, and field identification numbers. Flounder and lobster samples were individually analyzed. Bivalve samples were generally composited to provide sufficient tissue for analysis.

Sample Preparation

In the laboratory, the samples were thawed to the point where they could be handled but remained sufficiently frozen to retain body fluids. Samples were cleaned of any surface debris before tissues were removed. Flounder samples consisted of the combined dorsal and ventral fillets. Lobster samples included tail meat, claw meat, tomalley (hepatopancreas), and gonads. Bivalve samples included all soft body parts normally considered edible.

All knives, forceps, scissors, and blender cups used to prepare samples were composed of stainless steel and were thoroughly cleaned and rinsed in acetone. The removal of tissues was done on acetone-rinsed aluminum foil over a wooden platform that was placed in a stainless steel drainage trough. Edible tissues were defined as follows.

> *Flounder:* Filleted and skinned using a filleting knife and spatula. Both fillets were transferred to a blender cup and homogenized.
> *Lobster:* Tissues were picked from open lobster with surgical scissors, forceps, and spatula, transferred to a blender cup, and homogenized.
> *Bivalves:* Shucked directly into a blender cup and homogenized.

Samples were refrozen in plastic bags and thawed as needed for chemical analysis.

Chemical Analyses

Cadmium, chromium, copper, lead, and zinc analyses were performed by digesting tissue samples with hot nitric acid and 30% hydrogen peroxide, followed by determination of metal concentrations by flame atomic absorption spectrophotometry with a Perkin-Elmer model 3030B atomic absorption spectrophotometer (AAS). The AAS was optimized according to the manufacturer's specifications to the primary absorption wavelength. Calibration standards (Fisher Certified Atomic Absorption Reference Standards) were prepared in 2% vol/vol nitric acid (70.0–71.0% 'Baker Instra–analyzed' with 2 megohm purity deionized water). Calibration standards encompassing the expected range in samples were used to define an analysis curve for each element, and accuracy was verified by comparison with at least two U.S. Environmental Protection Agency (U.S. EPA) Water Pollution Laboratory Performance Evaluation (WP Series) trace metal samples. Mean sample absorbance based on three machine readings was recorded and used in subsequent calculations. See Duston et al. (1990) for details.

Quality control reagent blanks were analyzed at least three times during analysis of each batch of samples: before the first sample, after the last sample, and after every 10% of samples in each sample batch. No more than 36 samples were included in each batch. EPA Water Pollution Quality Control samples of trace metals in fish were used as the standard reference material (SRM). Spiked reagent blanks accompanied each batch. At least 10% of the sample load consisted of sample triplicates and spiked tissue samples. Recoveries of 80–120% of expected value were considered acceptable, and sample batch groups with recoveries outside this range were re-extracted. Metal concentrations, including mercury, in tissues were corrected for blanks accompanying each batch of samples but were not corrected for percentage of recovery. Log books were maintained for all samples as they were received, prepared, digested, and analyzed. Raw AAS data were stored on machine printouts. All calculations, including mercury, were stored in notebooks. Some data had been entered into computer files; however, this process was incomplete at the time of writing.

Mercury analysis was performed by digesting tissue samples in a solution of concentrated sulfuric and nitric acid for two to three days at room temperature, followed by a two-day hot digestion. Mercury concentrations were determined by manual cold vapor technique, using a Perkin-Elmer model MHS-10 hydride accessory attached to the AAS. The system was optimized to the manufacturer's specifications for the 253.7-nm wavelength. A prereductant solution of hydroxylamine sulfate/sodium chloride was added to the samples, followed by a sodium borohydride/sodium hydroxide reductant. A 1-ppm mercury standard (Fisher Certified Atomic Absorption Reference Standard) was prepared in 2% vol/vol nitric acid in deionized water. Calibration standards, prepared from aliquots of the 1-ppm mercury solution and added to digestion vessels, were processed the same as tissue samples.

Three SRMs were digested and analyzed to verify for completeness of digestion and accuracy of machine calibration for the mercury analysis. The SRMs were the National Institute of Science and Technology (NIST) Oyster Tissue (NIST SRM #1566a), the National Research Council of Canada (NRC) Marine Analytical Chemistry Standards Program Dogfish Muscle Reference Material (NRC DORM-1), and NRC Lobster Hepatopancreas Marine Reference Material (NRC TORT-1). Additional quality control included digestion and analysis of reagent blanks, and at least 10% of tissue samples as duplicate/spiked tissue samples. Values for calibration standards and SRMs were checked at frequent intervals. If calibration and SRM values deviated from expected results by more than 10%, the tissue samples in batches assigned to these quality control samples were reanalyzed. Recoveries of 80–120% of expected values for both SRMs and sample spikes were considered acceptable, and 84% of these analyses had recoveries between 90% and 110%. Ninety-one percent of the mercury analyses had recoveries between 90% and 110%.

Detection Limits and Frequency of Nondetection

The detection limits for all analyses were established at three times the standard deviation of the amount in reagent blanks with each batch (Keith et al.

1983). Among all metals, nondetection occurred frequently with lead, chromium, and cadmium in winter flounder, and lead in lobster and surf clam. Metal data below the detection limit were reviewed in accordance with NOAA protocol (NOAA 1989). In all cases, below-detection determinations were assumed to be zero when calculating means. Whenever more than half of the data for a sample were nondetect values, those data were not used for tests of significance and data comparisons. Descriptive statistics for cadmium, chromium, and lead in winter flounder and lead in lobster and surf clam are provided; however, statistical interpretation was not performed. All metal concentrations are reported as parts per million (ppm) wet weight (μg/g wet wt.). Descriptive statistics are contained in Appendix A, all metal data are in Appendix B, station coordinates are in Appendix C, and station maps are in Appendix D. (Data appendices are not included in this chapter, but a complete copy of the report can be obtained from the authors.)

Tests of Significance and Data Comparisons

The procedure for statistical comparisons was the same as the procedure used for PCB data (Schwartz et al. 1991). The mean concentration for each metal in flounder or lobster samples was compared spatially between coastal Massachusetts, Boston Harbor, and Salem Harbor. Seasonal comparisons within each study area were based on samples collected from February to July grouped as spring, and compared with samples collected from August to January grouped as fall/winter. Yearly (February through January) mean concentrations in each study area were compared to detect changes with time. Statistical differences using the Kruskal-Wallis (K-W) test were considered significant at the level of probability of $p < .05$. K-W tests were performed on a computer using software from NH Software, Inc. (Roseville, MN). When a significant difference between yearly means was detected, the data were further analyzed by linear regression to determine if any trend was significant—i.e., metal concentrations increasing or decreasing over time at the probability level of $p < .05$. Statistical comparisons by K-W tests for bivalve samples were limited to group means of metals in soft shell clam between Boston Harbor and Salem Harbor and seasonal comparisons for those species with sufficient sample size. Interspecies statistical comparisons among bivalve molluscs were not performed per NOAA's recommendation (NOAA 1992) because the ability of molluscs to incorporate and discharge metals is highly species-specific and results in wide variations in tissue concentrations.

Previously published reports of metal data for marine species from Massachusetts and other coastal and offshore areas around New England were cited for comparison (nonstatistically) to the data in our survey. When previous metals data were reported as parts per million dry weight, we assumed an 80% moisture content and noted this conversion. For previous reports that provided the dry weight/wet weight ratio, the data were converted to wet weight concentrations. Comparisons of our lobster composite analyses with metal concentrations in lobster from previous surveys were not possible because different surveys separately analyzed either lobster muscle (tail and claw meat), hepatopancreas (tomalley), or both. Each tissue type is identified in the text along with the citation.

Findings

Winter Flounder

Most of the winter flounder samples, regardless of the sampling area, had cadmium concentrations below the limits of detection. The range of detection limits was 0.014–0.054 ppm. The range of detection limits encompassed cadmium concentrations previously reported for winter flounder from Massachusetts and New England. Cadmium levels in Boston Harbor flounder were reported as <0.001 ppm by the U.S. EPA (1988) and listed between 0.03 ppm and 0.15 ppm by Capuzzo et al. (1986). Long Island Sound flounder contained 0.02–0.06 ppm cadmium according to the Connecticut Department of Environmental Protection (CT DEP 1987), and samples from Georges Bank contained cadmium concentrations between 0.050 ppm and 0.086 ppm (Hall et al. 1978). The highest cadmium concentration recorded in the present study was 0.096 ppm in Boston Harbor.

The range of detection limits for chromium in winter flounder was 0.00–0.14 ppm. The majority of flounder samples, regardless of sampling area, contain chromium concentrations less than 0.10 ppm. The highest chromium concentration in flounder was 0.24 ppm in Boston Harbor. The majority of flounder samples from Boston Harbor collected by the U.S. EPA (1988) contained chromium concentrations between 0.00 ppm and less than 0.01 ppm, with one high reading of 0.34 ppm. The CT DEP (1987) reported chromium levels in flounder from Long Island Sound between 0.00 ppm and 0.08 ppm, but concentrations between 0.42 ppm and 0.56 ppm were also reported in flounder from Long Island Sound (Capuzzo et al. 1986; Reid et al. 1982). Chromium concentrations in winter flounder from Georges Bank were between 0.1 ppm and 0.3 ppm (Hall et al. 1978).

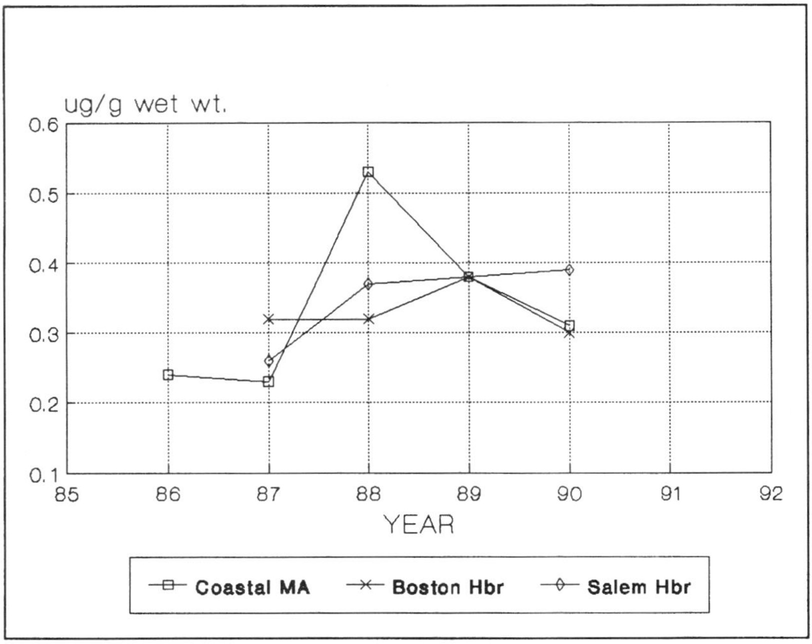

Figure 15-1. Copper in winter flounder.

Mean copper concentrations in winter flounder were not significantly different between Boston Harbor (0.33 ppm), Salem Harbor (0.34 ppm), and coastal Massachusetts (0.34 ppm). The range of concentrations was 0.10–2.10 ppm. The highest copper concentration was collected in coastal Massachusetts. In coastal winter flounder, there was a significant difference in mean copper concentrations between collection years (Figure 15-1), but no significant change over time. Coastal flounder collected in spring had a significantly higher mean copper concentration (0.39 ppm) than did fall/winter samples (0.26 ppm). Capuzzo et al. (1986) listed copper concentrations in flounder from Boston Harbor between 0.7 ppm and 10 ppm. Concentrations in Long Island Sound flounder were generally between 0.08 ppm and 0.26 ppm (Reid et al. 1982), but one reported concentration near the mouth of the Housatonic River was 3.60 ppm (CT DEP 1987). Copper concentrations in Georges Bank flounder were between 0.2 ppm and 0.4 ppm (Hall et al. 1978).

The range of detection limits for lead in winter flounder was 0.08–0.77 ppm. The highest lead concentration in flounder was 0.90 ppm in a sample from coastal Massachusetts. Previous studies indicate the range of lead con-

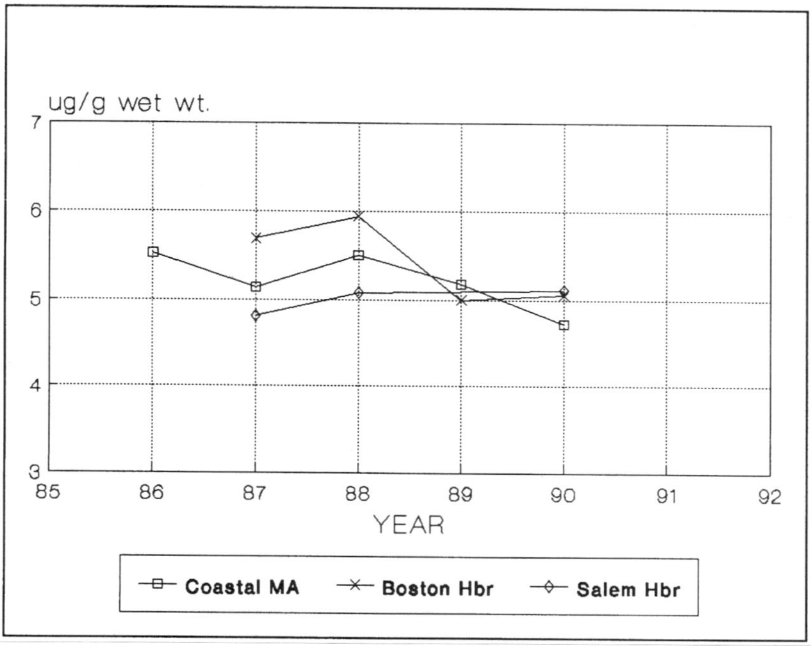

Figure 15-2. Zinc in winter flounder.

centrations in Boston Harbor flounder can be 0.07–0.65 ppm (Capuzzo et al. 1986). The majority of lead concentrations in flounder reported by the U.S. EPA (1988) were less than 0.02 ppm. Reid et al. (1982) reported lead concentrations in Long Island Sound flounder as <0.06 ppm and the CT DEP (1987) reported a range of lead concentrations from 0.16 ppm to 0.44 ppm. The majority of Georges Bank flounder contained lead concentrations between 0.3 ppm and 0.4 ppm, but some samples exceeded 0.6 ppm (Hall et al. 1978).

The mean zinc concentration in winter flounder from Boston Harbor (5.60 ppm) was significantly higher than in flounder from Salem Harbor (5.00 ppm) or coastal Massachusetts (5.12 ppm). However, the absolute difference between the highest and lowest mean value was only 12%. Mean zinc concentrations in coastal and Boston Harbor flounder were significantly different between years (Figure 15-2) but did not significantly change over time. Flounder collected in the spring in Boston Harbor had significantly higher mean zinc concentration (5.24 ppm) compared with those collected in fall-winter (4.97 ppm). The range of zinc concentrations was 3.38–9.03 ppm. In one study, the range of zinc concentrations in floun-

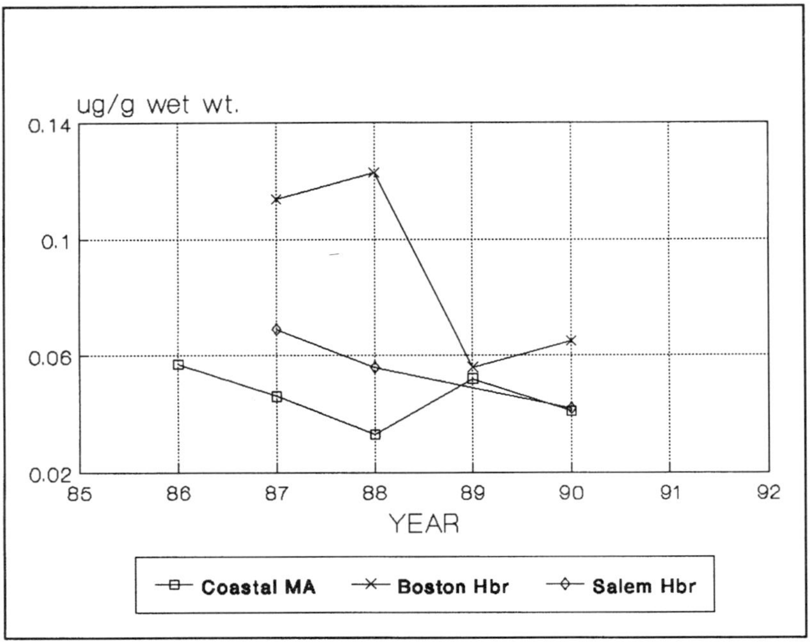

Figure 15-3. Mercury in winter flounder.

der from Long Island Sound was 2.94–4.96 ppm (Reid et al. 1982), whereas the CT DEP (1987) reported zinc concentrations in flounder from Long Island Sound from 3.9 ppm to 19.0 ppm. Hall et al. (1978) reported that flounder from Georges Bank contained zinc concentrations from 1.63 ppm to 12.50 ppm.

The mean mercury concentration (Figure 15-3) in Boston Harbor winter flounder (0.101 ppm) was significantly higher (approximately 100%) than Salem Harbor (0.057 ppm) and coastal flounder (0.045 ppm). In all three study areas, there were significant differences in yearly mean mercury concentrations. Mean mercury concentrations in Boston Harbor flounder significantly decreased over time. There were no significant seasonal differences in mean mercury concentrations among Boston Harbor or coastal flounder. Capuzzo et al. (1986) listed a range of mercury concentrations in Boston Harbor flounder from 0.04 ppm to 0.5 ppm. The U.S. EPA (1988) reported a mean mercury concentration of 0.03 ppm in Boston Harbor flounder. The majority of mercury concentrations in Georges Bank flounder were below 0.1 ppm; the range was 0.000–0.540 ppm (Hall et al. 1978).

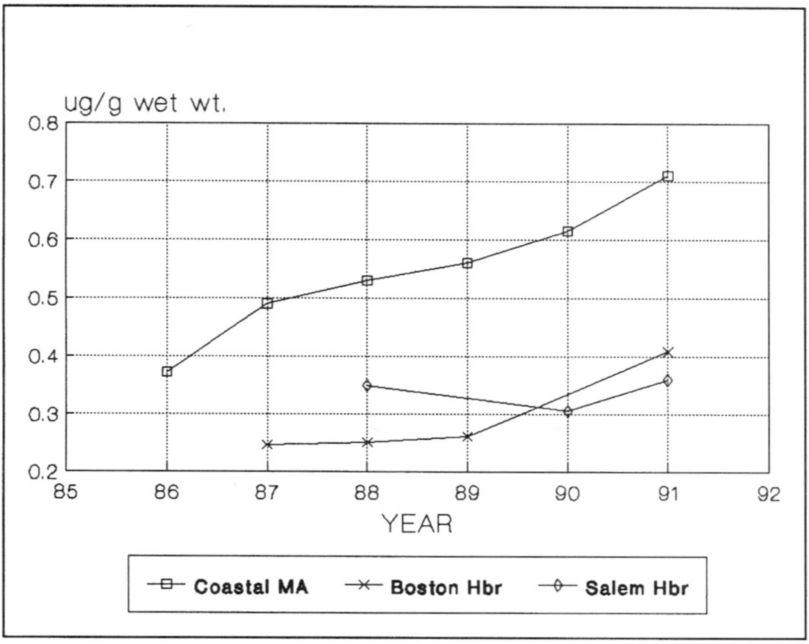

Figure 15-4. Cadmium in American lobster.

American Lobster

The mean cadmium concentration in coastal lobster (0.558 ppm) was significantly higher than in lobster from either Boston Harbor (0.256 ppm) or Salem Harbor (0.306 ppm). The yearly mean cadmium concentration in coastal lobster (Figure 15-4) significantly increased over time, from 0.372 ppm in 1986 to 0.711 ppm in 1991. There was no significant seasonal difference in mean cadmium concentrations among lobster in any of the three study areas. The range of cadmium concentrations was from less than 0.029 ppm to 2.195 ppm. Hall et al. (1978) reported a range of cadmium concentrations in Georges Bank lobster tails and claws from 0.070 ppm to 0.695 ppm, with most of these lobster containing from 0.2 ppm to 0.3 ppm. Wallace et al. (1988) reported mean cadmium concentrations in lobster muscle tissue from Boston Harbor and Salem Harbor of 0.0043 ppm and 0.0052 ppm, respectively, and the U.S. EPA (1988) reported a mean of 0.002 ppm in lobster muscle and 1.391 ppm in hepatopancreas from Boston Harbor.

Lobster from Salem Harbor contained a significantly higher mean chromium concentration (0.21 ppm) than lobster from Boston Harbor (0.09

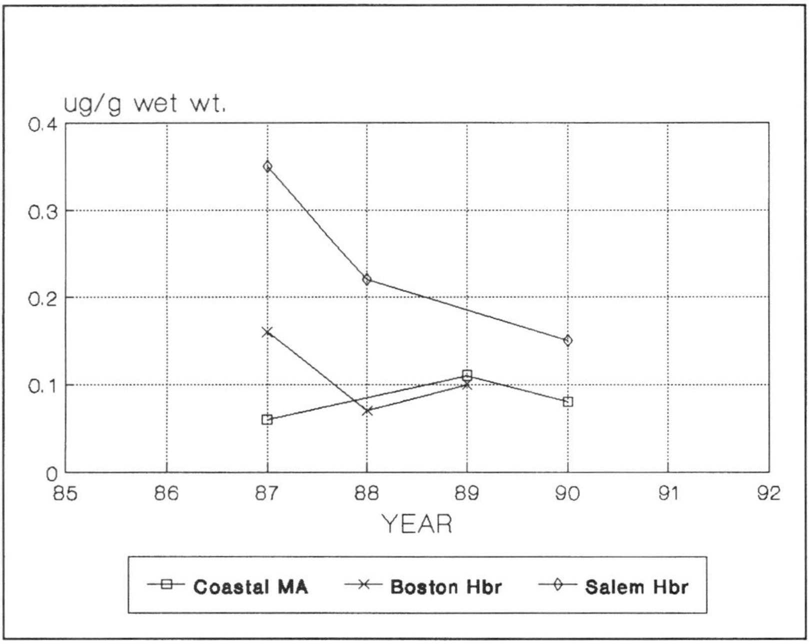

Figure 15-5. Chromium in American lobster.

ppm) or coastal Massachusetts (0.08 ppm). Lobster collected in the spring in Salem Harbor had a significantly higher mean chromium concentration (0.27 ppm) than those collected in fall-winter (0.10 ppm), but in approximately 44% of the fall/winter samples in Salem Harbor the concentration was below detection—i.e., statistically equivalent to zero. There was no significant linear regression for yearly mean chromium concentrations in Salem Harbor lobster although chromium levels appear to decrease over time (Figure 15-5). Lobster tail and claw meat from Georges Bank typically contained chromium concentrations between 0.1 ppm and 0.2 ppm, but concentrations could vary between 0.046 ppm and 0.690 ppm (Hall et al. 1978). Wallace et al. (1988) reported mean chromium concentrations in lobster muscle tissue from Boston Harbor and Salem Harbor of 0.048 ppm and <0.048 ppm, respectively, and the U.S. EPA (1988) reported a mean chromium concentration of 0.02 ppm in lobster muscle and 0.73 ppm in hepatopancreas from Boston Harbor.

The mean copper concentration in Salem Harbor lobster (22.56 ppm) was significantly lower than in both coastal (25.56 ppm) and Boston Harbor lobster (27.02 ppm). The absolute difference between the highest

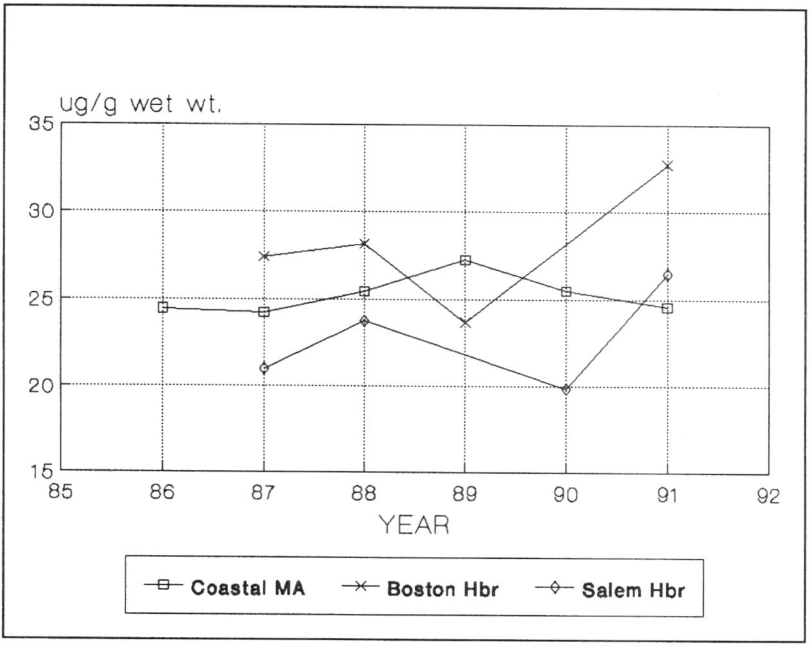

Figure 15-6. Copper in American lobster.

and lowest mean value was approximately 16%. Coastal lobster collected in spring had a higher mean copper concentration (27.38 ppm) than lobster collected in fall-winter (22.39 ppm). Yearly mean copper concentrations in Salem Harbor lobster were significantly different (Figure 15-6) but did not significantly change over time. The range of copper concentrations among all samples was 7.15–61.68 ppm. The range of copper concentrations in tail and claw meat of lobster from Georges Banks was 5.75–33.75 ppm (Hall et al. 1978). Wallace et al. (1988) reported mean copper concentrations in lobster muscle tissue from Boston and Salem harbors of 14.4 ppm and 9.9 ppm, respectively, and the U.S. EPA (1988) reported a mean copper concentration of 4.09 ppm in lobster muscle and 145 ppm in hepatopancreas from Boston Harbor.

The range of detection limits for lead in lobster was 0.10–0.24 ppm. Very few samples exceeded their detection limit. Of those samples that did exceed their detection limit, two coastal lobster samples had anomalously high lead concentrations that were confirmed by repeat analysis of the sample homogenates (samples PM2295 and PM2406). These samples were processed in accordance with field and laboratory quality control proce-

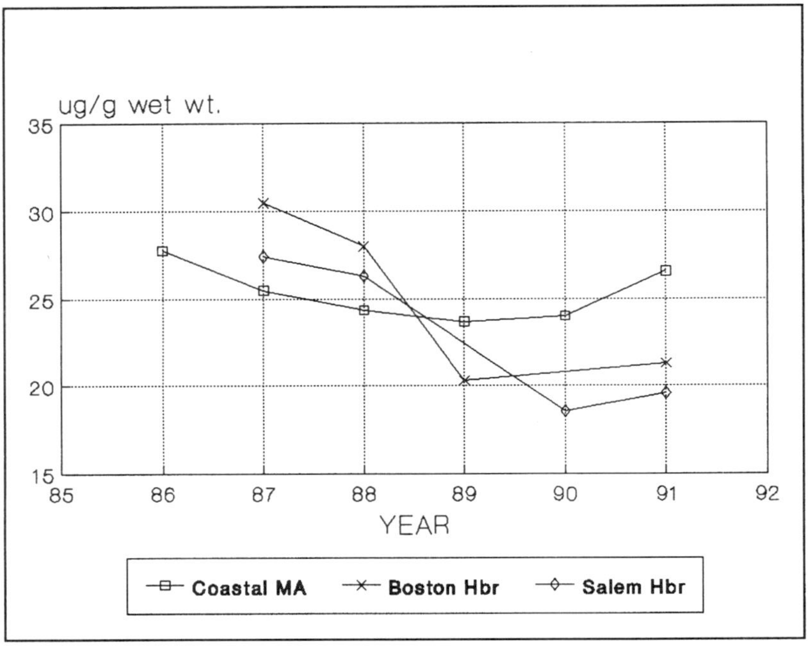

Figure 15-7. Zinc in American lobster.

dures; however, the high concentrations suggest the samples were possibly contaminated before our analysis. The majority of lead concentrations in tail and claw meat of Georges Bank lobster were 0.5–0.6 ppm (Hall et al. 1978). Wallace et al. (1988) reported mean lead concentrations in muscle tissue of Boston Harbor and Salem Harbor lobster of less than 0.024 ppm and less than 0.064 ppm, respectively. The U.S. EPA (1988) reported a mean lead concentration of 0.17 ppm in lobster muscle and 0.01 ppm in hepatopancreas from Boston Harbor.

The mean zinc concentration in Salem Harbor lobster (22.95 ppm) was significantly lower than in lobster from Boston Harbor (25.92 ppm) or coastal Massachusetts (24.95 ppm). The greatest difference between the highest and lowest mean value between study areas was only 11%. The range of zinc concentrations among all samples in our survey was 11.41–53.88 ppm. In Boston Harbor, Salem Harbor, and coastal Massachusetts, mean zinc concentrations were higher in spring (26.66 ppm, 24.31 ppm, and 26.92 ppm, respectively) compared with fall/winter (21.91 ppm, 20.47 ppm, and 21.51 ppm, respectively); in all three study areas, there was a significant difference in yearly mean zinc concentrations (Figure 15-7) but no significant change over time. The

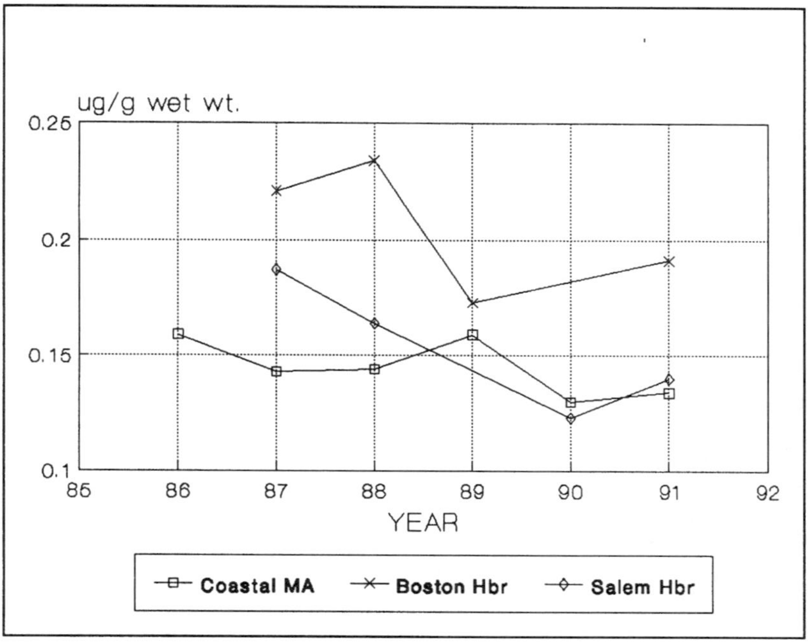

Figure 15-8. Mercury in American lobster.

majority of zinc concentrations in tail and claw meat of Georges Bank lobster were between 20.00 ppm and 30.00 ppm, but the range was 10.00–183.45 ppm (Hall et al. 1978). Wallace et al. (1988) reported mean zinc concentrations in lobster muscle tissue from Boston Harbor and Salem Harbor of 17.6 ppm and 21.3 ppm, respectively.

The mean mercury concentration in Boston Harbor lobster (0.213 ppm) was significantly higher than in lobster from Salem Harbor (0.151 ppm) or coastal Massachusetts (0.144 ppm). Mercury concentrations in coastal lobster were significantly higher in spring compared with fall/winter. Yearly mean mercury concentrations in lobster were significantly different in all three study areas (Figure 15-8) but did not significantly change over time. The majority of mercury concentrations in tail and claw meat of Georges Bank lobster were 0.2–0.3 ppm (Hall et al. 1978). Wallace et al. (1988) reported mean mercury concentrations in muscle tissue of Boston Harbor and Salem Harbor lobster of 0.141 ppm and 0.148 ppm, respectively. The U.S. EPA (1988) reported a mean mercury concentration of 0.085 ppm in lobster muscle and 0.024 ppm in hepatopancreas from Boston Harbor.

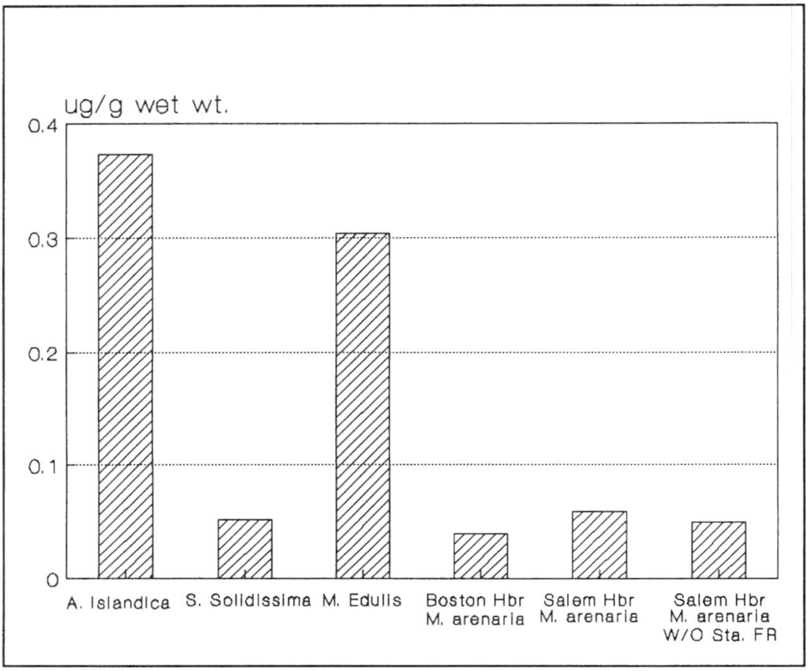

Figure 15-9. Cadmium in bivalve molluscs.

Bivalve Molluscs

The mean cadmium concentration (Figure 15-9) in Salem Harbor soft shell clam (0.060 ppm) was significantly higher ($p < 0.048$) than in Boston Harbor clams (0.040 ppm). Samples from one of the three stations in Salem Harbor (FR, Forest River) had approximately twice the mean cadmium concentration of samples from the other two stations. Historical records indicate that the location of the Forest River sampling area is in the vicinity of the site of a lead mill that existed from about 1826–1920 (Essex Institute 1971; Directory for Salem and Beverly 1920). Cadmium is commonly found in lead deposits (May and McKinney 1981). Additionally, the Forest River sampling station is downstream from an industrial glass landfill (EOEA 1991) with relatively high levels of metals, including cadmium, in abutting river marsh sediments. Since either source could potentially influence cadmium levels in Forest River clam samples, the samples from the two remaining Salem Harbor stations were compared with Boston Harbor soft shell clam. There was no significant difference in cadmium concentrations between Boston Harbor and Salem

Harbor with the elimination of the Forest River station. There was no seasonal difference in cadmium concentrations in Salem Harbor soft shell clam with or without the elimination of clams from Forest River. Wallace et al. (1988) reported a mean cadmium concentration of 0.060 ppm for Boston Harbor clams, and the U.S. FDA (1988) reported a cadmium concentration for Massachusetts soft shell clam (including samples from Boston Harbor) of 0.044 ppm. The U.S. EPA (1988) reported a mean cadmium concentration in Boston Harbor soft shell clam of 0.02 ppm.

The mean cadmium concentration in ocean quahog was 0.373 ppm. The range of cadmium concentrations reported for Georges Bank and coastal New England was 0.27–0.70 ppm, converted from dry weight (Steimle et al. 1986). The mean cadmium concentration for surf clam was 0.053 ppm. The mean cadmium concentration reported by Hall et al. (1978) for surf clam from Block Island Sound was 0.110 ppm. There was no seasonal difference in mean cadmium concentrations among surf clam in our study.

The mean cadmium concentration in blue mussel was 0.304 ppm. The National Oceanic and Atmospheric Administration (NOAA 1987) reported blue mussel cadmium concentrations of 0.15–0.35 ppm (converted from dry weight) for Boston Harbor and 0.13–0.26 ppm for Buzzards Bay. There was no seasonal difference in mean cadmium concentrations among blue mussel in our study.

The mean chromium concentration (Figure 15-10) in Salem Harbor soft shell clam (1.22 ppm) was significantly higher than in Boston Harbor clams (0.30 ppm). In Salem Harbor, clam samples collected in spring had significantly higher chromium concentrations (1.42 ppm) than those collected in fall-winter (0.98 ppm). Wallace et al. (1988) reported a mean chromium concentration of 0.720 ppm for Boston Harbor clams, and the U.S. FDA (1988) reported a chromium concentration for Massachusetts soft shell clam (including samples from Boston Harbor) of 0.48 ppm. The U.S. EPA (1988) reported a mean chromium concentration in Boston Harbor soft shell clam of 0.21 ppm.

The mean chromium concentration in ocean quahog was 0.92 ppm. The range of concentrations for this species collected on Georges Bank and offshore areas of southern New England was 0.44–0.61 ppm, converted from dry weight (Steimle et al. 1986). The mean chromium concentration in surf clam was 0.23 ppm. The range of chromium concentrations reported by Hall et al. (1978) for surf clam from Block Island Sound was 0.51–0.71 ppm. Surf clam collected in spring in our survey had significantly higher chromium concentrations (0.32 ppm) than fall/winter samples (0.16 ppm).

The mean chromium concentration in blue mussel was 0.32 ppm. NOAA (1987) reported blue mussel chromium concentrations of 0.32–0.53 ppm

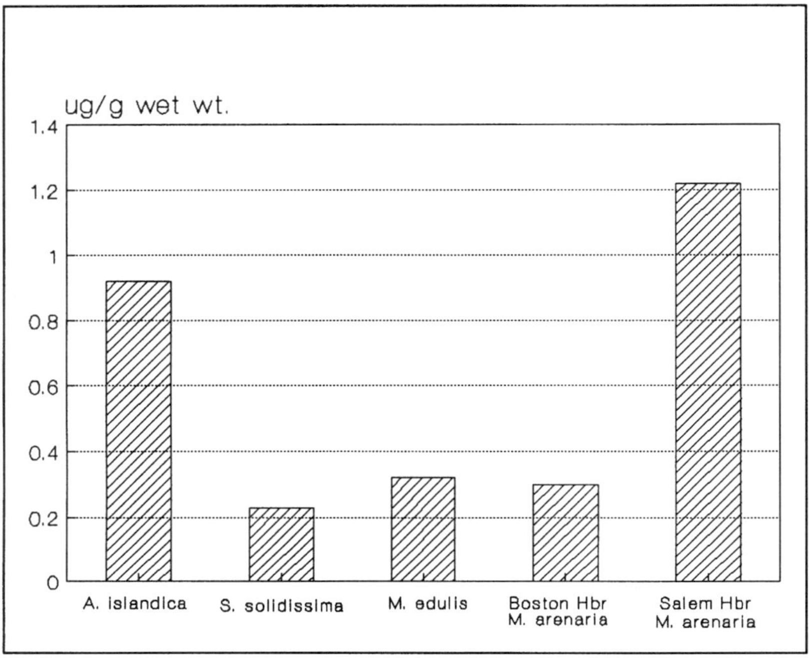

Figure 15-10. Chromium in bivalve molluscs.

(converted from dry weight) for Boston Harbor and 0.06–0.29 ppm for Buzzards Bay. NOAA (1987) noted that blue mussel may have an affinity for accumulating chromium. There was no seasonal difference in mean chromium concentrations among blue mussel in our study. Chromium levels of 2.8 ppm and 4.0 ppm were listed by Capuzzo et al. (1986) for blue mussel near Salem Harbor.

The mean copper concentration (Figure 15-11) in Boston Harbor soft shell clam (4.10 ppm) was significantly higher than in Salem Harbor clams (2.18 ppm). In Salem Harbor, the soft shell clam samples collected in spring had a significantly higher copper concentration (2.50 ppm) than samples from fall/winter (1.79 ppm). Wallace et al. (1988) reported a mean copper concentration of 6.22 ppm for Boston Harbor clams, and the U.S. FDA (1988) reported a copper concentration for Massachusetts soft shell clam (including samples from Boston Harbor) of 4.1 ppm. The U.S. EPA (1988) reported a mean copper concentration in Boston Harbor soft shell clam of 1.86 ppm.

The mean copper concentration in ocean quahog was 2.06 ppm. The range of copper concentrations for this species collected on Georges Bank and offshore areas of southern New England was 1.3–2.6 ppm, converted from dry

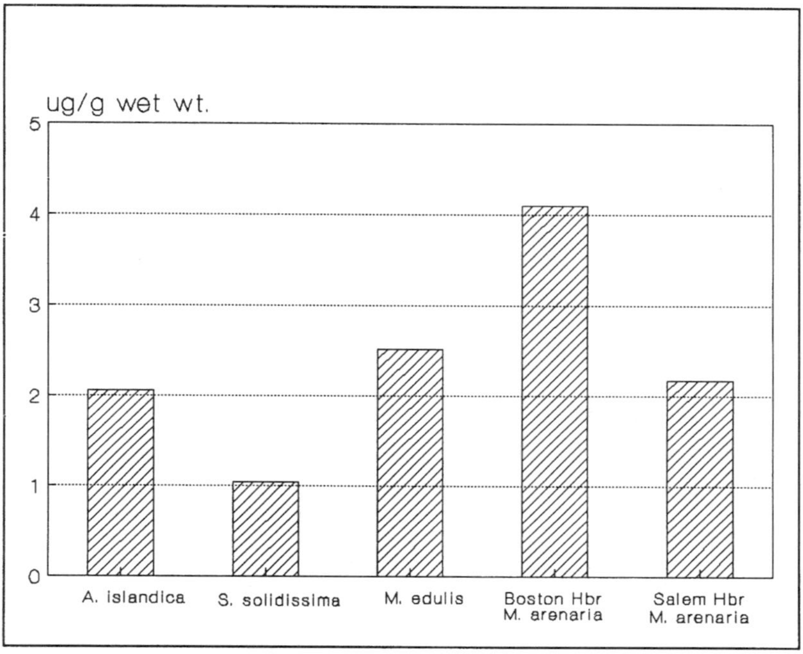

Figure 15-11. Copper in bivalve molluscs.

weight (Steimle et al. 1986). The mean copper concentration in surf clam was 1.04 ppm. The range of copper concentrations reported by Hall et al. (1978) for surf clam from Block Island Sound was 1.57–3.31 ppm. Surf clam collected in spring in our survey had significantly higher copper concentrations (1.35 ppm) than fall/winter samples (0.84 ppm).

The mean copper concentration in blue mussel was 2.52 ppm. NOAA (1987) reported blue mussel copper concentrations of 2.05–2.66 ppm (converted from dry weight) for Boston Harbor and 1.12–2.01 ppm for Buzzards Bay. There was no seasonal difference in mean copper concentrations among blue mussel in our study.

Soft shell clam samples from the Forest River in Salem Harbor had a mean lead concentration of 5.98 ppm, which was approximately one order of magnitude higher than that at the other two Salem Harbor sampling stations. The highest lead concentration we recorded was 24.94 ppm from Forest River. As previously noted, this shoreline station in Salem Harbor is in the vicinity of a lead mill that existed from around 1826 (Essex Institute 1971) until around 1920 (Directory of Salem and Beverly 1920). Another potential source of lead is a landfill upstream from the Forest River sampling station (EOEA 1991).

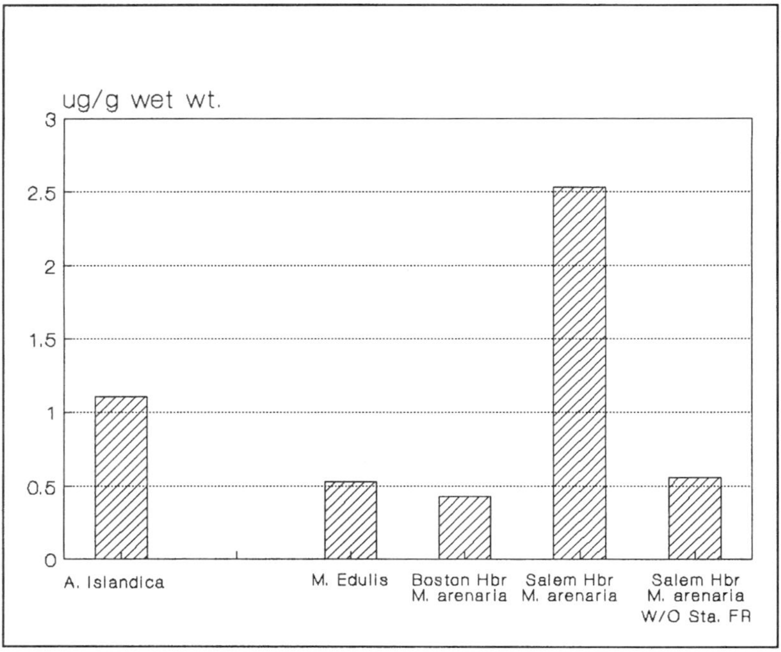

Figure 15-12. Lead in bivalve molluscs.

Without the FR station, the mean lead concentrations for Salem Harbor and Boston Harbor (Figure 15-12) were 0.56 ppm and 0.43 ppm, respectively, with no significant difference. Soft shell clam samples from Salem Harbor (without station FR) collected in spring had significantly more lead (0.68 ppm) than samples from fall-winter (0.39 ppm). Wallace et al. (1988) reported a mean lead concentration of 1.37 ppm for Boston Harbor clams, and the U.S. FDA (1988) reported a lead concentration for Massachusetts soft shell clam (including samples from Boston Harbor) of 0.43 ppm. The U.S. EPA (1988) reported a mean lead concentration in Boston Harbor soft shell clam of 0.45 ppm.

The mean lead concentration in ocean quahog was 1.04 ppm. The range of lead concentrations for this species collected on Georges Bank and offshore areas of southern New England was 0.60–2.21 ppm, converted from dry weight (Steimle et al. 1986). The range of detection limits for lead concentrations in surf clam was from less than 0.08 ppm to less than 0.30 ppm, with one sample containing 0.45 ppm. The range of lead concentrations reported by Hall et al. (1978) for surf clam from Block Island Sound was 0.875–1.250 ppm.

The mean lead concentration in blue mussel was 0.53 ppm. NOAA (1987) reported blue mussel lead concentrations of 0.89–2.60 ppm (converted from

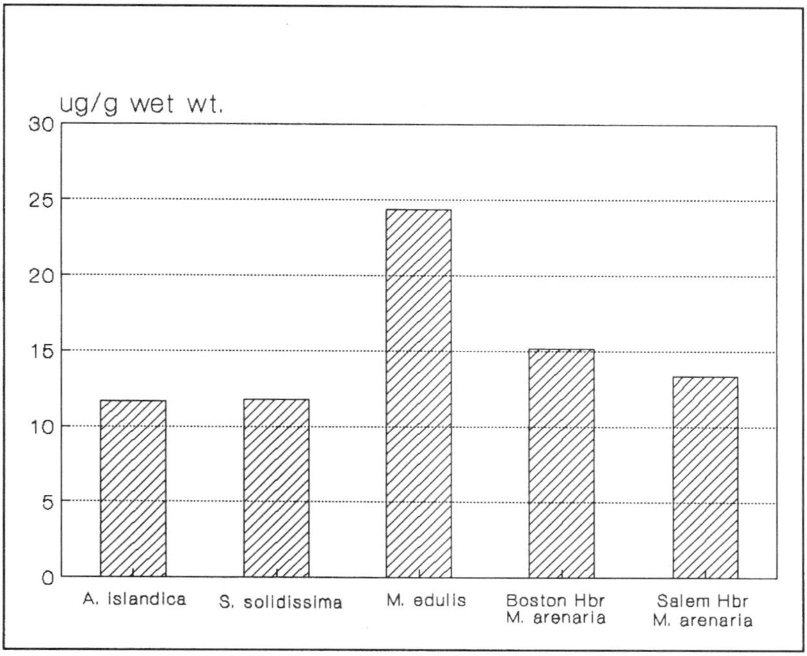

Figure 15-13. Zinc in bivalve molluscs.

dry weight) from Boston Harbor, and 0.19–0.92 ppm from Buzzards Bay. NOAA (1987) indicated that blue mussel may possess a particular affinity for accumulating lead. There was no seasonal difference in mean lead concentrations among blue mussel in our study.

There was no significant difference in zinc concentrations (Figure 15-13) between Boston Harbor (mean = 15.17 ppm) and Salem Harbor (mean = 13.34 ppm) soft shell clam. There was no seasonal difference among Salem Harbor soft shell clam. Wallace et al. (1988) reported a mean zinc concentration of 15.6 ppm for Boston Harbor clams, and the U.S. FDA (1988) reported a zinc concentration for Massachusetts soft shell clam (including samples from Boston Harbor) of 15 ppm.

The mean zinc concentration in ocean quahog was 11.70 ppm. The range of zinc concentrations for this species collected on Georges Bank and offshore areas of southern New England was 12.4–30.8 ppm, converted from dry weight (Steimle et al. 1986). The mean zinc concentration in surf clam was 11.77 ppm. The range of zinc concentrations reported by Hall et al. (1978) for surf clam from Block Island Sound was 15.27–18.75 ppm. There was no seasonal difference in mean zinc concentrations among surf clam in our study.

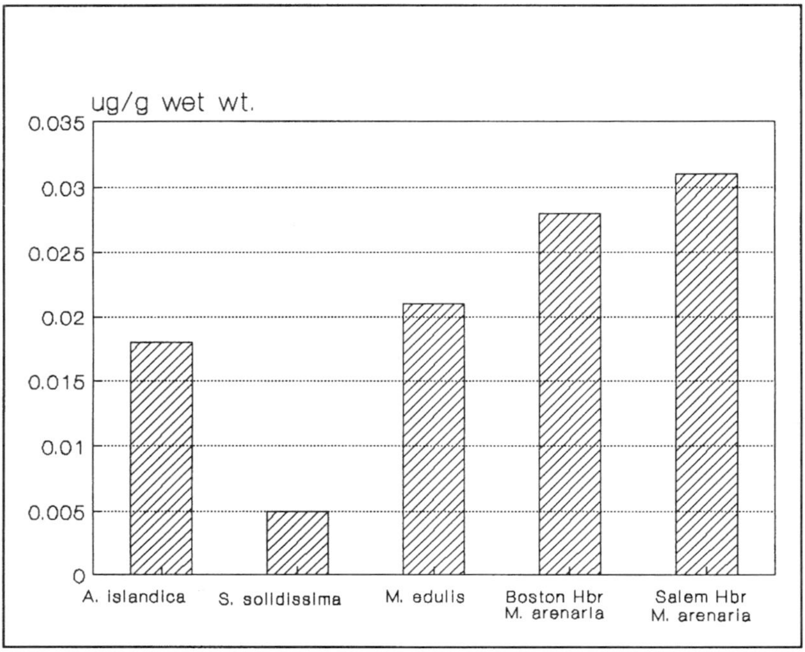

Figure 15-14. Mercury in bivalve molluscs.

The mean zinc concentration in blue mussel was 24.37 ppm. NOAA (1987) reported blue mussel zinc concentrations of 22.0–27.2 ppm (converted from dry weight) from Boston Harbor, and 10.6–21.4 ppm from Buzzards Bay. There was no seasonal difference in mean zinc concentrations among blue mussel in our study.

There was no significant difference in mean mercury concentrations (Figure 15-14) between Boston Harbor (0.028 ppm) and Salem Harbor (0.031 ppm) soft shell clam. There was no seasonal difference among Salem Harbor soft shell clam. Wallace et al. (1988) reported a mean mercury concentration of 0.053 ppm for Boston Harbor clams. The U.S. EPA (1988) reported a mercury concentration of less than 0.002 ppm in soft shell clam. The range of mercury concentrations for Massachusetts (including Boston Harbor samples) listed by Capuzzo et al. (1986) was 0.01–0.04 ppm.

The mean mercury concentration in ocean quahog was 0.018 ppm. The mean mercury concentration in surf clam was 0.006 ppm. The range of mercury concentrations reported by Hall et al. (1978) for surf clam from Block Island Sound was 0.020–0.050 ppm. There was no seasonal difference in mean mercury concentrations among surf clam in our study.

The mean mercury concentration in blue mussel was 0.021 ppm. NOAA (1987) reported blue mussel mercury concentrations of 0.038–0.066 ppm (converted from dry weight) from Boston Harbor and 0.012–0.022 ppm from Buzzards Bay. There was no seasonal difference in mean mercury concentrations among blue mussel in our study.

Summary

Unlike human-made contaminants, such as PCBs, the trace metals in this study can be found as naturally occurring elements throughout the marine environment. Some of these elements, such as copper and zinc, are essential elements for normal biological functions. Concentrations of these and other trace metals in marine crustaceans and bivalve molluscs are strongly influenced by changes in the physiological state of the organism that occur with molting, growth and reproductive cycles, and feeding (Engel 1988). Furthermore, different tissues accumulate trace metals at different rates (U.S. EPA 1988). Metal concentrations in both fish and invertebrates will not always be elevated by high metal concentrations in underlying sediments, even when the site is impacted by mine tailings (Harding and Goyette 1989).

Environmental factors such as coastal upwelling of cadmium-rich deep ocean water was postulated by NOAA (1992) as the source for high cadmium levels in mussels off northern California because of the absence of an industrial or urban source. Engel (1988) similarly hypothesized that cadmium transferred from cadmium-rich deep ocean waters through the food chain resulted in surprisingly high levels in the liver and kidneys of beaked whales. Although biological and environmental variables cannot be controlled, their influence must be considered when assessing spatial and temporal trends of trace metal concentrations in marine species as part of a contaminant monitoring program (Engel 1988).

Our results indicate that the metal concentrations we detected, overall, are not substantially different from concentrations we previously reported (Duston et al. 1990) or from several reports for trace metals for fish and invertebrates from Massachusetts and other locales around New England cited in the text. The difficulties we encountered determining statistical trends both spatially and over time were also encountered by the mussel watch project of the National Status and Trends Program. NOAA (1992) was unable to clearly establish long-term trends after monitoring organic and inorganic (trace metal) contaminants for 5 years. This was attributed to the sum of biological and environmental variables discussed above and the relatively long time span needed to detect statistically significant trends. Additional improvements in trace metal detection limits using analytical techniques such as inductively coupled plasma (ICP), graphite furnace atomic absorption (GFAA), or both, will increase our ability to resolve statistical trends for some trace metals. Despite these variables and analytical limitations, some spatial and temporal attributes of metal concentrations in Massachusetts marine species do emerge from the data base and are summarized as follows:

1. Mercury concentrations in Boston Harbor winter flounder were higher than mercury concentrations in flounder in Salem Harbor or coastal Massachusetts. The mercury level in Boston Harbor flounder decreased by approximately 50% over a four-year period. Zinc concentrations in Boston Harbor flounder were significantly elevated compared with those in Salem Harbor or coastal flounder.
2. Among all samples in the database, coastal winter flounder had higher copper and zinc concentrations in spring compared with fall-winter.
3. Coastal lobster had significantly higher cadmium concentrations compared with lobster from Boston Harbor or Salem Harbor. The cadmium concentration in coastal lobster significantly increased over a six-year period.
4. Chromium concentrations in Salem Harbor lobster were more than twice the concentrations in lobster from Boston Harbor or coastal Massachusetts. Chromium levels in Salem Harbor lobster appear to decrease over time, but the concentrations are too variable to support a statistically significant trend. Copper and zinc concentrations in Salem Harbor lobster were slightly but statistically lower compared with lobster from Boston Harbor or coastal Massachusetts.
5. Concentrations of the following metals in lobster from different study areas were all significantly higher in spring compared with fall/winter: chromium (Salem Harbor), copper (coastal), zinc (coastal, Boston Harbor, and Salem Harbor), and mercury (coastal).
6. Copper concentrations in Boston Harbor soft shell clam were nearly twice the concentrations of copper in Salem Harbor soft shell clam.
7. Chromium concentrations of Salem Harbor soft shell clam were four times higher than concentrations in Boston Harbor soft shell clam. In Salem Harbor soft shell clam, the concentrations of chromium, copper, and lead were all significantly higher in spring compared with fall/winter.
8. Concentrations of chromium and copper in surf clam were significantly higher in spring compared with fall-winter.

The relatively higher chromium concentrations in Salem Harbor lobster and soft shell clam, combined with high chromium levels in Salem Harbor sediments reported by NOAA (1988), strongly suggest that chromium levels we observed in Salem Harbor are the result of human activity. Otherwise, there was no corroborating information to equate metal concentrations in sediments with metal concentrations in the species we analyzed. For example, mercury concentrations in Boston and Salem Harbor sediments are approximately the same (NOAA 1988), yet substantially higher mercury concentrations were observed in Boston Harbor winter flounder compared with Salem Harbor flounder.

We are unaware of any human activity that could be a source explaining the increased cadmium concentrations in coastal lobster. It has been suggested that cadmium-rich deep ocean water imparts higher cadmium concentrations to coastal marine species in other parts of the United States (Engel 1988; NOAA 1992). In coastal Massachusetts, cadmium concentrations in ocean quahog and blue mussel were five times higher than cadmium concentrations in soft shell clam from Boston or Salem harbors. Surf clam, however, do not contain high levels of cadmium. Interspecies differences in the uptake and concentration of metals among bivalve molluscs prevents direct comparison of coastal bivalve species to soft shell clam in the harbors. Although the possible effect of deep ocean water in increasing the cadmium concentrations in coastal marine species should not be ignored, we are unable to explain the high levels of cadmium in coastal samples compared with the harbor samples.

Not every metal had a seasonal change in concentration among every species, but all the seasonal differences in metal concentrations that were detected indicated higher metal concentrations in spring compared with fall/winter. These seasonal differences were very likely the result of biological cycles that regulated the uptake, accumulation, and discharge of trace metals (Engel 1987, 1988; Engel and Brouwer 1984, 1987; Fowler et al. 1986). Future contaminant monitoring efforts could utilize the spring sampling period to determine an annual upper estimate of metal concentrations in marine species, thus eliminating the need to sample twice annually.

The absence of statistical changes over time for all metal concentrations in all marine species does not preclude the occurrence of anthropogenic changes but does indicate that these changes will not become statistically apparent without continued monitoring. Temporal trends of metal concentrations in winter flounder and lobster from Salem and Boston harbors, for example, indicate either a significant decrease, such as mercury in Boston Harbor flounder, or no change over time. We found no temporal increase in metal concentrations among the harbor samples. In some cases, such as chromium and mercury in lobster, metal concentrations appear nonstatistically to be decreasing. This suggests that efforts to control and reduce the input of metals to the harbors could be a reason for reductions in metal concentrations in marine species. Further monitoring will be necessary to discern between natural and anthropogenic concentrations of trace metals in Massachusetts marine species.

Data Appendices

Data appendices have not been included. A complete copy of the original report can be obtained from the authors: Division of Marine Fisheries, 100 Cambridge Street, Boston, MA 02202.

References

Capuzzo, J.M., McElroy, A., and Wallace, G.T. 1986. Fish and Shellfish in New England Waters: An Evaluation and Review of Available Data on the Distribution of Chemical Contaminants. Prepared for the Coastal Alliance, Washington, DC.

Connecticut Department of Environmental Protection (CT DEP). 1987. Report on Long Island Sound Activities. Hartford: Connecticut Department of Environmental Protection.

Directory of Salem and Beverly. 1920. Naumkeag Series No. 27. Salem, MA: Henry M. Meek Publishing. [Available at the Essex Institute Library, Salem, MA.]

Duston, N. M., Batdorf, C. A., and Schwartz, J. P. 1990. Metal Concentrations in Marine Fish and Shellfish from Boston and Salem Harbors, and Coastal Massachusetts: Progress Report. Mass. Div. Marine Fish., Boston, MA. Publ. #16-378-129-200-7-90-C.R.

Engel, D. W. 1987. Metal regulation in the blue crab, *Callinectes sapidus*: Copper, zinc, and metallothionein. Biol. Bull. 172:69–82.

Engel, D. W. 1988. The effect of biological variability on monitoring strategies: Metallothioneins as an example. Water Resources Bull. 24(5):981–987.

Engel, D. W., and Brouwer, M. 1984. Trace metal binding proteins in marine molluscs and crustaceans. Mar. Environ. Res. 13:177–194.

Engel, D. W., and Brouwer, M. 1987. Metal regulation and molting in the blue crab, *Callinectes sapidus*: Metallothionein function in metal metabolism. Biol. Bull. 173:239–251.

Essex Institute. 1971. 1868—Essex Institute Historical Collections, Second Series; Vol. I, Part II, p. 65. (Reprinted 1971.) Salem, MA: Essex Institute Press.

Executive Office of Environmental Affairs (EOEA). 1991. Draft Environmental Impact Report, Glass Landfill Closure, GTE Products Corporation, Salem, Massachusetts. EOEA No. 8437. Boston: Massachusetts Executive Office of Environmental Affairs.

Fowler, B. A., Engel, D. W., and Brouwer, M. 1986. Purification and characterization studies of cadmium-binding proteins from the American oyster, *Crassostrea virginica*. Environ. Health Perspect. 65:63–69.

Hall, R. A., Zook, E. G., and Meaburn, G. M. 1978. National Marine Fisheries Service Survey of Trace Elements in the Fishery Resource. NOAA Tech. Rpt. NMFS/SSRF-721. Charleston, SC: National Oceanic and Atmospheric Administration.

Harding, L., and Goyette, D. 1989. Metals in northeast Pacific coastal sediments, and fish, shrimp, and prawn tissues. Mar. Poll. Bull. 20(4):187–189.

Keith, L. H., Crummett, W., Deegan, J., Jr., et al. 1983. Principles of environmental analysis. Anal. Chem. 55(14):2210–2218.

May, T. W., and McKinney, G. L. 1981. Cadmium, lead, mercury, arsenic and selenium concentrations in freshwater fish, 1976–1977—National Pesticide Monitoring Program. Pest. Monit. J. 15(1):14–38.

National Oceanic and Atmospheric Administration (NOAA). 1984. National Status and Trends Program: Progress Report and Preliminary Assessment of Findings of the Benthic Surveillance Program—1984. Rockville, MD: National Oceanic and Atmospheric Administration, Offfice of Oceanography and Marine Assessment.

National Oceanic and Atmospheric Administration (NOAA). 1987. National Status and Trends Program for Marine Environmental Quality—Progress Report: A Summary of Selected Data on Chemical Contaminants in Tissues Collected During 1984, 1985, 1986, and 1987. NOAA Tech. Mem. NOS OMA 38. Rockville, MD: National Oceanic and Atmospheric Administration.

National Oceanic and Atmospheric Administration (NOAA). 1988. National Status and Trends Program for Marine Environmental Quality—Progress Report: A Summary of Selected Data on Chemical Contaminants in Sediments Collected During 1984, 1985, 1986, and 1987. NOAA Tech. Mem. NOS OMA 44. Rockville, MD: National Oceanic and Atmospheric Administration.

National Oceanic and Atmospheric Administration (NOAA). 1989. National Status and Trends Program for Marine Environmental Quality—Progress Report: A Summary of Data on Tissue Contamination from the First Three Years (1986–1988) of the Mussel Watch Project. NOAA Tech. Mem. NOS OMA 49. Rockville, MD: National Oceanic and Atmospheric Administration.

National Oceanic and Atmospheric Administration (NOAA). 1992. Recent Trends in Coastal Environmental Quality: Results from the First Five Years of the NOAA Mussel Watch Project. Silver Spring, MD: Coastal Monitoring Branch, Coastal Monitoring and Bioeffects Assessment Division, NOAA/NOS, SSMCIV, ORCA2.

Reid, R. N., O'Reilly, J. E., and Zdanowicz, V. S. (eds.). 1982. Evaluation of New York Bight and Long Island Sound Sediments and Demersal Species, and Contaminant Effects on Benthos, Summer 1980. NOAA Tech. Mem. NMFS-F/NEC-16. Sandy Hook, NJ: National Oceanic and Atmospheric Administration.

Schwartz, J. P., Duston, N. M., and Batdorf, C. A. 1991. PCBs in Winter Flounder, American Lobster, and Bivalve Molluscs from Boston Harbor, Salem Harbor, and Coastal Massachusetts: 1984–1989. Mass. Div. Marine Fish., Boston, MA. Publ. #16,966-63-250-10-91-C.R.

Steimle, F. W., Boehm, P. D., Zdanowicz, V. S., and Bruno, R. A. 1986. Organic and trace metal levels in ocean quahog, *Arctica islandica linne*, from the northwestern Atlantic. Fish. Bull. 84(1):133–140.

U.S. Environmental Protection Agency (U.S. EPA). 1988. A Histopathological and Chemical Assessment of Winter Flounder, Lobster, and Soft-shelled Clam Indigenous to Quincy Bay, Boston Harbor and an In Situ Evaluation of Oysters Including Sediment (Surface and Cores) Chemistry (G. R. Gardner and R. J. Pruell, Principal

Investigators). Narragansett, RI: U.S. Environmental Protection Agency, Environmental Research Laboratory.

U.S. Food and Drug Administration (U.S. FDA). 1988. Lead, Cadmium, and Other Elements in Domestic Shellfish: Summary of FY'85/86 Field Program (S.G. Capar, Principal Investigator). Washington, DC: U.S. Food and Drug Administration, Division of Contaminants Chemistry, Center for Food Safety and Applied Nutrition.

Wallace, G. T., Egenhouse, R. P., Pitts, L. C., and Gould, B. R. 1988. Analysis of Contaminants in Marine Resources. (Prepared for the Massachusetts Department of Environmental Quality Engineering, Contract # 85-18, and the U.S. Environmental Protection Agency, Project # 812527-01-0.) Boston: Environmental Science Program, University of Massachusetts–Boston.

16

Riverine Contributions to Heavy Metal Inputs to the Northeast Shelf Ecosystem

Herbert L. Windom

Abstract

Major rivers discharging into the Northeast Shelf include the Roanoke, James, Rappahannock, Potomac, Susquehanna, Delaware, Hudson, Connecticut, Kennebec, Penobscot, and St. John. The dissolved and particulate fluxes of Cd, Co, Cu, Ni, Pb, and Zn and the dissolved fluxes of As and Sb in these and other rivers have been estimated, and the results are extrapolated to the entire Northeast Shelf watershed. Dissolved metal concentrations in rivers are evaluated in relation to general water chemistry in an attempt to identify anthropogenic influences. The differentiation of anthropogenic contributions to the particulate flux is based on elemental abundances in natural riverine suspended sediment.

The Northeast Continental Shelf receives annually about 180 km^3 of runoff from the United States (Curtis et al. 1973) and an additional 10 km^3 from Canada. This runoff is transported to the coast primarily in 11 major rivers (the St. John, Penobscot, Kennebec, Connecticut, Hudson, Delaware, Susquehanna, Potomac, Rappahannock, James, and Roanoke). The watersheds of these rivers represent about two-thirds of the approximately 400,000 km^2 drainage area of the Middle Atlantic Bight and Gulf of Maine (Table 16-1).

In this chapter, we draw on results of a general study of the chemistry of 18 East Coast rivers (Figure 16-1) to estimate the riverine flux of trace metals to the Northeast Continental Shelf. This study involved the collection of samples from rivers at locations near their mouths, but above any influence of salt water, so that they better reflect the mobilization and transport characteristics of the watersheds.

Table 16-1. Northeast watersheds

River system	Drainage area (km^2)	Discharge ($m^3 \cdot s^{-1}$)
Medway	1,390	56
St. John	39,900	1,030
Penobscot	17,920	360
Kennebec	7,040	121
Connecticut	25,000	445
Hudson	21,000	360
Delaware	17,600	327
Susquehanna	66,500	1,100
Potomac	29,900	310
Rappahannock	4,090	50
James	16,200	200
Roanoke	21,800	230
Total	268,340	

Because of the variability of constituent concentration and discharge with time, a statistically good estimate of the riverine flux of a substance requires that sufficient data on concentration be gathered over the annual hydrography of a river to allow the annual load to be calculated using an appropriate numerical interpolation procedure (Table 16-2). Although the main objectives of our river study did not include the estimation of riverine fluxes, the data obtained can be used for such purposes providing that the large uncertainties in the estimates are recognized.

Methods

The details of our sampling and analytical methodology are described elsewhere (Windom 1990; Windom et al. 1991) but are briefly summarized here.

Water samples were collected from small boats, from bridges or, in a few cases, from riverbanks by using silicone tubing attached to a PVC pole and a hand-powered peristaltic pump. Samples for dissolved trace metal analysis were filtered through Gelman Minicapsule cartridge filters (0.45 μm) attached to the outlet end of the silicone tubing, directly into precleaned polyethylene bottles.

Unfiltered samples for particulate trace metal analysis were pumped directly into 4-liter precleaned polyethylene bottles. Within 8 hours after sampling, particulate trace metals were collected on precleaned 0.4-μm Nucleopore filters held in precleaned polycarbonate filter holders.

Figure 16-1. East Coast watersheds included in the general study.

Table 16-2. Numerical procedures for calculating riverine fluxes or loads

$$\text{Total load} = K\left(\sum_{i=1}^{n} C_i / n\right)\left(\sum_{i=1}^{n} Q_i / n\right)$$

$$\text{Total load} = K\bar{Q}_r\left(\sum_{i=1}^{n} C_i / n\right)$$

$$\text{Total load} = K\sum_{i=1}^{n}\left(C_i Q_i / n\right)$$

$$\text{Total load} = K\sum_{i=1}^{n}\left(C_i \overline{Q_p}\right)$$

$$\text{Total load} = \frac{K\sum_{i=1}^{n}\left(C_i Q_i\right)}{\sum_{i=1}^{n} Q_r}$$

K = Conversion factor to account for period of record; C_i = instantaneous concentration associated with individual samples; Q_i = instantaneous discharge at time of sampling; Q_r = mean discharge for period of record; Q_p = mean discharge for interval between samples; n = mumber of samples.

Dissolved Cd, Co, Cu, Ni, Pb, and Zn were concentrated by extraction onto 8-hydroxyquinoline immobilized on silica gel before analysis by graphite furnace atomic absorption spectrophotometry. Dissolved As and Sb were analyzed using a hydroxide generation procedure followed by quartz furnace atomic absorption spectrophotometry (Byrd 1990).

Particulate samples were totally digested in polytetrafluoroethylene (Teflon) bombs using HF and HNO_3. Subsequent analyses of Cd, Co, Cu, Ni, Pb, and Zn were accomplished using graphite furnace atomic absorption spectrophotometry.

Results

Samples for dissolved and particulate trace metal analyses were collected in triplicate from the rivers listed in Table 16-1 during April–May 1987 and July–August 1988. These sampling times were chosen because they generally represent high and low runoff conditions, respectively. Although two sampling periods are insufficient to provide a reliable estimate of material flux in rivers, the conditions during which we collected samples probably do represent extremes of transport. Therefore, fluxes, F, calculated using the equation:

$$F = Q(\bar{c}) \tag{1}$$

Table 16-3. Mean and range of dissolved trace metal concentrations in Northeast rivers

	High runoff			Low runoff		
Metal	Mean	Minimum	Maximum	Mean	Minimum	Maximum
Arsenic (nmol · l^{-1})	1.9	0.9	3.5	5.2	1.7	4
Antimony (nmol · l^{-1})	2.6	0.2	7.7	1.3	0.3	2.8
Cadmium (pmol · l^{-1})	89	28	230	43	13	110
Cobalt (nmol · l^{-1})	1.1	0.5	2.6	0.9	0.3	2.3
Copper (nmol · 1^{-1})	16	7	28	19	1	64
Nickel (nmol · l^{-1})	11	1	51	9	1	22
Lead (pmol · l^{-1})	137	33	255	98	50	230
Zinc (nmol · l^{-1})	11	1	36	3.8	1.4	9.2
TDS (mg · l^{-1})	49	24	97	88	27	200

where Q is the annual river discharge and $\bar{c}$ is the average concentration for samples collected during the two sampling periods, are probably the best estimates that can be made based only on two samples.

Fluxes calculated in this way do provide a reasonable basis for comparison of watershed delivery characteristics. The collective data may also be used to develop a better regional understanding of sources and processes responsible for trace metal mobilization and transport.

In the following, we consider dissolved and particulate fluxes separately, since their natural components are controlled by different weathering processes. Their unnatural or anthropogenic compounds may also be controlled by different processes.

Dissolved Fluxes

The dissolved concentrations of the trace metals studied generally differ between low and high runoff (Table 16-3). One group of metals, Sb, Cd, Pb, and Zn, has higher concentrations during high runoff. Another group, Co, Cu, and Ni, has more similar mean concentrations in the two sample sets. Arsenic concentrations are higher during low runoff, similar to total dissolved solids.

Using metal concentrations and discharge and drainage area, we calculated the flux of each metal per unit area for each watershed (Table 16-4). We also calculated the average flux for the sum of the watersheds studied (see Table 16-1), which we will use later to estimate total flux for the entire watershed of the Northeast Shelf.

The fluxes from each watershed can be compared with the average flux (see Table 16-4) to assess relative efficiencies of mobilization and transport. A few watersheds have conspicuously higher fluxes for a given metal. The Medway

Table 16-4. Dissolved trace metal flux (g·km^{-2})

River system	As	Sb	Cd	Co	Cu	Ni	Pb	Zn
Medway	349	110	8	87	352	104	43	831
St. John	189	34	5	43	487	177	13	263
Penobscot	128	27	4	18	289	145	12	373
Kennebec	172	39	6	26	518	127	16	514
Connecticut	97	70	5	25	1,173	189	20	403
Hudson	232	164	8	13	1,602	265	21	208
Delaware	63	82	6	13	394	129	10	233
Susquehanna	80	140	2	67	661	1,112	7	84
Potomac	232	121	8	35	323	327	8	91
Rappahannock	39	188	5	29	380	122	5	35
James	248	189	4	23	321	199	12	30
Roanoke	57	189	5	19	447	105	8	431
Average flux*	126	107	4	37	630	460	12	220

*Sum of total fluxes from each watershed divided by sum of the watershed areas.

watershed has significantly higher fluxes of most of the metals (i.e., As, Cd, Co, Pb, and Zn). This is surprising considering that the Medway is the smallest watershed and probably the least developed.

The ability to assess what portion of the estimated fluxes is the result of natural mobilization and transport and what portion is the result of anthropogenic influences was a goal. The watersheds differ somewhat geologically and geomorphologically, but there is enough similarity between groups of watersheds that natural weathering processes should produce some similarities or patterns in the metal concentrations in natural weathering solutions. The fact that considerable differences in metal concentrations are observed suggests that natural processes, which should have similar effects in similar watersheds, do not dominate. Anthropogenic inputs or alterations to the natural weathering regime are obviously suspected.

Natural chemical weathering is normally dominated by H_2CO_3 as the primary source of protons for the dissolution of solid phases. Organic acids produced during organic matter decomposition and sulfuric acid resulting from the oxidation of reduced sulfur compounds within watersheds are also involved in weathering. Even so, the cation charge of natural surface weathering solutions (i.e., river water) is balanced primarily by HCO_3. In rivers in the Northeast, this does not appear to be the case. Most of the samples taken from the rivers had SO_4/alkalinity mole ratios considerably higher than major world rivers (Table 16-5), which reflect weathering solutions on continental scales.

Annual atmospheric sulfate deposition associated with acid rain for four sampling sites in Massachusetts, Pennsylvania, Delaware, and North Carolina

Table 16-5. Sulfate/alkalinity ratio (moles per equivalent)

	Northeast rivers		Large world rivers*	
River system	High runoff	Low runoff		
Medway	1.32	2.07	Chiang	0.07
St. John	0.13	0.11	Huang He	0.21
Penobscot	0.41	1.15	Brahmaputra	0.11
Kennebec	0.64	0.45	Mekong	0.04
Connecticut	0.37	0.31	Zaire	0.06
Hudson	0.16	—*	Amazon	0.06
Delaware	0.50	0.39		
Susquehanna	0.45	0.94		
Potomac	0.23	0.43		
Rappahannock	0.37	0.12		
James	0.11	0.27		
Roanoke	0.19	0.48		

*Influenced by sea water.
Source: M.-H. Hu, R. F. Stallard, and J. M. Edmond. (1982). Major ion chemistry of some large Chinese rivers. Nature 298:550–553.

averages about 21 mmole · m^{-2} during the period 1979–1983. Similar inputs (i.e., approximately 19 mmole · $m^{-2} \cdot y^{-1}$) of nitric acid were also estimated (Samson 1987).

Although other sources, both natural and anthropogenic, may account for a portion of observed sulfate concentrations over expected levels, a large portion is clearly the result of atmospheric deposition associated with air pollution. Thus, higher sulfate/alkalinity ratios in rivers may also reflect increased heavy metal inputs associated with atmospheric contamination. This could explain the relatively high metal fluxes from a relatively undeveloped watershed such as the Medway (see Table 16-4).

Metal inputs from anthropogenic sources other than through the atmosphere obviously occur, but inputs from natural sources may also be enhanced by the additional acid deposition. Increased weathering within watersheds would be expected to increase the mobilization of metals, and the increased acidity of weathering solution, resulting from increased acid deposition, should influence metal partitioning between solid and dissolved phases. This appears to be the case for Pb, Cd, and Zn (Figure 16-2) and suggests that acid deposition enhances the mobilization and transport of these metals.

Because of the variability of unnatural sources for metal inputs to watersheds and the apparent altered weathering regime, as discussed above, it is difficult to assess the natural and anthropogenic components of the dissolved metal fluxes in rivers in the Northeast.

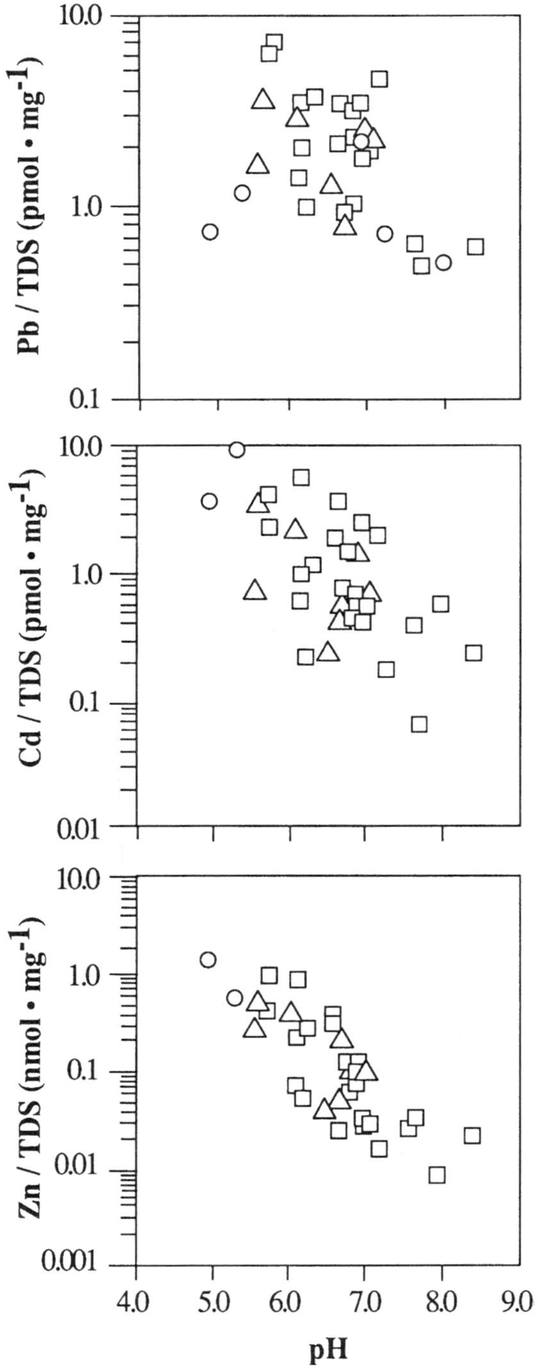

Figure 16-2. Dissolved Pb, Cd, and Zn normalized to total dissolved solids (TDS) versus pH. Open circles represent Coastal Plain rivers, open triangles represent Piedmont rivers, and open squares represent Northeast rivers.

Table 16-6. Total particulate metal flux

River system	Cd ($g \cdot km^{-2}$)	Co ($g \cdot km^{-2}$)	Cu ($g \cdot km^{-2}$)	Ni ($g \cdot km^{-2}$)	Pb ($g \cdot km^{-2}$)	Zn ($g \cdot km^{-2}$)	TSS (10^3 $kg \cdot km^{-2}$)
Medway	1.6	37	640	1,300	90	1,100	2.4
St. John	4.3	60	130	208	70	650	2.4
Penobscot	4.9	48	190	100	85	1,100	2.2
Kennebec	2.8	75	190	230	120	790	2.7
Connecticut	9.4	85	450	690	290	1,300	4.0
Hudson	8.2	170	440	390	440	2,500	4.9
Delaware	2.0	7	58	180	34	460	1.4
Susquehanna	11.0	1,000	690	1,600	330	3,400	4.0
Potomac	3.5	120	170	290	110	1,200	1.4
Rappahannock	3.8	70	240	270	110	900	1.9
James	3.7	70	170	170	90	640	2.3
Roanoke	3.6	140	350	400	180	1,200	3.9
Average flux	6.4	320	350	640	200	1,700	2.7

Particulate Fluxes

Suspended sediment mobilization and transport depend on weathering within watersheds and their geomorphology. These characteristics along with the chemistry of the suspended particles (controlled by the geology of the watershed) obviously determine the natural flux of trace metals, assuming the suspended particles are not contaminated. In rivers in the Northeast, however, particles are expected to contain an anthropogenic component, as well as the natural component, since most trace metals are particle-reactive (i.e., they will become associated with particles subsequent to their input). Using Equation 1, we can determine a total particulate metal flux from data on the total concentration of metals in suspended sediment (Table 16-6). With the exception of copper, most of the total metal flux from Northeast watersheds is in particulate phases.

What portion of the particulate flux is natural and what portion anthropogenic? To account for the natural concentrations of trace metals in suspended sediments, we calculated the enrichment factor for each sample, which is the observed metal/aluminum ratio divided by the metal/aluminum ratio reported for average crustal material (Taylor and McLennan 1981). It is not surprising that the enrichment factor is relatively constant in samples taken from rivers having total suspended solids above approximately 15 $mg \cdot l^{-1}$ (Figure 16-3) and tends to be higher in samples from rivers containing low total suspended solids. Clearly, the anthropogenic contribution to suspended sediments in samples with high suspended solids is masked by natural contri-

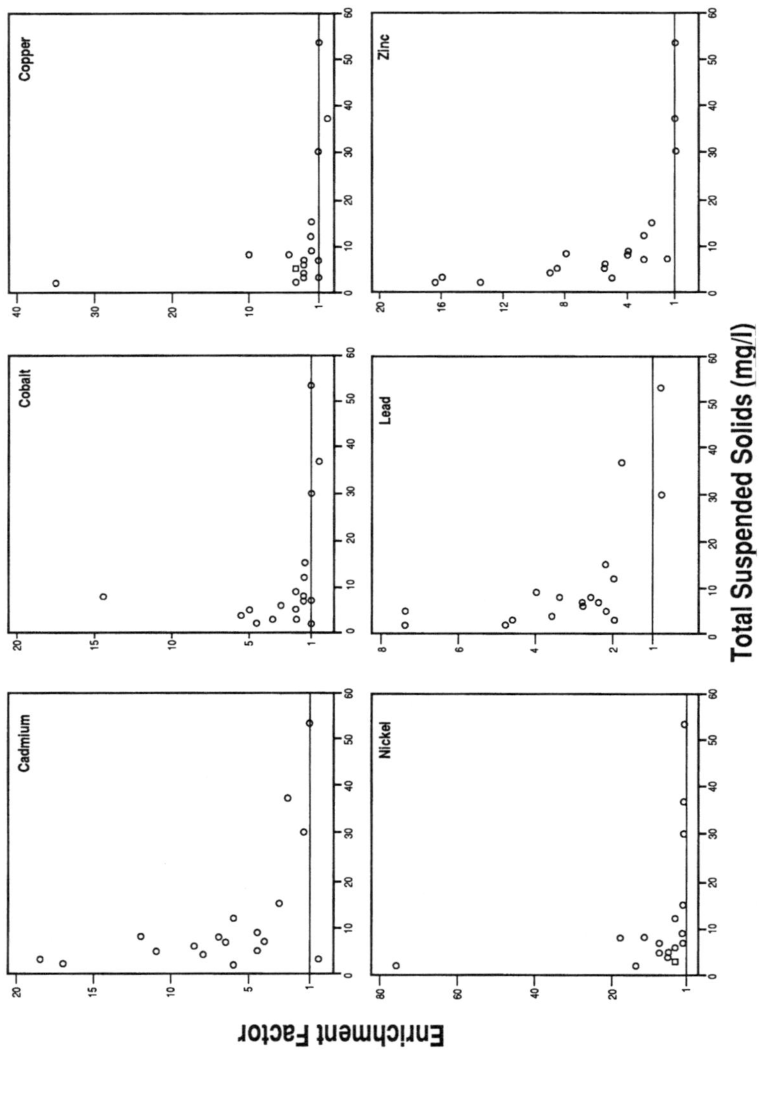

Figure 16-3. Enrichment factors for trace metals in suspended sediments in Northeast rivers, calculated using Equation 2.

Table 16-7. Excess particulate metal flux

River system	Cd (g·km^{-2})	Co (g·km^{-2})	Cu (g·km^{-2})	Ni (g·km^{-2})	Pb (g·km^{-2})	Zn (g·km^{-2})
Medway	1.4	26	610	1,280	75	1,000
St. John	–	29	72	150	28	460
Penobscot	4.4	25	140	58	53	1,000
Kennebec	2.1	40	110	160	75	560
Connecticut	6.6	25	220	520	110	530
Hudson	5.6	49	150	140	270	1,700
Delaware	1.9	2	44	170	26	430
Susquehanna	9.4	930	520	1,480	230	3,000
Potomac	3.1	100	120	240	80	1,000
Rappahannock	2.9	30	150	190	57	640
James	3.3	50	120	130	67	520
Roanoke	1.3	18	60	140	16	350
Average flux	4.6	260	220	530	120	1,300
% of total particulate flux	71	81	63	82	60	76

butions. When total suspended solids are low, however, the anthropogenic fraction is more evident.

The metal/aluminum ratio in average crustal material is an assumed approximation of the natural composition of suspended sediment in Northeast rivers. A better approximation is probably derived from the average metal/aluminum ratios in samples from rivers having total suspended solids greater than 15 mg · l^{-1}. The metal/aluminum ratios of these samples are generally similar to those reported for average crustal material (Taylor and McLennan 1981). By use of these ratios, the excess metal above assumed natural concentrations in suspended sediments, M_x, presumably resulting from the anthropogenic inputs, can be estimated by the following equation:

$$[M]_x = [M]_s - [Al]_s([M]:[Al])_{15+} \quad (2)$$

where $[M]_s$ and $[Al]_s$ are the total metal and aluminum concentrations in the sample, respectively, and $([M]:[Al])_{15+}$ is the mean metal/aluminum ratio in all samples we analyzed from Northeast rivers with total suspended solids greater than 15 mg · l^{-1}.

We can calculate the excess metal flux on particles using Equation 1 and the excess metal concentrations $[M]_x$ calculated from Equation 2. The average excess metal flux on particles ranges from about 60% (Pb) to about 81% (Co) of the total particulate metal flux (Table 16-7).

Table 16-8. Summary of riverine trace metal fluxes

	Annual flux in metric tons							
	As	Sb	Cd	Co	Cu	Ni	Pb	Zn
Dissolved	50	43	1.6	15	250	180	5	90
Particulate	5*	3*	2.6	130	140	260	80	680
Total	55	46	4.2	145	390	440	85	770
% Dissolved	91	93	38	11	64	41	6	12
Unnatural flux	—	—	3	110	245	360	51	585

*Estimate based on assumption As and Sb concentrations are suspended particles of 5 and 2.5 $\mu g \cdot g^{-1}$, respectively (Martin and Whitfield 1983), and on suspended sediment flux given in Table 16-6.

Table 16-9. Excess trace metal flux in Northeast rivers compared with 1988 U. S. consumption

Metal	Consumption* (metric tons)	Excess flux (% of consumption)
Cadmium	3,600	0.08
Cobalt	8,000	1.4
Copper	2,200,000	0.01
Nickel	120,000	0.3
Lead	1,200,000	0.004
Zinc	1,300,000	0.05

*Data from U.S. Bureau of Mines (1989).

Summary

The results presented above can be scaled up to estimate the total trace metal flux, in both dissolved and particulate form, from the entire 400,000 km^2 watershed of the Northeast Continental Shelf (Table 16-8). An unnatural or excess flux estimate is made on the following assumptions: (1) there is an excess particulate flux equal to that given in Table 16-7, and (2) there is excess dissolved flux that is of the same proportion to the total dissolved flux as the excess particulate flux to the total particle flux.

The significance of "excess metal fluxes," which can be assumed to represent maximum anthropogenic fluxes, can be assessed by comparing them with the annual U.S. consumption of these metals (Table 16-9). For example, cobalt and nickel have the highest excess fluxes relative to their annual consumption. These two metals are primarily used in the steel and alloy manufacturing industries, which are common in the Northeast, particularly in the Susquehanna drainage basin.

References

Byrd, J. T. 1990. Comparative geochemistries of arsenic and antimony in rivers and estuaries. Sci. Total Environ. 97/98:301–314.

Curtis, W. F., Culbertson, J. K., and Chase, E. B. 1973. Fluvial-Sediment Discharge to the Oceans from the Conterminous United States. In U. S. Geological Survey: A Contribution to the International Hydrologic Decade. Circular 670. Washington, DC: U. S. Geological Survey, pp. 1–17.

Hu, M.-H., Stallard, R. F., and Edmond, J. M. 1982. Major ion chemistry of some large Chinese rivers. Nature 298:550–553.

Martin, J. M., and Whitfield, M. 1983. The Significance of the River Input of Chemical Elements to the Ocean. In C. S. Wong, E. Boyle, K. W. Bruland, J. D. Burton, and E. D. Goldberg (eds.), Trace Metals in Sea Water. NATO Conference Series IV: Marine Sciences. pp. 265–296.

Samson, P. J. 1987. Estimating the atmospheric input of pollutants into a watershed. Water Air Soil Poll. 35:7–17.

Taylor, S. R., and McLennan, S. M. 1981. The composition and evolution of the continental crust: Rare earth element evidence from sedimentary rocks. Phil. Trans. R. Soc. Lond. A301:381–399.

U. S. Bureau of Mines. 1989. Minerals Yearbook, 1988. Washington, DC: U.S. Government Printing Office.

Windom, H. L. 1990. Flux of particulate metals between East Coast North American rivers and the North Atlantic Ocean. Sci. Total Environ. 97/98:115–124.

Windom, H. L., Byrd, J. T., Smith, R. G., Jr., and Huan, F. 1991. Inadequacy of NASQAN data for assessing metal trends in the nation's rivers. Environ. Sci. Technol. 25:1137–1142.

17

Trends in Nutrient Loading and Eutrophication: A Comparison of the Chesapeake Bay and the Hudson River Estuarine Systems

Thomas C. Malone and Daniel J. Conley

Abstract

Estuaries and their coastal plumes influence the quantity and quality of nutrients transported from land to the sea through processes that make them among the most productive ecosystems on earth. The response of phytoplankton to nitrogen and silicon inputs plays a central role in this regard. Seasonal scale comparison of the Hudson River and Chesapeake Bay systems illustrates the effects of variable inputs of new nutrients on the production and fate of phytoplankton biomass. Exports of dissolved inorganic nitrogen and silicon appear to be small relative to inputs and to dissolved inorganic phosphorus, largely as a consequence of phytoplankton uptake in the coastal plume of the Hudson and of uptake and subsequent transformations within the Chesapeake Bay. Contrasts in phytoplankton dynamics of the two systems reflect differences in nutrient input patterns and dilution rates.

Phytoplankton production in estuaries has been shown to be enhanced by nitrogen loading (Nixon 1981; Boynton et al. 1982), which has increased by as much as an order of magnitude during this century (Walsh et al. 1981). In the Middle Atlantic Bight, the delivery of nitrate by rivers and estuaries increased by about 50% during 1974–1981 as a consequence of inputs related to agricultural activities (animal waste and fertilizers), atmospheric deposition (acid rain), and sewage wastes (Smith et al. 1987). Partially stratified estuaries, such as the salt-intruded reaches of the Chesapeake Bay and Hudson River systems, are particularly responsive to nutrient enrichment as a consequence of two-layered flow regimes and benthic–pelagic coupling, which increase the retention and recycling of nutrients (Ketchum 1967; Kemp and Boynton 1984). Officer et al. (1984) argued that increases in nutrient loading and phytoplankton production during 1950–1980 caused the volume of oxygen-depleted bottom water to increase in the mesohaline reach of Chesapeake Bay.

High and variable nutrient inputs can destabilize plankton communities, resulting in large accumulations of phytoplankton biomass (Caperon et al. 1971). Such instabilities may cause shifts in the flow of energy and nutrients from metazoan food webs, which support high fish production, to microbial food webs, which lead to greater decomposition and oxygen consumption (e.g., Verity 1987). If anoxia occurs, the release of gaseous nitrogen via coupled nitrification–denitrification is inhibited, promoting the regeneration of ammonium and yet higher phytoplankton production (Kemp et al. 1990). In addition to stimulating phytoplankton production, inputs of anthropogenic nutrients may lead to changes in floristic composition by altering the relative availability of essential nutrients. For example, nitrogen (N) and phosphorus (P) loading can cause dissolved silicate (DSi) depletion, which may also cause changes in the patterns of energy flow and nutrient cycling (Schelske and Stoermer 1972; Officer and Ryther 1980).

Our goal here is to evaluate the effects of nutrient inputs in terms of the production and fate of phytoplankton in estuarine ecosystems. For this purpose, we compare phytoplankton responses to seasonal variations in nutrient input for different years in the mesohaline reach of Chesapeake Bay and in two contrasting estuarine systems, the Chesapeake Bay and Hudson River estuaries and their coastal plumes (Figure 17-1). We examine the relationships between annual phytoplankton cycles and seasonal variability in the magnitude and timing of allochthonous nutrient inputs and the implications of these effects in terms of coastal eutrophication.

Nutrient Inputs and Distributions

The salt-intruded reaches of the Hudson River and Chesapeake Bay are partially stratified estuaries with coastal plumes formed by the discharge of buoyant, low salinity water onto the continental shelf. Nutrient inputs to the two estuaries are similar. Total N and P inputs to Chesapeake Bay and its tribu-

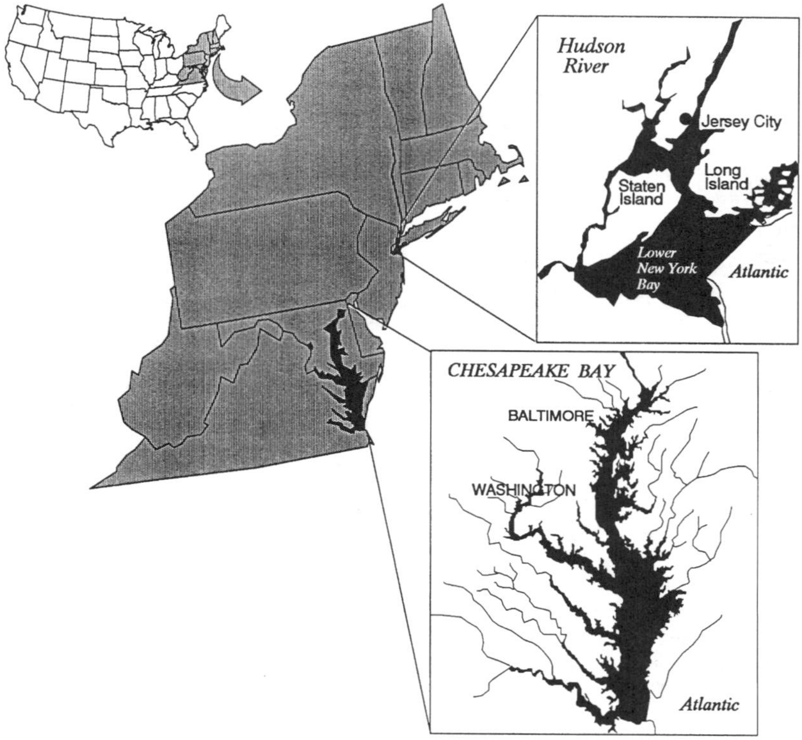

Figure 17-1. The Hudson estuary and Chesapeake Bay ecosystems.

taries are currently estimated to be about 152×10^6 kg N yr^{-1} and 11×10^6 kg P yr^{-1} (Boynton et al. 1995) compared with 182×10^6 kg N yr^{-1} and 14×10^6 kg P yr^{-1} to the Hudson estuary (Malone 1982, 1984). The fate and effects of nutrient inputs to each estuary differ largely because of differences in flushing rate and the location, chemical form, and temporal patterns of nutrient input.

River flow and the location of inputs are important variables of both nutrient input and residence time. The Susquehanna and Hudson Rivers account for most freshwater input to the Chesapeake Bay and the Hudson estuary, respectively. As is typical of mid-latitude rivers, the annual cycles of freshwater discharge exhibit a spring maximum and a summer minimum (Figure 17-2). Based on the ratio of estuarine volume to river flow (Q_f), the upper limit for freshwater residence time in the Hudson estuary ranges from approximately 1 week (spring) to 1 year (summer) with a mean of 1 month. The freshwater residence time in the Chesapeake Bay ranges from approximately 3 months (spring) to 7 years (summer) with a mean of 1 year. Thus,

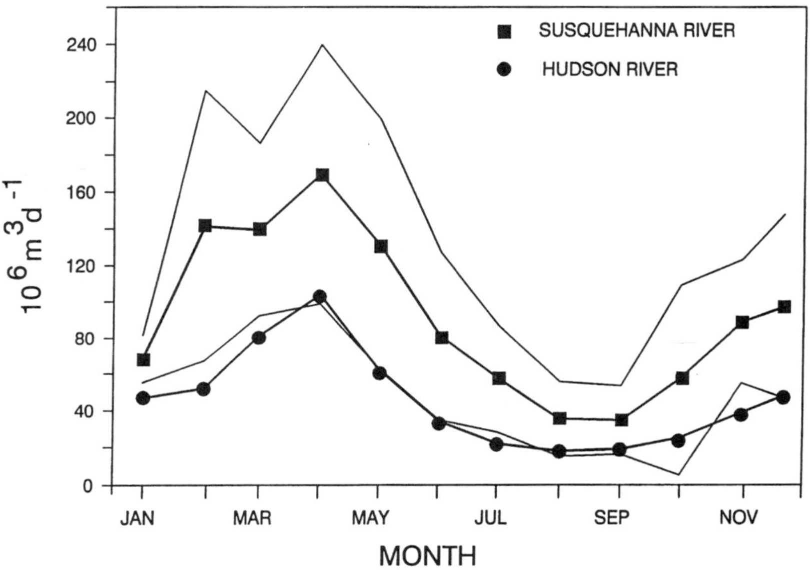

Figure 17-2. Monthly mean freshwater flow of the Susquehanna River (annual mean for 1984–1990 with ±1 standard deviation) and the Hudson River (1973–1976) corresponding to the periods during which phytoplankton data were collected.

the time scale for nutrient residence time is on the order of 1 month in the Hudson compared with 1 year in the Bay.

The effects of these differences are amplified by the location of nutrient inputs. Sewage discharges near the mouth of the estuary account for most of the allochthonous inputs of nitrogen (N) and phosphorus (P) to the Hudson, whereas nonpoint agricultural inputs associated with freshwater runoff near the head of the Bay and its subtributaries account for most nutrient input to the Chesapeake (Garside et al. 1976; Smullen et al. 1982; Schubel and Pritchard 1986). In addition, because seasonal variations in sewage discharge are small relative to freshwater runoff, inputs of N and P to the Hudson estuary are comparatively stable (Simpson et al. 1977; Malone 1982) while inputs to the Chesapeake exhibit large seasonal variations that are correlated with river flow (Malone et al. 1988). Discharge from the Susquehanna River accounts for approximately 70% of dissolved inorganic N (DIN) and 55% of dissolved inorganic P (DIP), with over 50% of the input occurring during the spring period of maximum flow (Schubel and Pritchard 1986; Boynton et al. 1995).

Differences in nutrient sources are also reflected in the chemical quality of dissolved and particulate nutrient inputs (Table 17-1). Ammonium and DIP

Table 17-1. Nutrient concentrations and ratios of (1) source inputs to Chesapeake Bay (Susquehanna River) and the Hudson estuary (ambient concentrations near the point source input), (2) surface water at the mouth of Chesapeake Bay, and (3) surface water in their coastal plumes

Chesapeake system nutrients	Source 1982–83[a]	Source 1989–90[b]	Outflow 1989–90[b]	Plume 1982–83[a]	Plume 1985–86[c]
TN (μM)	129	128	20	18	6
% DIN	71	68	14	15	14
% DON	—	22	47	—	—
% Ph-N	3	4	19	10	41
% TON	28	32	86	85	86
TP (μM)	1.4	—	—	1.4	—
% DIP	27	—	—	31	—
% TOP	73	—	—	67	—
DIP (μM)	0.4	0.2	0.3	0.4	0.5
DIN/DIP	235	900	18	6	2
DSi/DIP	160	629	55	14	3
Hudson system nutrients	Source 1973–74[d]	Plume 1973–74[d]			
DIP (μM)	2.7	1.8			
DIN/DIP	23	7			
DSi/DIP	8	3			

[a]Fisher et al. (1988).
[b]Glibert and Fisher (unpublished data).
[c]Malone and Ducklow (1990).
[d]Malone et al. (1985).

account for most sewage-derived inputs of N and P in the Hudson (Garside et al. 1976; Simpson et al. 1977), whereas nitrate and particulate P account for most input to the mesohaline Bay (Fisher et al. 1988), and concentrations of major nutrient elements (P, N, and Si) differ relative to each other. In the region of point source inputs to the lower Hudson estuary, nutrient ratios are close to Redfield ratios,* with the ratio of dissolved inorganic N to dissolved inorganic P (DIN:DIP) tending to be at or above Redfield and the ratio of dissolved inorganic Si to phosphorus (DSi:DIP) at or below Redfield. In contrast, nutrient ratios in the riverine input to Chesapeake Bay are considerably higher than Redfield.

Within the Hudson estuary, nutrients generally distribute conservatively with salinity upstream and downstream from point source discharges; most

*The Redfield ratio is the elemental ratio of C:N:P in nutrient-sufficient phytoplankton; it averages 106:16:1 (Redfield 1958). Similarly, Si:P averages 16:1 (Brzenzinski 1985).

nutrient input is exported to the coastal plume (Ketchum 1969; Garside et al. 1976; Simpson et al. 1977; Malone 1982) where phytoplankton uptake is high and DSi and DIN are depleted relative to DIP (Table 17-1; Figure 17-3). Most nutrient input to Chesapeake Bay is taken up in the mesohaline reach (see Figure 17-3), and DIN:DIP and DSi:DIP ratios decline during transport through the Bay (see Table 17-1). Relationships between nutrient distributions and salinity indicate that DIN and DSi are depleted within both plumes (Malone 1982; Glibert et al. 1991). Thus, although inorganic nutrient inputs to the Hudson plume are substantially higher than to the Chesapeake plume, nutrient distributions and ratios in both plumes indicate that DIN and DSi are depleted by phytoplankton uptake before DIP.

Differences in the spatial and temporal patterns of nutrient input and nutrient residence times are reflected in the levels of phytoplankton production and biomass that characterize the estuaries and their coastal plumes (Table 17-2). In the Chesapeake Bay, annual production and mean annual biomass (chlorophyll *a* concentration) are high in the mesohaline portion of the Bay and low in the coastal plume. In contrast, production and biomass are low in the Hudson estuary and high in its coastal plume, reflecting the rapid export of inorganic nutrients from the estuary. In general, export of dissolved inorganic nutrients from Chesapeake Bay appears to be low compared with export from the Hudson estuary. The distributions of N and P among dissolved and particulate pools in the Chesapeake plume suggest that most nitrate input to the Bay is exported to the coastal plume in organic form, with little net change between P inputs and exports.

Estuarine Productivity and Biomass

Annual cycles of phytoplankton biomass (CHL) and productivity (PP) in Chesapeake Bay differ from those in the Hudson estuary (Figure 17-4). In the Bay, CHL and PP exhibit annual maxima during spring and summer, respectively. Phytoplankton productivity also peaks during the summer in the Hudson estuary, but CHL is comparatively low and stable throughout the year. Although annual cycles of CHL differ, both estuaries exhibit similar transitions from spring to summer phytoplankton distributions as surface temperature exceeds 20°C during a 2- to 3-week period in late May to early June. The transition coincides with major changes in the distribution, floristic composition, and size structure of the plankton community (Malone 1977, 1991; Malone et al. 1980). During spring, CHL is typically higher below the euphotic zone than in the overlying surface layer, and phytoplankton biomass is

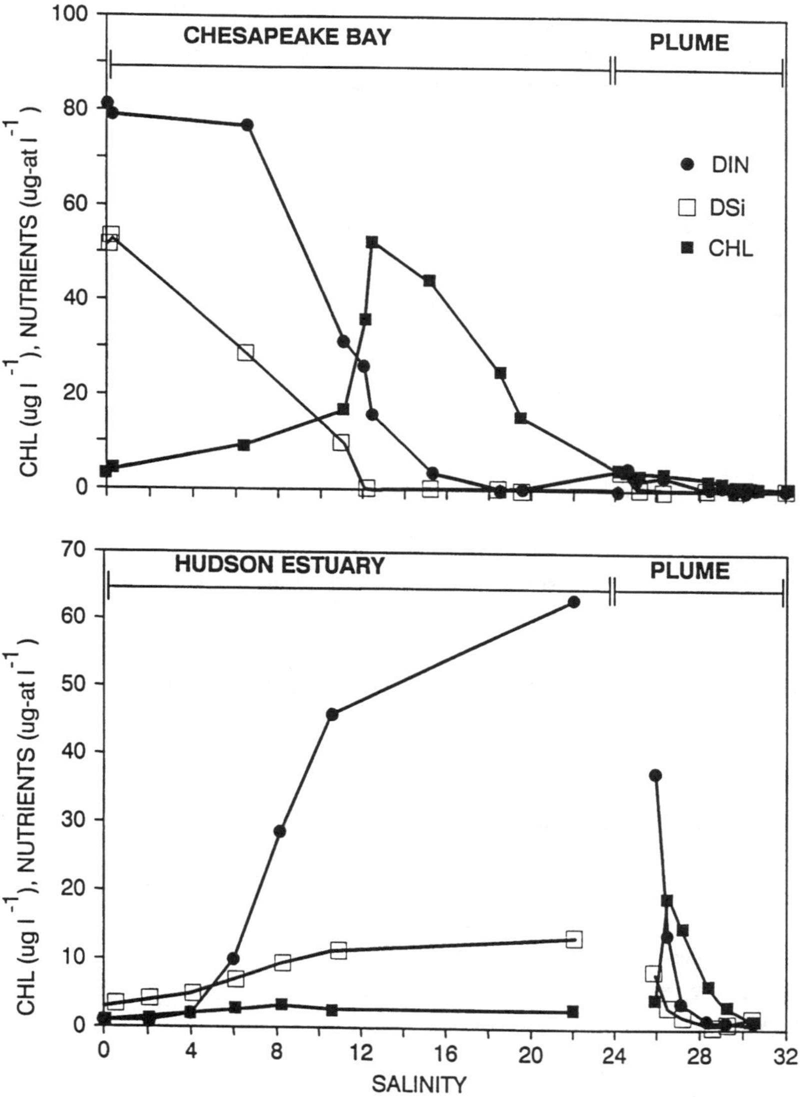

Figure 17-3. Surface distributions of chlorophyll a (μg l^{-1}) and dissolved inorganic nitrogen and silicon (μg-at l^{-1}) along the salinity gradients of the Chesapeake and Hudson estuarine systems under high flow conditions.

Table 17-2. Annual phytoplankton production (PP; g C m^{-2}), mean water column chlorophyll a concentration (CHL; mg m^{-2}), and chlorophyll specific productivity (P_{CHL}; g C g Chl^{-1} d^{-1}) of the mesohaline reach of Chesapeake Bay, the salt-intruded reach of the Hudson estuary, and their respective coastal plumes

	PP	CHL	P_{CHL}
Estuary			
Mesohaline Bay[a]	630	180	17
Hudson River[b]	220	25	36
Plume			
Chesapeake[c]	320	48	35
Hudson[d]	590	80	30

[a]1984–1989: Mean; Malone (unpublished data).
[b]1973–1974: Malone (1977).
[c]1984–1985: Malone (unpublished data).
[d]1973–1978: Malone et al. (1985).

dominated by chain-forming diatoms with occasional intrusions of large dinoflagellates. After the transition, high CHL is restricted to the surface layer, PP increases to its seasonal maximum, and small, solitary cyanobacteria, chrysophytes, cryptophytes, and chlorophytes dominate. As suggested by these distributions, phytoplankton sedimentation rates are highest during spring, when chain-forming diatoms dominate (Kemp and Boynton 1991).

Dilution Rate and Growth

The spring–summer transition coincides with a rapid increase in CHL-specific productivity (P_{CHL}), with P_{CHL} being higher in the Hudson throughout most of the year (Figure 17-4). High P_{CHL} and low CHL during summer are indicative of rapid biomass turnover. The phasing of annual cycles of CHL and P_{CHL} suggests a seasonal progression from high productivity relative to biomass losses during spring (Chesapeake Bay only) to a more closely balanced coupling between productivity and losses during summer (both estuaries).

The relationship between phytoplankton growth and dilution rate resulting from nontidal flow provides useful insights to CHL dynamics in the two estuaries (Malone 1977; Malone et al. 1988). Mean euphotic growth rates (μ, calculated from P_{CHL} in Figure 17-4 and C:CHL = 50) in the Hudson estuary range from a winter minimum of 0.1 d^{-1} to a summer maximum of 1.8 d^{-1} compared with surface layer dilution rates (D, calculated from Q_f in Figure 17-2 and salt distributions) of 2.4 d^{-1} to 0.2 d^{-1} during spring and summer, respectively. Likewise, phytoplankton growth in the mesohaline Chesapeake ranges seasonally from 0.1 to 0.8 d^{-1} compared with dilution rates of 0.1 d^{-1}

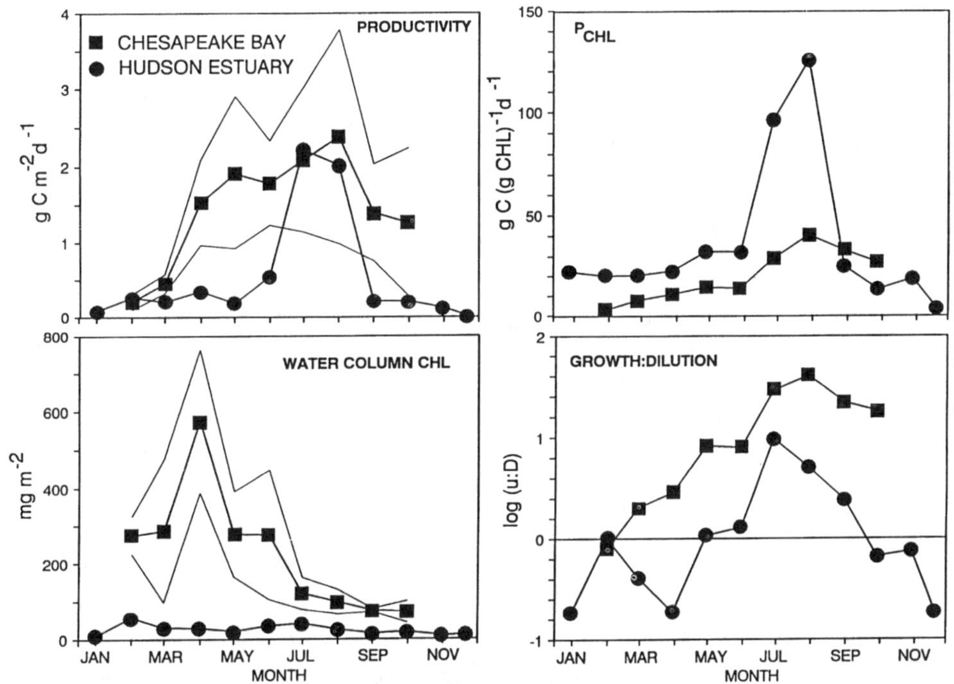

Figure 17-4. Annual cycles of phytoplankton productivity (PP) (euphotic zone); chlorophyll a (CHL) (total water column); chlorophyll specific productivity (P_{CHL}) (euphotic zone); and the ratio of carbon specific growth rate (μ) to dilution rate (D) of the euphotic zone (3–10 m in depth and generally confined to the surface layer in both estuaries). $\mu = 0.02 \times P_{/CHL}$, where 0.02 = CHL:C; $D = Q_s/V_s$, where V_s = volume of the surface layer; Q_s = volume transport of the surface layer calculated from $Q_f(S_b/[S_b - S_s])$; Q_f = river flow, S_b = mean salinity of the bottom layer, and S_s = mean salinity of the surface layer.

to 0.04 d^{-1}. Comparison of μ:D indicates that although growth rates appear to be higher in the Hudson estuary than in Chesapeake Bay (see Figure 17-4), the potential for increases in CHL because of local growth is lower in the Hudson, where μ:D is clearly greater than 1 only during summer (see Figure 17-4). As discussed by Malone et al. (1980), CHL variability in the Hudson estuary is driven by advective inputs from adjacent coastal waters during winter–spring and by local growth during summer. In contrast, most phytoplankton biomass in Chesapeake Bay is produced within the estuary throughout the year. During summer, when the potential for rapid increases in biomass is greatest and sedimentation rates appear to be low (Malone et al. 1980; Kemp and Boynton 1991), biomass appears to be kept in check by grazing in both estuaries (Malone and Chervin 1979; Malone et al. 1988).

Biomass Responses to Seasonally Variable Nutrient Input

The seasonal pulsing of nutrient input and long nutrient residence times in the Chesapeake lead to higher and less stable levels of CHL and the frequent occurrence of summer anoxia in bottom waters of the Bay (Malone 1991), a phenomenon that has not occurred in the Hudson estuary in recent decades (cf., Garside et al. 1976; Malone 1982). The accumulation of phytoplankton biomass during spring in the Chesapeake appears to be a response to nutrient input associated with freshwater flow and to two-layered, estuarine circulation, which concentrates and retains biomass in the mesohaline reach of the Bay (Malone et al. 1988).

Water column CHL begins to increase in late winter as flow increases and peaks about 1 month after the spring freshet. Assuming a 1-month lag to account for the delayed response of the mesohaline Bay relative to the gauged flow of the Susquehanna River (Malone et al. 1988; Boicourt 1991), monthly mean CHL content of the mesohaline reach of the Bay during spring is significantly correlated ($p < 0.001$) with monthly mean Q_f (Figure 17-5). This suggests that the magnitude of the spring bloom is regulated by the input of allochthonous nutrients associated with freshwater input. It is noteworthy in this context that, although the concentration of DIP in freshwater entering the Bay has decreased steadily since 1982 (Figure 17-6), mean CHL content of the water column in April–May actually increased from 220 mg m^{-2} in 1985 to 340 mg m^{-2} in 1988. Given these opposing trends and the pattern of DSi and DIN depletion along the salinity gradient, it appears that the accumulation of phytoplankton biomass is related to the inputs of both DIN and DSi, with DIN regulating the accumulation of total phytoplankton biomass and DSi regulating the accumulation of diatom biomass (Conley and Malone 1992).

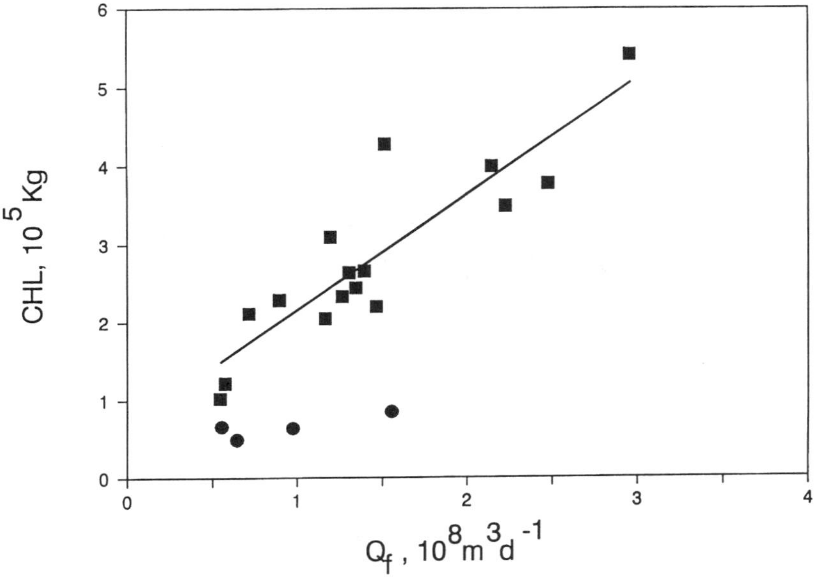

Figure 17-5. Relationship between the chlorophyll a content of the mesohaline reach (5–20 ppt) of the main stem of the Chesapeake Bay (CHL) during the spring bloom and volume transport of the Susquehanna River (Q_f) lagged by 1 month ($r^2 = 0.78$, $n = 16$; solid circles = data from 1989 not used in the correlation as discussed in the text).

The seasonal accumulation of CHL depends on both the magnitude and timing of the spring freshet relative to the spring–summer transition. Spring CHL and Q_f were low in 1989 (see Figure 17-5), and the spring bloom failed to develop (Figure 17-7). In contrast to 1985, 1986, 1987, 1988, and 1990, when Q_f generally peaked in April, Q_f in 1989 did not peak until May (see Figure 17-7). Consequently, maximum riverborne nutrient input to the mesohaline Bay was not achieved until after the transition from spring to summer plankton communities. Although no CHL response to the freshet was observed, P_{CHL} was higher than the monthly mean in June (see Figure 17-7). The absence of a spring bloom and high P_{CHL} subsequent to the spring–summer transition may be indicative of changes in the fate of phytoplankton biomass dependent on the timing of the spring freshet. During most years, nutrient inputs peak before the transition from spring to summer plankton communities, phytoplankton biomass accumulates, and benthic sedimentation is high. If new nutrient inputs peak after the transition, it appears that the close coupling between production and grazing prevents biomass from accumulating.

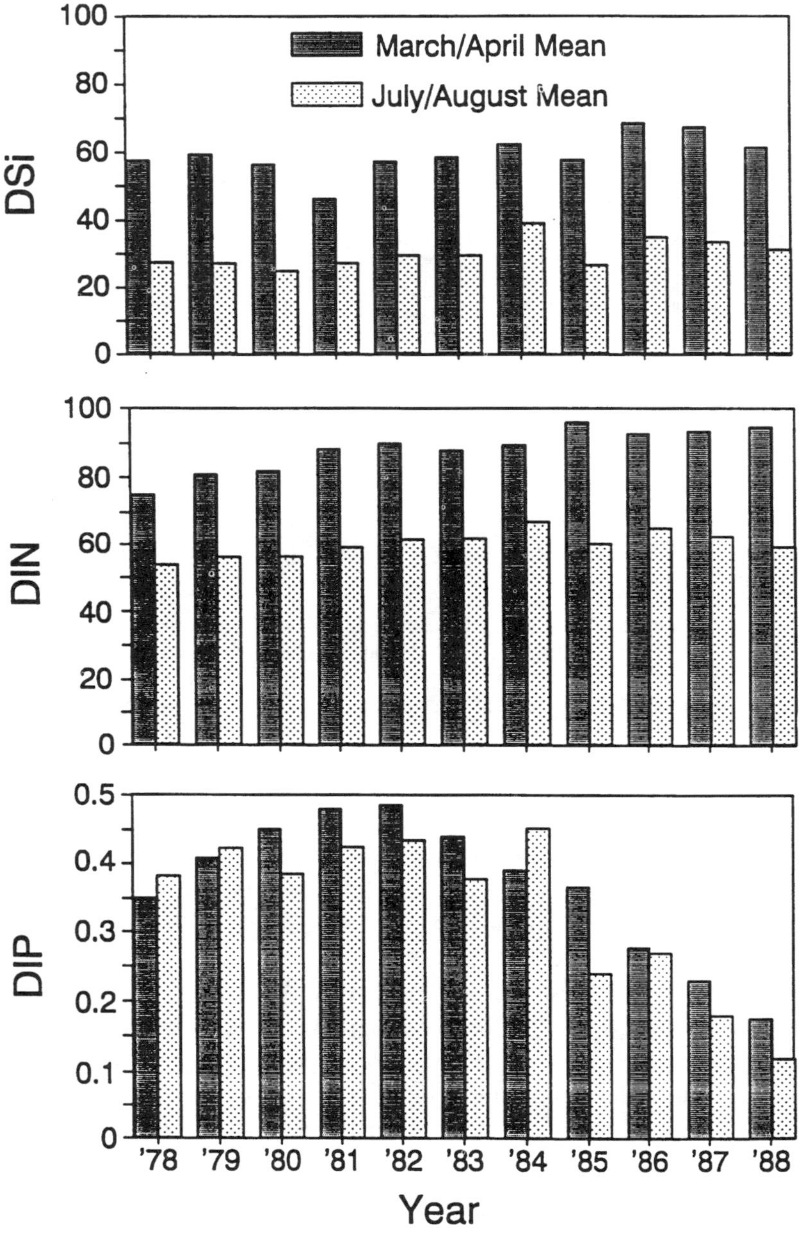

Figure 17-6. Mean spring and summer nutrient concentrations of DSi, DIN, and DIP in freshwater entering the Chesapeake Bay from the Susquehanna River. (Data kindly provided by the Maryland Department of the Environment.)

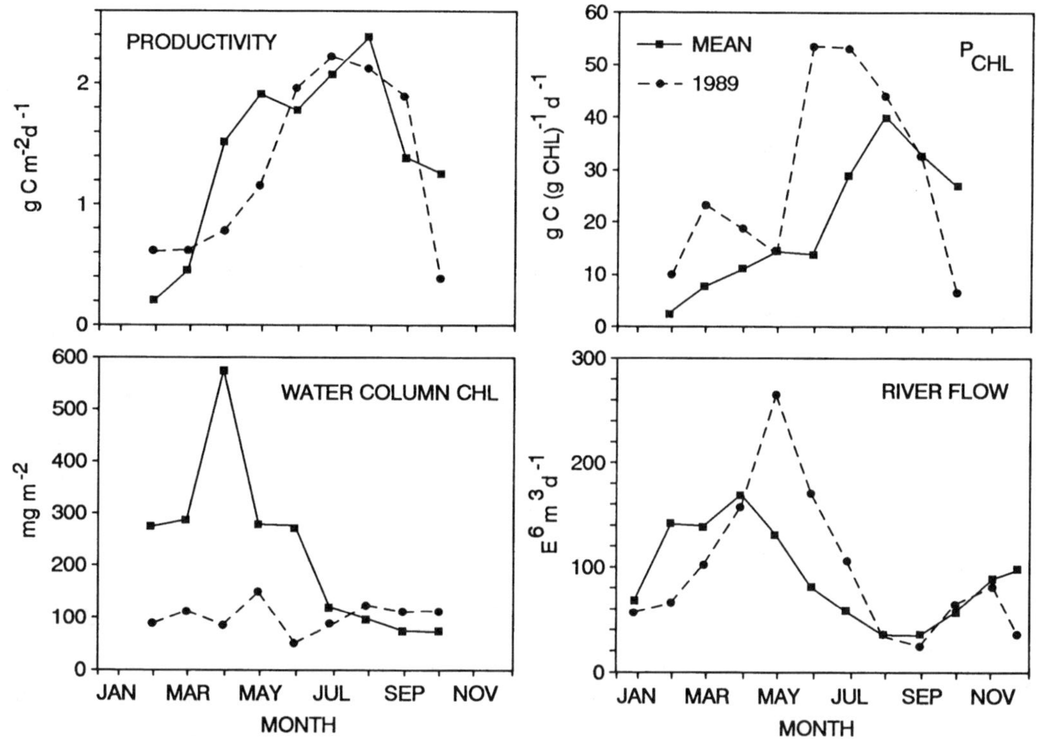

Figure 17-7. Comparison of 1989 annual cycles of productivity, chlorophyll a, and chlorophyll specific productivity with the mean annual cycle for 1985, 1986, 1987, 1988, and 1990.

Productivity Responses to Light and Nutrient Recycling

Annual cycles in phytoplankton productivity are driven by incident photosynthetically active radiation (I_o^{PAR}, E m^{-2} d^{-1}). Cole and Cloern (1984) have shown that seasonal variations of PP in nutrient-rich, temperate estuaries are significantly correlated ($r^2 = 0.82$, n = 211, $p < 0.001$) with the product of I_o^{PAR} and euphotic zone CHL (mg m^{-2}) as described by the least square regression (model II).

$$PP = 146 + 0.73\,(I_o^{PAR})(CHL)$$

Similar analyses for the mesohaline Bay ($r^2 = 0.34$, n = 87; pooled data from 1984–1988) and the Hudson estuary ($r^2 = 0.89$, n = 15, pooled data from 1973–1976) give the following regressions (± SE):

$$\text{Chesapeake } PP = 630 \pm 805 + (0.31 \pm 0.05)\,(I_o^{PAR})(CHL)$$

$$\text{Hudson } PP = 65 \pm 123 + (0.73 \pm 0.07)\,(I_o^{PAR})(CHL)$$

These results indicate a strong influence of incident radiation and CHL on seasonal scale variations in phytoplankton productivity. The slope and r^2 of the relationship for the nutrient-rich Hudson estuary are high and similar to results reported by Cole and Cloern (1984). The low r^2 and large scatter of the Chesapeake Bay relationship (Figure 17-8) indicate that much of the variability in productivity is related to other factors, especially during the summer when productivity is highest. Nutrient limitation is suggested by the low slope of the regression line.

Although DIN inputs to the two estuaries are similar, annual phytoplankton production in the Bay is higher by a factor of nearly 3 (see Table 17-2). Assuming a C:N ratio of 5.7 (wt/wt), these productions are equivalent to 8% and 350% of annual DIN inputs to the Hudson and Chesapeake estuaries, respectively. High phytoplankton production relative to allochthonous N input in estuaries has been attributed to nutrient recycling (Nixon and Pilson 1983). In this regard, phytoplankton production per unit nitrogen loading in the mesohaline Chesapeake Bay is exceptionally high (Boynton et al. 1982).

High turnover rates of phytoplankton biomass during summer in Chesapeake Bay require high nutrient fluxes to the euphotic zone at a time when riverborne nutrient input is low. Much of this flux probably occurs via heterotrophic plankton (bacteria to macrozooplankton) within the euphotic zone. However, nutrient inputs from external sources must also occur to

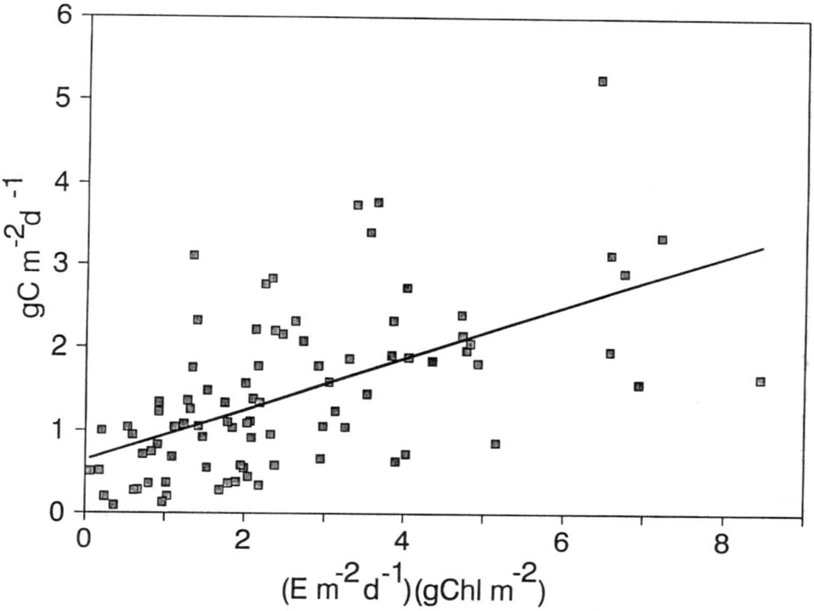

Figure 17-8. Relationship between euphotic zone phytoplankton productivity and the product of I_0^{PAR} (E m^{-2} d^{-1}) and CHL (gCHL m^{-2}).

maintain the integrity of the plankton community. Seasonal variations in the vertical distribution of nutrients (Figure 17-9) indicate that benthic regeneration may be a major source of nutrients required to support high summer productivity, a conclusion consistent with the annual cycles of benthic N and P fluxes, which also peak during the summer (Kemp et al. 1990; Kemp and Boynton 1991). These observations suggest a seasonal coupling between spring riverborne nitrate input and summer benthic ammonium fluxes. The accumulation, sedimentation, and decomposition of phytoplankton biomass during the spring bloom provides the linkage (Kemp and Boynton 1984; Malone et al. 1988). Such a seasonal coupling between new nutrient input during spring and the development of the summer productivity maximum does not occur in the Hudson estuary.

If summer phytoplankton productivity is limited by vertical fluxes of nutrients released from the sediments, vertical mixing might be expected to be a controlling factor. This hypothesis is supported by the relationship between summer productivity (July–August mean) and density stratification (Figure 17-10). Under nutrient-saturated conditions, increases in stratification should have a positive effect on productivity. An inverse relationship suggests that the

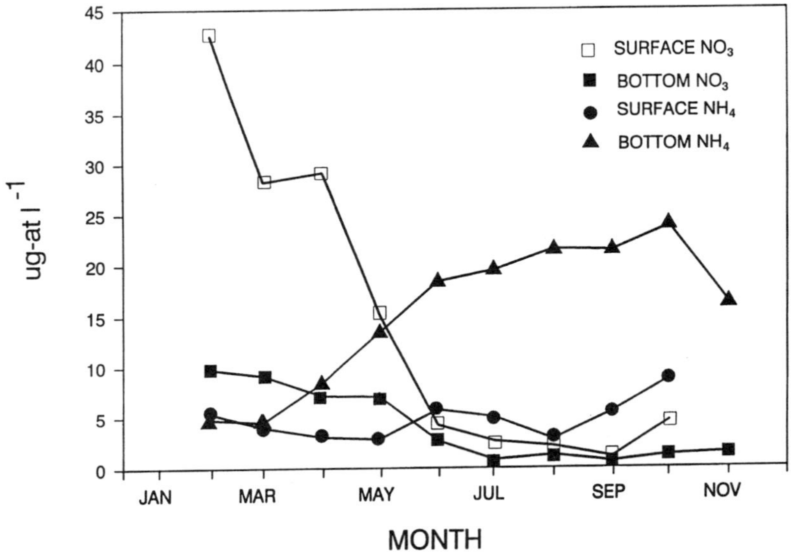

Figure 17-9. Mean (1985–1990) annual cycles of nitrate and ammonium in surface and bottom water of the mesohaline reach of Chesapeake Bay.

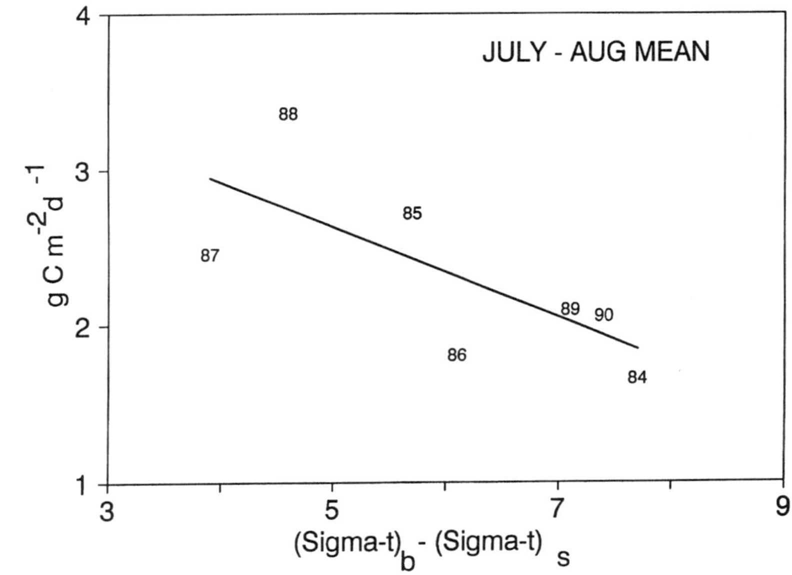

Figure 17-10. Relationship between mean euphotic zone productivity (g C m^{-2} d^{-1}) and the vertical density gradient across the pycnocline in the mesohaline reach of Chesapeake Bay during the summers (July–August) of 1984 through 1990.

occurrence and magnitude of the summer productivity maximum depends on the efficiency with which nutrients delivered to the Bay during spring are recycled between the benthos and surface layer of the estuarine water column. The scatter and low slope in the relationship between PP and I_0CHL for the Bay relative to the Hudson estuary may reflect this effect. The lack of a declining trend in summer productivity also suggests that phytoplankton production has not been responsive to the declining trend in DIP input.

Phytoplankton Production and Biomass in Coastal Plumes

Total nitrogen export from the two estuaries appears to be similar. Recent mass balance calculations suggest that total nitrogen export from the Chesapeake Bay is about 50% of that exported from the Hudson River (Malone 1984; Boynton et al. 1995). Most N export from the Hudson is DIN, whereas export from the Bay is mainly in organic form, with as much as 60% being dissolved (see Table 17-1; McCarthy et al. 1977). These differences are reflected in annual phytoplankton production and mean CHL (see Table 17-2), as well as in the annual cycles of productivity and biomass (Figure 17-11).

Phytoplankton production and CHL in the Hudson plume exhibit similar annual cycles, with spring maxima and winter minima. Variations in PP are significantly correlated (r^2 = 0.90, n = 15) with the composite variable (I_0^{PAR})(CHL) by the regression (± SE)

$$PP = 71 \pm 138 + (0.97 \pm 0.06)\,(I_0^{PAR})(CHL).$$

In contrast, the annual cycles of PP and CHL in the Chesapeake plume appear to be similar to the Bay, with CHL peaking during winter–spring and PP peaking during the summer. As in their respective estuaries, annual cycles of P_{CHL} exhibit winter minima and summer maxima (Figure 17-10). However, the relationship between PP and (I_0^{PAR})(CHL) in the Chesapeake plume varies seasonally (winter–spring: r^2 = 0.67, n = 8; summer: r^2 = 0.74, n = 11):

Winter–spring $\quad PP = 170 \pm 120 + (0.17 \pm 0.05)\,(I_0^{PAR})(CHL)$

Summer $\quad PP = 430 \pm 430 + (1.24 \pm 0.24)\,(I_0^{PAR})(CHL)$

The summer relationship is similar to that for the Hudson plume, where inorganic nutrient input is high throughout the year.

These differences may reflect contrasting patterns of new nutrient input and the relative importance of nutrient regeneration. In the Hudson plume,

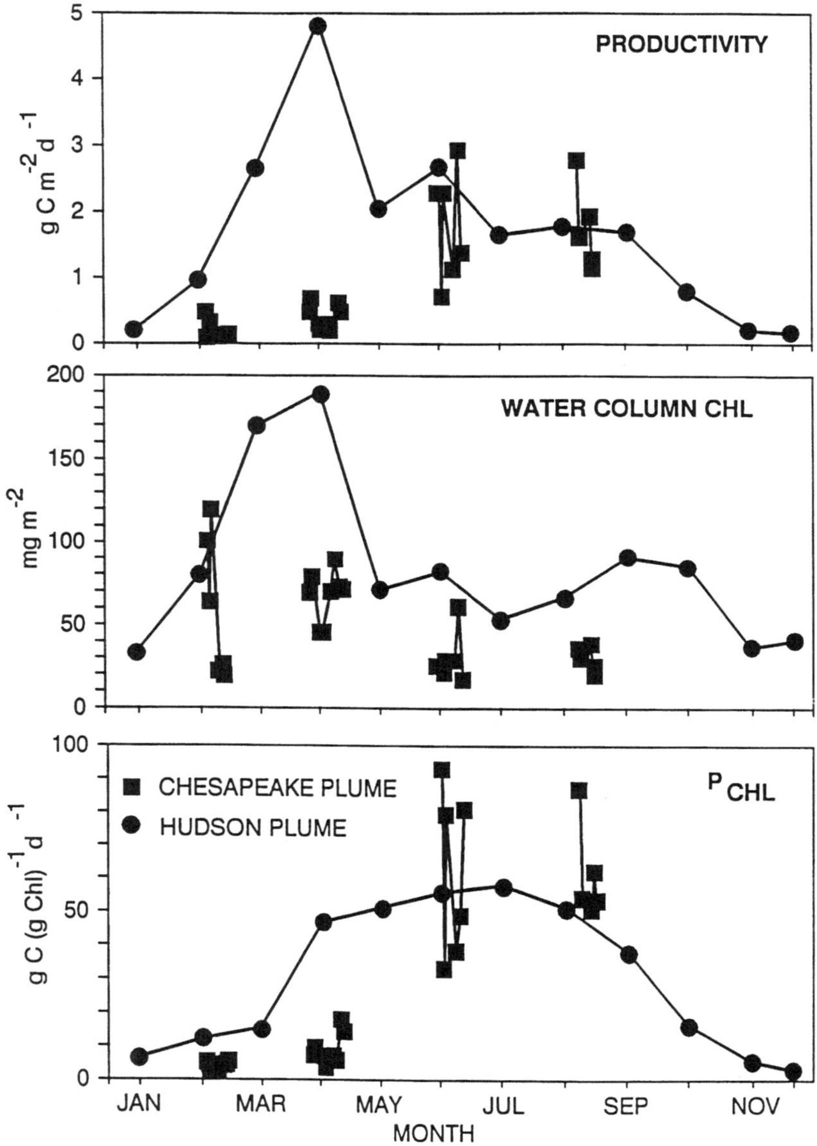

Figure 17-11. Annual cycles of phytoplankton productivity (g C m^{-2} d^{-1}), chlorophyll a (mg m^{-2}), and chlorophyll specific productivity (g C [g Chl]$^{-1}$ d^{-1}) in the coastal plumes of the Hudson River (mean 1973–1976) and Chesapeake Bay (daily time series in February, June, and August 1985, and April 1986).

estuarine export supports about 50% of new production during winter–spring and 100% of new production during summer, as regenerated production increases from about 10% of total production during winter–spring to 80% during summer (Malone 1984). In the Chesapeake plume, most nutrient export from the Bay is in organic form, and phytoplankton production is more dependent on nutrient regeneration. Thus, regenerated production increases with increasing temperature from about 70% of total production during winter to about 90% during summer (Glibert et al. 1991). This increase coincides with an increase in bacterioplankton biomass relative to phytoplankton biomass from <10% during winter to approximately 100% during summer (Malone and Ducklow 1990). Malone and Ducklow (1990) and Glibert et al. (1991) interpret these patterns to indicate a grazing-enhanced flow of phytoplankton carbon to bacteria and associated increases in nitrogen regeneration.

Conclusions

Inputs of DIN and DSi are assimilated within both the Chesapeake and Hudson estuarine systems. Nutrient residence times are short in the Hudson estuary, and most uptake occurs in the coastal plume. In contrast, most nutrient uptake in the Chesapeake Bay occurs in the mesohaline reach of the estuary. Consequently, N available for export to waters of the continental shelf is largely in organic form, with approximately 50% in the dissolved organic N pool.

The seasonal coupling between the spring peak in allochthonous nitrate input and summer maxima in benthic ammonium regeneration and phytoplankton productivity in the Chesapeake Bay underscores the functional limitations of the distinction between new and regenerated production based on nutrient oxidation states (Malone et al. 1988) as the coupling between benthic and pelagic processes becomes closer with decreasing depth (Kemp et al. 1990). Perhaps the distinction between new and regenerated nutrients should be based on time lags between phytoplankton production and nutrient regeneration relative to the generation times of the primary producers. Thus, ammonium in Chesapeake Bay is regenerated N if it is recycled within the euphotic zone but new N if it is regenerated below the euphotic zone with a lag time longer than the generation time of phytoplankton.

Responses of phytoplankton productivity and biomass to new nutrient input in temperate estuarine ecosystems differ depending on the timing of the input relative to annual cycles of temperature and incident radiation. As indicated by the annual cycles of CHL in the Chesapeake Bay and the coastal plumes, increases in biomass are driven by new production. The lack of a CHL response is the result of high dilution rates in the Hudson and grazing in the Chesapeake (1989), both of which effectively prevent accumulations of CHL. Seasonal variations in P_{CHL} appear to be more closely related to regenerated production in both estuaries and their plumes.

With the exception of Chesapeake Bay and its plume during spring, the slopes of regressions of PP on (I_o^{PAR})(CHL) are similar to each other and to the results of Cole and Cloern (1984). This suggests that phytoplankton production in the Hudson system and in the Chesapeake plume during summer approaches nutrient saturation on a seasonal scale. In the Bay and in the Bay's plume during winter–spring, low slopes may be indicative of nutrient-limited production, a consequence of low dilution relative to growth in the Bay and of low DIN export and regenerative fluxes in the Chesapeake plume during winter. The phasing of annual cycles of CHL and PP in the Bay–plume system suggests that biomass is nutrient-limited during winter–spring and that P_{CHL} is nutrient-limited during summer. As argued by Caperon et al. (1971), such responses to nutrient inputs provide a means of quantifying eutrophication in terms of ecosystem dynamics. In this model, oligotrophic ecosystems are characterized by stable phytoplankton biomass because of the close coupling between phytoplankton production and losses. Increases in nutrient input result in increases in P_{CHL}. In eutrophic ecosystems, phytoplankton biomass is unstable because of the uncoupling of phytoplankton production and losses; increases in nutrient input are responded to by an increase in CHL. Thus, the eutrophication of Chesapeake Bay is largely a consequence of nonpoint N inputs that occur during winter–spring before the transition from spring to summer phytoplankton communities.

ACKNOWLEDGMENTS

This work was supported by grants from the National Science Foundation and the National Oceanic and Atmospheric Administration, and by the Sea Grant Program of the University of Maryland System. Data for Figure 17-6, collected as part of the Chesapeake Bay Monitoring Program, were kindly provided by the Maryland Department of the Environment. William Boicourt, Walter Boynton, and Michael Kemp provided valuable discussion.

References

Boicourt, W. C. 1992. The Influence of Circulation Processes on Dissolved Oxygen in Chesapeake Bay. In D. Smith, M. Leffler, and G. Mackiernan (eds.), Oxygen Dynamics in Chesapeake Bay: A Synthesis of Research. College Park, MD: University of Maryland Sea Grant Publ., pp. 7–60.

Boynton, W. R., Garber, J. H., Summers, R., and Kemp, W. M. 1995. Inputs, Transformations, and Transport of Nitrogen and Phosphorus in Chesapeake Bay and Selected Tributaries. Estuaries 18:285–314.

Boynton, W. R., Kemp, W. M., and Keefe, C. W. 1982. A Comparative Analysis of Nutrients and Other Factors Influencing Estuarine Phytoplankton Production. In V. S. Kennedy (ed.), Estuarine Comparisons. San Diego, CA: Academic Press, pp. 69–90.

Brzenzinski, M. A. 1985. The Si:C:N ratio of marine diatoms: Interspecific variability and the effect of some environmental variables. J. Phycol. 21:347–357.

Caperon, J., Cattell, S. A., and Krasnick, G. 1971. Phytoplankton kinetics in a subtropical estuary: Eutrophication. Limnol. Oceanogr. 16:599–607.

Cole, B. E., and Cloern, J. E. 1984. Significance of biomass and light availability to phytoplankton productivity in San Francisco Bay. Mar. Ecol. Prog. Ser. 17:15–24.

Conley, D. J., and Malone, T. C. 1992. The annual cycle of dissolved silicate in Chesapeake Bay: Implications for the production and fate of phytoplankton biomass. Mar. Ecol. Prog. Ser. 81:121–128.

Fisher, T. R., Harding, L. W., Stanley, D. W., and Ward, L. G. 1988. Phytoplankton, nutrients, and turbidity in the Chesapeake, Delaware and Hudson Estuaries. Estuar. Coast. Shelf Sci. 27:61–93.

Garside, G., Malone, T. C., Roels, O. A., and Sharfstein, B. A. 1976. An evaluation of sewage-derived nutrients and their influence on the Hudson Estuary and New York Bight. Estuar. Coast. Mar. Sci. 4:281–289.

Glibert, P. M., Garside, C., Fuhrman, J. A., and Roman, M. R. 1991. Time-dependent coupling of inorganic and organic nitrogen uptake and regeneration in the plume of the Chesapeake Bay estuary and its regulation by large heterotrophs. Limnol. Oceanogr. 36:895–909.

Kemp, W. M., and Boynton, W. R. 1984. Spatial and temporal coupling of nutrient inputs to estuarine production: The role of particulate transport and decomposition. Bull. Mar. Sci. 35:242–247.

Kemp, W. M., and Boynton, W. R. 1992. Benthic-pelagic Interactions: Nutrient and Oxygen Dynamics. In D. Smith, M. Leffler, and G. Mackiernan (eds.), Oxygen Dynamics in Chesapeake Bay: A Synthesis of Research. College Park, MD: University of Maryland Sea Grant Publ., pp. 149–222.

Kemp, W. M., Sampou, P., Caffrey, J., et al. 1990. Ammonium recycling versus denitrification in Chesapeake Bay sediments. Limnol. Oceanogr. 35:1545–1563.

Ketchum, B. H. 1967. Phytoplankton Nutrients in Estuaries. In G. H. Lauff (ed.), Estuaries. AAAS Publ. No. 83. Washington, DC: American Association for the Advancement of Science, pp. 329–335.

Ketchum, B. H. 1969. Eutrophication in Estuaries. In Eutrophication: Causes, Consequences, and Correctives. Proceedings of a Symposium of the National Academy of Sciences. Washington, DC: National Academy of Sciences, pp. 187–209.

Malone, T. C. 1977. Environmental regulation of phytoplankton productivity in the lower Hudson Estuary. Estuar. Coast. Mar. Sci. 5:157–171.

Malone, T. C. 1982. Factors Influencing the Fate of Sewage-derived Nutrients in the Lower Hudson Estuary and New York Bight. In G. F. Mayer (ed.), Ecological Stress and the New York Bight: Science and Management. Columbia, SC: Estuarine Research Federation, pp. 389–400.

Malone, T. C. 1984. Anthropogenic Nitrogen Loading and Assimilation Capacity of the Hudson River Estuarine System, USA. In V. S. Kennedy (ed.), The Estuary as a Filter. San Diego, CA: Academic Press, pp. 291–311.

Malone, T. C. 1991. River Flow, Phytoplankton Production and Oxygen Depletion in Chesapeake Bay. In R. V. Tyson and T. H. Pearson (eds.), Modern and Ancient Continental Shelf Anoxia. Geol. Soc. Spec. Publ. No. 58. Washington, DC: Geological Society, pp. 83–93.

Malone, T. C., and Chervin, M. B. 1979. The production and fate of phytoplankton size fractions in the plume of the Hudson River, New York Bight. Limnol. Oceanogr. 27:683–696.

Malone, T. C., Chervin, M. B., Stepien, J. C., et al. 1985. Synoptic Investigation of Nutrient Cycling in the Coastal Plume of the Hudson and Raritan Rivers: Plankton Dynamics. NOAA Spec. Rept. Rockville, MD: National Oceanic and Atmospheric Administration.

Malone, T. C., Crocker, L. H., Pike, S. E., and Wendler, B. W. 1988. Influences of river flow on the dynamics of phytoplankton production in a partially stratified estuary. Mar. Ecol. Prog. Ser. 48:235–249.

Malone, T. C., and Ducklow, H. W. 1990. Microbial biomass in the coastal plume of Chesapeake Bay: Phytoplankton-bacterioplankton relationships. Limnol. Oceanogr. 35:296–312.

Malone, T. C., Neale, P. J., and Boardman, D. 1980. Influences of Estuarine Circulation on the Distribution and Biomass of Phytoplankton Size Fractions. In V. S. Kennedy (ed.), Estuarine Perspectives. San Diego, CA: Academic Press, pp. 249–262.

McCarthy, J. J., Taylor, W. R., and Loftus, M. E. 1977. Nitrogenous nutrition of the plankton in the Chesapeake Bay. I. Nutrient availability and phytoplankton preferences. Limnol. Oceanogr. 22:996–1011.

Nixon, S. W. 1981. Remineralization and Nutrient Cycling in Coastal Marine Ecosystems. In B. J. Neilson and L. E. Cronin (eds.), Estuaries and Nutrients. Totowa, NJ: Humana Press, pp. 111–138.

Nixon, S. W., and Pilson, M. G. 1983. Nitrogen in Estuarine and Coastal Marine Ecosystems. In E. J. Carpenter and D. G. Capone (eds.), Nitrogen in the Marine Environment. San Diego, CA: Academic Press, pp. 565–648.

Officer, C. B., Biggs, R. B., Taft, J. L., et al. 1984. Chesapeake Bay anoxia: Origin, development, and significance. Science 223:22–27.

Officer, C. B., and Ryther, J. H. 1980. The possible importance of silicon in marine eutrophication. Mar. Ecol. Prog. Ser. 9:91–94.

Redfield, A. C. 1958. The biological control of chemical factors in the environment. Am. Sci. 46:205–221.

Schelske, C. L., and Stoermer, E. F. 1972. Phosphorus, Silica and Eutrophication of Lake Michigan. In G. Likens (ed.), Nutrients and Eutrophication: The Limiting Nutrient Controversy. Am. Soc. Limnol. Oceanogr. Spec. Symp. Vol. 1. American Society of Limnology and Oceanography, pp. 157–171.

Schubel, J. R., and Pritchard, D. W. 1986. Responses of upper Chesapeake Bay to variations in discharge of the Susquehanna River. Estuaries 9:236–249.

Simpson, H. J., Williams, S. C., Olsen, C. R., and Hammond, D. E. 1977. Nutrient and Particulate Matter Budgets in Urban Estuaries. In Estuaries, Geophysics, and the Environment. Washington, DC: National Academy of Sciences, pp. 94–103.

Smith, R. A., Alexander, R. B., and Wolman, M. G. 1987. Water-quality trends in the Nation's rivers. Science 235:1607–1615.

Smullen, J. T., Taft, J. L., and Macknis, J. 1982. Nutrient and Sediment Loads to the Tidal Chesapeake Bay System. In U.S. Environmental Protection Agency, Chesapeake Bay Technical Studies: A Synthesis. Washington, DC: U.S. Environmental Protection Agency, pp. 147–262.

Verity, P. G. 1987. Factors Driving Changes in the Pelagic Trophic Structure of Estuaries, with Implications for the Chesapeake Bay. In M. P. Lynch and E. C. Krome (eds.), Perspectives on the Chesapeake Bay. CRC Publ. No. 127. Chesapeake Bay Research Consortium, pp. 35–56.

Walsh, J. J., Rowe, G. T., Iverson, R. E., and McRoy, C. P. 1981. Biological export of shelf carbon is a sink of the global CO_2 cycle. Nature 291:196–201.

18

Preliminary Estimates of the Pollutant Loads and Fluxes into the Northeast Shelf Ecosystem

Norbert A. Jaworski and Robert Howarth

The pollutant loadings entering the Northeast Shelf large marine ecosystem (LME) are being estimated to quantify the impact of natural and anthropogenic stresses on the health of the LME and its estuaries. At present, no comprehensive estimates of the pollutant loadings and fluxes for the coastal watershed exist for the Northeast Shelf LME. We are currently compiling a comprehensive estimate of pollutant loadings from point and nonpoint sources in the coastal watersheds of the Northeast Shelf LME. The objective of this chapter is to provide a status report on this effort and to present a preliminary estimate of pollutant loadings and fluxes.

The Large Marine Ecosystem Concept

LMEs as defined by Sherman (1991, 1992a) are extensive areas of ocean, approximately 200,000 km^2 or greater, characterized by distinct bathymetry, oceanography, productivity, and trophically dependent populations. Increasing attention has been focused on the synthesis of environmental data affecting the productivity, diversity, and yields of these LMEs and identification of factors causing major shifts in species composition and abundance.

Natural and anthropogenic pollutants originating from watersheds and coastal urban areas can have a great impact on the water and benthic quality of these LMEs. Of the 49 LMEs in the global coastal ecosystem, Sherman (1992b) has suggested that 50% have felt the impact of pollutant stresses.

Study Area

The Northeast Shelf LME encompasses about 258,000 km^2, extending from the Gulf of Maine in the north to Cape Hatteras in the south. It is one of the most productive in the world, with a fishery yield worth about $1 billion annually (Sherman et al. 1988). The LME has been further divided into four geographical areas: (1) the Gulf of Maine, (2) Georges Bank, (3) Southern New England, and (4) the Middle Atlantic Bight (see Chapter 7, Figure 7-1).

The coastal zone draining into the Northeast Shelf LME has an area of 478,787 km^2 and supports a population of about 55 million, at an average density of about 115 people per square kilometer. Our preliminary estimates of wastewater flow from municipal and industrial treatment facilities is about 27.2 billion liters per day (7.2 billion gallons per day).

Geographic data for the LME, clustered into the four subareas (Gulf of Maine, Georges Bank, Southern New England, and Middle Atlantic Bight) are presented in Table 18-1. For all practical purposes, the Georges Bank area has no coastal watershed and is therefore omitted from further analysis here. The Middle Atlantic Bight watershed is the largest in drainage area, estuarine surface area, population, and wastewater flow. The Southern New England coastal watershed, including Narragansett Bay and Buzzards Bay, has the highest population density, about 722 people/km^2.

Data Sources

The primary data sources for the point source loadings and, to a lesser extent, non-point sources, are the numerous characterization studies of the various National Estuary Programs (NEPs) of the U.S. Environmental Protection Agency (EPA NEP 1992). These have been done for Casco Bay, Massachusetts Bay, Buzzards Bay, Narragansett Bay, Long Island Sound, New York–New Jersey Harbor, Delaware Bay, and the Chesapeake Bay (by a forerunner of the NEP). Most of the non-point loadings for metals were derived from a study by Windom (see Chapter 16). Most of the nutrient non-point loadings came from the various NEP reports and from an unpublished study by Jaworski et al. (1991).

Our current effort (Howarth and Jaworski, in preparation) will describe in detail extrapolation methods used to estimate pollutant loadings and flux-

Table 18-1. Geographic data on the Northeast Shelf ecosystem

Subarea	Size (km^2)	Watershed area (km^2)	Estuary area (km^2)	Estimated population and density (people per km^2)
Gulf of Maine (including Massachusetts Bay)	98,000	175,520	6,134	5,000,000 (28.5)
Georges Bank	418,000	0	0	0
Southern New England (Buzzards Bay, Narragansett Bay, and Long Island Sound)	59,900	48,016	4,387	10,757,000 (722.9)
Middle Atlantic Bight (Hudson, Delaware, and Chesapeake)	58,300	255,251	19,747	39,155,000 (153.3)
Total	258,000	478,787	30,268	54,912,000 (114.8)

Source: Reprinted from National Oceanic and Atmospheric Administration. A Special NOAA 20th Anniversary Report: Estuaries of the United States, Vital Statistics of a National Resource Base. Rockville, MD: U.S. Department of Commerce, 1990.

es. All data sources are being reviewed and more refined estimates of loadings will be made.

Preliminary Estimates

Preliminary estimates of total organic carbon (TOC), total nitrogen (TN), and total phosphorus (TP) are presented in Table 18-2. Of the 1,111.9 × 10^6 kg/yr of TOC, about 60% was from sources in the Middle Atlantic Bight watershed. For TN, about 68% of the estimated 477.6 × 10^6 kg/yr for the entire LME came from the Middle Atlantic Bight watershed. Likewise, for TP, 60% of the 70.9 × 10^6 kg/yr originates from the Middle Atlantic Bight watershed. Although the Middle Atlantic has the highest loadings, the flux rates for TOC and TP per watershed area were the highest for Southern New England watersheds because of the high population density (see Table 18-1). The high TN flux rate for the Middle Atlantic Bight watersheds was a result of large contributions from animal waste and high atmospheric deposition (Jaworski et al. 1992).

Total metal loadings to the LME were estimated (Table 18-3). Of the five metals considered, zinc was the highest, followed by copper, nickel, lead, and cadmium. The loadings and fluxes were for the total of the dissolved and particulate bound components. The high flux rates for waters of Southern New England watersheds were indicative of the population and manufacturing base in that area.

Table 18-2. Preliminary estimates of wastewater flows (liters × 10^6/day), nutrient loading (kg × 10^6/yr), and flux rates (kg/km^2/yr) from the coastal watersheds

Coastal area	Waste-water flow, l × 10^6/day	TOC, kg × 10^6/yr (kg/km^2/yr)	TN, kg × 10^6/yr (kg/km^2/yr)	TP, kg × 10^6/yr (kg/km^2/yr)
Gulf of Maine	2,592	293.0 (1,674)	96.1 (548)	11.5 (66)
Southern New England	5,136	145.5 (3,020)	53.8 (1,104)	16.8 (350)
Middle Atlantic Bight	19,435	673.4 (2,639)	327.7 (1,282)	42.6 (167)
Total for LME	27,163	1,111.9 (2,324)	477.6 (997)	70.9 (148)

TOC = total organic carbon; TN = total nitrogen; TP = total phosphorus.

Table 18-3. Preliminary estimates of metal loadings (kg × 10^3/yr) and flux rates (kg/km^2/yr) from the coastal watershed

	Metal loadings, kg × 10^3/yr (flux rate, kg/km^2/yr)				
Coastal area	Copper	Cadmium	Lead	Zinc	Nickel
Gulf of Maine	227 (1.3)	14 (0.08)	337 (1.9)	656 (3.7)	124 (0.7)
Southern New England	282 (5.9)	9 (0.19)	119 (2.5)	1,079 (22.2)	505 (10.5)
Middle Atlantic Bight	912 (3.6)	24 (0.09)	364 (1.4)	1,819 (7.1)	743 (2.9)
Total for LME	1,421 (2.9)	47 (0.09)	820 (1.7)	3,554 (7.4)	1,372 (2.8)

Discussion

It should be noted that these are preliminary estimates and they will be refined as this effort continues. The present estimates are annual averages, and vary mainly as a function of climate. For example, the measured nutrient fluxes for the Upper Potomac River Basin watershed (Jaworski et al. 1992) have an annual variability up to a factor of 3, mainly because of rainfall–runoff conditions.

The average flux rates for TOC, TN, and TP (in kg/km^2/yr) for the total LME and Chesapeake Bay watersheds are compared in Table 18-4. The LME watershed flux rates are very similar to those estimated for the Chesapeake Bay watershed (Jaworski et al. 1991), except for TP. The lower flux rate for phosphorus in the Chesapeake Bay watershed reflects the removal of phosphorus by wastewater treatment plants discharging into the Potomac estuary and the use of low phosphorus detergents.

The estimated carbon, nitrogen, and phosphorus water surface area loadings to the Northeast Shelf LME (Table 18-5) reflect the impacts of large pop-

Table 18-4. Comparison of watershed flux rates for total organic carbon (TOC), total nitrogen (TN), and total phosphorus (TP) (kg/km²/yr)

Pollutant	Entire LME watershed	Chesapeake Bay watershed
TOC	2,324	2,240
TN	997	920
TP	148	70

Table 18-5. Preliminary estimates of organic carbon, nitrogen, and phosphorus loadings to the Northeast Shelf LME from the coastal watersheds

Coastal area	Total organic carbon (g/m²/yr)	Total nitrogen (g/m²/yr)	Total phosphorus (g/m²/yr)
Gulf of Maine	2.989	0.980	0.117
Southern New England	2.429	0.898	0.280
Middle Atlantic Bight	11.550	5.620	0.730
Area weighted average	4.309	1.851	0.275

ulation centers in the Northeast and agricultural practices. The loading rates per unit area of coastal water for nitrogen and phosphorus for the Middle Atlantic Bight, 5.62 g/m²/yr and 0.73 g/m²/yr, respectively, were similar to the loadings for the upper Potomac River Basin watershed from all sources (including point, animal waste, fertilizer, atmospheric depositions, etc.) of 5.14 g/m²/yr and 0.98 g/m²/yr, as reported by Jaworski et al. (1991). Only about 17% of the nitrogen and 9% of the phosphorus was exported out of the watershed by the Potomac River, with the remaining nutrients retained within the watershed. These are high nutrient loadings that we would expect to support a high rate of primary productivity. Sherman et al. (1988) has estimated that the Middle Atlantic Bight has annual primary production rates of 260–505 g C/m²/yr.

The metal flux rates as reported by Windom (see Chapter 16) indicated considerable variability from watershed to watershed. Within the three coastal watershed geographical regions, the flux data (see Table 18-3) varied, at most, by a factor of 15 for a given metal. Stigliani et al. (1993) estimated the loadings for cadmium, lead, and zinc from the Rhine Basin, which has a population of 50 million and a watershed area of 200,000 km². The Rhine Basin flux rates for these three metals were very similar to those of the Northeast Shelf LME (Table 18-6), a similarity that will be investigated further by the authors.

Table 18-6. A comparison of total metal fluxes (kg/km^2/yr) for the Northeast Shelf LME and the Rhine Basin

Metal	Rhine Basin	Northeast Shelf LME
Cadmium	0.06	0.09
Lead	2.00	1.70
Zinc	12.14	7.40

Future Effort

The loadings and flux rates presented in this chapter are preliminary. As more data are obtained and extrapolation methods improved, the estimates will be refined. The three coastal geographic areas will be subdivided further into 14 coastal watershed areas. Future efforts will attempt to differentiate between point and non-point sources. The contribution from atmospheric deposition also will be estimated.

Future efforts will include comparison of metal loadings (normalized by estuarine area) and metal concentrations in the sediments for the 14 subareas. Similar comparisons will be made for nutrients and state of eutrophication.

References

Howarth, R. W., and Jaworski, N. A. (In preparation.) Pollutant Loadings to the Northeast Shelf LME.

Jaworski, N. A., Groffman, P. M., Keller, A. A., and Prager, J. C. 1992. A watershed nitrogen and phosphorus balance: The upper Potomac River basin. Estuaries 15:83–95.

Jaworski, N. A., Howarth, R. W., and Linker, L. C. 1991. A comparison of total organic carbon fluxes from various land uses in the Potomac, Susquehanna, and Hudson River basin watersheds. Unpublished.

National Oceanic and Atmospheric Administration (NOAA). 1990. A Special NOAA 20th Anniversary Report: Estuaries of the United States, Vital Statistics of a National Resource Base. Rockville, MD: U.S. Dept. Commerce, National Oceanic and Atmospheric Administration.

Sherman, K. 1991. The large marine ecosystem concept: Research and management strategy for living marine resources. Ecol. Appl. 1(4):349–360.

Sherman, K. 1992a. Large Marine Ecosystems. In W. A. Nierenberg (ed.), Encyclopedia of Earth System Science (Vol. 2). San Diego, CA: Academic Press, pp. 653–673.

Sherman, K. 1992b. Monitoring and Assessment of Large Marine Ecosystems: A Global and Regional Perspective. In D. H. McKenzie, D. E. Hyatt, and V. J. McDonald (eds.), Ecological Indicators. London: Elsevier, Ch. 59, pp. 1041–1074.

Sherman, K., Grosslein, M., Mountain, D., et al. 1988. The Continental Shelf Ecosystem Off the Northeast Coast of the United States. In H. Postma and J. J. Zijlstra (eds.), Ecosystems of the World 27: Continental Shelves. Amsterdam: Elsevier, pp. 279–337.

Stigliani, W. M., Jaffé, P. R., and Anderberg, S. 1993. Heavy metal pollution in the Rhine Basin. Environ. Sci. Technol. 27:786–793.

U. S. Environmental Protection Agency National Estuary Program [EPA NEP]. 1992. The National Estuary Program After Four Years; A Report to Congress. EPA 503/9-92/007. Washington, DC: U.S. Environmental Protection Agency, Office of Water.

V

Habitat Degradation

19

Habitat Mitigation

Thomas E. Bigford

Sustainable fisheries require healthy stocks, which are strongly dependent on a delicate balance of fishing pressure, species productivity, and environmental conditions. When that balance is imperiled, resource managers typically respond with a flurry of regulations and recommendations. Whereas stock managers react by adjusting harvest allocations or gear specifications, habitat managers historically have balanced impacts through a five-step "mitigation" process. This chapter discusses how managers mitigate or offset habitat threats in the Northeast Shelf large marine ecosystem (LME).

The Management/Legal/Bureaucratic Arena

The Northeast Shelf LME is a dynamic arena, affected continually by human activities and natural cycles. For management purposes, we must expand our perspectives to include habitats from river headwaters to the continental shelf break. As resources are harvested and essential habitats are affected by the cumulative effects of humankind, the ecosystem changes. Impacts may be measured by quality (e.g., relative productivity, preferred species, overall health) or quantity (e.g., acreage, harvests). Any approach is open to intense scrutiny and political interpretations. Meteorological and physical oceanographic factors complicate the issue further. Long-term sustainability hinges on our collective ability to operate within this complex arena of science and management.

For the habitat portion of the fishery management equation, regulators address the environmental consequences of human activities by mitigating impacts through a process established in regulations of the Council on

Environmental Quality (CEQ) used in implementing the National Environmental Policy Act (43 FR 55978–56007, November 29, 1978; 40 CFR 1500–1508):

> 1508.20 Mitigation
>
> "Mitigation" includes:
>
> (a) *Avoiding* the impact altogether by not taking a certain action or parts of an action;
>
> (b) *Minimizing* impacts by limiting the degree or magnitude of the action and its implementation;
>
> (c) *Rectifying* the impact by repairing, rehabilitating, or restoring the affected environment;
>
> (d) *Reducing or eliminating* the impact over time by preservation and maintenance operations during the life of the action; and
>
> (e) *Compensating* for the impact by replacing or providing substitute resources or environments. [italics added]

Although this sequence has never been codified in other relevant laws, the philosophy is embodied in the operating procedures of most state and federal agencies.

Government decisions associated with a permit, management plan, or policy have environmental consequences and trigger the mitigation sequence. Managers regularly seek to avoid and minimize before resorting to restoration or the other options for balancing lost habitat value. Scientific knowledge supports that sequence.

Avoidance and minimization are significantly more achievable than the other mitigation options. Careful site design and stringent engineering controls can help trustees avoid or minimize most impacts on resources and their habitat. Policies such as "water dependency," "buffer strips," and "no net loss" emphasize the first two steps in mitigation. However, the remaining "unavoidable" impacts must be offset by a suite of skills best described as "evolving art" rather than "tested science." Many of the more controversial mitigation debates affecting Northeast Shelf LME environs are complicated by the lack of science and engineering experience. Managers simply do not yet have the ability to offset all unavoidable habitat impacts associated with human activities.

Habitat Threats to the Northeast Shelf Large Marine Ecosystem

The Northeast Shelf LME extends from Cape Hatteras northward to the Canadian maritimes, lying adjacent to the industrialized northeast corridor.

Habitats have been significantly altered by generations of human activity, ranging from coastal development to offshore resource extraction. The priority threats to shelf habitats shift from watersheds to the continental shelf break, with chemical and physical degradation being the two recurring themes:

Coastal fringe
Chemical contamination that degrades habitat health and functions (e.g., contaminants).
Physical interference that eliminates natural habitats (e.g., filling, water diversions, waterway blockages, shading).

Shallow waters
Chemical contamination (e.g., piped discharges and non-point sources).
Physical alterations (e.g., dredging, filling, fishing/mining activities).

Shelf waters
Chemical discharges (e.g., ocean disposal and industrial effluents).
Physical alterations (e.g., ocean disposal, fishing/mining activities).

Programs to Mitigate Losses

As habitat losses have mounted, governments have responded with a series of scientific and management conferences, policy pronouncements, and administrative fixes, including many involving or affecting the National Oceanic and Atmospheric Administration/National Marine Fisheries Service (NOAA/NMFS):

- President George Bush's statement in the February 1989 State of the Union Address that his goal was for "no net loss of wetlands";
- U.S. Army Corps of Engineers–U.S. Environmental Protection Agency agreement under the Clean Water Act, Section 404(b)(1), implemented in February 1990 to clarify mitigation approaches;
- NOAA Conference on Habitat Restoration, September 1990, Washington, DC, the first assembly of restoration experts;
- U.S. Army Corps of Engineers–NOAA/NMFS agreement to "restore and create fish habitat," signed in January 1991;
- NOAA Damage Assessment and Restoration Program, established in 1991 with units addressing legal, damage assessment, and restoration issues.

In total, these and many other programs represent a significant groundswell of environmental concern, public opinion, and government response. Natural resource managers will be required to adapt to the evolving requirement that degradation associated with human activities be balanced by some measure of mitigation. This mandate applies equally to fishery management, dredged material management, water quality management, and every other human activity that affects a component of the shelf ecosystem.

How Can We Mitigate Habitat Losses?

Confined by the present state of knowledge, we can only hope to avoid and minimize, although even those proven steps in the CEQ sequence require more political willpower than most agencies can muster. While we might be unable to avoid or minimize, managers are routinely expected to mitigate for lost habitat values. With very little scientific justification, resource managers are required by legislative mandates and program regulations to render decisions based on the best available information. Unfortunately, that translates into ill-fated attempts to rectify, replace, or compensate for unavoidable losses.

Some mitigation successes have been documented, but often progress has been incomplete—usually limited to only several habitat values, not the full suite of values that contribute to ecosystem health. For example, replacement eelgrass plugs may form the rudimentary beginnings of a vegetated habitat, but indigenous species may not return in the natural ratios or densities.

The Mitigation Challenge

The challenge to all resource managers is to advance our knowledge of species, habitats, and human activities so we can make intelligent mitigation decisions. Before we can hope to offset lost habitat values, we must understand the ecological connections of species to the environment, and how human activities can shift that balance. Basic ecological information is sorely lacking. Engineering experience is practically nonexistent.

Despite these realities, development pressures along the Northeast Shelf LME continue to mount; the need for defensible science and consistent management will be paramount. In the northeast, various pressures promise to complicate our efforts:

- Burgeoning populations in coastal counties;

- Mounting pollution throughout watersheds, with cumulative impacts affecting habitats from the shore to the shelf;
- Continued development pressures to transform functional marine habitats into human habitats, and to use waterfront property for non–water-dependent activities;
- Diminished population health of important species, especially harvested fish and protected marine mammals;
- Increased perception that seafood quality is being compromised by human activities; and
- Greater chances for special management problems associated with protected species, air quality, drinking water, recreational enjoyment, and other issues of prime social importance.

Among the specific habitat impacts expected in the Northeast Shelf LME are:

- Dredged material disposal sites off Maine, Massachusetts, Connecticut, New York/New Jersey, and Maryland/Virginia, with implications for the water column, bottom sediments, and associated species;
- Municipal wastewater outfall pipes ranging from the massive MWRA/Boston Metropolitan Area outfall into Massachusetts Bay to dozens of smaller discharges from sewerage districts;
- Expanding plumes of pollutants from rivers, urban centers, and other discrete areas; and
- Continued pressure to remove resources from the ecosystem, including,
  - Freshwater (e.g., waterfowl impoundments in the Chesapeake Bay area, drinking water reservoirs in the Virginia tidewater region, hydropower in New England, irrigation);
  - Fish, resulting in a species assemblage shift affecting ecosystem quality, or in direct impacts that decrease habitat suitability or productivity;
  - Petroleum from the Middle Atlantic;
  - Construction material (e.g., sand and gravel from Georges Bank and the New York Bight); and
  - Others (e.g., water from Stellwagen Bank).

These challenges will demand solid scientific support, a strong management will, and convictions to meet the fullest intent of the CEQ's mitigation definition.

20

Status, Trends, and Health of Wetlands: A 200-Year Overview and Perspective

James Postles Thomas

Abstract

Large marine ecosystems (LMEs), such as the U.S. Northeast Continental Shelf, are considered to include adjacent estuaries and their wetlands. These provide habitat for estuarine and coastal living marine resources (LMRs), some of which may support fisheries of the U.S. Exclusive Economic Zone (EEZ). This chapter describes the history of attempts, in an evolutionary sense, to determine the extent, health, and rate of change in the size and functioning (i.e., health) of wetlands, particularly marine coastal wetlands, in the United States. Included in this history is some discussion of the strengths and weaknesses of the various programs as they have evolved. What is apparent is that whereas wetlands continue to be lost, the rate of loss is decreasing.

Determining the health of wetlands via remote sensing is only now beginning. The recent advent of landscape ecology, Landsat Thematic Mapper imagery, and the use of geographical information systems presents the potential for determining the relationship of changes in land cover of adjacent uplands to changes in coastal wetlands and the LMRs dependent on them. If we are to appropriately manage LMEs such as the Northeast Shelf, it is essential that we consider the effects of changes in coastal land cover on the estuarine and coastal components of such systems, the linkage between these components and uplands on the one hand, and the linkage between these components and the Continental Shelf on the other.

Recent newspaper headlines have documented the intense interest by society in the status (i.e., extent), trends, and health (i.e., functioning) of our coastal ecosystems and their fish and shellfish. The *Baltimore Sun* (The Sun 1991) published an article on 16 July 1991 entitled, "Drop in Fish Population in U.S. Coastal Waters Alarms Researchers." The article stated, "...many fish species are threatened by overfishing and an advancing tide of pollutants and coastal development that have degraded habitats and wetlands that fish use as spawning and migrating grounds." The *New York Times* (Gold 1991) on 6 August 1991, in an article entitled, "Pollution's Clench Could Strangle L.I. Bays," stated, "Long Island's coastal pollution mirrors what is happening along much of the nation's shoreline.... The trend isn't getting any better.... The problem is we, the people—the higher the density, the more pollution."

The issue is not unique to the United States. A United Nations Environment Program (IMO/FAO/UNESCO/WHO/IAEA/UN/UNEP 1990) report on the state of the marine environment stated, "The coastal strip, encompassing the shallow-water and intertidal area along with the immediately adjacent land, is clearly the most vulnerable as well as the most abused marine zone. Its sensitivity is directly tied to the diversity and intensity of activities which take place there, and the threat to its future is related to the increasing concentration of the world population in this area. The consequences of coastal development are thus of the highest concern. They arise not only from the variety of contaminating inputs associated with great concentrations of people, commerce and industry, but also from the associated physical changes in natural habitats, especially salt marshes, seagrass beds, coral reefs and mangrove forests."

In fact, one of the important trends observed in the marine environment, particularly in coastal areas of the United States, is a decline in abundance of certain fish and shellfish stocks (NOAA 1991), and, in some cases, a change in species composition from more desirable to so-called "less-desirable" species (Sherman 1988). At the same time, a dramatic increase in human population in coastal areas is evident (Culliton et al. 1990). Concomitant with this population increase and fisheries decline has been a broad-scale degradation of the coastal habitat of living marine resources (Murchelano and Wolke 1985; McCain et al. 1988; U.S. Congress 1989; Chambers 1991, 1992). While an inference is drawn here, it is realized that populations of living marine resources (LMRs) also are affected by natural environmental changes and overutilization by recreational and commercial fishermen.

These coastal areas, their status, trends, and health, are important to the long-term, sustainable productivity (economic and ecological) of the coastal lands and oceans. Over two-thirds of the commercially and recreationally important marine fish of the United States are estuarine-dependent (McHugh 1966). Estuarine areas, which provide critical habitat for spawning, feeding, and survival of fish and shellfish, are and have been affected by human activities occurring within the estuary proper as well as on adjacent upland areas (Pearce 1991). As these critical habitats are affected, so too are the LMRs using them, as well as recruitment to estuarine and in some cases coastal (i.e., U.S. Exclusive Economic Zone) fisheries. The process of changes in human demographic patterns and resultant land-use changes is affecting habitat and LMRs faster than we presently are monitoring it. If human presence and activity could be related in a holistic and understandable way to changes in LMRs, using research, data integration, and modeling in a geographical information system (GIS) to detect changes in land cover, land use (e.g., from natural forest to agricultural or urbanized use), and fishery habitat, then long-term planning and regulation could be instituted to sustain the productivity of the coastal land–ocean complex.

In this chapter, I first describe the history of our attempts, in an evolutionary sense, to determine the extent, health, and rate of change in the size of wetlands, particularly coastal wetlands, in the United States. Included in this history will be discussion of the strengths and weaknesses as various programs have evolved. Next, I will suggest what is needed. Finally, I hope to make the topic relevant to the U.S. Northeast Continental Shelf in the context of large marine ecosystems (LMEs).

Status, Trends, and Health of Coastal Wetlands

General

This discussion concentrates on estuarine wetlands, because palustrine and lacustrine habitat information, although relevant, is not available in a holistic (i.e., quantified in a standard way by drainage basin), readily interpretable (i.e., ecologically significant) way for anadromous species. Additionally, estuarine wetlands, a known critical habitat for LMRs, are under major threat from human development in the coastal zone. This does not diminish the importance of freshwater habitats to anadromous species many miles upstream and inland, nor the contribution of adjacent uplands through physical, chemical, and biological processes to the water quality encountered by estuarine and anadromous species. Thus, these upland areas, as well as certain defined palustrine and lacustrine habitats of ecological significance to LMRs, need to be delineated.

Although the original quantity of wetlands in the United States is uncertain, several reasonably reliable estimates exist and may be used as points of reference. Roe and Ayres (1954) estimated that the original total agricultural lands requiring reclamation by drainage (i.e., wetlands) was about 87.4 million hectares (ha) in the contiguous United States (Alaska and Hawaii excluded).

The Soil Survey Staff (1975) estimated that originally 85.4 million ha of hydric soils (i.e., wetlands) existed in this area, with an additional 66.8 million ha in Alaska (152.2 million ha in toto). Other estimates for the contiguous United States, 86.2 million ha (Pavelis 1987) and 87.85 million ha (USDA 1989), were made by determining agricultural lands in drainage in 1950 combined with comparable estimates of wetlands from Frayer et al. (1983) or the National Resources Inventory (Iowa State University Statistical Laboratory 1984). Dahl (1990) estimated the original quantity of wetlands as 89.47 million ha in the lower 48 states, 68.83 million ha in Alaska, and 23,887 ha in Hawaii, for a 50-state total of 158.3 million ha. More recently, Hall et al. (1994) estimated the present quantity of wetlands in Alaska to be 70.75 million ha. No comparable estimates have been calculated for estuarine or coastal wetlands.

The earliest known census of wetlands in the United States was made in 1906 (Wright 1907). The U.S. Department of Agriculture was asked by Congress to determine how much wetland could be reclaimed for agriculture. The response was 31.98 million ha; this was, however, incomplete in that all tidal wetlands were excluded from the census, as were the wetlands in Alaska, Hawaii, and eight states of the arid west.

The second inventory of wetlands, conducted in 1922 by the Bureau of Agricultural Economics of the U.S. Department of Agriculture (Gray et al. 1923), was based on data from (1) the U.S. Bureau of Public Roads, (2) soil-survey reports, (3) topographic maps of the U.S. Geological Survey, (4) state reports, and (5) results of the 1920 census of drainage. This inventory estimated the presence of 37,062,000 ha of wetlands in the lower 48 states, of which 2,981,000 ha were listed as tidal marsh. Of the total, an estimated 30,360,000 ha had potential for conversion to agriculture.

In 1940, the Soil Conservation Service of the U.S. Department of Agriculture estimated 39,406,000 ha of wetlands in the lower 48 states, exclusive of drainage enterprises (Wooten and Purcell 1949). The drainage enterprises included lands (wetlands and non-wetlands) being drained or planned to be drained. For instance, the non-wetland category of drainage enterprises included irrigated lands drained to remove alkali salts. Again, because of such a pointed focus and the exclusion of certain wetlands, the overall estimate remained incomplete.

In June 1954, the Fish and Wildlife Service of the Department of the Interior completed its first inventory of wetlands of the contiguous United States (Shaw and Fredine 1956). The purpose was to estimate wetlands available to waterfowl and wildlife. In addition to field work with state agencies, they made use of aerial photographs, topographic maps of the U.S. Geological Survey, charts of the U.S. Coast and Geodetic Survey (now NOAA), maps of the U.S. Forest Service, soil maps and land-use maps of federal and state agencies, and county highway maps. They believed that 90% of the wetlands used significantly by waterfowl were included in their inventory. Their estimate was 30.12 million ha, of which 3.77 million ha were tidally influenced (listed in their report as Coastal Fresh and Coastal Saline, according to the classification scheme of Martin et al. [1953]). Their national totals (in millions of ha) were: Inland Fresh, 25.7; Inland Saline, 0.6; Coastal Fresh, 1.6; and Coastal Saline, 2.1. It should be noted that open water areas within the Coastal Fresh and Coastal Saline categories were included in the acreages because these waters were considered shallow enough to be diked and filled for reclamation and agricultural or industrial uses. If coastal open fresh water (0.08 million ha) and coastal sounds and bays (0.45 million ha) are excluded, the total for tidally influenced wetlands becomes 3.2 million ha. However, tidal fresh wetlands in the classification scheme of Cowardin et al. (1979) are considered palustrine. If all of the Coastal Fresh category (1.6 million ha) plus the Coastal Saline subcategory of sounds and bays (0.45 million ha) are excluded, the resulting acreage (1.7 million ha) more nearly matches the estuarine intertidal wetlands class as defined by Cowardin et al. (1979). Additionally, if Coastal Salt Meadows (0.39 million ha), Irregularly Flooded Salt Marshes (0.28 million ha), and Regularly Flooded Salt Marshes (0.64 million ha) are added together, an estimate of Estuarine Intertidal Emergent (1.31 million ha) may be obtained.

In 1974, the U.S. Fish and Wildlife Service (FWS) undertook a review of existing state and local wetland surveys conducted from 1965 to 1975 (U.S. Department of the Interior 1976). They found that individual states had unique classification schemes for land/habitat cover categories; legal definitions for wetlands varied from one state to the next. Additionally, surveys were not repeated, or were repeated without standard methodology. Thus, no national, regional, or interjurisdictional data could be developed. The results confirmed that a national wetlands inventory (NWI) was required to ensure a standard methodology for mapping wetlands in the United States.

In 1979, the U.S. Fish and Wildlife Service's NWI began applying a standard methodology for determining the status (quantity) of U.S. wetlands (Wilen 1990). The methodology is based on interpretation of aerial photographs and the use of the Cowardin classification scheme (Cowardin et al. 1979) to depict the various classes observed on 1:24,000 scale hard-copy maps. Production of the maps, a labor-intensive and time-consuming process, is rigorously controlled, and their quality is generally excellent. Approximately 3,200 maps at 1:24,000 scale are produced per year for the lower 48 states, and 60 maps at 1:63,360 scale are produced for Alaska. This is a rate of 5% and 2%

per year, respectively, and theoretically means that it should take 20 and 50 years, respectively, for complete coverage. Passage of the Emergency Wetlands Resources Act of 1986 mandated completion of NWI mapping of wetlands for the contiguous United States by 1998. More recent amendments require the NWI to complete mapping the wetlands of Alaska by the year 2000.

Estuarine Loss

In 1983, the U.S. Fish and Wildlife Service produced a report entitled "Status and Trends of Wetlands and Deepwater Habitats in the Coterminous United States: 1950's to 1970's" (Frayer et al. 1983). Trend analysis was introduced for the first time. Acreages for analysis were determined based on stratified random samples selected from within each of the 35 physical subdivisions (i.e., strata) described by Hammond (1970). For each four-square-mile (1,036 ha) sample unit randomly selected from within a stratum, aerial photographs from the 1950s (median year 1954) and 1970s (median year 1974) were obtained and interpreted. Ninety-eight percent of the aerial photographs were made within 5 years of the median years, and the median, mode, and average interval was 20 years. The units were interpreted in entirety for the 1950s; only the changes were interpreted for the 1970s. Results were scaled up to the stratum level. Within the lower 48 states, 43.77 million ha of wetlands existed during the 1950s; by the 1970s, only 40.08 million ha remained. This equates to an approximate average annual loss of 184,211 ha. Of the above totals, approximately 2,270,931 ha of Estuarine Intertidal wetlands (including 1,731,174 ha of Emergent wetlands, 239,717 ha of Forested and Scrub/Shrub wetlands, and 300,000 ha of Non-Vegetated wetlands) existed in the 1950s; 2,122,389 ha (including 1,588,178 ha of Emergents, 231,984 ha of Forested and Scrub/Shrub, and 302,227 ha of Non-Vegetated) existed in the 1970s. The fact that the numbers do not add up suggests uncertainty in the estimates of Frayer et al. (1983). Of the 148,542 ha of Estuarine Intertidal wetlands lost, most went to Estuarine Subtidal areas and to urban development. The report, because it is derived from stratified random samples and not a comprehensive survey, presents these data only in tabular and graphical form, by class at the national level, and for certain states or areas where the losses are particularly dramatic. Thus, where the losses occurred geographically cannot be seen (as with a map).

Loss of estuarine wetlands (i.e., all wetlands semi-enclosed by land and with ocean-derived salinity greater than 0.5%) during this time was greatest in Louisiana and Florida, followed by Texas, New Jersey, and California. Louisiana was estimated to be losing 40 square miles or 10,121 ha of coastal marshes a year (Fruge 1982). Others, however, have estimated 50 square miles or 12,955 ha a year (Day and Craig 1981). These losses were attributed to a combination of natural and man-induced effects—i.e., channelization and levee construction along the Mississippi River, canal dredging in the marshes, and subsidence of the marshes themselves (Craig et al. 1980). The channelization, canal dredging, and subsidence resulted in erosion and salt water intrusion, whereas levee construction prevented sediment needed to sustain the wetlands from reaching them. According to Gosselink and Baumann (1980), losses outside of Louisiana were directly related to population density. In south Florida, major losses were caused by the building of levees and canals to prevent flooding, channel waters, and drain wetlands for urban and agricultural expansion (Yates 1982). These activities blocked the sheet flow of freshwater, so critical for the Everglades, and upset the distribution of salinity in estuarine nursery areas. Major losses elsewhere were associated with dredge-and-fill activities for urban and industrial development.

Note that Estuarine Subtidal wetlands are excluded from the estimates of Frayer et al. (1983) because they include water depths below extreme low water of spring tide or water depths greater than 2 m, below the maximum depth at which emergent plants normally grow. Also note that the acreage for Coastal Saline wetlands (sans sounds and bays), 1.7 million ha in 1954 (Shaw and Fredine 1956), does not match the acreage for Estuarine Intertidal wetlands (2.27 million ha) in the 1950s (Frayer et al. 1983). Differences result because the two studies used different classification schemes and field methods. Shaw and Fredine (1956) used the classification scheme of Martin et al. (1953), whereas Frayer et al. (1983) used the scheme of Cowardin et al. (1979). Cowardin et al. (1979) compared a number of different classification schemes, including Martin et al. (1953), and the differences are apparent. Additionally, Shaw and Fredine (1956) concentrated on wetland areas of known importance to waterfowl, and consequently may have missed other wetlands of supposedly lesser importance to waterfowl.

In 1984, the NWI with financial assistance from the U.S. Environmental Protection Agency instituted the first intensive study of a region, the five states of Delaware, Maryland, Pennsylvania, Virginia, and West Virginia, to determine the status (i.e., extent) and trends (i.e., changes in extent) of wetlands in the Chesapeake drainage area and other wetlands in the five states (Tiner and Finn 1986; Tiner 1987). The NWI maps were used to determine the most recent (1978–1981) wetland acreages. Trend analysis, using aerial photographs from 1955–1957 and 1978–1981, was determined as before (Frayer et al. 1983), but the number of strata for these states was increased from 17 to 29, and enough additional plots were sampled within each stratum to estimate totals for major wetland types in each state, with a standard error of less than 20% of the mean (Tiner and Finn 1986).

Whereas Tiner and Finn (1986) present a number of detailed classes, I have chosen to focus on Estuarine Vegetated wetland (i.e., Estuarine Intertidal Emergent and Estuarine Intertidal Scrub–Shrub wetlands) and Palustrine Vegetated Wetland (i.e., Palustrine Forested, Palustrine Scrub–Shrub, and Palustrine Emergent wetlands). The Chesapeake Bay watershed had 59,537 ha of Estuarine Intertidal Vegetated wetlands in 1956, and 54,442 ha in 1979, a loss rate of 221 ha/yr. This was attributed to coastal impoundments, dredging, sea-level rise, and urban development. For Palustrine Vegetated wetlands, the watershed had 391,836 ha in 1956, and 369,734 ha in 1979, a loss rate of 961 ha/yr. These losses were the result of agriculture, pond construction, and, to a lesser extent, urban development. Similar data for the states of Delaware, Maryland, and Virginia are presented in Table 20-1.

The losses of Estuarine Vegetated wetlands observed for Delaware and Virginia were attributed mostly to urban development and saltwater impoundment; less important factors were dredge-and-fill operations, sea-level rise, and, particularly for Delaware, conversion to freshwater impoundments. Losses in Maryland were primarily caused by saltwater impoundment and, to a lesser extent, sea-level rise and urban development. For Palustrine Vegetated wetlands, losses were attributed to agriculture, impoundment, and urban development. It was not until 1989 that comparable data became available for wetlands of the California Central Valley (Frayer et al. 1989), and 1991 for the wetlands of Florida (Frayer and Hefner 1991) and the states of the East and Gulf Coasts of the United States (Tiner 1991).

In 1986, the National Oceanic and Atmospheric Administration (NOAA), Department of Commerce, initiated a coastal wetlands project to develop a digital data base that could be used for making decisions about coastal wetlands. No digital data base then existed for coastal wetlands. In the first attempt (Field et al. 1988), data were obtained from 22 coastal states (contiguous United States). Federal data sets were also examined, but not used; the National Wetlands Inventory maps of the U.S. Fish and Wildlife Service were incomplete for the coastal area, and the U.S. Geological Survey's Land Use Data Analysis Program only had two categories of wetlands, forested and non-forested. Hence, state data (1950–1982) were acquired, evaluated, categorized (salt marsh, fresh marsh, tidal flats, and swamp), and summarized. Based on these data, an estimated 4.45 million ha of coastal wetlands existed, including 1.8 million ha of salt marsh. The latter figure for salt marsh may be compared with the 1,588,178 ha of Estuarine Intertidal Emergents estimated via stratified random samples by Frayer et al. (1983). The discrepancy is attributed to the variable comparability

Table 20-1. Areal changes (1950s–1970s) in estuarine and palustrine vegetated wetlands for Delaware, Maryland, and Virginia

Wetland type and state	Hectares	Hectares	Hectares lost/year
Estuarine intertidal vegetated wetlands			
Delaware	27,250 (1955)	25,680 (1981)	60
Maryland	48,693 (1955)	44,635 (1978)	176
Virginia	56,906 (1956)	54,838 (1977)	98
Palustrine vegetated wetlands			
Delaware	74,067 (1955)	58,714 (1981)	591
Maryland	109,364 (1955)	103,270 (1978)	265
Virginia	327,846 (1956)	304,753 (1977)	1,100

Note: Numbers in parentheses are dates of aerial photography.
Source: Modified from R. W. Tiner, Jr., and J. T. Finn. 1986. Status and Recent Trends of Wetlands in Five Mid-Atlantic States: Delaware, Maryland, Pennsylvania, Virginia, and West Virginia. Newton Corner, MA: U.S. Fish and Wildlife Service, Region 5, National Wetlands Inventory Project, and Philadelphia: U.S. Environmental Protection Agency, Region III.

and completeness in the state inventories, as well as differences between the state data and Frayer et al. (1983). The states used different wetland definitions, wetland classifications, census methods, and census dates (Alexander et al. 1986).

As a result of this analysis of state and federal data, a workshop was held in April 1986 to review the current status of wetland information and to make recommendations for developing a digital data base (NOAA 1986). The workshop established the following: (1) a digital data base would be valuable, (2) the NWI represented the most complete and standardized wetland information available, (3) the NWI was only partially digitized, with no plans to complete the very expensive (up to $1,000/map) and time-consuming (4–5 days/map) effort, and (4) in lieu of digitizing NWI maps, a 45-acre (18.22 ha) cell-size grid sampling of NWI maps (5,500 in the coastal area) should be undertaken by NOAA after statistical validity and adequacy is demonstrated (accomplished 20 May 1986). It was acknowledged that rare and possibly important wetland categories would be missed by this approach.

In June 1986, to develop a digital data base, NOAA began the 18.22-ha cell-size grid sampling of the NWI maps available for the contiguous U.S. coastal region (Field et al. 1990; 1991). This region includes the area composed of the land and water component of an entire watershed that most directly affects an estuary, called the estuarine drainage area (EDA), and counties adjacent to marine waters not in an EDA. The NWI maps sampled depicted wetlands as they existed, based on aerial photographs, from 1972 to 1985. Sampling of the 5,290 NWI maps available was completed in October 1989. The NWI coverage of all EDAs was not complete, particularly for the inland portions. Data were entered into a geographical information system (i.e., the

Spatial Analysis System). Regional reports were produced for the Gulf of Mexico (Reyer et al. 1988), Mid-Atlantic (Reyer et al. 1990b), and West Coast (Reyer et al. 1990a). Comparisons between the wetland acreages obtained by the NOAA grid-sampling method and by computer sampling of digitized NWI maps are included in each regional report. In general, the more abundant a wetland type, the better the agreement between the two. Wetland types greater in abundance than 1,000–2,000 ha have a predicted error in agreement between the two methods of less than 10%.

Data for all regions, including the Southeast Region (not available previously), are published in a national summary (Field et al. 1991). These data are considered acceptable determinations of the abundances of the various wetland types at the national, regional, and estuarine levels of analysis. At the national level, compared with 40.08 million ha of wetlands in the 1970s for the contiguous United States (Frayer et al. 1983), or 42.87 million ha for the mid-1970s (Dahl and Johnson 1991) and 41.82 million ha for the mid-1980s (Dahl and Johnson 1991), approximately 11 million ha of coastal wetlands existed for this same period (Field et al. 1991). The most abundant class of these coastal wetlands was the Coastal Forested and Scrub–Shrub (7.0 million ha), followed by Coastal Fresh Marsh (1.8 million ha), Salt Marsh (1.6 million ha), and Tidal Flats (0.5 million ha). In terms of distribution, most of the coastal wetlands (51–56%) were in the Gulf of Mexico region, 30–33% in the South Atlantic, 9–10% in the Mid-Atlantic, and about 0.5% each in the Pacific and North Atlantic (Gulf of Maine) regions. Between 10% and 19% of the coastal wetlands of the United States were adjacent to the Northeast Shelf, depending on whether and how the coastal wetlands of southern Virginia and northern North Carolina were included (Tables 20-2 and 20-3). Grid sampling based solely on drainage basins (i.e., EDAs) omitted certain coastal wetlands, specifically, those wetlands adjacent to marine waters that are not part of an EDA (see Table 20-2). In contrast, grid sampling of the entirety of each county for all counties included, in toto or in part, within the EDAs, plus those counties not in an EDA but adjacent to marine waters, led to an overestimation of coastal wetlands (see Table 20-3). In either case, most of the coastal wetlands adjacent to the Northeast Shelf were located in the mid-Atlantic region.

The different classes of coastal wetlands adjacent to the Northeast Shelf are presented in Tables 20-4 and 20-5. The most abundant class was the Coastal Forested and Scrub–Shrub, followed by Salt Marsh, Tidal Flats, and Coastal Fresh Marsh. Again, these data have been presented as summaries of EDA data (Table 20-4), and by counties (Table 20-5), in order to show both under- and overestimates as mentioned above for Tables 20-2 and 20-3. What is missing are Canadian data for coastal wetlands of the Gulf of Maine. No

Table 20-2. U.S. coastal wetlands by portion of region within estuarine drainage areas adjacent to the Northeast Shelf

Region	Hectares
North Atlantic region (U.S. Gulf of Maine)	328,543
Mid-Atlantic region	933,239
South Atlantic region (Albemarle/Pamlico Sound)	458,057
Total	1,719,839

Note: Data based on grid sampling of estuarine drainage areas.
Source: Modified from D. W. Field, A. J. Reyer, P. V. Genovese, and B. D. Shearer. 1991. Coastal Wetlands of the United States: An Accounting of a Valuable National Resource. Washington, DC: U.S. Department of Commerce, NOAA, National Ocean Service.

Table 20-3. Extent of U.S. coastal wetlands in counties of estuarine drainage areas (EDAs) and coastal counties not in EDAs but adjacent to marine waters of states along the Northeast Shelf

State	Hectares
Maine	307,247
New Hampshire	34,858
Massachusetts	189,717
Rhode Island	26,802
Connecticut	61,822
New York	44,170
New Jersey	331,417
Pennsylvania	8,259
Delaware	90,891
Maryland	218,421
District of Columbia	121
Virginia	369,757
North Carolina (Albemarle/Pamlico Sound)	456,599
Total	2,140,081

Note: Data based on grid sampling of counties in Estuarine Drainage Area, in whole or in part, plus those coastal counties adjacent to marine waters.
Source: Modified from D. W. Field, A. J. Reyer, P. V. Genovese, and B. D. Shearer. 1991. Coastal Wetlands of the United States: An Accounting of a Valuable National Resource. Washington, DC: U.S. Department of Commerce, NOAA, National Ocean Service.

data are presented for seagrass beds, which are important fishery habitats. Additionally, upland areas that affect wetlands and LMR habitat in estuarine and coastal areas via runoff and water quality are not included. Since this effort, a 1992 amendment to the Emergency Wetlands Resources Act of 1986 requires the U.S. Fish and Wildlife Service to produce a digital wetlands data base for the United States, based on NWI maps, by 2004. To date (April 1995), the entire coastal areas of Maine, New York, New Jersey, Delaware, Maryland,

Table 20-4. Extent of U.S. coastal wetland types by region in estuarine drainage areas adjacent to the Northeast Shelf

Wetland type and region	Hectares	Total
Salt marsh		
U.S. Gulf of Maine	13,644	
Mid-Atlantic	225,466	
South Atlantic	63,806	
		302,916
Coastal fresh marsh		
U.S. Gulf of Maine	14,170	
Mid-Atlantic	50,202	
South Atlantic	14,777	
		79,149
Coastal forested and scrub-shrub		
U.S. Gulf of Maine	249,474	
Mid-Atlantic	568,340	
South Atlantic	366,883	
		1,184,697
Tidal flats		
U.S. Gulf of Maine	51,255	
Mid-Atlantic	89,190	
South Atlantic	12,591	
		153,036
Grand total		1,719,798

Note: Data based on grid sampling of estuarine drainage areas.
Source: Modified from D. W. Field, A. J. Reyer, P. V. Genovese, and B. D. Shearer. 1991. Coastal Wetlands of the United States: An Accounting of a Valuable National Resource. Washington, DC: U.S. Department of Commerce, NOAA, National Ocean Service.

North Carolina, Florida, Washington, Hawaii, Puerto Rico, and the Virgin Islands, and portions of the coastal areas of Massachusetts, Rhode Island, Connecticut, Virginia, South Carolina, Georgia, Alabama, Mississippi, Louisiana, Texas, California, Oregon, and Alaska have been digitized by the NWI (T. Dahl, personal communication).

In 1991, the National Wetlands Inventory produced a report on the status and trends of wetlands in the contiguous United States from the mid-1970s to the mid-1980s (Dahl and Johnson 1991). It provides the most authoritative information to date, and will be updated every ten years as required by the Emergency Wetlands Resources Act of 1986. The methodology used was the same as that in Frayer et al. (1983), except that the photography used was from the mid-1970s (1971–1978; mean year 1974) and mid-1980s (1981–1986; mean year 1983). During the mid-1970s, an estimated 42.87 million ha of wetlands existed in the contiguous United States; by the mid-1980s, 41.82 million ha existed. These figures differ from those presented by Frayer et al. (1983) for the

Table 20-5. Extent of U.S. coastal wetland types by region in counties of estuarine drainage areas (EDAs) and coastal counties not in EDAs but adjacent to marine waters along the Northeast Shelf

Wetland type and region	Hectares	Total
Salt marsh		
U.S. Gulf of Maine	27,571	
Mid-Atlantic	279,190	
South Atlantic	54,251	
		361,012
Coastal fresh marsh		
U.S. Gulf of Maine	23,603	
Mid-Atlantic	48,907	
South Atlantic	15,304	
		87,814
Coastal forested and scrub-shrub		
U.S. Gulf of Maine	375,547	
Mid-Atlantic	623,522	
South Atlantic	452,713	
		1,451,782
Tidal flats		
U.S. Gulf of Maine	57,814	
Mid-Atlantic	103,684	
South Atlantic	6,397	
		167,895
Grand total		2,068,503

Note: Data based on grid sampling of counties in Estuarine Drainage Areas, in whole or in part, plus those coastal counties adjacent to marine waters.
Source: Modified from D. W. Field, A. J. Reyer, P. V. Genovese, and B. D. Shearer. 1991. Coastal Wetlands of the United States: An Accounting of a Valuable National Resource. Washington, DC: U.S. Department of Commerce, NOAA, National Ocean Service.

mid-1970s (40.08 million ha). The difference is attributable to (1) an expansion of the coastal zone stratum to include coastal wetlands seaward of the estuarine–palustrine break on NWI maps, previously omitted, and (2) interpretation differences between the lesser quality 1:80,000-scale black-and-white aerial photographs of the 1970s and the better quality 1:58,000-scale color infrared photographs used in the 1980s. Use of the better film allowed correction of the 1970s data base. Average loss of wetlands during this time interval was approximately 116,667 ha/yr, down from the approximate annual average losses between the mid-1950s and the mid-1970s of 184,211 ha/yr (Frayer et al. 1983).

This decrease in loss represents the results of federal, state, and local efforts to conserve wetlands. Most loss occurred in freshwater wetlands of the southeastern United States, caused primarily by agriculture and conversion to other land uses and secondarily by urban development.

For estuarine wetlands (i.e., Estuarine Intertidal Unconsolidated Shore, Estuarine Intertidal Aquatic Beds, Estuarine Intertidal Emergent, and Estuarine Intertidal Forested and Scrub/Shrub), 2,239,717 ha (or 2,122,389 ha per Frayer et al. 1983) existed in the mid-1970s; by the mid-1980s, 2,215,668 ha remained. It should be noted that Estuarine Intertidal Aquatic Beds were not systematically surveyed according to accepted methods as defined by Orth et al. (1991) and Ferguson and Wood (1990) and therefore are underestimated. The underestimation comes about because of the use of high-altitude aerial photographs obtained early in the growing season at various tide, wind, and turbidity conditions. The problem is that aquatic beds may be seen in photographs obtained at one time and missed in photographs obtained at another, or may be partially seen or missed altogether.

For the dominant estuarine wetlands type, Estuarine Intertidal Emergent wetlands, 1,678,107 ha (Frayer et al. 1983; 1,588,178 ha) were present in the mid-1970s; by the mid-1980s, 1,649,437 ha remained, a loss rate of approximately 3,185 ha/yr. These losses were attributed mostly to sea-level rise, subsidence, and erosion, particularly in the Gulf Coast area, and secondarily to impoundments, agriculture, and urban development. Note that losses of Estuarine Intertidal Emergent wetlands are greater than losses of total estuarine wetlands. This is the result of conversion from one estuarine wetland type to another, so that net loss of estuarine wetland may not necessarily occur. Also note that this loss rate (3,185 ha/yr) is less than half of what it was between the mid-1950s and mid-1970s (7,150 ha/yr). Urban development of these wetlands decreased precipitously, from 2,156 ha/yr between the mid-1950s and mid-1970s to 203 ha/yr between the mid-1970s and mid-1980s (Tiner 1991).

Although no data are presented at the state level, Dahl and Johnson (1991) noted that certain states lost more than 50% of their wetlands between the 1780s and the mid-1980s. For coastal states, these include California (91% loss), Texas (52%), Mississippi (59%), Alabama (50%), Maryland (73%), Delaware (54%), New York (60%), and Connecticut (74%). The latter four states are adjacent to the Northeast Shelf.

Tiner (1991), using the same data as Dahl and Johnson (1991), breaks out coastal wetlands for each of the Atlantic and Gulf Coast states. Except for state reports mentioned earlier, these are the first data presented by state. Of particular interest are the status of estuarine wetlands in the mid-1980s (Table 20-6), and the status and trends of estuarine vegetated wetlands (i.e., Estuarine Intertidal Emergents, Estuarine Intertidal Aquatic Beds, Estuarine Intertidal Forested, and Scrub/Shrub wetlands) from the mid-1970s to the mid-1980s (Table 20-7) for those states along the Northeast Shelf. Tiner (1991) stated that when the percentage of standard error is less than 25%, the

Table 20-6. Estuarine wetlands during mid-1980s for states along the Northeast Shelf

State	Hectares	% standard error
Maine	32,994	39.5
New Hampshire	891	92.4
Massachusetts	29,913	31.2
Rhode Island	6,649	74.6
Connecticut	4,990	35.3
New York	68,849	46.2
New Jersey	76,211	14.7
Pennsylvania	No data	No data
Delaware	22,186	34.2
Maryland	29,845	23.5
District of Columbia	No data	No data
Virginia	39,889	29.6
North Carolina	65,781	18.6
Total (includes North Carolina)	378,198	
Total (excludes North Carolina)	312,417	

Source: RW Tiner, Jr. 1991. Recent Changes in Estuarine Wetlands of the Coterminous United States. Proceedings of the Coastal Wetlands, Coastal Zone '91 Conference—American Society of Civil Engineers, Long Beach, CA, July 1991. New York: American Society of Civil Engineers.

Table 20-7. Status and trends of estuarine vegetated wetlands for states adjacent to the Northeast Shelf

State	Mid-1970s hectares	Mid-1980s hectares	Ha lost/yr 1974–1983
Maine	8,864	8,864	0
New Hampshire	115	115	0
Massachusetts	15,122	15,057	7.2
Rhode Island	4,730	4,719	1.2
Connecticut	4.425	4,421	0.4
New York	54,485	54,444	4.6
New Jersey	75,122	74,657	51.7
Pennsylvania	No data		
Delaware	21,516	21,456	6.7
Maryland	29,421	29,317	11.6
District of Columbia	No data		
Virginia	38,957	38,953	0.4
North Carolina	64,363	64,326	4.1
Totals (including North Carolina)	317,120	316,329	87.9
Totals (excluding North Carolina)	252,757	252,003	83.8

Note: For comparison, losses of estuarine marsh in Louisiana were estimated to be 10,121 ha/yr between the mid-1950s and mid-1970s by Fruge (1982), and 12,955 ha/yr for the same interval by Day and Craig (1981). Tiner (1991) estimated losses of estuarine marsh in Louisiana to be 2,568 ha/yr between the mid-1970s and mid-1980s.
Source: RW Tiner, Jr. 1991. Recent Changes in Estuarine Wetlands of the Coterminous United States. Proceedings of the Coastal Wetlands, Coastal Zone '91 Conference—American Society of Civil Engineers, Long Beach, CA, July 1991. New York: American Society of Civil Engineers.

estimate is considered reliable. Even though a number of the estimates in Table 20-6 have a percentage of standard error in excess of 25%, these estimates are the best available. Between the mid-1970s and mid-1980s, Louisiana was losing estuarine vegetated wetlands at an approximate rate of 2,568 ha/yr (Tiner 1991), compared with earlier estimates of 10,121 ha/yr (Fruge 1982) and 12,955 ha/yr (Day and Craig 1981). Comparison of Tables 20-1 and 20-7 shows that rate of loss of estuarine vegetated wetlands has decreased substantially between the earlier period (mid-1950s to late 1970s) and more recently (mid-1970s to mid-1980s). Such a statement cannot include Estuarine Intertidal Aquatic Beds, however, since they have been poorly censused.

In 1977, and subsequently in 1982, 1987, and 1992, the Soil Conservation Service (SCS), now the National Resources Conservation Service (NRCS), of the U.S. Department of Agriculture (USDA) produced National Resources Inventories (NRIs). The Rural Development Act of 1972 authorizes the SCS to conduct the NRI every 5 years to provide status, condition, and trend information on soil, water, and related resources for all non-federal, rural lands in the United States. The first trend analysis was completed shortly after the 1987 NRI (SCS 1989). During June and July 1991, a special, partial NRI was conducted to update information on wetland conversion rates, to be used for the continued development of soil and water conservation programs (SCS 1991). Alaska was not part of these surveys. Wetland classes were compared using the classification scheme of Shaw and Fredine (1956). The data (Table 20-8) represent estimates derived from sample points within primary sampling units statistically selected from within strata. For comparison, the NWI estimated conversion of wetlands to agriculture for the contiguous United States to be approximately 161,000 ha/yr between 1954 and 1974 (Frayer et al. 1983), and 64,000 ha/yr from 1974 to 1983 (Dahl and Johnson 1991). Notice in Table 20-8 that the trend in loss of wetlands to agriculture appears to continue downward: 20,000 ha/yr between 1982 and 1987, and 12,000 ha/yr between 1987 and 1991. However, the fact that only non-federal, rural lands were surveyed, and that different classification schemes [i.e., Shaw and Fredine (1956) versus Cowardin et al. (1979)] were used, at best confounds and at worst may negate such intercomparison. Using only the Cowardin et al. (1979) classification scheme for comparison, the most recent report (SCS 1995) shows 12,515 ha/yr of wetlands lost to agriculture between 1982 and 1992. It also indicates that estuarine wetland losses of all types were approximately 2,321 ha/yr in that same time span. For the states adjacent to the Northeast Shelf, the loss rates for estuarine wetlands are presented in Table 20-9. Except for

Table 20-8. Conversion of wetlands on non-federal, rural land in the United States, excluding Alaska

Losses to	1982–1987 (ha/yr)	1987–1991 (ha/yr)	1982–1992 (ha/yr)
Agriculture	20,000	12,000	12,515
Development	23,000	24,000	35,879
Other (open water)	10,000	8,000	Not specified
Total	53,000	44,000	63,233

Note: For comparison, the National Wetlands Inventory estimated conversion of wetlands to agriculture for the coterminous United States to be approximately 161,000 ha/yr between 1954 and 1974, and 64,000 ha/yr from 1974 to 1983. These losses become significant when compared with the losses of wetlands listed in Table 20-12.
Source: Soil Conservation Service (1991, 1995).

Table 20-9. Status and trends of estuarine wetlands for states adjacent to the Northeast Shelf

State	1982 Hectares	1992 Hectares	Ha lost/yr 1982–1992
Maine	5,306	5,306	0
New Hampshire	4,212	4,212	0
Massachusetts	19,481	19,481	0
Rhode Island	29,079	29,079	0
Connecticut	6,602	6,602	0
New York	25,313	25,313	0
New Jersey	101,615	101,250	36.5
Pennsylvania	0	0	0
Delaware	60,872	61,155	+28.3
Maryland	231,174	230,648	52.6
District of Columbia	No data		
Virginia	229,311	229,595	+28.4
North Carolina	636,822	636,174	64.8
Total (including North Carolina)	1,349,787	1,348,815	97.2
Total (excluding North Carolina)	712,965	712,641	32.4

Source: Soil Conservation Service (1995).

Maryland and North Carolina, the rates of loss appear to be less from 1982 to 1992 (see Table 20-9) than from the mid-1970s to 1980s (see Table 20-7). However, notice the large differences in quantity of estuarine wetlands between Tiner (1991) (see Table 20-6) and the Soil Conservation Service (1995) (see Table 20-9). Total wetland losses for coastal states between 1982 and 1992 were approximately 28,731 ha/yr, whereas non-coastal states gained approximately 3,151 ha/yr. The losses were most notable in North Carolina (12,243 ha/yr), Florida (7,112 ha/yr), Oregon (5,816 ha/yr), and Louisiana (4,577 ha/yr).

A Recent Program

A relatively new program, the Coastal Change Analysis Program (C-CAP) is of interest regarding development of status and trends data on wetlands within a landscape approach. In 1990, NOAA initiated C-CAP to monitor coastal wetlands, including SVRP, and adjacent upland cover and change in the coastal region of the United States (Thomas and Ferguson 1990; Thomas et al. 1991; Ferguson et al. 1992a). The long-term goal of C-CAP is to determine how land cover and changes in land cover affect living marine resources—their abundance, distribution, and health. To do this, NOAA is developing a comprehensive, nationally standardized information system for land cover and change in the coastal region of the United States, making use of satellite imagery, aerial photography, surface level, and other collateral data within a geographical information system (GIS) context. The project is intended to be a cooperative effort with other federal and state agencies. As a consequence, the first 4 years were devoted primarily to developing a standardized protocol based on a series of regional workshops and smaller working group meetings held around the country with other federal, state, and academic personnel (Dobson and Bright 1991; Orth et al. 1991; Haddad 1992; Dobson et al. 1995). Additional research and development is continuing in areas such as accuracy assessment, classification, tidal effects, and modeling.

The coastal region covered by C-CAP includes those land and water components of the various watersheds within the United States, its possessions, and territories that most directly influence estuarine and coastal marine habitats used by LMRs. The land cover includes those classes of vegetation and physical cover of ecological significance to LMRs and their habitats (Klemas et al. 1993; Dobson et al. 1995). The major classes include wetlands, uplands, and water and submerged land cover. Satellite imagery is the primary data source for coastal wetlands and adjacent uplands. Aerial photography is the primary source for determining abundance and distribution of SRVPs. The planned time interval for repeated looks at the coastal region of the United States is every 1–5 years. Regions with little change or interest will be monitored every 5 years, and areas of intense development every 2 or 3 years; areas disturbed by extreme events (e.g., oil spills, hurricanes) will be censused annually. Data will be collected as synoptically as possible to facilitate change analysis.

Additionally, a component of C-CAP is being developed so that not just areal coverage is determined, but also functional health, whereby a decline in the functioning of a coastal habitat could be observed before any loss in its area. To accomplish this, NOAA is undertaking research to relate condition of the wetlands to spectral radiances, as measured by satellite or airborne sensors

(Hardisky et al. 1984, 1986, 1992; Gross et al. 1987, 1991; Patience and Klemas 1993). Should the research effort be successful, it would allow large areas of wetlands to be surveyed and assessed much more rapidly and easily.

C-CAP has released several maps depicting the distribution of SRVPs (Ferguson et al. 1989, 1991) and change in the distribution of SRVPs in Core Sound, North Carolina between 1985 and 1988 (Ferguson et al. 1992b, 1993). This portion of C-CAP is working with the U.S. Fish and Wildlife Service, and states in the Gulf of Mexico region and along the Atlantic and Pacific coasts to map the extent and change of SRVPs. The SRVP data being produced by the Chesapeake Bay program (Orth et al. 1990) will be utilized by C-CAP. The plan is to integrate SRVP data with coastal wetland and adjacent upland data to produce a holistic data set for the coastal region of the United States.

In December 1992, the NOAA National Oceanographic Data Center (NODC 1992) announced the first availability of a C-CAP data set for coastal wetlands and adjacent uplands showing land cover and changes in land cover between 1984 and 1988–89 for the Chesapeake Bay estuarine drainage area, based on four Landsat Thematic Mapper scenes for each time period. The data were prepared by Jerome E. Dobson and Edward A. Bright (Dobson and Bright 1994) at the Oak Ridge National Laboratory, Oak Ridge, Tennessee, and have been extensively ground truthed in cooperation with the states of Maryland and Virginia, several universities, and federal agencies (i.e., EPA, USDI/FWS, USDI/USGS, USDA/SCS) (Shapiro 1995). The data are also available from the National Geophysical Data Center on a CD-ROM (Hittelman et al. 1994). What is of particular interest about the data set is the loss of upland forest (approximately 14,300 ha/yr) to crop/grassland and development. By comparison, wetlands losses and gains were relatively minor. Palustrine wetlands sustained losses of approximately 700 ha/yr, whereas estuarine marsh gained about 400 ha/yr.

These data are reasonably in line with the data of Tiner (1987) and Tiner and Finn (1986) who, for the period between 1956 and 1979, presented loss rates of 961 ha/yr and 221 ha/yr for palustrine wetlands and estuarine marsh, respectively. More important, the data set places changes in upland landscape cover in context with changes in wetlands. In contrast to Brady and Flather (1994), who found losses in forested wetlands to be greater than losses of upland forest for the coterminous United States from 1982 to 1987, Dobson and Bright (1994) found losses of upland forest in the Chesapeake Bay area to be much greater. The cumulative effects of all these changes, uplands and wetlands alike, result in significant but as yet undefined changes in the water quality of estuarine and coastal habitats used by LMRs (Thomas 1995). For this reason, C-CAP believes a holistic view of the coastal region of the United

States, beyond just wetlands, is extremely important for developing the understanding necessary for the protection, maintenance, and restoration of estuarine and coastal habitats and the LMRs dependent on them.

Summary and Recommendations

The way in which people have considered wetlands has not only determined what has happened to these lands, but also how and to what extent information was collected about them. Historically, people thought of wetlands as undesirable tracts of land that caused disease (i.e., malaria) or interfered with development. Wetlands were considered wastelands unless they could be drained for agriculture. Early in the history of the United States, the federal government, through the Federal Swampland Acts of 1845, 1850, and 1860, ceded to the states all land unfit for cultivation because of flooding. These acts enabled states to build levees to contain flood waters as well as to drain wetlands for other purposes. Such early uses of wetlands included peat mining; flooding, diking, and draining for salt production, as was done in San Francisco (South) Bay from 1856 to 1985; hay production and cattle grazing; rice production; and growing sugar cane, to mention only a few. Among the other early influences on wetlands were dams built in the 1830s and later on the upper Susquehanna River and elsewhere. These dams not only blocked fish migration, but also changed water flow, sediment loading, and, ultimately, water quality downstream. Beginning about 1881, extensive areas of Florida were drained for agriculture (Light and Dineen 1994). As human population growth continued, wetlands were dredged or filled for development or used for sanitary landfills, among the other uses already mentioned. Beginning at least as far back as the 1950s, wetlands were impounded for waterfowl, and later for aquaculture of food species. Between 1962 and 1973, Florida undertook an additional, extensive canal building program to further drain wetlands for agriculture and urbanization.

Gradually, based on research in the 1950s and 1960s, wetlands began to be considered for other values (i.e., flood protection, water quality, and habitat). Tidal influence on estuarine wetlands was considered as good as a tractor and fertilizer for maintaining their productivity (Odum 1971). The states, beginning with Massachusetts in 1963 (followed by Rhode Island in 1965; Connecticut in 1969; New Jersey in 1970; Maryland in 1970; Georgia in 1970; New York in 1972; Delaware in 1973; and others since), began mandating protection of wetlands. Before this time, most wetlands had passed into private ownership, and these owners could do with their lands as they wished, so long as their actions did not interfere with navigation as covered by the Rivers and Harbors Act of 1899. The Federal Clean Water Act of 1977 extended and strengthened existing controls over wetlands. Whereas estuarine wetlands seemed to be increasingly protected, palustrine wetlands were not. In response, the states began passing laws to address the issue of palustrine wetlands loss. Through the Food Security Act of 1985, the federal government extended protection of wetlands by discouraging farmers from converting them to agricultural lands. Most recently an awareness in landscape ecology has begun to encourage integration of collection and analysis of information, and its application for management by drainage basin or other management unit using a GIS. This is resulting in the need for machine-manipulable data (i.e., digital data) on temporal and spatial scales appropriate for the research and management issues being addressed. Finally, it seems true that with the more rapidly increasing growth of human populations in coastal areas, the demand for more up-to-date information has also accelerated.

The emphasis of early inventories was on determining wetlands available for agricultural development (i.e., cropland or pasture). Emphasis on waterfowl and wildlife came much later. As a consequence, estuarine wetlands often were not surveyed. In spite of more recent efforts to determine the status and trends of wetlands in the United States, the effort is still incomplete. Wetlands of Alaska's coastal regions are only partially mapped. Those of U.S. possessions and territories in the Pacific still require mapping. Certain critical habitats for LMRs (e.g., seagrass) have yet to be surveyed systematically. Different methods and classification schemes have been used, making comparison from one survey to the next difficult, if not impossible. As a consequence, the numbers presented in the various publications and over time appear to be disparate, even though they exhibit internal consistency (Tables 20-10 and 20-11). Adding to the potential uncertainty is the fact that the data have been quantified by statistical methods rather than by actual census, both because of the magnitude and cost of the task, and because of the tools that historically have been available.

Changes in acreage between one time and another have been estimated by comparing data collected over a span of years with data collected over a second span of years, rather than by comparing data from a single year with data from a different single year. Lack of digitized data has made integration and manipulation of numerous, large, and different data sets quite difficult, often hindering our ability to conceptually understand and deal with issues.

Upland areas that affect coastal habitat have not been integrated into surveys of the coastal region at all. Each state and federal agency has collected and will continue to collect data based on its own interests and mandates.

Table 20-10. Wetlands of the contiguous United States

Survey year	Reference	Area (millions of Ha)
1780	Roe and Ayers (1954)	87.40
1780	Soil Survey Staff (1975)	85.40
1780	Pavelis (1987)	86.20
1780	USDA (1989)	87.80
1780	Dahl (1990)	89.47
1906	Wright (1907)	31.98
1922	Gray et al. (1923)	37.06
1940	Wooten and Purcell (1949)	39.41
1954	Shaw and Fredine (1956)	30.12
1954	Frayer et al. (1983)	43.77
1974	Frayer et al. (1983)	40.08
1974	Dahl and Johnson (1991)	42.87
1983	Dahl and Johnson (1991)	41.82

Expressed differently, each state and federal agency quantifies land cover only for those lands for which it has interest (e.g., forested lands in the case of the USDA/Forest Service; nonfederal, rural lands for the USDA/NRCS). Such collection efforts, while appropriate for individual agencies, may not be supportive of the solution for broader issues (e.g., the effects of development in the coastal region on productivity of the coastal zone and coastal ocean). Even if the various data bases from agencies were put together, there would be gaps, since not all lands are surveyed for land cover; there would also be inconsistencies in the data, because different classification schemes and methods of assessment are used. As a consequence, except for the Chesapeake Bay area and a few smaller areas, there is no holistic examination of land cover in the coastal region of the United States. Additionally, no recent (i.e., collected within the last 7 years) land cover change data exist for coastal wetlands and adjacent uplands, except for the Chesapeake Bay area and perhaps a few smaller areas. Finally, perhaps only in Florida (Haddad and McGarry 1991) have SRVPs been integrated with coastal wetlands and adjacent uplands.

Despite these problems certain trends are evident in the data presented in this paper. Wetlands in the United States have declined in areal coverage (see Tables 20-10 and 20-11). The apparent increase in acreage with time (see Table 20-10) exhibited in the data between certain publications is not seen in the data within an individual publication. This apparent disparity is the result of different methods being used.

Table 20-11. Tidal wetlands, estuarine intertidal wetlands, and estuarine intertidal emergent wetlands for the contiguous United States

Survey year	Reference	Area (millions of Ha)	Type
1922	Gray et al. (1923)	2.98	Tidal
1954	Shaw and Fredine (1956)	3.77	Tidal
1954	Shaw and Fredine (1956)	1.70	Estuarine intertidal
1954	Shaw and Fredine (1956)	1.31	Estuarine intertidal emergent
1954	Frayer et al. (1983)	2.27	Estuarine intertidal
1974	Frayer et al. (1983)	2.12	Estuarine intertidal
1954	Frayer et al. (1983)	1.73	Estuarine intertidal emergent
1974	Frayer et al. (1983)	1.59	Estuarine intertidal emergent
1950–1982	Field et al. (1988)	1.80	Estuarine intertidal emergent
1972–1985	Field et al. (1991)	1.60	Estuarine intertidal emergent
1974	Dahl and Johnson (1991)	2.24	Estuarine intertidal
1983	Dahl and Johnson (1991)	2.22	Estuarine intertidal
1974	Dahl and Johnson (1991)	1.68	Estuarine intertidal emergent
1983	Dahl and Johnson (1991)	1.65	Estuarine intertidal emergent

In contrast, annual rates of loss are decreasing for all wetlands, including Estuarine Intertidal wetlands, and Estuarine Intertidal Emergent wetlands (Tables 20-1, 20-7, and 20-12). The literature indicates that the largest losses of wetlands have been and continue to be palustrine wetlands. Overall, however, recent state and federal efforts appear to be effectively reducing the annual rates of loss for wetlands in the contiguous United States. These efforts should be continued. Finally, although in general uplands have not been inventoried together with coastal wetlands, such data for the Chesapeake Bay region suggest that changes in uplands may be of greater importance than heretofore thought relative to areal changes in wetlands. It must be remembered that uplands are a significant determinant of the water quality affecting wetland habitats and their living resources.

What is needed is a database that will allow a holistic look at the coastal region of the United States. It should include all significant components—submerged rooted vascular plants (SRVPs), coastal wetlands, and adjacent uplands (perhaps entire drainage basins). The data must be machine manipulable because of the volume of data and complexity of analyses to be accomplished. This means that the data must be digital, and must be entered into

Table 20-12. Annual loss rates of wetlands in the contiguous United States

Time period	Reference	Ha loss/yr	Type
1954–1974	Frayer et al. (1983)	184,211	Wetlands
1974–1983	Dahl and Johnson (1991)	116,667	Wetlands
1982–1987	SCS (1991)	53,000	Wetlands
1987–1991	SCS (1991)	44,000	Wetlands
1982–1992	SCS (1995)	63,233	Wetlands
1954–1974	Frayer et al. (1983)	7,427	Estuarine intertidal
1974–1983	Dahl and Johnson (1991)	2,672	Estuarine intertidal
1954–1974	Frayer et al. (1983)	7,150	Estuarine intertidal emergent
1974–1983	Dahl and Johnson (1991)	3,185	Estuarine intertidal emergent

Note: Data from Soil Conservation Service (1991), includes only non-federal, rural lands, and uses Shaw and Fredine (1956) classification scheme. Other studies include Federal lands and use the Cowardin et al. (1979) classification scheme.

and used in a GIS. The data could come from single or multiple sources, but must be compatible, if the latter. The integration of data from different sources would require Federal and state interagency coordination not only on data standards and classification schemes, such as are being worked on by the Federal Geographic Data Committee under the Office of Management and Budget Circular A-16 (FGDC 1991), but also on data collection, processing, and analysis, as C-CAP is attempting to do. Ideally, data from programs that provide detailed, local coverage should be integrated with those that provide broad and rapid but less detailed coverage. This would allow high accuracy and precision as well as the ability to update changes rapidly. Where possible, the data base should be extended backward in time to gain perspective, as Costa (1988) did for seagrass in Massachusetts. Such backward extension might facilitate development of models linking landscape changes with changes in coastal habitats and their LMRs. The data base also should be extended forward, with periodic updates to include data less than 5 years old. These updates would enable management to be more proactive in preventing losses of critical habitat and sustaining the long-term productivity (both economic and ecological) of the coastal zone and coastal ocean.

Finally, even though wetlands are not as dominant along the U.S. Northeast Continental Shelf as they are along the Southeast and Gulf coastal areas, and loss rates for wetlands have decreased dramatically, the Northeast Shelf does receive the input of a number of major streams, and is impacted by the waste loads and activities of 39 million people living in the adjacent coastal region (Culliton et al. 1990). Some mechanism is needed to integrate and periodically update these cumulative impacts (see Pearce 1991) on the Northeast Shelf. The sources of data can be multiple, but must be digital and compatible within a GIS framework. Examination of land cover and change in land cover, encompassing SRVPs, coastal wetlands, and adjacent uplands, should be incorporated into any over-arching research and management scheme for the Northeast Shelf. Such inclusion would allow a way to consider the major impacts from land on adjacent shelf ecosystems.

If we are to appropriately manage large marine ecosystems such as the Northeast Shelf, it becomes essential that we include consideration of the effects of changes in land cover on the estuarine and coastal components of such systems, and the linkage between these components and uplands on the one hand, and linkage between these components and the Continental Shelf on the other. Such linkages should include those estuarine and coastal species that support fisheries in the U.S. Exclusive Economic Zone.

ACKNOWLEDGMENTS

I thank Ford A. Cross, Jerome E. Dobson, John B. Pearce, Victor V. Klemas, and Donald W. Field for advice and comment on this paper.

References

Alexander, C. E., Broutman, M. A., and Field, D. W. 1986. An Inventory of Coastal Wetlands of the USA. Mimeogr. Rept. Rockville, MD: National Oceanic and Atmospheric Administration, Strategic Assessment Branch.

Brady, S. J., and Flather, C. H. 1994. Changes in wetlands on non-federal rural land of the coterminous United States from 1982 to 1987. Envir. Manag. 18(5):693–705.

Chambers, J. R. 1991. Habitat Degradation and Fisheries Declines in the U.S. In H. S. Bolton (ed.), Coastal Wetlands. Coastlines of the World. Coastal Zone 91, Seventh Symposium on Coastal and Ocean Management. New York: American Society of Civil Engineers, pp. 46–60.

Chambers, J. R. 1992. Coastal Degradation and Fish Population Losses. In R. H. Stroud (ed.), Stemming the Tide of Coastal Fish Habitat Loss. Marine Recreational Fisheries Symposium 14. Savannah, GA: National Coalition for Marine Conservation, pp. 45–51.

Costa, J. E. 1988. Eelgrass in Buzzards Bay: Distribution, Production and Historical Changes in Abundance. EPA 503/4-88-002. Washington, DC: U.S. Environmental Protection Agency, Office of Marine and Estuarine Protection.

Cowardin, L. M., Carter, V., Golet, F. C., and LaRoe, E. T. 1979. Classification of Wetlands and Deepwater Habitats of the United States. FWS/OBS-79/31. Washington, DC: U.S. Fish and Wildlife Service.

Craig, N. J., Turner, R. E., and Day, J. W., Jr. 1980. Wetland losses and their consequences in coastal Louisiana. Z. Geomorph. N.F. Suppl. Bd. 34:225–241.

Culliton, T. J., Warren, M. A., Goodspeed, T. R., et al. 1990. 50 Years of Population Change Along the Nation's Coasts, 1960–2010. Coastal Trends Series, Rept. #2. Rockville, MD: National Oceanic and Atmospheric Administration.

Dahl, T. E. 1990. Wetlands Losses in the United States, 1780's to 1980's. Washington, DC: U.S. Fish and Wildlife Service.

Dahl, T. E., and Johnson, C. E. 1991. Wetlands Status and Trends in the Coterminous United States, mid-1970s to mid-1980s. Washington, DC: U.S. Fish and Wildlife Service.

Day, J. W., Jr., and Craig, N. J. 1981. Comparison of Effectiveness of Management Options for Wetlands Loss in the Coastal Zone of Louisiana. In D. F. Boesch (ed.), Proceedings of the Conference on Coastal Erosion and Wetland Modification in Louisiana: Causes, Consequences, and Options. Oct. 5–7, 1981, Baton Rouge, LA. FWS/OBS-82/59. Washington, DC: U.S. Fish and Wildlife Service, pp. 232–239.

Dobson, J. E., and Bright, E. A. 1991. CoastWatch—Detecting change in coastal wetlands. Geo. Info. Systems. January/February 1991. pp. 36–40.

Dobson, J. E., and Bright, E. A. 1994. Large Area Change Analysis: The CoastWatch Change Analysis Project (C-CAP)—Land Information from Space-based Systems. Proceedings of the Pecora 12 Conference, August 24–26, 1993. Bethesda, MD: American Society for Photogrammetry and Remote Sensing. pp. 73–81.

Dobson, J. E., Ferguson, R. L., Field, D. W., et al. 1995. NOAA Coastal Change Analysis Project: Guidance for Regional Implementation. NOAA Tech. Rept. NMFS 123. Rockville, MD: National Oceanic and Atmospheric Administration.

Ferguson, R. L., Rivera, J. A., and Wood, L. L. 1989. Seagrasses in Southern Core Sound, North Carolina. NOAA Coastal Ocean Program, Submerged Aquatic Vegetation Study. [Chart.] Beaufort, NC: NOAA/National Marine Fisheries Service, Beaufort Laboratory, SEFSC.

Ferguson, R. L., and Wood, L. L. 1990. Mapping Submerged Aquatic Vegetation in North Carolina with Conventional Aerial Photography. In S. J. Kiraly, F. A. Cross, and J. D. Buffington (eds.), Federal Coastal Wetland Mapping Programs. Biol. Rept. 90(18). Washington, DC: U.S. Fish and Wildlife Service, pp. 125–132.

Ferguson, R. L., Wood, L. L., and Pawlak, B. T. 1991. SAV Habitat from Drum Inlet to Ocracoke Inlet, North Carolina. NOAA, Coastal Ocean Program, Submerged Aquatic Vegetation Study. [Chart.] Beaufort, NC: NOAA/National Marine Fisheries Service, SEFSC, Beaufort Laboratory.

Ferguson, R. L., Wood, L. L., and Graham, D. B. 1992a. Detection of Change in Submerged Coastal Habitat. Proceedings American Society for Photogrammetry and Remote Sensing/American Congress on Surveying and Mapping/Resource Technology '92 Convention, Washington, DC, August 1992. Bethesda, MD: American Society for Photogrammetry and Remote Sensing/American Congress on Surveying and Mapping, pp. 70–79.

Ferguson, R. L., Wood, L. L., and Pawlak, B. T. 1992b. SAV Habitat in 1985 and 1988: Cape Lookout to Drum Inlet, N.C. NOAA Coastal Ocean Program, Submerged Aquatic Vegetation Study. [Chart.] Beaufort, NC: NOAA/National Marine Fisheries Service, SEFSC, Beaufort Laboratory.

Ferguson, R. L., Wood, L. L., and Graham, D. B. 1993. Monitoring spatial change in seagrass habitat with aerial photography. (Special issue on Monitoring and Mapping Global Change.) Photogram. Engin. Remote Sensing. 59(6):1033–1038.

Federal Geographic Data Committee (FGDC). 1991. OMB [Office of Management and Budget] revises Circular A-16; Federal Geographic Data Committee organized. FGD

Newsletter. ISSN 1055-8357, Newsletter No. 1, Spring 1991. Reston, VA: U.S. Department of the Interior, U.S. Geological Survey.

Field, D. W., Alexander, C. E., and Broutman, M. 1988. Toward developing an inventory of U.S. coastal wetlands. Mar. Fish. Rev. 50(1):40–46.

Field, D. W., Reyer, A. J., Alexander, C. E., et al. 1990. National Oceanic and Atmospheric Administration's Coastal Wetlands Inventory. In S. J. Kiraly, F. A. Cross, and J. D. Buffington (eds.), Federal Coastal Wetland Mapping Programs. Biol. Rept. 90(18). Washington, DC: U.S. Fish and Wildlife Service, pp. 39–49.

Field, D. W., Reyer, A. J., Genovese, P. V., and Shearer, B. D. 1991. Coastal Wetlands of the United States: An Accounting of a Valuable National Resource. Washington, DC: U.S. Department of Commerce, NOAA, National Ocean Service.

Frayer, W. E., and Hefner, J. M. 1991. Florida Wetlands, Status and Trends, 1970s to 1980s. Atlanta: U.S. Fish Wildlife Service, Southeast Region.

Frayer, W. E., Monahan, T. J., Bowden, D. C., and Graybill, F. A. 1983. Status and Trends of Wetlands and Deepwater Habitats in the Coterminous United States: 1950's to 1970's. Ft. Collins, CO: Department of Forest Wood Sciences, Colorado State University.

Frayer, W. E., Peters, D. D., and Pywell, H. R. 1989. Wetlands of the California Central Valley, Status and Trends, 1939 to mid-1980s. Portland, OR: U.S. Fish and Wildlife Service, Region 1.

Fruge, D. W. 1982. Effects of Wetland Deterioration on the Fish and Wildlife Resources of Coastal Louisiana. In D. F. Boesch (ed.), Proceedings of the Conference on Coastal Erosion and Wetland Modification in Louisiana: Causes, Consequences, and Options. FWS/OBS-82/59. Washington, DC: U.S. Fish and Wildlife Service, pp. 99–107.

Gold, A. R. 1991. Pollution's clench could strangle L.I. bays. New York Times (Aug. 6, 1991) CXL(48,684):B-1.

Gosselink, J. G., and Baumann, R. H. 1980. Wetland inventories: Wetland loss along the United States coast. Z. Geomorph. N.F. Suppl. Bd. 34:173–187.

Gray, L. C., Baker, O. E., Marschner, F. J., Weitz, B. O., et al. 1923. The Utilization of Our Lands for Crops, Pasture, and Forests. In Yearbook 1923. Washington, DC: U.S. Department of Agriculture. pp. 415–506.

Gross, M. F., Hardisky, M. A., Klemas, V., and Wolf, P. L. 1987. Quantification of biomass of the marsh grass *Spartina alternaflora* Loisel using Landsat Thematic Mapper imagery. Photogram. Engin. Remote Sensing 53(11):1577–1583.

Gross, M. F., Hardisky, M. A., Wolf, P. L., and Klemas, V. 1991. Relationship between aboveground and belowground biomass of *Spartina alternaflora* (Smooth Cordgrass). Estuaries 14(2):180–191.

Haddad, K. D. 1992. CoastWatch Change Analysis Program (C-CAP) Remote Sensing and GIS protocols. In Proceedings of the American Society for Photogrammetry and Remote Sensing/American Congress on Surveying and Mapping/Resource Technology '92 Convention, Washington, DC, August 1992. Bethesda, MD: American Society for Photogrammetry and Remote Sensing/American Congress on Surveying and Mapping. pp. 58–69.

Haddad, K. D., and McGarry, G. A. 1991. Marine Resource Mapping and Monitoring in Florida. Coastal Wetlands Coastal Zone '91 Conference—American Society of Civil Engineers, Long Beach, California, July 1991. New York: American Society of Civil Engineers. pp. 347–361.

Hall, J. V., Frayer, W. E., and Wilen, B. O. 1994. Status of Alaskan Wetlands. Anchorage, AK: U.S. Fish and Wildlife Service, Alaska Region.

Hammond, E. H. 1970. Physical Subdivisions of the United States. National Atlas of the United States. Washington, DC: U.S. Geological Survey, pp. 61–64.

Hardisky, M. A., Daiber, F. C., Roman, C. T., and Klemas, V. 1984. Remote sensing of biomass and annual net aerial primary productivity of a salt marsh. Remote Sensing of Environment 16:91–106.

Hardisky, M. A., Gross, M. F., and Klemas, V. 1986. Remote sensing of coastal wetlands. Bioscience 36(7):453–460.

Hardisky, M. A., Wolf, P. L., and Gross, M. F. 1992. Remote Sensing of Biomass and Stress in Coastal Wetlands. [Abstract.] Proceedings of the First Thematic Conference on Remote Sensing for Marine and Coastal Environments, New Orleans, June 1992.

Hittelman, A. M., Iredale, H., Hollinger, M., and Field, D. 1994. Coastwatch Change Analysis Project: Chesapeake Bay Region. Boulder, CO: National Oceanic and Atmospheric Administration, National Geophysical Data Center.

International Maritime Organization/Food and Agriculture Organization/United Nations Educational Scientific and Cultural Organization/World Health Organization/International Atomic Energy Agency/United Nations/United Nations Environment Programme. 1990. Joint Group of Experts on the Scientific Aspects of Marine Pollution (GESAMP): The State of the Marine Environment. Rept. Stud. GESAMP No. 39. Nairobi, Kenya: United Nations Environment Programme.

Iowa State University Statistical Laboratory. 1984. National Resources Inventory—A Guide for Users of 1982 NRI Data Files. Washington, DC: U.S. Department of Agriculture.

Klemas, V. V., Dobson, J. E., Ferguson, R. L., and Haddad, K. D. 1993. A coastal land cover classification system for the NOAA CoastWatch Change Analysis Project. J. Coast. Res. 9(3):862–872.

Light, S. S., and Dineen, V. W. 1994. Water Control in the Everglades: A Historical Perspective. In S. M. Davis and V. C. Ogden (eds.), Everglades: The Ecosystem and Its Restoration. Delray Beach, FL: St. Lucie Press. pp. 47–84.

Martin, A. C., Hotchkiss, N., Uhler, F. M., and Bourn, W. S. 1953. Classification of wetlands of the United States. Special Sci. Rept.—Wildlife No. 20. Washington, DC: U. S. Fish and Wildlife Service.

McCain, B. B., Brown, D. W., Krahn, M. M., et al. 1988. Marine pollution problems, North American West Coast. Aquat. Toxicol. 11:143–162.

McHugh, J. L. 1966. Management of Estuarine Fishes. Spec. Publ. No. 3. Bethesda, MD: American Fish Society, pp. 133–154.

Murchelano, R., and Wolke, R. 1985. Epizootic carcinoma in winter flounder, *Pseudopleuronectes americanus*. Science 228:587–589.

National Oceanic and Atmospheric Administration (NOAA). 1986. Summary of Proceedings: NOAA Coastal Wetlands Workshop. (Mimeogr. Rept.) Rockville, MD: National Oceanic and Atmospheric Administration, Strategic Assessment Branch.

National Oceanic and Atmospheric Administraiton (NOAA). 1991. Status of the Fishery Resources Off the Northeastern United States for 1990. NOAA Tech. Mem. NMFS-F/NEC-81. Woods Hole, MA: National Oceanic and Atmospheric Administration, Northeast Fisheries Center.

National Oceanographic Data Center (NODC). 1992. CoastWatch Change Analysis Project (C-CAP): Chesapeake Bay Land Cover Classification Data, 1984 and 1988–89. NODC Environmental Information Bull. No. 92-3. Washington, DC: U.S. Department of Commerce, NOAA, National Oceanographic Data Center.

Odum, E. P. 1971. Fundamentals of Ecology (3rd ed.). Philadelphia: W. B. Saunders.

Orth, R. J., Moore, K. A., and Nowak, J. C. 1990. Monitoring Seagrass Distribution and Abundance Patterns: A Case Study from the Chesapeake Bay. In S. J. Kiraly, F. A. Cross, and J. D. Buffington (eds.), Federal Coastal Wetland Mapping Programs. U.S. Fish Wildl. Serv., Biol. Rept. 90(18). Washington, DC: U.S. Fish and Wildlife Service, pp. 112–124.

Orth, R. J., Ferguson, R. L., and Haddad, K. D. 1991. Monitoring seagrass distribution and abundance patterns. Proceedings of the Coastal Wetlands, Coastal Zone '91 Conference—American Society of Civil Engineers, Long Beach, CA, July 1991. New York: American Society of Civil Engineers. pp. 281–300.

Patience, N., and Klemas, V. 1993. Wetland Functional Health Assessment Using Remote Sensing and Other Techniques: Literature Search. NOAA Tech. Memo. NMFS-SEFC-319. Rockville, MD: National Oceanic and Atmospheric Administration.

Pavelis, G. A. (ed.). 1987. Farm Drainage in the United States: History, Status, and Prospects. Misc. Publ. No. 1455. Washington, DC: U.S. Department of Agriculture, Economic Research Service.

Pearce, J. B. 1991. Collective effects of development on the marine environment. Proceedings of the International Colloquium on the Environment of Epicontinental Seas, Lille, France, March 20–22, 1990. Oceanologica Acta Vol. sp. 11:287–298.

Reyer, A. J., Field, D. W., Cassells, J. E., et al. 1988. The Distribution and Areal Extent of Coastal Wetlands in Estuaries of the Gulf of Mexico. (Mimeogr. Rept.) Rockville, MD: National Oceanic and Atmospheric Administration, Strategic Assessment Branch.

Reyer, A. J., Shearer, B. D., Genovese, P. V., et al. 1990a. The Distribution and Areal Extent of Coastal Wetlands in Estuaries of the West Coast Region. (Mimeogr. Rept.) Rockville, MD: National Oceanic and Atmospheric Administration, Strategic Assessment Branch.

Reyer, A. J., Shearer, B. D., Genovese, P. V., et al. 1990b. The Distribution and Areal Extent of Coastal Wetlands in Estuaries of the Mid-Atlantic Region. (Mimeogr. Rept.) Rockville, MD: National Oceanic and Atmospheric Administration, Strategic Assessment Branch.

Roe, H. B., and Ayres, Q. C. 1954. Engineering for Agricultural Drainage. New York: McGraw-Hill.

Soil Conservation Service (SCS). 1989. Summary Report, 1987, National Resources Inventory. Statistical Bull. No. 790. Washington, DC: U.S. Department of Agriculture, SCS, and Iowa State Statistical Laboratory.

Soil Conservation Service (SCS). 1991. 1991 Update of National Resources Inventory. Wetlands Data for Non-Federal Rural Lands. Washington, DC: U.S. Department of Agriculture, SCS.

Soil Conservation Service (SCS). 1995. Summary Report: 1992 National Resources Inventory. Issued July 1994, Revised January 1995. Washington, DC: U.S. Department of Agriculture, SCS.

Shapiro, C. 1995. Coordination and integration of wetland data for status and trends and inventory estimates: Progress report. Reston, VA: Federal Geographic Data Committee Wetlands Subcommittee, U.S. Geological Survey.

Shaw, S. P., and Fredine, C. G. 1956. Wetlands of the United States, Their Extent and Their Value to Waterfowl and Other Wildlife. Circular 39. Washington, DC: U.S. Fish and Wildlife Service.

Sherman, K. 1988. Large Marine Ecosystems as Global Units for Recruitment Experiments.

In B. J. Rothschild (ed.), Toward a Theory on Biological-Physical Interactions in the World Ocean. NATO ASI Series. Series C: Mathematical and Physical Sciences, Vol. 239, Dordrecht, The Netherlands: Kluwer Academic, pp. 459–476.

Soil Survey Staff. 1975. Soil Taxonomy: A Basic System of Soil Classification for Making and Interpreting Soil Surveys. Washington, DC: U.S. Department of Agriculture, Soil Conservation Service.

The Sun. 1991. Drop in fish population in U.S. coastal waters alarms researchers. Baltimore (Maryland), The Sun (July 16, 1991) 309(52):1.

Thomas, J. P. 1995. Remote sensing and relating coastal development to living marine resources and their habitats. Natural Areas J. 15:21–36.

Thomas, J. P., and Ferguson, R. L. 1990. NOAA's Habitat Mapping Under the Coastal Ocean Program. In S. J. Kiraly, F. A. Cross, and J. D. Buffington (eds.), Federal Coastal Wetland Mapping Programs. Biol. Rept. 90(18). Washington, DC: U.S. Fish and Wildlife Service, pp. 44–54.

Thomas, J. P., Ferguson, R. L., Dobson, J. E., and Cross, F. A. 1991. NOAA's CoastWatch: Change Analysis Program. Proceedings of the Coastal Wetlands Coastal Zone '91 Conference—American Society of Civil Engineers, Long Beach, CA, July 1991. New York: American Society of Civil Engineers. pp. 259–267.

Tiner, R. W., Jr. 1987. Mid-Atlantic Wetlands: A Disappearing Natural Treasure. Newton Corner, MA: U.S. Fish and Wildlife Service, Region 5, National Wetlands Inventory Project, and Philadelphia: U.S. Environmental Protection Agency, Region III.

Tiner, R. W., Jr. 1991. Recent Changes in Estuarine Wetlands of the Coterminous United States. Proceedings of the Coastal Wetlands, Coastal Zone '91 Conference—American Society of Civil Engineers, Long Beach, CA, July 1991. New York: American Society of Civil Engineers. pp. 100–109.

Tiner, R. W., Jr., and Finn, J. T. 1986. Status and Recent Trends of Wetlands in Five Mid-Atlantic States: Delaware, Maryland, Pennsylvania, Virginia, and West Virginia. Newton Corner, MA: U.S. Fish and Wildlife Service, Region 5, National Wetlands Inventory Project, and Philadelphia: U.S. Environmental Protection Agency, Region III.

U. S. Congress. 1989. Coastal Waters in Jeopardy: Reversing the Decline and Protecting America's Coastal Resources. Oversight Report of the Committee on Merchant Marine and Fisheries. Serial 100-E. Washington, DC: U.S. Government Printing Office.

U. S. Department of Agriculture (USDA). 1989. Wetland priority analysis—Lower 48 States. Washington, DC: U.S. Department of Agriculture, Economic Research Service.

U. S. Department of the Interior. 1976. Existing State and Local Surveys 1965–75. Washington, DC: U.S. Fish and Wildlife Service.

Wilen, B. O. 1990. The U.S. Fish and Wildlife Service's National Wetlands Inventory. In S. J. Kiraly, F. A. Cross, and J. D. Buffington (eds.), Federal Coastal Wetland Mapping Programs. Biol. Rept. 90(18). Washington, DC: U.S Fish and Wildlife Service, pp. 9–20.

Wooten, H. H., and Purcell, M. R. 1949. Farm Land Development: Present and Future by Clearing, Drainage, and Irrigation. Circular 825. Washington, DC: U.S. Department of Agriculture.

Wright, J. O. 1907. Swamp and Overflowed Lands in the United States. Circular 76. Washington, DC: U.S. Department of Agriculture.

Yates, S. 1982. Florida's broken rain machine. The Amicus Journal 4:48–55.

21

Relating Habitat Stress to Fish Productivity: Problems and Approaches

David S. Peters and Ford A Cross

In most large marine ecosystems, including the Northeast Shelf, the variety of habitats and fishery species is so large and the ecological processes that influence fish production are so complex that it is currently not possible to accurately predict or measure impacts of environmental stress on fish production. However, that does not mean that stress is absent or that impacts on production are unimportant. Because there is no generally accepted method of relating stress to production, the assigned topic is not an easy one, and we have chosen to discuss it from a conceptual point of view.

Before doing so, we need to clarify the definition of two terms: habitat and stress. Use of the term *habitat* is highly variable, ranging from the general—i.e., the area where an animal is found (Odum 1971)—to the more restrictive—i.e., that local structural component of the environment that attracts organisms and serves as a center of biological activity (Ryder and Kerr 1989). To avoid any confusion, we will use the term *environmental stress* rather than *habitat stress*, with *environment* defined as the total physical, chemical, and biological surroundings of an organism, including habitat and other biota.

Defining stress is not easy. As Levin (1985) stated, "one who undertakes the task of defining stress either has an enormous ego, is immeasurably stupid or is totally mad." Not wanting to be classified as any of those, we use the term as it is commonly accepted by those who work with fish (Heath 1990)—i.e., stress refers to "diminished function, which implies potential death of the individual or depressed reproduction leading to a declining population." The difficulty with this use of the term *environmental stress* in a fisheries context is that population impacts that are a direct result of fishery harvest cannot be easily distinguished from those that are an indirect result of alteration of the environment. Distinguishing these impacts is complicated by four major problems inherent in the state of our knowledge or the way we conduct fisheries and environmental studies.

One problem is that environmental research and fisheries research often are conducted on different scales, both geographic and biological. For example, fishery scientists tend to collect information from the entire range occupied by fish large enough to be harvested. For many species (e.g., Atlantic menhaden, striped bass, bluefish, weakfish, and summer flounder), this range exceeds the limits of the Northeast Shelf ecosystem. On the other hand, scientists working on the effects of environmental stress on fish usually focus on much smaller geographic scales, often only in a single bay or estuary or the immediate vicinity of a spill or contaminant. Because anthropogenic environmental impacts are usually concentrated close to land, such impacts are usually studied in estuaries or very close to shore, environments that may not represent the average environment occupied by a particular species. The estuaries and areas near shore are more likely to be juvenile nurseries, so that young fish are more apt to be subject to environmental impacts, whereas fishery data tend to emphasize older fish. If nearshore degradation is severe, some nursery areas may contribute little to any fishery, and if they are separate from the primary adult habitat, they may not be included in areas represented by the fishery data base.

A second problem in relating environmental stress to fish productivity is that environmental impacts that are most easily and often measured occur below the organism level and are not useful as indicators of effects at higher levels of biological organization. Fishery data deals with population level variables (e.g., stock size, age structure, mortality, and growth rates), whereas environmental stress measurements often deal with observation at sub-organism levels (e.g., biochemical, histological, or physiologic). Although we agree that biochemical or physiologic responses may be indicative of environmental health and may ultimately provide the key to understanding why some process fails at a higher level of organization (Figure 21-1), they are of limited value in measuring or predicting population or ecosystem response. If we want to be able to infer the effects of environmental stress on populations, then we must focus on population characteristics such as abundance, recruitment, and distribution. Research should not progress from one level of biological organization to another but must be carried out at all levels simultaneously and in an integrated manner.

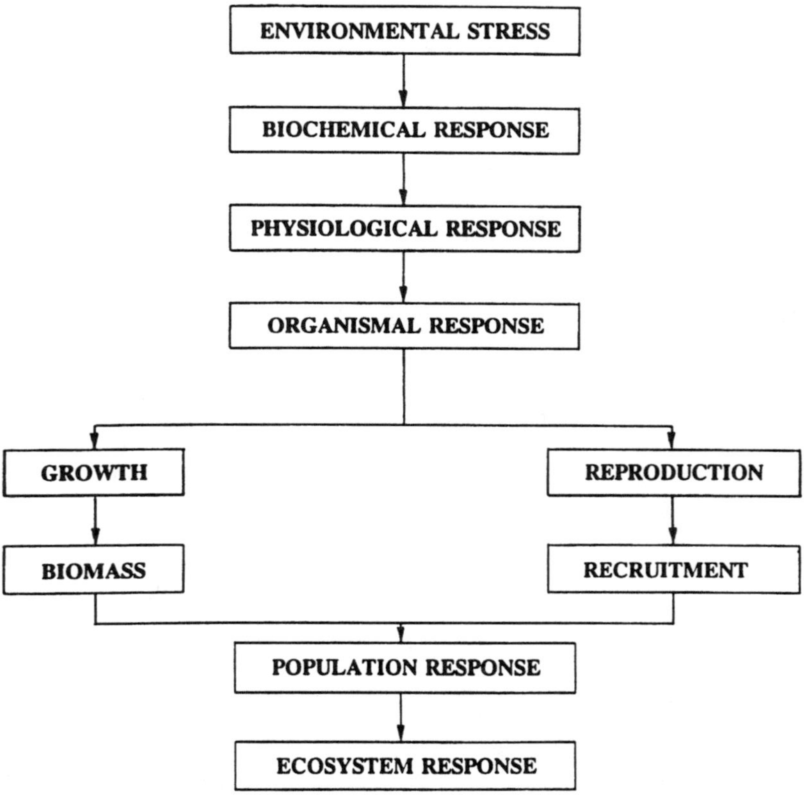

Figure 21-1. Levels of biological response to environmental stress.

A third problem is that temporal and spatial variation in the abundance of marine organisms makes it extremely difficult to separate natural variation from anthropogenic impacts (Jones 1982). The magnitude of this problem is illustrated by Vaughan et al. (1986), who estimated that if a single year-class of Atlantic menhaden suffered a 71% decline in recruitment to age 1 or a 42% decline in catch at age 2, there is only a 50% chance that it could be detected with any degree of statistical confidence. Obviously, environmental stress will rarely cause such extreme mortality, except perhaps in a localized habitat where a large fraction of a population aggregates during some phase of its life history (Schaaf et al. 1987).

Another major reason why it is difficult to measure environmental effects on fish populations is that these populations have an unquantified natural

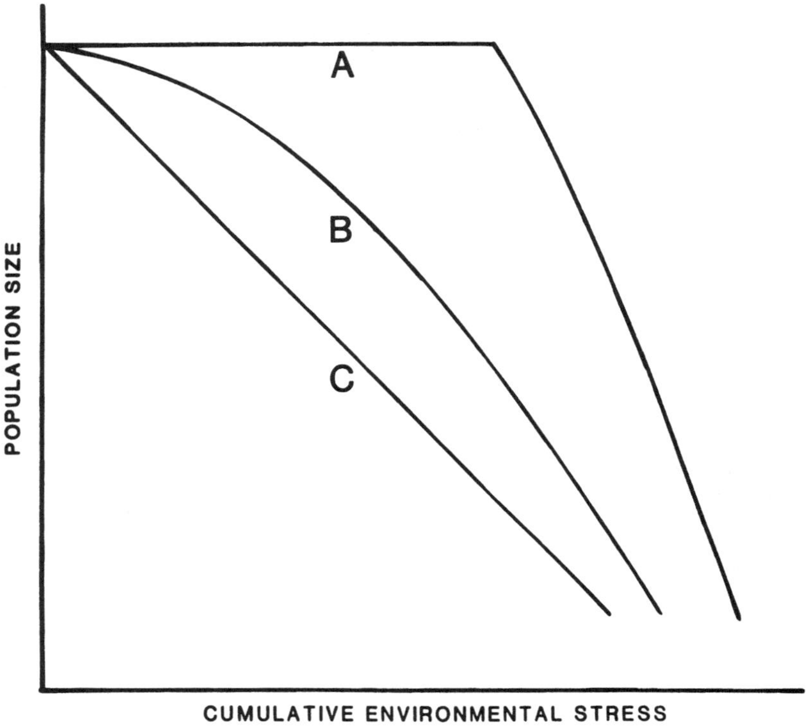

Figure 21-2. Population size as a function of stress level, with three hypothetical patterns of compensation: (A) complete compensation, then rapid decline, (B) partial compensation, and (C) no compensation.

capacity to compensate for increased mortality. As mortality from fishing and other human activities increases, fish can compensate in a variety of ways, such as increased growth and fecundity rates (Goodyear 1980). Because the compensatory reserve of a population cannot be measured, the amount of additional mortality that a population can accommodate without a severe reduction in size cannot be predicted. In other words, we do not know how populations respond to cumulative environmental stress. The relationship may be direct, there may be some partial compensation, or the population may fully compensate up to a point and then decline very rapidly (Figure 21-2). Because many fish populations already are heavily exploited, particularly in nearshore areas, their capacity to compensate for additional sources of stress may be limited (Vaughan 1981).

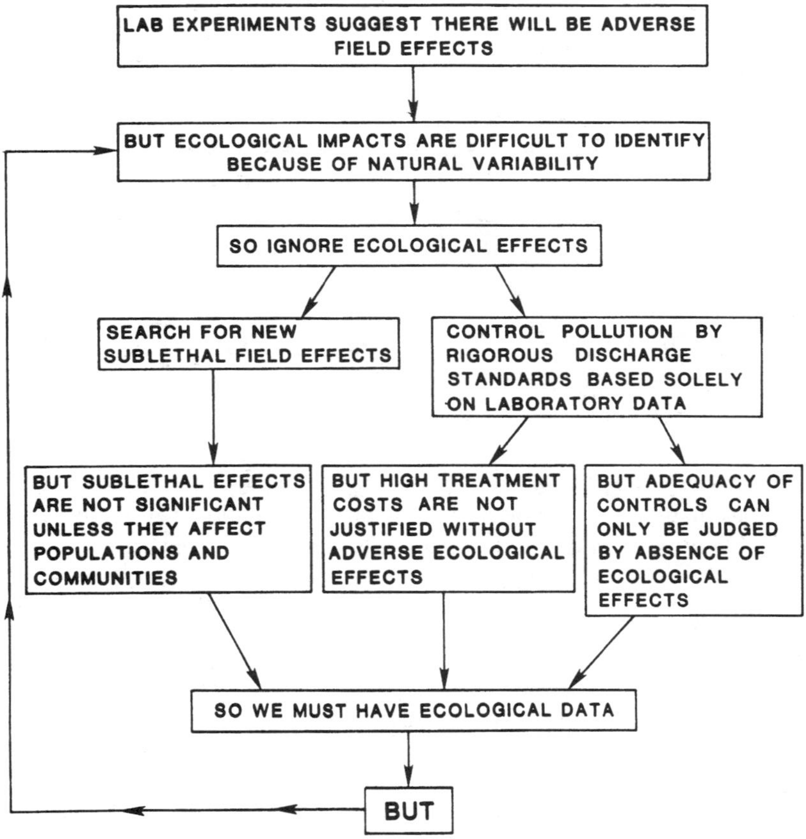

Figure 21-3. Conceptual diagram illustrating the difficulty in assessing effects of pollution in the natural environment. (Reprinted from F. A Cross, D. S. Peters, and W. E. Schaaf. Implications of waste disposal in coastal waters on fish populations. Aquatic Toxicology and Hazard Assessment: Seventh Symposium. ASTM STP 854. Philadelphia: American Society for Testing and Materials, 1985. pp. 383–399.)

We believe that the four problems described above are largely responsible for our lack of progress in assessing environmental stress; our current dilemma (Figure 21-3) calls for new approaches. We suggest that increased emphasis in three areas will be very helpful.

First, we need to document or measure environmental changes on appropriate geographic scales—i.e., on scales that relate to the problems in large marine ecosystems. As the U.S. population shifts from the interior part of our country to the coastal regions, it is accompanied by increased development (more pavement, more marinas, more sewage outfalls), which affects both the amount of habitat and environmental or water quality. This, in turn, may be having some unquantifiable effect on our coastal fisheries. This change is not localized but is occurring along our entire coastline; thus, it is essential that we conduct programs that can address environmental change on a regional scale. We believe that programs described elsewhere in this book (see, for example, Chapters 13 and 20) are an important first step in developing an environmental data base on an appropriate scale that we can relate to changes in fisheries.

The next step is to develop methods to infer the effects of environmental stress on fish populations. Recent examples of the effort to evaluate environmental effects on winter flounder populations include Boreman et al. (1993) and Ingham and Pawlowski (1992). Integrating pollution-related mortality terms into population models (Vaughan et al. 1984) and individually based modeling approaches being developed at Oak Ridge National Laboratory may (1) allow an estimation of the effects of environmental stress on fishery populations that may already be stressed by high exploitation; (2) permit assessment of multiple stresses, such as a substantial pollution event coincident with generalized habitat degradation; and (3) lead to the use of derived parameters that indicate the susceptibility of populations to additional stress (Schaaf et al. 1987, 1993). There are few studies that compare growth (e.g., Burke et al. 1993), fecundity and mortality (e.g., Slooff and de Zwart 1983) among polluted and less polluted waters; more are needed.

If our interest is in fish productivity, then we clearly need a better understanding of fish production systems—how they are structured, how they function and vary—in both natural and polluted or degraded environments. Empirical models (e.g., Peters and Schaaf 1991) are needed that include information such as species composition and magnitude of the fishery harvest, delineation of the major food webs, sources of primary production, and flux of organic matter through the populations of yield species. It is unlikely that such models will provide reliable estimates of expected change because of pollution over the next decade (e.g., will total yield decrease or remain stable but change in species composition?). Recognizing that several fisheries are involved in most systems, and that the number of compensatory mechanisms and interactions among the fishery species and their supporting food webs is large, mechanistic prediction of system response may not be practical. We are not promoting empirical modeling of marine ecosystems as a tool of immediate management utility, but it could serve as the basis for documenting variability within and among fishery systems, leading to inferences about cause and effect.

In conclusion, we believe that past progress in relating environmental stress to fish production was inhibited by too narrow an approach. However, this approach is broadening and we are beginning to acquire the necessary information. For example, programs being developed are beginning to obtain relevant ecological information over proper spatial and temporal scales and to find ways to relate this information to changes in populations and ecosystems. Also, increased understanding of natural fishery systems should allow us to interpret this information in a useful way. If we are to successfully address this complex issue, it is essential that we acquire long-term data sets for both fishery and environmental information, and that research, monitoring, and synthesis efforts be conducted on regional scales that approximate the geographical range of many economically important fisheries. This will require multidisciplinary efforts that integrate the fields of toxicology, population dynamics, and ecosystem dynamics.

References

Boreman, J., Correia, S. J., and Witheral, D. B. 1993. Effects of changes in age-0 survival and fishing mortality on egg production of winter flounder in Cape Cod Bay. American Fisheries Society Symposium 14:39–45.

Burke, J. S., Peters, D. S., and Hanson, P. J. 1993. Morphological indices and otolith microstructure of Atlantic croaker, *Micropogonias undulatus*, as indicators of habitat quality along an estuarine pollution gradient. Environ. Biol. Fish. 36:25–33.

Goodyear, C. P. 1980. Compensation in Fish Populations. In C. H. Hocutt and J. R. Stauffer (eds.), Biological Monitoring of Fish. Lexington, MA: D. C. Heath, pp. 253–280.

Heath, A. G. 1990. Summary and perspectives. American Fisheries Society Symposium 8:183–191.

Ingham, M. C., and Pawlowski, R. J. 1992. Modeling Environmental Effects on Winter Flounder Populations. Report of Northeast Environmental Assessment Workshop No. 2. Gloucester, MA: Nattional Marine Fisheries Service.

Jones, R. 1982. Population fluctuation and recruitment in marine populations. Philos. Trans. R. Soc. Lond. [B]297:353–368.

Levin, S. 1985. A Definition of Stress? In G. P. Moberg (ed.), Animal Stress. Bethesda, MD: American Physiological Society, pp. 51–69.

Lewis, J. R. 1980. Options and problems in environmental management and evaluation. Helgol. Wiss. Meeresunters. 33:452–466.

Odum, E. P. 1971. Fundamentals of Ecology. Philadelphia: W.B. Saunders.

Peters, D. S., and Schaaf, W. E. 1991. Empirical model of the trophic basis for fishery yield in coastal waters of the eastern USA. Trans. Am. Fish. Soc. 120:459–473.

Ryder, R. A., and Kerr, S. R. 1989. Environmental Priorities: Placing Habitat in Hierarchic Perspective. Spec. Publ. Can. J. Fish. Aquat. Sci. 105:2–12.

Schaaf, W. E., Peters, D. S., Coston-Clements, L., et al. 1993. A simulation model of how life history strategies mediate pollution effects on fish populations. Estuaries 16(4):697–702.

Schaaf, W. E., Peters, D. S., Vaughan, D. S., et al. 1987. Fish population responses to chronic and acute pollution: The influence of life history strategies. Estuaries 10:267–275.

Slooff, W., and de Zwart, D. 1983. The growth, fecundity and mortality of bream, *Abramis brama*, from polluted and less polluted surface waters in the Netherlands. Sci. Total Environ. 27:149–162.

Vaughan, D. S. 1981. An Age Structure Model of Yellow Perch in Western Lake Erie. In D. G. Chapman and V. F. Gallucci (eds.), Quantitative Population Dynamics. Fairland, MD: International Co-operative Publishing, pp. 189–216.

Vaughan, D. S., Merriner, J. V., and Schaaf, W. E. 1986. Detectability of a reduction in a single year-class of a fish population. J. Elisha Mitchell Sci. Soc. 102:122–128.

Vaughan, D. S., Yoshiyama, R. M., Breck, J. E., and DeAngelis, D. L. 1984. Modeling Approaches for Assessing the Effects of Stress on Fish Populations. In V. W. Cairns, P. V. Cairns, and J. O. Nriagu (eds.), Contaminant Effects on Fisheries. New York: John Wiley, pp. 259–278.

22

Biotoxins and the Health of Living Marine Resources of the Northeast Shelf Ecosystem

Alan W. White

Over the past 20 years or so, biotoxins produced by unicellular marine algae have become of increasing concern in terms of fisheries resources and public health. This is concomitant with an apparent global trend of intensification and spreading of toxic marine algal blooms and red tides (Anderson 1989; Smayda 1990). Marine algal toxins have an impact on the utilization of fisheries resources because they accumulate in some animals (e.g., molluscan shellfish, crustaceans, fish, etc.) to levels that jeopardize human health, thus requiring sanctions against harvesting. In addition, as emphasized by recent dramatic events worldwide, the toxins directly affect the health of marine animals, causing mass mortalities or debilitation of animals both in the wild and in aquaculture operations (Shumway 1990; White 1988). This chapter focuses on the effects of biotoxins, particularly "paralytic shellfish toxins," on the health of living marine resources in the northeast region of the United States.

Biotoxin Concerns in the Northeast Shelf Ecosystem

The Northeast Shelf ecosystem spans the region from the Gulf of Maine to Cape Hatteras. Within this region, the biotoxins of greatest impact on marine animals are the so-called *paralytic shellfish toxins* produced by the dinoflagellates *Alexandrium* (formerly *Gonyaulax*) *tamarense* and *A. fundyense*. Blooms and red tides of these toxic dinoflagellates occur nearly annually in the northeast region, and the toxins have been implicated in mass kills of fish, birds, and marine mammals, and in the illness of humans through paralytic shellfish poisoning (PSP).

A biotoxin of potential but not yet demonstrated impact on the health of marine animals in the Northeast is domoic acid, which is produced by various species of diatoms of the genus *Nitzschia* and recently has been found to be responsible for amnesic shellfish poisoning (ASP) in Prince Edward Island (Bates et al. 1989; Perl et al. 1990). In the fall of 1991, domoic acid caused mass mortalities of pelicans and cormorants in California when the birds ate anchovies contaminated with the toxin as a result of feeding on toxic diatoms (Work et al. 1993). Low levels of domoic acid have been detected recently in molluscan shellfish in the Northeast (J. Nassif, personal communication 1991), but whether this biotoxin affects the health of living marine resources in this region remains to be determined.

Brevetoxins, produced by the marine dinoflagellate *Gymnodinium breve* and long associated with mass kills of fish and with neurotoxic shellfish poisoning (NSP) in the Gulf of Mexico, first appeared along the North Carolina coast in the winter of 1987–1988, requiring the closure of important shellfisheries in that area (Tester et al. 1991). Brevetoxins in fish may have been related to the mass mortalities of bottlenose dolphin along the East Coast in 1987, and these toxins in sea squirts have been implicated in the death of manatees on Florida's Gulf Coast (Anderson and White 1989).

There are several other marine algae in the Northeast that can adversely affect marine animals, perhaps via toxins. The chrysophyte flagellate *Gyrodinium aureolum*, associated with the mass mortalities of marine animals in northwestern Europe, has formed intense blooms on several occasions in the Northeast, some of which have been implicated in animal deaths (Mahoney et al. 1990; Heinig and Campbell 1992). The brown tide organism, *Aureococcus anophagefferens*, which has caused extensive damage to the bay scallop industry on Long Island, may exert its action through a biotoxin (Cosper and Carpenter 1990). A strain of *Prorocentrum* spp. isolated from Long Island Sound causes a reduction in growth of hard clams and bay scallops, and there are indications that a toxin is involved (Wikfors and Smolowitz 1993). The first reported instance of diarrhetic shellfish poisoning (DSP) in North America occurred in August 1991 on the Atlantic coast of Nova Scotia as a result of a bloom of the dinoflagellate *Prorocentrum lima* (Marr et al. 1992). It is unknown whether the diarrhetic toxins, produced by several cosmopolitan coastal dinoflagellates, affect marine animals.

Finally, a newly discovered and as yet unnamed dinoflagellate has been determined to be the cause of many estuarine and coastal fish kills in the North Carolina area (Burkholder et al. 1992). Steidinger et al. (1996) describe this organism as a new species, *Pfeisteria piscicida*. The alga has a novel mode of action. Its excystment is stimulated in the presence of live fish. It then excretes a potent ichthyotoxin, feeds on the tissues of the dead fish, and then quickly encysts. The geographic distribution of this previously unknown organism is now being investigated along the East Coast.

Paralytic Shellfish Toxins

Paralytic shellfish toxins are produced by several species of red-tide dinoflagellates that occur in many locations around the world. In New England and eastern Canada, these toxins are produced by *Alexandrium tamarense* (=*excavata*) and *A. fundyense*. The toxins consist of a suite of structurally similar compounds closely related to and including saxitoxin. The toxins are modified purines. Their role in the physiology and ecology of the dinoflagellates is unknown. These compounds are among the most potent neurotoxins known, especially to vertebrate animals. The toxins block sodium channels in nerve and muscle membranes, causing paralysis. Death results from asphyxiation.

Paralytic Shellfish Toxins in the Northeast

For at least hundreds of years, *Alexandrium* blooms have occurred nearly annually in the Bay of Fundy and the St. Lawrence estuary, and the accompanying affliction of humans, PSP, has been recognized in these locations (Prakash et al. 1971). The consequences of PSP toxins for marine animals are not limited to inshore areas because the blooms and red tides of the toxic dinoflagellates occur over the entire central and southern Bay of Fundy, from Nova Scotia to New Brunswick (Martin and White 1988).

An intense bloom of *Alexandrium tamarense* was first observed along the coast of New England in the fall of 1972, extending southward as far as the south coast of Cape Cod and Narragansett Bay. Since then, toxic blooms have occurred in the coastal waters of New England nearly every year, as indicated by the occurrence of the toxins in inshore shellfish. The yearly blooms contaminate shellfish from Maine to northern Massachusetts, sometimes extending to areas south of Boston. On occasion the toxins have been detected, at low levels, in shellfish from the coasts of Connecticut and Long Island.

Until recently, little was known about the seaward extent of *Alexandrium* blooms in the Gulf of Maine. In 1989, PSP toxins were detected in surf clams and ocean quahogs on Georges Bank at levels exceeding the public health safety threshold, leading to the first shellfishery closure in offshore waters in the

Northeast (White et al. 1993). Another offshore bloom occurred in the spring of 1990. In June of that year, eight fishermen suffered PSP after eating mussels taken as bycatch from Georges Bank.

Paralytic Shellfish Toxins in the Food Web

Over the years, it has been learned that the toxins from *Alexandrium* spp. are transferred through the marine food web and can occur in a wide array of marine animals (Figure 22-1 and White 1984), sometimes with devastating effects. Although we are becoming increasingly aware that these toxin transfer events occur during and after *Alexandrium* blooms, we have little quantitative knowledge of the fate of the toxins in the marine food web.

Effects of Paralytic Shellfish Toxins on Marine Animals

The risk that paralytic shellfish toxins pose to humans, and the resulting cases of human poisoning, have been well documented. We are becoming increasingly aware that these toxins pose a significant risk to marine animals as well. Through the food web routes depicted in Figure 22-1, paralytic shellfish toxins have caused mass mortalities of fish and birds and have also been implicated in the mass mortalities of marine mammals.

In general, vertebrate animals are sensitive to paralytic shellfish toxins whereas invertebrates are relatively insensitive via oral exposure. Hence, invertebrates that ingest *Alexandrium* cells or toxin-containing animals are able to accumulate the toxins and then serve as toxin transvectors to sensitive animals.

MOLLUSCAN SHELLFISH. Filter-feeding molluscs are notorious for their ability to accumulate high levels of paralytic shellfish toxins on ingestion of *Alexandrium* cells, thus posing a danger to vertebrates that consume them. Marine molluscs are particularly insensitive to these toxins and even high levels do not kill them, although some physiological and metabolic effects may be detectable at high toxin levels (Gainey and Shumway 1988; Twarog and Yamaguchi 1975). Carnivorous molluscs, such as whelks and moonsnails, which acquire the toxins secondarily from filter feeders, are also able to accumulate high levels of the toxins. Mussels and sea scallops (the viscera) can contain extraordinary levels of these toxins, with little apparent effect. Many shellfish show no aversion to ingestion of *Alexandrium* cells, and, in fact, the blooms serve as a good source of food. Studies suggest that Alaska butter clams have evolved to exploit their ability to accumulate extremely high levels of the toxins in their siphons as a chemical defense mechanism against vertebrate predators, including siphon-nipping fish, sea otters, and gulls (Kvitek 1993).

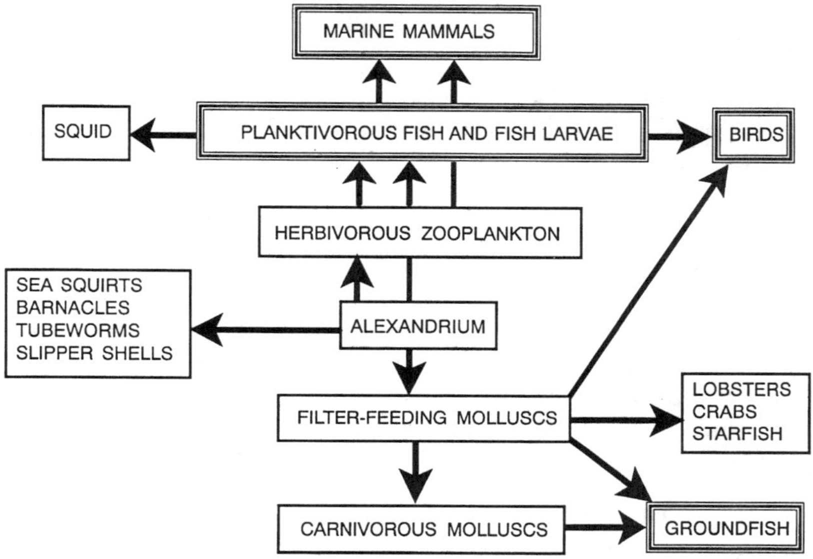

Figure 22-1. Pathways of transmission of Alexandrium toxins (= paralytic shellfish toxins) in the marine food web. Plain boxes represent animals that are relatively insensitive to the toxins via oral exposure. Framed boxes represent animals that are sensitive to the toxins; mass kills of planktivorous fish, birds, and marine mammals have occurred.

CRUSTACEAN SHELLFISH. Recently, paralytic shellfish toxins have been detected at moderate levels in the viscera of lobsters in the Northeast (J. Hurst, personal communication 1991) and in crabs from Puget Sound (Jonas-Davies and Liston 1985). In Japan, the presence of toxins in various crabs has been known for many years (Hashimoto et al. 1967). The toxins seem to have no effect on lobsters and crabs via the oral route (Yentsch and Balch 1975; Foxall et al. 1979). This insensitivity is not related to insensitivity of the nerves, because these animals (and shrimp) are very sensitive to the toxins on injection directly into their tissues (Foxall et al. 1979; Hwang et al. 1990). The larvae of lobsters and crabs are apparently also immune to the toxins when feeding on *A. tamarense* (Robineau et al. 1991a; Yazdandoust 1985). It appears, then, that PSP toxins have little direct consequence for the health of crustacean shellfish.

ZOOPLANKTON. Zooplankton are major grazers of dinoflagellate blooms. Field and laboratory studies have shown that a wide variety of zooplankters, including tintinnids, copepods, cladocerans, nauplii, and pteropods, feed on toxic *Alexandrium* cells and accumulate the toxins (White 1977, 1979, 1980, 1981a, 1984). The effects of the toxins on zooplankton require elucidation. Some studies report no apparent effects on various invertebrate zooplankters from ingestion of toxic *Alexandrium* cells (Robineau et al. 1991b; White 1981a); other studies show a decreased level of ingestion of the toxic cells by certain copepods (Huntley et al. 1986; Ives 1985).

During *Alexandrium* blooms, these cells can overwhelmingly dominate the phytoplankton community for several weeks. During this period, herbivorous zooplankton, including fish larvae that are sensitive to the toxins (see below), have little alternative but to eat *Alexandrium* cells and thereby acquire the toxins. These contaminated zooplankton then serve as vectors of PSP toxins to zooplanktivorous fish (such as herring and mackerel) and other planktivorous animals, including certain marine mammals, such as right whales. Field evidence indicates that this phenomenon of toxin transfer into the zooplankton community accompanies each *Alexandrium* bloom (White 1979, 1984), and calculations suggest that only a few grams of contaminated zooplankton may serve as a lethal meal for planktivorous fish (White 1984).

FISH. Fish in general are sensitive to paralytic shellfish toxins, and exposure to relatively low levels of the toxins can cause their death. The lethal oral and intraperitoneal doses for fish are not much different than for warm-blooded vertebrates (Evans 1972; White 1981b). Within minutes of exposure, fish show symptoms of poisoning, including loss of equilibrium, irregular swimming, and finally immobilization and labored, arrhythmic breathing. For other vertebrates, death is the result of asphyxiation. There is evidence that some fish may be able to detect the toxins in their prey and thus avoid poisoning (Kvitek 1993; Yamamori et al. 1988).

Paralytic shellfish toxins have caused massive kills of adult herring, menhaden, and sand lance in the Northeast. Hundreds of tons of adult herring died in the Bay of Fundy during the *Alexandrium* bloom in the summer of 1976 as a result of toxin transfer through pteropods (White 1977). The toxins also caused herring kills in the Bay of Fundy in 1979, this time involving the cladoceran, *Evadne nordmanni*, as a transvector (White 1979).

In 1978, PSP toxins were associated with a kill of sand lance, bluefish, dogfish, skates, and monkfish off Wellfleet, Massachusetts (Nisbet 1983). The toxins were also associated with a menhaden kill off the coast of Maine in 1979 (J. Hurst, personal communication 1980), probably resulting from the direct ingestion of the toxic dinoflagellates by these phytoplanktivorous fish. In this case, the toxins were also detected in the viscera of squid that had been feeding on the menhaden.

From the point of view of human health, there is limited concern about paralytic shellfish toxins in fish because the toxins do not accumulate in fish flesh; fish die when traces of the toxins reach their muscle tissues (White 1981b). However, the toxins can occur in the viscera of fish, and in this way present a danger to animals (such as other fish, marine mammals, and, in some cases, humans) that eat whole fish, including the viscera. In this connection, it should be mentioned that mackerel throughout much of the Northeast have recently been found to contain low to moderate levels of the toxins in their livers (Geraci et al. 1989; Haya et al. 1990; C. Martin, personal communication 1988), raising questions regarding the source or sources of the toxins and the ability of the mackerel to accumulate the toxins without harm.

In sum, PSP toxins generally have a lethal effect on fish. We need to learn more about the dynamics of this process to understand why blooms sometimes result in fish kills and sometimes not. We also need to examine the consequences of chronic exposure to sublethal toxin doses for the health of fish populations. For example, the temporary debilitation that results from sublethal doses (White 1981b) of the toxins may be associated with higher predation rates. Chronic exposure to sublethal levels of PSP toxins may relate to the unexplained mechanism by which Atlantic mackerel in the Northeast store the toxins in their livers.

FISH LARVAE. Perhaps the greatest impact of PSP toxins on the health of fish populations concerns their effects on fish larvae and juveniles. In the midst of toxic *Alexandrium* blooms, fish larvae have little alternative but to eat *Alexandrium* cells or zooplankton that have eaten the toxic cells. Laboratory studies show that marine fish larvae (including herring, mackerel, capelin, cod, winter flounder, red sea bream, and Japanese anchovy) are killed by eating toxic *Alexandrium* cells or by eating zooplankton that have eaten the toxic cells (Gosselin et al. 1989; Mills and Klein-Macphee 1979; Robineau et al. 1991a, b; White 1982; White et al. 1989). Calculations suggest that ingestion of only one or a few toxic cells is sufficient to affect first-feeding larvae.

It appears likely, then, that *Alexandrium* blooms and paralytic shellfish toxins can influence the survival of fish larvae and juveniles. The significance of these toxic events to fish populations may be considerable, especially considering the recurrent nature of the blooms and the possible cumulative effects on recruitment. Assessing the overall impact of these toxic blooms on fish populations, however, is difficult (White 1982, 1984).

BIRDS. Paralytic shellfish toxins have caused massive kills of marine birds on a number of occasions. In 1968 off the northeast coast of England, nearly 80% of the breeding population of shags and other birds died after eating sand lance containing the toxins acquired from an *Alexandrium* bloom (Adams et al. 1968; Coulson et al. 1968). During the first recorded toxic red tide in New England, in 1972, more than 1,600 black ducks and 600 waterfowl, gulls, and shorebirds died after eating filter-feeding molluscs (Sasner et al. 1975). In 1978 off Wellfleet (Cape Cod), Massachusetts, a kill of breeding terns and other birds occurred after they ate sand lance containing paralytic shellfish toxins (Nisbet 1983).

Seabirds may be able to learn to avoid the toxins. Kvitek (1993) found that wild glaucous-winged gulls regurgitated highly toxic butter clams, but not non-toxic clams, and that at a chronically toxic site, gulls took significantly fewer butter clams than at a non-toxic site. Also, in the Bay of Fundy and St. Lawrence estuary area, where toxic *Alexandrium* blooms have been a persistent feature for at least hundreds of years, there are no reports of mass mortalities of seabirds. It may be that birds quickly learn to avoid the toxins, and thus, massive kills of birds occur only in instances where the blooms are rather new events.

MARINE MAMMALS. The herring kills in the Bay of Fundy in the 1970s and the discovery that the zooplankton community became contaminated with *Alexandrium* toxins each year raised the question of whether marine mammals that eat large quantities of zooplankton or whole fish can be affected by these toxins (White 1984). An answer was soon to come. In the late fall of 1987 off Provincetown, Massachusetts (the tip of Cape Cod), 14 humpback whales died and were washed ashore. Paralytic shellfish toxins were detected in the stomachs, livers, and kidneys of the whales, and also in Atlantic mackerel that the whales had eaten (Anderson and White 1989; Geraci et al. 1989). Further tests found moderate levels of the toxins in mackerel over a wide area in the Northeast (Geraci et al. 1989; C. Martin, personal communication 1988). The evidence and resulting calculations point to the toxins, accumulated in mackerel viscera, as the likely cause of the kill.

It appears, then, that the health of marine mammals can be endangered by PSP toxins in the food web. We are just at the first stage of knowledge in this area. Relatively low doses of the toxins may prove fatal for marine mammals if the loss of equilibrium interferes with the animal's ability to orient to the surface for breathing. With the toxic blooms perhaps expanding into offshore waters, we may see more biotoxin events involving marine mammals. However, there is some evidence that marine mammals, like birds, may learn to avoid the toxins. Sea otters appear to be able to learn to switch prey from highly toxic butter clams or to discard the highly toxic parts (the siphons) of the clams (Kvitek 1993).

Conclusion

Biotoxins from the New England red tide dinoflagellate *Alexandrium tamarense* accumulate in a wide variety of animals within the marine food web and, through a number of food web routes, can cause mass kills of marine vertebrates. There is little information on the distribution of *Alexandrium* cells and the dynamics of the toxic blooms in the central and offshore waters of the Gulf of Maine. Recent events involving the widespread occurrence of the toxins in mackerel throughout the Northeast, the implication of the toxins in a kill of humpback whales, and the apparently first occurrence of the toxins in Georges Bank shellfish suggest that the toxic blooms are now more prevalent in the offshore waters of the Gulf of Maine than in the past, and that paralytic shellfish toxins are having an increasing impact on the health of living marine resources throughout much of the Northeast Shelf ecosystem.

References

Adams, J. A., Seaton, D. D., Buchanan, J. B., and Longbottom, M. R. 1968. Biological observations associated with the toxic phytoplankton bloom off the East Coast. Nature 220:24–25.

Anderson, D. M. 1989. Toxic Algal Blooms and Red Tides: A Global Perspective. In T. Okaichi, D. M. Anderson, and T. Nemoto (eds.), Red Tides: Biology, Environmental Science, and Toxicology. New York: Elsevier, pp. 11–16.

Anderson, D. M., and White, A. W. (eds.). 1989. Toxic Dinoflagellates and Marine Mammal Mortalities. Woods Hole Oceanog. Inst. Tech. Rept. WHOI-89-36 (CRC-89-6). Woods Hole, MA: Woods Hole Oceanographic Institute.

Bates, S. S., Bird, C. J., deFreitas, S. W., et al. 1989. Pennate diatom *Nitzschia pungens* as the primary source of domoic acid, a toxin in shellfish from eastern Prince Edward Island, Canada. Can. J. Fish. Aquat. Sci. 46:1203–1215.

Burkholder, J. M., Noga, E. J., Hobbs, C. H., and Glasgow, H. B. 1992. New "phantom" dinoflagellate is the causative agent of major estuarine fish kills. Nature 358:407–410.

Cosper, E. M., and Carpenter, E. J. 1990. Novel "Brown Tide" Blooms in Long Island Embayments: A Search for the Causes. In E. Granéli, B. Sundström, L. Edler, and D. M. Anderson (eds.), Toxic Marine Phytoplankton. New York: Elsevier, pp. 17–28.

Coulson, J. C., Potts, G. R., Deans, I. R., and Fraser, S. M. 1968. Exceptional mortality of shags and other seabirds caused by paralytic shellfish poison. Brit. Birds 61:381–404.

Evans, M. H. 1972. Tetrodotoxin, Saxitoxin, and Related Substances: Their Applications in Neurobiology. In C. C. Pfeiffer and J. R. Smythies (eds.), International Review of Neurobiology (Vol. 15). New York: Academic Press, pp. 83–166.

Foxall, T. L., Shoptaugh, N. H., Ikawa, M., and Sasner, J. J. 1979. Secondary Intoxication with PSP in *Cancer irroratus*. In D. L. Taylor and H. H. Seliger (eds.), Toxic Dinoflagellate Blooms. New York: Elsevier/North-Holland, pp. 413–418.

Gainey, L. F., and Shumway, S. E. 1988. A compendium of the responses of bivalve molluscs to toxic dinoflagellates. J. Shellfish Res. 7:623–628.

Geraci, J. R., Anderson, D. M., Timperi, R. J., et al. 1989. Humpback whales (*Megaptera novaeangliae*) fatally poisoned by dinoflagellate toxin. Can. J. Fish. Aquat. Sci. 46:1895–1898.

Gosselin, S., Fortier, L., and Gagné, J. A. 1989. Vulnerability of marine fish larvae to the toxic dinoflagellate *Protogonyaulax tamarensis*. Mar. Ecol. Prog. Ser. 57:1–10.

Hashimoto, Y., Konosu, S., Yasumoto, T., et al. 1967. Occurrence of toxic crabs in Ryukyu and Amami Islands. Toxicon 5:85–90.

Haya, K., Martin, J. L., Waiwood, B. A., et al. 1990. Identification of Paralytic Shellfish Toxins in Mackerel from the Southwest Bay of Fundy, Canada. In E. Granéli, B. Sundström, L. Edler, and D. M. Anderson (eds.), Toxic Marine Phytoplankton. New York: Elsevier, pp. 350–355.

Heinig, C. S., and Campbell, D. E. 1992. The environmental context of a *Gyrodinium aureolum* bloom and shellfish kill in Maquoit Bay, Maine, September 1988. J. Shellfish Res. 11:11–122.

Huntley, M., Sykes, P., Rohan, S., and Marin, V. 1986. Chemically mediated rejection of dinoflagellate prey by the copepods *Calanus pacificus* and *Paracalanus parvus*: Mechanism, occurrence and significance. Mar. Ecol. Prog. Ser. 28:105–120.

Hwang, D. F., Chueh, C. H., and Jeng, S. S. 1990. Susceptibility of fish, crustacean, and mollusk to tetrodotoxin and paralytic shellfish poison. Nippon Suisan Gakkaishi 65:337–343.

Ives, J. D. 1985. The Relationship Between *Gonyaulax tamarensis* Cell Toxin Levels and Copepod Ingestion Rates. In D. M. Anderson, A. W. White, and D. G. Baden (eds.), Toxic Dinoflagellates. New York: Elsevier, pp. 413–418.

Jonas-Davies, J., and Liston, J. 1985. The Occurrence of PSP Toxins in Intertidal Organisms. In D. M. Anderson, A. W. White, and D. G. Baden (eds.), Toxic Dinoflagellates. New York: Elsevier, pp. 467–472.

Kvitek, R. G. 1993. Paralytic Shellfish Toxins as a Chemical Defense in the Butter Clam (*Saxidomus giganteus*). In T. J. Smayda and Y. Shimizu (eds.), Proceedings of the Fifth International Conference on Toxic Marine Phytoplankton. Amsterdam: Elsevier, pp. 407–412.

Mahoney, J. B., Olsen, P., and Cohn, M. 1990. Blooms of a dinoflagellate *Gyrodinium* cf. *aureolum* in New Jersey coastal waters and their occurrence and effects worldwide. J. Coastal Res. 6:121–135.

Marr, J. C., Jackson, A. E., and McLachlan, J. L. 1992. Occurrence of *Prorocentrum lima*, a DSP toxin-producing species from the Atlantic coast of Canada. J. Appl. Phycol. 4:17–24.

Martin, J. L., and White, A. W. 1988. Distribution and abundance of the toxic dinoflagellate *Gonyaulax excavata* in the Bay of Fundy. Can. J. Fish. Aquat. Sci. 45:1968–1975.

Mills, L. J., and Klein-Macphee, G. 1979. Toxicity of the New England Red Tide Dinoflagellate to Winter Flounder Larvae. In D. L. Taylor and H. H. Seliger (eds.), Toxic Dinoflagellate Blooms. New York: Elsevier/North-Holland, pp. 389–394.

Nisbet, I.C.T. 1983. Paralytic shellfish poisoning: Effects on breeding terns. Condor 85:338–345.

Perl, T. M., Bédard, L., Kosatsky, T., et al. 1990. An outbreak of toxic encephalopathy caused by eating mussels contaminated with domoic acid. N. Engl. J. Med. 322:1775–1780.

Prakash, A., Medcof, J. C., and Tennant, A. D. 1971. Paralytic shellfish poisoning in eastern Canada. Bull. Fish. Res. Board Can. 177:87.

Robineau, B., Gagné, J. A., Fortier, L., and Cembella, A. D. 1991a. Potential impact of a toxic dinoflagellate (*Alexandrium excavatum*) bloom on survival of fish and crustacean larvae. Mar. Biol. 108:293–301.

Robineau, B., Fortier, L., Gagné, A., and Cembella, A. D. 1991b. Comparison of the response of five larval fish species to the toxic dinoflagellate *Alexandrium excavatum* (Braarud) Balech. J. Exp. Mar. Biol. Ecol. 152:225–242.

Sasner, J. J., Ikawa, M., and Barrett, B. E. 1975. The 1972 Red Tide in New Hampshire. In V. R. LoCicero (ed.), Proceedings of the First International Conference on Toxic Dinoflagellate Blooms. Wakefield, MA: Massachusetts Science and Technology Foundation, pp. 517–523.

Shumway, S. E. 1990. A review of the effects of algal blooms on shellfish and aquaculture. J. World Aquaculture Soc. 21:65–104.

Smayda, T. J. 1990. Novel and Nuisance Phytoplankton Blooms in the Sea: Evidence for a Global Epidemic. In E. Granéli, B. Sundström, L. Edler, and D. M. Anderson (eds.), Toxic Marine Phytoplankton. New York: Elsevier, pp. 29–40.

Steidinger, K. A., Burkholder, J. M., Glasgow, H. B., Jr., et al. 1996. *Pfeisteria piscicida* gen et. sp. nov. (Pfeisteriaceae fam. nov.), a new toxic dinoflagellate with a complex life cycle and behavior. J. Phycology 32:157–164.

Tester, P. A., Stumps, R. P., Vukovich, S. M., et al. 1991. An expatriate red tide bloom: Transport, distribution, and persistence. Limnol. Oceanogr. 36:1053–1061.

Twarog, B.M., and Yamaguchi, H. 1975. Resistance to Paralytic Shellfish Toxins in Bivalve Molluscs. In V. R. LoCicero (ed.), Proceedings of the First International Conference on Toxic Dinoflagellate Blooms. Wakefield, MA: Massachusetts Science and Technology Foundation, pp. 381–393.

White, A. W. 1977. Dinoflagellate toxins as probable cause of an Atlantic herring (*Clupea harengus harengus*) kill, and pteropods as apparent vector. J. Fish. Res. Board Can. 34:2421–2424.

White, A. W. 1979. Dinoflagellate Toxins in Phytoplankton and Zooplankton Fractions During a Bloom of *Gonyaulax excavata*. In D. L. Taylor and H. H. Seliger (eds.), Toxic Dinoflagellate Blooms. New York: Elsevier/North-Holland, pp. 381–384.

White, A. W. 1980. Recurrence of kills of Atlantic herring (*Clupea harengus harengus*) caused by dinoflagellate toxins transferred through herbivorous zooplankton. Can. J. Fish. Aquat. Sci. 37:2262–2265.

White, A. W. 1981a. Marine zooplankton can accumulate and retain dinoflagellate toxins and cause fish kills. Limnol. Oceanogr. 26:103–109.

White, A. W. 1981b. Sensitivity of marine fishes to toxins from the red-tide dinoflagellate *Gonyaulax excavata* and implications for fish kills. Mar. Biol. 65:255–260.

White, A. W. 1982. The scope of impact of toxic dinoflagellate blooms on finfish in Canada. Can. Tech. Rept. Fish. Aquat. Sci., No. 1063. St. Andrews, New Brunswick: Canadian Dept. of Fisheries and Oceans.

White, A. W. 1984. Paralytic Shellfish Toxins and Finfish. In E. P. Ragelis (ed.), Seafood Toxins. American Chemical Society Symposium Series, No. 262. Washington, DC: American Chemical Society, pp. 171–180.

White, A. W. 1988. Blooms of Toxic Algae Worldwide: Their Effects on Fish Farming and Shellfish Resources. In Proceedings of the International Conference on Impact of Toxic Algae on Mariculture. Aqua-Nor '87 International Fish Farming Exhibition, 13–18 August 1987, Trondheim, Norway, pp. 9–14.

White, A. W., Fukuhara, O., and Anraku, M. 1989. Mortality of Fish Larvae from Eating Toxic Dinoflagellates or Zooplankton Containing Dinoflagellate Toxins. In T. Okaichi, D. M. Anderson, and T. Nemoto (eds.), Red Tides: Biology, Environmental Science, and Toxicology. New York: Elsevier, pp. 395–398.

White, A. W., Nassif, J., and Shumway, S. E. 1993. Recent Occurrence of Paralytic Shellfish Toxins in Offshore Shellfish in the Northeastern United States. In T. J. Smayda and Y. Shimizu (eds.), Proceedings of the Fifth International Conference on Toxic Marine Phytoplankton. Amsterdam: Elsevier, pp. 435–440.

Wikfors, G. H., and Smolowitz, R. M. 1993. Detrimental Effects of a *Prorocentrum* Isolate upon Hard Clams and Bay Scallops in Laboratory Feeding Studies. In T. J. Smayda and Y. Shimizu (eds.), Proceedings of the Fifth International Conference on Toxic Marine Phytoplankton. Amsterdam: Elsevier, pp. 447–452.

Work, T. M., Beale, A. M., Fritz, L., et al. 1993. Domoic Acid Intoxification of Brown Pelicans and Cormorants in Santa Cruz, California. In T. J. Smayda and Y. Shimizu (eds.), Proceedings of the Fifth International Conference on Toxic Marine Phytoplankton. Amsterdam: Elsevier, pp. 643–649.

Yamamori, K., Nakamura, M., Matsui, T., and Hara, T. J. 1988. Gustatory responses to tetrodotoxin and saxitoxin in fish: A possible mechanism for avoiding marine toxins. Can. J. Fish. Aquat. Sci. 45:2182–2186.

Yazdandoust, M. H. 1985. *Cancer* Crab Larvae and Goby Fish: Vector and Victim of Paralytic Shellfish Poisons. In D. M. Anderson, A. W. White, and D. G. Baden (eds.), Toxic Dinoflagellate Blooms. New York: Elsevier, pp. 419–424.

Yentsch, C. M., and Balch, W. 1975. Lack of secondary intoxication by red-tide poison in the American lobster, *Homarus americanus*. Environ. Lett. 9:240–254.

23

Emergent Stressors and Public Health Implications in Large Marine Ecosystems: An Overview*

Paul R. Epstein

Curious changes have been taking place, with many animals invading this cold-temperate zone from the south and pushing up through Maine and even up into Canada. This new distribution is, of course, related to the widespread change of climate that seems to have set in about the beginning of the century and is now well recognized—a general warming-up noticed first in arctic regions, then in subarctic, and now in the temperate areas of northern states.

Rachel Carson, The Edge of the Sea

Abstract

From the rocky northern coasts and mid-latitude sandy beaches to Florida's coral-ringed beaches, the Mississippi Delta, and the Gulf of Mexico, North America's marine ecosystems are endangered by local and global human activities. Excess nutrient effluents, overharvesting

*This report was prepared for the Climate and Policy Assessment Division of the U.S. Environmental Protection Agency's Office of Policy, Planning, and Evaluation.

of marine resources, habitat degradation, and climate changes affecting sea surface temperatures and ocean circulation are altering the composition of large marine ecosystems. These changes are manifested in a "global epidemic" of coastal algal blooms of greater persistence and toxicity to fish, shellfish, and humans (Smayda 1990).

Toxic chemicals such as polychlorinated biphenyls are monitored in many near-coastal systems, but coastal micro-ecosystems involve chemicals and trace metals, nutrients, algae, bacteria, and viruses. Recently a set of new seafood biotoxins have emerged. These include domoic acid-induced amnesic shellfish poisoning on both coasts of North America, an "ambush predator" of fish spreading north- and southward from the Pimlico estuary, and a brown tide off Long Island that first crippled the scallop industry and recently became toxic, with a viral agent suspected. Elsewhere, the explosive outbreak of cholera (using algae as a reservoir) along the west coast of Latin America in 1991, and the appearance of a new, environmentally hardy cholera variant ("0139 Bengal") in Bangladesh in 1993, are indicative of increasingly vulnerable coastal ecosystems worldwide.

Algal mats, besides amplifying infectious pathogens and permitting the development of new biotoxins, also cause anoxia in fish, shellfish, marine mammal, and seabird nurseries, thus influencing future stocks. Finally, coastal algal blooms are increasingly fouling tourist and recreation areas in many nations. Algae may be seen as key biological indicators of ecosystem health, responding rapidly to a multiplicity of environmental changes. The condition of the viral-bacterial-algal microsystem that forms the base of the marine food web is crucial to the health of all the other components in the system. The financial costs of the growth in toxic blooms and the species shifts within them (associated with warming and stratification in the water column) are now being felt in terms of health outcomes and health care expenditures, surveillance and inspection systems, food safety, food commerce and allied food industries, and tourism. Improved ecosystem management, including control of agricultural and industrial discharges in watershed areas, preservation of wetlands, establishment of marine reserves, and prudent climate change policies, is necessary to protect coastal marine ecosystems, preserve their future productivity, and prevent an acceleration in hazards to public health.

Large marine ecosystems (LMEs) along continental coasts worldwide provide nearly 95% of the living marine resources used annually by humans (Sherman 1993). Primary production is concentrated in LMEs, where it is consistently greater than in the open-ocean (pelagic) areas of the globe. Large marine ecosystems are large areas, on the order of 200,000 km^2, and are defined by distinct bathymetry, hydrography, productivity, and trophically dependent populations. The LMEs along the North American coast include (1) the Eastern Bering Sea, (2) the Gulf of Alaska, (3) the California Current, (4) the Gulf of California, (5) the Gulf of Mexico, (6) the Southeast U.S. Continental Shelf, (7) the Northeast U.S. Continental Shelf, (8) the Scotian Shelf, and (9) the Newfoundland Shelf (see Figure 1-1, Chapter 1). Most North American LMEs subsume coastal waters of more than one state; coordination among states and through federal agencies is therefore necessary for resource monitoring and management.

The food web within marine ecosystems includes nutrients, phytoplankton, zooplankton, macro-algae and numerous intermediate-sized forms, shellfish, molluscs, finfish, sea mammals, and marine birds. Perturbations anywhere along the food chains can have cascading reverberations throughout the web. Multiple perturbations to a system may profoundly affect ecosystem structure, stability, resilience, and productivity. While the health and integrity of LMEs are threatened by the "global epidemic" of coastal algal blooms (Smayda 1991; Anderson 1992), fresh water bodies are also experiencing an overgrowth of micro-algae (including cyanobacteria) and macro-algae (seaweeds).

Environmental Factors

Anthropogenic inputs and resource depletion are the driving forces for environmental and climate change affecting LMEs. Repetitively, local perturbations affect ecosystems; when these activities involve biotic feedbacks and occur on a global scale (e.g., fossil fuel pollutants and deforestation), they affect the climate system. The risk assessment approach for this overview combines climate impacts, which can result in changes in algal abundance and species composition, with evaluation of ecosystem vulnerability to these changes.

Sunlight, pH, ocean currents, winds, and river runoff govern the precise location and timing of plankton blooms. The major anthropogenic influences are (1) eutrophication and pollutants, (2) overharvesting, (3) loss of habitat, (4) atmospheric/oceanic warming and algal growth, (5) climate variability and storm frequency, (6) changing winds and currents, (7) ultraviolet light and plankton, (8) chemical composition of the sea, (9) sea level rise, (10) increased ultraviolet-B radiation, (11) mariculture, and (12) ballast water. This section focuses on issues pertinent to North America, but includes others important to tropical marine ecosystems.

Eutrophication and Pollutants

Excessive sewage and fertilizers from domestic, agricultural, and industrial sources are the primary sources of marine eutrophication; soil erosion, chemicals (from agricultural deforestation) and acid precipitation add additional nitrogen and phosphates. Excessive phosphates biologically used to make ATP ("energy packets") may encourage the more active algal groups (e.g., dinoflagellates). Toxic chemicals such as polychlorinated biphenyls (PCBs), heavy metals, and pesticides bioaccumulate as they move up the food chain. In addition to the direct pathophysiological effects on organisms, contaminants harm constituents differentially along the chain, and alter ecosystem structure and equilibrium. Oil slicks and solid plastics endanger sea mammals and marine birds, thus altering predation pressures.

Overharvesting

The overharvesting of finfish (e.g., cod off eastern Canada and New England; salmon off Alaska, British Columbia, Oregon, and Washington) and shellfish reinforces algal growth by reducing algivorous grazing. Nine of the world's 17 major fisheries are in serious decline, with those in the northeast and northwest LMEs of North America among the most disrupted. Contrary to the impact of natural selection on fish populations and communities, the initial symptoms of overexploitation usually include a decline in average fish size and the disappearance of larger species (Bohnsack 1993).

Loss of Habitat

Wetlands ("nature's kidneys") filter nitrogen and phosphorus, store carbon, and support fish and sea bird colonies. Coastal and inland salt marshes and seagrasses are disappearing through development practices such as drainage and landfill; over the past several decades, California has lost 90% of its coastal wetlands. Mangroves in many nations are disappearing due to mariculture and coastal development. Coral reefs, sources of enormous biodiversity, are being widely mined for construction materials. Sea surface warming and other stresses directly harm coral reefs, causing the expulsion of the algal symbiont and leaving bleached polyps behind (Hayes and Goreau 1991). Coral reefs close to Africa, Southeast Asia, and Florida and in the Caribbean are in critical condition. Bleaching during warm El Niño years has been especially extensive worldwide. In the summer of 1995, the Belize Barrier reef experienced bleaching for the first time, as SSTs exceeded 29°C (Hayes and Goreau 1996). Subsurface habitats such as seagrasses can be disrupted by pollution and by warmer ocean temperatures (see below).

Atmospheric/Oceanic Warming and Algal Growth

As early as 1955, Rachel Carson (*The Edge of the Sea*) presented data on the northward movement of marine species because of sea surface warming occurring this century. Although there is a debate concerning sea surface temperatures (SSTs) based on "seatruth" data (ship sampling via the Combined Ocean Atmospheric Data Set—COADS), multispectral sea surface temperature (MSSST) data obtained by satellite imaging, and expected data from the acoustic temperature ocean experiment (ATOC) of Scripps Oceanographic Institution, there is general agreement that sea surface temperature has risen 0.5°C since 1850, mirroring the same degree rise in global atmospheric temperature. In the Atlantic, Pacific, and Indian oceans, there is preliminary evidence of warming at deeper levels (Parrilla et al. 1994; Thwaites 1994; Travis 1994).

Warming SSTs reduce dissolved oxygen and, within limits, stimulate photosynthesis, metabolism (two- to sixfold for each 10°C increase), and growth. (Excessive warming can kill off some families or species of algae entirely.) Moreover, warming and stratification in the water column appears to favor shifts toward pico-phytoplankton—the cyanobacteria and dinoflagellates—groups containing members more toxic to predators (Valiela 1984). The resultant decrease in predation pressure may add a further positive feedback (increase) on algal growth, whereas the shift in species structure within the marine community can lead to more expected toxic species and greater numbers of biotoxic events.

Warming and increased SSTs affect the global hydrological cycle. It is projected that a doubling of CO_2 will increase evaporation and precipitation 7–15%, and recent evidence suggests that atmospheric water vapor is up (IPCC 1995). Because water vapor is the chief greenhouse gas, increased atmospheric moisture will increase tropospheric warming for a further positive feedback.

Climate Variability and Storm Frequency

The Intergovernmental Panel on Climate Change (IPCC) (Houghton et al. 1990) projects more frequent and more intense extreme weather events with climate change. Similar projections are made by three coupled atmospheric-oceanic general circulation models (Meehl and Washington 1993; Manabe and Stouffer 1993; Bengtsson et al. 1993). There is now evidence that extreme events in the United States have increased in the past decade (Karl et al. 1995). In addition, recently extracted ice-core records (GRIP 1993) indicate that climate instability, variability, and "jumps" to new regimes (i.e., variations around new means) are possible within just a few years.

Climate instability and changes in storm patterns may present the greatest threat to coastlines and coastal ecosystems. The chief climate signal associated with El Niño–Southern Oscillation (ENSO) phenomena is a warm SST anomaly of up to 2–3°C, maintained for up to one year. These oceanic anomalies may affect the jet stream, and thus global weather patterns of precipitation—floods, droughts, and storms—and temperatures. The increase in the number and duration of ENSO events since 1983 has had profound impact on Pacific marine life (Wooster and Fluharty 1985), and the unexpected rewarming of the Pacific in austral spring, 1993 (Climate Analysis Center 1994) and jet stream changes may paradoxically have been related to the sustained arctic air mass that enveloped North America in January, 1994. The El Niño lasting 5 years and 8 months (1990–1995) is unprecedented in the historical record (Trenberth and Hoar 1996).

Changing Winds and Currents

In addition to altered SSTs, the increasing contrasts between land and sea temperatures have been and will continue to be associated with changes in alongshore coastal winds, sea breezes, and ocean circulation (Bakun 1993). Along western continental boundaries, upwelling currents will increase equatorially, whereas along eastern seaboards, winds and currents are expected to increase poleward. Multidecadal records show an increasing trend in upwelling and along-shore (inward-blowing or advecting) wind stress already occurring in the Californian, Peruvian, Iberian, and Benguelan LMEs. Changes in ocean circulation may have been responsible for past climate shifts (Broecker 1987). The shift in the Mediterranean circulation in 1995 must be considered as a dramatic change worthy of intense inspection (Roether 1996).

Chemical Composition of the Sea

As a whole, the oceans contain 50 times as much carbon (C) as the atmosphere, and 17 times the C in the terrestrial biosphere. In a warmer ocean, pCO_2 of the water will increase, causing CO_2 outgassing. For a temperature change of +1°C, the change in pCO_2 is on the order of 10 μatm, a decreased CO_2 uptake by the ocean that could amplify future atmospheric CO_2 by 5% (IPCC 1994). Oceanic CO_2 concentrations also affect the acid–base balance of the ocean and could reduce the amount of dissolved calcium carbonate available to reef-building corals.

Changes in ocean circulation with either greater stratification or increased upwelling (there is great uncertainty in this area) may not disrupt the "Redfield ratio" of nutrients (C:N:P in a molar ratio of 106:16:1), but delivery of such nutrients in a different composition from land effluents and atmospheric deposition may substantially influence species composition at the base of the food chain. With changes in net ecosystem production (i.e., the ratio of respiration to production), the production of trace gases such as dimethyl sulfide (which is important in cloud formation, precipitation, and—along with industrial sulfate aerosols—tropospheric temperature control); and concentrations of N_2O, CO, NH_3, and chlorofluorocarbons may vary. With warming, the composition and distribution of some species (e.g., Coccolithophoridae, salps) may change, as well as the fluxes to the ocean of iron, molybdenum, and other bio-essential trace metals. Enhanced decomposition of dissolved organic C (a large oceanic reservoir, on the order of 1,000 gigatons) may occur. Finally, any changes in freshwater runoff will alter ocean salinity (normally 32 parts per thousand), thus potentially affecting regional marine communities.

Carbon Dioxide in Freshwater

In many ponds and some lakes, devoid of dissolved carbonates, CO_2 becomes a limiting factor for algal growth. Thus, increased atmospheric CO_2 may fertilize aquatic organisms having chlorophyll, just as it does on land, when other nutrients are in abundance (water being unlimited). This may have an impact on the growth of cyanobacteria and other toxic phytoplankton, as well as the algal reservoir for human pathogens in sewage lagoons.

Increased Ultraviolet B Radiation

Ultraviolet radiation can cause mutations in DNA. Additionally, a series of experiments demonstrate that phytoplankton are capable of producing pigments (e.g., chlorophyll and carotenes) that provide protection against DNA damage from ultraviolet-B radiation (Bothwell et al. 1994; Carreto and Benevides 1993). This makes sense from a historical perspective, as cyanobacteria are at least 3.45 billion years old, according to Australian fossil records—that is, they evolved before the ozone shield that is a by-product of tropospheric O_2. Increased ultraviolet radiation may therefore affect fish larvae and zooplankton differentially, thus further destabilizing ecological relationships, perhaps increasing phytoplankton populations due to decreased grazing.

Sea Level Rise

Sea level rises of 5 cm (with a range of 1–9 cm) per decade may have significant impacts on coastal marine ecosystems, the most important repercussion being the increased loss of wetlands essential for filtering effluents and toxins and storing carbon. Wetlands also serve as nurseries for fish, shellfish, and marine birds. Gradual changes, combined with potential increases in storm surges, may influence the salinity of groundwater and would affect the extent and duration of coastal algal blooms and related health impacts.

Mariculture

Mariculture, an industry of growing importance to many nations for food consumption and export, contributes an additional set of problems. Mangrove forests are destroyed to create controlled and concentrated breeding grounds, and nutrients and antibiotics (in an attempt to control contamination) are discharged excessively into the systems. Red tides are frequent in such altered ecosystems and contaminate the mariculture products. Many systems are "crashing" due to several viruses and *Vibrio* bacteria. The systems themselves are highly vulnerable, and there are questions as to their long-term sustainability.

Ballast Water

The transport of marine organisms through ballast water is another human factor that plays a role in the dispersion of potentially disease-causing oceanic flora and fauna (Rigby et al. 1993; Carlton and Geller 1993). Where conditions permit the persistence of introduced organisms, such "hierarchical" dispersion of toxic algae and bacterial symbionts into new settings is an additional public health hazard. This mechanism may have been involved in seeding the rich coastal plankton blooms off Peru with *Vibrio cholerae* in 1991. Monitoring of a nation's coastal ecosystems thus intersects with issues of international transport, regulation of which falls under the purview of the International Maritime Organization (IMO).

Summary of Stressor Events

During the last decade, several "natural experiments" have occurred. Periodic warming events driven by regional climate shifts, often associated with El Niño events, have favored recruitment, survival, and consequent population increases of fish species in the South Atlantic. In the 1987 ENSO, for example, shifts in advecting winds and changes in SSTs in the Benguela Current LME off the west coast of Africa were associated with a toxic phytoplankton bloom, an extensive fish kill, and cascading effects on marine mammals and sea birds (Crawford et al. 1989). Other ecosystems where the fish community appears to have responded to changes in temperature signals are in the North Atlantic: the Iberian LME and the Canary Current LME (Bakun 1993).

In the Pacific, anomalous sea surface temperatures and an altered pattern of advecting winds and currents overwhelmed the cool, rich, upwelling Humboldt current. This affected Peruvian sardine and anchovy yields for years afterward (Alheit and Bernal 1993).

The 1986–1987 El Niño was also associated with new growth patterns of tropical and temperate species in higher northern and southern latitudes. In 1987, after a shoreward intrusion of Gulf Stream eddies, the dinoflagellate *Gymnodinium breve*, previously found as far north as the Gulf of Mexico, bloomed off Cape Hatteras, North Carolina, where it has persisted at low levels ever since. In the same year, anomalous rain patterns (high sediment runoff) and warm Gulf Stream eddies sweeping unusually close to Prince Edward Island, combined with other local factors to generate a bloom of the pennate diatom (*Nitzschia pungens* f. *multiseries*), and for the first time, domoic acid was produced, killing five Canadian mussel consumers and causing at least 107 cases of amnesic shellfish poisoning. Several persons were left with permanent short-term memory loss. Also in 1987, massive seagrass die-offs began in the central part of Florida Bay (Barley 1993), while on the beaches of the North Atlantic coast, numerous dolphins and whales died, with algal NSP (neurological shellfish poisoning) toxins (via mackerel?), PCBs, and warming possibly combining to reduce immunity to viruses. This convergence of events in 1987 suggests a strong role for climatic signals. Ongoing studies are reviewing these global algal data and climate signals (M. Altalo, P. Epstein, personal communication 1994).

The following summary of disease and other adverse events in United States LMEs provides examples of continued ecosystem degradation and vulnerability to environmental stressors.

Scotian Shelf

Closures of commercial shellfish harvesting areas because of the accumulation of PSP (paralytic shellfish poisoning) toxins are an annual occurrence in the Bay of Fundy and are caused there by *Alexandrium fundyense* blooms. In 1988, for the first time, ASP (amnesic shellfish poisoning) toxin was detected in shellfish in the southwest Bay of Fundy. Highest densities of *Pseudonitzschia delicatissima* (which produces the domoic acid responsible) were observed in four consecutive years, during September of 1987 through 1990, and in June of 1991 and 1992 (Martin et al. 1993).

In 1992, PSP was reported from the Grande Entrée Lagoon and the Îles-de-la-Madeleine in the Gulf of St. Lawrence, the major mussel farming sites in Quebec, and an area traditionally free of phycotoxins. The causative organisms, *Alexandrium* spp. with PSP toxins, were probably introduced through ballast water, since they were also found in the ballast tanks of ships docked at Mienes Selein Pier in June (Gosselin et al. 1993).

A new seafood, the Northern Moonshell, growing in the southern part of the Gulf of St. Lawrence has been found to contain PSP toxins since 1989 (Worms et al. 1993). *Buccinum undatum,* another carnivorous mollusc on Georges Bank between Nova Scotia and Massachusetts, has also been found to be toxic (White 1993). These toxins were unknown in that region before 1988.

Northeast U.S. Continental Shelf

Extensive biomass flips have been reported, including depletion of herring and mackerel, with replacement by sand eels in the 1970s and subsequent recovery of herring and mackerel in the 1991–1993 period, after a major decrease in fishing effort from the early 1980s to the present. The groundfish species of high economic value, including cod, haddock, and flounder, have been seriously overexploited since the mid-1970s, and are now at the lowest levels of record (see Chapters 9 and 27). The biofeedback response of the fish community within the ecosystem has been an unprecedented increase in small elasmobranchs (spiny dogfish and skates), for which only a very limited market demand exists (Sherman 1993; see also Chapter 28).

PARALYTIC SHELLFISH POISONING. In New England in 1972, red tides associated with paralytic shellfish poisoning (PSP) initiated much of the worldwide, modern era concern over algal blooms. Since then, periodic PSP outbreaks along the Northeast Shelf have occurred; monitoring is carried out by the Woods Hole Oceanographic Institution and marine stations in Maine, with sampling guided by satellite images from the Advanced Very High Resolution Radiometer (AVHRR) instruments on NOAA satellites (Keafer and Anderson 1993). For the past few years, commercially valuable shellfish resources in the Georges Bank region have been consistently contaminated with PSP saxitoxins, resulting in a continuous ban on the harvesting of shellfish from this productive offshore region. Dock-side testing has been proposed as a means of monitoring (White 1993).

In the summer of 1992, after a long, warm, calm period, *Alexandrium tamarense* blooms developed, causing PSP in Newfoundland (Hockin et al. 1993). In 1985, *A. tamarense* was detected in Long Island Sound, its most southern occurrence in eastern U.S. waters (Nuzzi and Waters 1993). New occurrences of PSP toxins have recently been reported in the southern Gulf of St. Lawrence, Canada (Worms et al. 1993). In 1990, PSP toxins were found in sufficiently toxic levels in crabs from the United Kingdom that their sale was banned after complaints of illness. In the same year, hepatopancreases of lobsters from the Atlantic Coast of Canada, mainly in the Gaspé area of Quebec, were found to contain high levels of PSP toxins (Todd 1993b; Desbiens and Cembella 1993). In 1993, PSP toxins were also found in British Columbia rock scallops (IOC 1992).

Fish as well as shellfish are subject to toxicity. For instance, in 1991 in Quebec, a small outbreak of illness traceable to consumption of swordfish occurred. Amines related to histamine were found in the fish, presenting a potential new area of concern (Todd and Holmes 1993).

TOXIC BROWN TIDE. Brown tide has existed in Peconic Bay at the eastern tip of Long Island since the mid-1980s, where it has crippled the scallop industry, killed eelgrass beds, and stunted the growth of clams (Nuzzi and Waters 1993). In the fall of 1993, it was reported at the Environmental Management of Enclosed Coastal Seas conference in Baltimore that the brown tide had become toxic. A virus was implicated (Cosper 1993). In the summer of 1993, brown tide was reported on Cape Cod beaches.

MARINE MAMMALS. In the past decade, there have been numerous whale and dolphin kills along the Northeast coast associated with a combination of viruses, algal toxins, and chemicals (e.g., PCBs), the last causing immune suppression. These die-offs of marine mammals were especially high in 1987 (an El Niño year), coincident with the translocation of *Gymnodinium breve* and the appearance of domoic acid amnesic shellfish poisoning in Canada. Warming may cause immune suppression and/or change distribution of food sources.

AMNESIC SHELLFISH POISONING. In November and December of 1987, Prince Edward Island (PEI), Canada reported its first confirmed outbreak of amnesic shellfish poisoning (ASP), attributed to the consumption of cultured mussels contaminated with domoic acid (an amino acid causing gastroenteritis and central nervous system damage). The event resulted in at least 107 poisonings, three deaths, and some persons left with permanent memory loss. In October 1991, a small bloom of *Pseudonitzschia delicatissima* or *multiseries* was recorded from northern PEI for the first time. In 1987 (an El Niño year), heavy rains led to increased sediments and effluents, and warm eddies of the Gulf Stream stayed closed to the coast. Toxin-producing blooms have also been documented from the coasts of Maine and Texas (Todd 1993a).

Southeast U.S. Continental Shelf

NEUROLOGICAL SHELLFISH POISONING. Since 1987, *Gymnodinium breve* organisms associated with neurological shellfish poisoning (NSP) have been present in the Cape Hatteras region (Tester et al. 1991). This transport and persistence marks a translocation of a tropical organism into higher, temperate latitudes.

Such species movement is also occurring elsewhere in the world. In 1991, paralytic shellfish toxins appeared for the first time in the Straits of Magellan, at the southern extreme of South America (Cosper 1993). In 1993, multiple shellfish toxins affected humans for the first time in New Zealand (Chang 1993). Many other areas, including France, Scandinavia, and the Philippines, have experienced such phenomena in the past 5 years.

THE AMBUSH PREDATOR. In 1992, a new phenomenon resulting in large fish kills (>1,000) occurred in the Pimlico estuary. The species responsible, *Pfiesteria piscimorte,* has 15 stages in its complex lifestyle. As an amoeboid form in bottom mud, it scavenges dead and decaying biological material. At a water temperature increased experimentally to 15°C (a condition found in shallow, turbid estuaries), the organism becomes a dinoflagellate that surfaces and causes disorientation, skin loss, subcutaneous hemorrhaging, and apparent suffocation in fish (Burkholder et al. 1993). The "ambush predator" has recently been associated with fish kills near Delaware (J. Burkholder, personal communication 1996).

Fish kills in the Chesapeake Bay are also associated with other algal species. In July 1992, a widespread bloom of *Ceratium furca* developed over 272 km^2 of the bay and persisted for a month. Algal blooms are, in general, becoming more extensive and long-lasting in the Chesapeake Bay. Traditional crab populations in the Bay have all but disappeared, and all marine resources are threatened by prolonged hypoxic conditions caused by surface algal mats (Marshall 1993).

In September 1994, an outbreak of *Gymnodinium breve* and resultant extensive fish kills began on Florida beaches. In February 1995, Karen Steidinger, of Florida's Department of Environmental Protection, reported an unusual winter outbreak that traveled with the Gulf Stream into the Atlantic Ocean (Reuters 1995).

Gulf of Mexico and Caribbean Sea

The Gulf of Mexico and Florida waters harbor *Gymnodinium breve* and 25 other species of dinoflagellates, seven species of *Chloromodas,* three species of other flagellates, and potentially toxic diatoms. Tampa Bay has 11 toxic or potentially toxic dinoflagellate species, in addition to *Chloromodas* and other flagellates. Neurological shellfish poisoning and ciguatera poisoning have been confirmed in Florida waters; *Heterosigna chattonela* is repeatedly found during summer months in brackish water environments and most likely contributes to fish kills. Toxic benthic (bottom) species, especially dinoflagellates, are associated with inshore tropical areas and reefs where both coral and micro-algal substrates are found. Thus, the Gulf of Mexico and Florida waters, with their coral reefs, nutrient-enriched canals, and associated flora and fauna, offer a full mosaic of environments within which many toxic species exist (Tomas et al. 1993). Satellite images demonstrate that runoff from the 1993 Mississippi floods is now being carried by the Gulf Stream up the Atlantic coast; the impact of the organisms and chemicals carried in this assault may not have been felt until the spring of 1994.

VIBRIO SPECIES. Shellfish-associated outbreaks of gastroenteritis have implicated a variety of pathogens, including *Vibrio* spp., *Salmonella typhi, Campylobacter* spp., hepatitis A, and Norwalk-like viruses. Those who are immunocompromised or who have other chronic problems (e.g., alcoholism; hepatic, gastrointestinal, or hematologic disorders; cancer; diabetes; kidney disease) may be susceptible to severe disease.

Vibrio cholerae 01, biotype El Tor, associated with the seventh pandemic of cholera and the explosive outbreak in Latin America beginning in January 1991 (Epstein 1993), and non-01s have been isolated from ballast water and from oysters originating in the Gulf of Mexico (CDC 1993a). DNA fingerprinting has demonstrated genetic identity with those strains causing the outbreak in Latin America (DePaola et al. 1992). Several cases of *V. cholerae* non-01, choleratoxin+ (dubbed "0139 Bengal") that emerged in Asia in 1993 have been imported into the United States, but there is no evidence of endemicity.

Vibrio vulnificus has also been found in oysters from the Gulf of Mexico. This bacteria has been associated with illness and deaths in persons with preexisting liver disease (CDC 1993b).

VIRAL GASTROENTERITIS. In 1993, the consumption of oysters harvested in the Gulf of Mexico caused an outbreak of 180 cases of viral gastroenteritis and led to a massive recall by the United States and Canada of oysters originating from two specific coral reefs (CDC 1993c). On November 16, 1993, the Louisiana Department of Health and Hospitals identified the Grande Pas Cabbage Reef harvesting areas off the Louisiana coast as the source of oysters associated with outbreaks of viral gastroenteritis in Louisiana, Mississippi, Maryland, and North Carolina. (Cases in Florida were linked to consumption of oysters from harvest areas other than the Louisiana coast.) By December 2, 23 separate outbreaks involving 180 people and four states had occurred. Because oysters from the beds implicated in this outbreak had been shipped to at least 14 states as well as Canada, this involved a major recall.

CIGUATERA FISH POISONING. Numerous cases of ciguatera fish poisoning (CFP) are reported regularly from the Caribbean, and the number is increasing. Ciguatera fish poisoning is associated with algal toxins traversing five trophic levels to reach humans. The World Health Organization reports over 200,000 cases of ciguatera in humans annually, primarily from the Caribbean and French Polynesia.

NEUROLOGIC SHELLFISH POISONING. In North America, neurologic shellfish poisoning (NSP) has largely been confined to the Gulf of Mexico, but since 1987, NSP has also been found in North Carolina (see Southeast U.S. Continental Shelf system) (Tester et al. 1993).

Gulf of California

A large number of dinoflagellates (*Gonyaulax polyedra, Gymnodinium splendens, Prorocentrum micans, Dinophysis* spp.) and diatoms (*Nitzschia pungens* and *Eucampia zodiacus*) have been present for years between Baja California and Mexico. Explosive reproduction favored by the presence of warm temperatures and the availability of organic and inorganic nutrients results in "hot spots," which often dissipate when wind strength is enough to break the stratification of warm surface water.

PARALYTIC SHELLFISH POISONING. In the fall of 1989, red tides of *Pyrodinium bahamense* var. *compressum* and several other variants occurred on the Pacific coast of Mexico, causing paralytic shellfish poisoning (PSP). Sixty-three persons in Mexico were intoxicated by eating contaminated shellfish, and three died. Large oysters (*Ostrea iridescens*), mussels (*Choromytylus palliopunctatus*), and barnacles (*Pollicipes polinerus*) were found to be contaminated with the saxitoxins (Navarro and Cuéllar 1993).

California Current

In October 1991, razor clams (*Siliqua patula*) from the Washington and Oregon coasts were found to be contaminated with domoic acid, an excitoxic amino acid causing the newly emerged syndrome of amnesic shellfish poisoning (see Scotian Shelf ecosystem). In early December, domoic acid was also isolated from the viscera of Dungeness crabs (*Cancer magister*) in Pacific coastal waters. Recreational and commercial harvest of the clams was prohibited in October, and the commercial fisheries closed in December. Domoic acid levels remained high until at least May 1992.

The presence of potentially toxic diatoms (*Pseudonitzschia peragallo, P. pungens* f. *multiseries* Hasle and *P. australis frenguelli*) signals a new problem with regard to toxic phytoplankton and public health on the U.S. West Coast (Horner et al. 1993). *Pseudonitzschia* spp. have been found in Monterey Bay, California, but were not reported to adversely affect wildlife or humans until the fall of 1991. In 1991, a massive bloom of *P. australis* led to acute accumulation of domoic acid in anchovies and the deaths of pelicans and cormorants, the first reported domoic acid toxic event in western North America (Worms et al. 1993). Domoic acid from *Pseudonitzschia* spp. may have occurred earlier in Southern California waters and gone undetected; 270 ppm domoic acid was found in the viscera of frozen anchovies caught in April 1991 between Newport and Los Angeles.

Gulf of Alaska

In September 1992, large numbers of dead king crabs (*Paralithodes camischatica*) were found along the beach of Captains Bay, on Unalaska Island, Alaska. The previous year, a large fish kill (>25,000 fish) and a small crab kill occurred in the same area. The 1992 event was found to be associated with phytoplankton blooms and hypoxic bottom conditions, along with pulses of heated, organic-laden water from a recently constructed fish processing plant. Examination of gills revealed the diatom *Chaetoceros convolutus;* the same organism was identified in surface phytoplankton blooms. Gill damage, caused by the spines of *C. convolutus* and exacerbated by hypoxic bottom water conditions, probably contributed to the king crab kill and to the fish kill reported the year before (Tester and Mahoney 1993).

Salmon populations in Northwest waters have declined significantly. Damming of rivers, deforestation of tributary watersheds, dioxin from paper mills, and overharvesting are the chief driving factors. Increased SSTs could severely dislocate remaining stocks, whereas other climatic changes in the hydrological cycle and storm patterns would further affect crucial spawning grounds.

Economic Analysis

Five types of costs are associated with seafood toxins: (1) those associated with human illness, (2) the costs of inspection and analysis of food products and monitoring of potentially toxic areas, (3) the losses of product sales as well as losses to associated industries because of closure of harvest areas or embargo of imported products, (4) the loss of long-term markets, and (5) reduction in tourism and recreation.

Direct Costs Linked to Foodborne Illness

Illness costs include health care, emergency transport, hospitalization, laboratory testing, epidemiologic investigation, and lives lost; individual costs include lost opportunities for work and leisure. Commercial costs include lost productivity, bans on harvesting, embargos, recalls and destruction of product, lost business, and losses to related industries (restaurants, hotels, food processors, catering companies, and institutions). Loss of product sales and closure of harvest areas is perhaps the most significant cost, and can result in widescale unemployment in coastal and fishing areas (e.g., the U.S. Northeast and almost all of the Alaskan coastline). Loss of long-term markets and export revenues can be extremely significant if seafood producing regions become identified as sources of toxins, and market substitutions are made by consuming nations (e.g., reduction of sales of Dungeness crabs from the U.S. West Coast to Japan because of domoic acid levels).

Loss of fish as food can be significant for many nations. In Indonesia, for example, fish supplies 60% of the protein consumed; for many small island nations, the percentage is close to 100%.

The estimated annual costs of seafood poisoning in Canada were calculated (Todd et al. 1992; Todd 1993a, b, c; Todd and Holmes 1993). The direct cost for 150 paralytic shellfish poisoning cases was over $225,000 in 1988, while monitoring amounted to $300,000. A 1987 episode of amnesic shellfish poisoning resulted in a cost of $57,700 per case for 145 cases, for a total loss of $8.4 million; the cost of implementing a domoic acid monitoring program in 1988 was an additional $1,390,000. Three hundred scombroid poisonings cost over $150,000 annually, and the expense of analyzing suspect lots of fish for the spoilage amines responsible was $27,000. For a possible 325 ciguatera poisonings each year, the annual cost is $350,000, and costs for diarrhetic and tetramine poisoning amount to $95,000 annually. Thus, the annual cost associated with an estimated 845 illnesses caused by seafood toxins is about $820,000, and an additional $4.7 million for product monitoring and analysis (Todd 1993c, 1994).

In the United States in 1989, losses for the estimated 58,200 cases of illness from PSP, scombroid poisoning, and ciguatera poisoning were $125 million; for 42,000 estimated *Vibrio* infections, it was $50 million (Todd 1989). Losses because of fish kills from algal biotoxins and temporary bans on harvesting are in the millions of dollars.

Costs to Tourism

Coastal regions used for recreation and tourism are threatened by coral reef damage and the expansion of algal blooms. Jamaica's northern coral reefs have virtually been destroyed and, worldwide, only reefs far from continents are neither threatened nor critical. The brown tide that has crippled the Long Island scallop industry may be spreading north to Cape Cod beaches. Expansive, toxin-emitting "red tides" have been reported off Florida and, increasingly, along the coastlines of Argentina, Chile, and Central America.

Tourist areas are vulnerable to terrestrial as well as marine disease vectors. In October 1993, for example, a sudden outbreak of 4,000 cases of dengue fever swept Costa Rica after record high temperatures and precipitation resulted in accelerated development and wider distribution of the vector mosquito, *Aedes aegypti* (L. Mata, personal communication 1994). Paleological records indicate that insects have flourished at times of climate transitions (Elias 1994). Costa Rica's tourism, allied services, industries, and travel are all put in jeopardy by the redistribution of disease vectors.

Conclusions

Changes in large marine ecosystems because of local practices and shifts in the world climate system are affecting marine resources, seafood safety, and health. The events described in this report require monitoring as well as intervention to support management decisions for ensuring the long-term development and sustainability of marine resources within these large ecosystems. The National Coastal Monitoring Act in the United States is one aspect of a national management system. The 1982 United Nations Convention on the Law of the Sea (UNCLOS) and the 1992 Declarations of the United Nations Conference on Environment and Development (UNCED) provide guidelines for improving degraded coastal ecosystems. Internationally, the Global Ocean Observing System (GOOS) of the International Oceanographic Commission of UNESCO is an effort of the international ocean science community to monitor marine ecosystems throughout the world.

The cost of the changing states of marine ecosystems to fish, shellfish, marine mammals, sea birds, and humans is increasing. Although small shifts may be controllable and livable, a large change, affecting multiple components of the marine ecosystem, damages resilience and may result in prolonged and perhaps permanent influences on fisheries (e.g., anchovies in Chile, cod and salmon in the United States) and consumer health (e.g., endemic cholera, new phytoplankton blooms). Local development policies that have an impact on these ecosystems include eutrophication, overharvesting of marine resources, and degradation of coastal habitats. These changes render our nation's LMEs more vulnerable to normal and variable climate factors. Changes in sea surface temperatures, in wind and current patterns, and in the patterns of storms are already influencing ecosystem dynamics; projected future changes in sea levels will exacerbate the threats examined in this report.

Health, seafood sources, commerce and livelihood, and tourism are all threatened by ecosystem degradation (NOAA 1992), compounded by changes in climate and in social/economic/political systems, with concomitant changes in population density, composition, and distribution. Insects, algae, and rodents are three rapidly responding and therefore key biological indicators of the structure and function of ecosystems. New seafood biotoxins, new viruses, and new forms of cholera are being increasingly reported. The growth in the number and extent of newly emerging diseases must be monitored.

Long-term integrated management of ecosystems is needed in order to plan adaptations and take more effective steps toward primary prevention. Necessary measures include (1) control of watershed effluents, (2) preservation and restoration of coastal wetlands, (3) establishment of more marine reserves, (4) maintenance of sustainable fish populations, and (4) implementation of policies to reduce the rate of climate change. The costs of "business as usual" are climbing.

References

(Note: A list of suggested additional reading is available from the author.)

Alheit, J., and Bernal, P. 1993. Effects of Physical and Biological Changes on the Biomass Yield of the Humboldt Current Ecosystem. In K. Sherman, L. M. Alexander, and B. D. Gold (eds.), Stress, Mitigation, and Sustainability of Large Marine Ecosystems. Washington, DC: AAAS Press, pp. 53–68.

Anderson, D. M. 1992. The fifth international conference on toxic marine phytoplankton: A personal perspective. Int. Mar. Sci. (UNESCO) 62(Harmful Algae News Suppl.):6–7.

Bakun, A. 1993. The California Current, Benguela Current, and Southwestern Atlantic Shelf Ecosystems: A Comparative Approach to Identifying Factors Regulating Biomass Yields. In K. Sherman, L. M. Alexander, and B. D. Gold (eds.), Stress, Mitigation, and Sustainability of Large Marine Ecosystems. Washington, DC: AAAS Press, pp. 199–221.

Barley, G. 1993. Integrated coastal management: The Florida Keys example from an activist citizen's point of view. Oceanus 36:15–18.

Bengtsson, L., Schlese, U., Roeckner, E., et al. 1993. A two-tiered approach to long-range climate forecasting. Science 261:1026–1029.

Bohnsack, J. A. 1993. Marine reserves: They enhance fisheries, reduce conflicts and protect resources. Oceanus 36:63–74.

Bothwell, M.L., Sherbot, D.M.J., and Pollock C.M. 1994. Ecosystem response to solar ultraviolet-B radiation: Influence of trophic-level interactions. Science 265:97.

Broecker, W. S. 1987. Unpleasant surprises in the greenhouse? Nature 328:123–126.

Burkholder, J. M., Glasgow, H. B., and Steidinger, K. A. 1993. Unraveling Environmental and Trophic Controls on Stage Transformations in the Complex Life Cycle of Anichthyotoxic "Ambush Predator" Dinoflagellate. Abstract presented at the Sixth International Conference on Toxic Marine Phytoplankton, 18–22 October 1993, Nantes, France.

Carlton, J. T., and Geller, J. B. 1993. Ecological roulette: The global transport of non-indigenous marine organisms. Science 261:78–82.

Carreto, J. I., and Benevides, H. R. 1993. World record of PSP in southern Argentina. Int. Mar. Sci. (UNESCO) 62(Harmful Algae News Suppl.):7.

Centers for Disease Control (CDC). 1993a. Isolation of *Vibrio cholerae* 01 from oysters: Mobile Bay, 1991–1992. MMWR 42:91–93.

Centers for Disease Control (CDC). 1993b. *Vibrio vulnificus* infections associated with raw oyster consumption: Florida, 1981–1992. MMWR 42:405–407.

Centers for Disease Control (CDC). 1993c. Multistate outbreak of viral gastroenteritis related to consumption of oysters: Louisiana, Maryland, Mississippi, and North Carolina, 1993. MMWR 42:945–948.

Chang, F.H. 1993. The First Records of *Gymnodinium* sp. *nov.* (cf. *breve*) and Other Harmful Phytoplankton Species in the Early 1993 Blooms in New Zealand. Abstract presented at the Sixth International Conference on Toxic Marine Phytoplankton, 18–22 October 1993, Nantes, France.

Climate Analysis Center. 1994. Long-lead Climate Forecasts. Washington, DC: National Weather Service, National Oceanic and Atmospheric Administration, U.S. Department of Commerce.

Cosper, E. 1993. Toxic Brown Tide Off Long Island. Presented at the Environmental Management of Enclosed Coastal Seas (EMECS) Conference, Baltimore, MD, November 1993.

Crawford, R. J. M., Shannon, L. V., and Shelton, P. A. 1989. Characteristics and Management of the Benguela as a Large Marine Ecosystem. In K. Sherman and L. M. Alexander (eds.), Biomass Yields and Geography of Large Marine Ecosystems. AAAS Selected Symposium 111. Boulder, CO: Westview Press, pp. 169–219.

DePaola, A., Capers, G. M., and Motes, M. L. 1992. Isolation of Latin American epidemic strain of *Vibrio cholerae* 01 from US Gulf Coast. Lancet 339:624.

Desbiens, M., and Cembella, A. D. 1993. Occurrence and Elimination Kinetics of PSP Toxins in the American Lobster (*Homarus americanus*). Abstract presented at the Sixth International Conference on Toxic Marine Phytoplankton, 18–22 October 1993, Nantes, France.

Elias, S. A. 1994. Quaternary Insects and Their Environments. Washington, DC: Smithsonian Institution Press.

Epstein, P. R. 1993. The role of algal blooms in the spread and persistence of human cholera. Biosystems 31:209–221.

Gosselin, S., Levasseur, M., and Gauthier, D. 1993. Transport and Introduction of Toxic Dinoflagellates via Ballast Water on the Eastern Atlantic Coast of North America. Abstract presented at the Sixth International Conference on Toxic Marine Phytoplankton, 18–22 October 1993, Nantes, France.

Greenland Ice-Core Project (GRIP). 1993. Climate instability during the last interglacial period recorded in the GRIP ice core. Nature 346:203–207.

Hayes, L., and Goreau, T. Coral bleaching in Belize; 1995. Presentation at the American Association for the Advancement of Science Annual Meeting, Baltimore, MD. Jan. 1996.

Hayes, L., and Goreau, T. J. 1991. The tropical coral reef ecosystem as a harbinger of global warming. World Resource Rev. 3:306–322.

Hockin, J., White, D., Spencer, K., et al. 1993. Paralytic shellfish poisoning: First cases in Newfoundland. Can. Dis. Wky. Rep. 9(7):25.

Horner, R. A., Postel, J. R., and Hinds, S. E. 1993. Harmful Diatoms in Western Washington Waters (USA). Abstract presented at the Sixth International Conference on Toxic Marine Phytoplankton, 18–22 October 1993, Nantes, France.

Houghton, J. T., Jenkins, G. L., and Ephraumis, J. J. (eds.). 1990. Climate Change: The IPCC Scientific Assessment. New York: Cambridge University Press.

Intergovernmental Oceanic Commission (IOC). 1992. Programme on Harmful Algal Blooms. IOC-SCOR Workshop, Newport, RI, USA, 2–3 November 1991. Workshop Report No. 80(SC92/WS-26), Annex VI. Paris: Intergovernmental Oceanographic Commission, UNESCO.

Intergovernmental Panel on Climate Changes (IPCC). 1994. Biotic Feedbacks in the Global Climate System: A Summary Statement from the Joint IPCC/Woods Hole Research Center Workshop, Woods Hole Research Center, Woods Hole, MA, 25–29 October, 1992.

Intergovernmental Panel on Climate Changes (IPCC). 1995: The Second Assessment Report. Cambridge, England: Cambridge University Press, 1995.

Karl, T.R., Knight, R.W., Easterling, D.R., and Quayle, R.G. 1995. Trends in U.S. climate during the 20th century. Consequences 1:3.

Karl, T.R., Knight, R.W., Plummer, N. 1995. Trends in high-frequency climate variability in the twentieth century. Nature 377:217.

Keafer, B. A., and Anderson, D. M. 1993. Use of Remotely-Sensed Sea Surface Temperatures in Studies of *Alexandrium tamarense* Bloom Dynamics. In T. J. Smayda and Y. Shimizu (eds.), Toxic Phytoplankton Blooms in the Sea. Amsterdam: Elsevier, pp. 763–768.

Manabe, S., and Stouffer, R. J. 1993. Century-scale effects of increased atmospheric CO_2 on the ocean–atmosphere system. Nature 364:215–218.

Marshall, H. G. 1993. The Seasonal Succession of Five Dinoflagellate Blooms in Chesapeake Bay, U.S.A. Abstract presented at the Sixth International Conference on Toxic Marine Phytoplankton, 18–22 October 1993, Nantes, France.

Martin, J. L., LeGresley, M. M., and Wildish, D. J. 1993. Environmental Factors and Occurrences of Organisms Responsible for PSP and ASP in Southwestern Bay of Fundy, Atlantic Canada. Abstract presented at the Sixth International Conference on Toxic Marine Phytoplankton, 18–22 October 1993, Nantes, France.

Meehl, G. A., and Washington, W. M. 1993. South Asian summer monsoon variability in a model with doubled atmospheric carbon dioxide concentration. Science 260:1101–1106.

National Oceanic and Atmospheric Administration (NOAA). 1992. Our Living Oceans: Report on the Status of U.S. Living Marine Resources. NOAA Tech. Memo NMFSF-SPO-2. Washington, DC: National Oceanic and Atmospheric Administration, U.S. Department of Commerce.

Navarro, O. S., and Dominguez Cuéllar, E. 1993. Toxic Red Tide of *Pyrodinium bahamense* var. *compressum* in the Tehuantepec Gulf, Mexico and the Central American Pacific System. Abstract presented at the Sixth International Conference on Toxic Marine Phytoplankton, 18–22 October 1993, Nantes, France.

Nuzzi, R., and Waters, R. M. 1993. The Occurrence of PSP Toxin in Long Island, New York, USA. In T. J. Smayda and Y. Shimizu (eds.), Toxic Phytoplankton Blooms in the Sea. Amsterdam: Elsevier, pp. 305–310.

Orellana-Cepeda, E., Morales-Zamorano, L. A., Castro-Castro, N. 1993. A Conceptual Model of Coastal Red Tides Off Baja California. Abstract presented at the Sixth International Conference on Toxic Marine Phytoplankton, 18–22 October 1993, Nantes, France.

Parrilla, G., Lavin, A., Bryden, H., et al. 1994. Rising temperatures in the sub-tropical North Atlantic Ocean over the past 35 years. Nature 369:48.

Reuters. 1995. "Red tide" tracked off Florida. *Boston Globe*, February 23, 1995, p. 17.

Rigby, G. R., Stevenson, I. G., Bolch, C. J., and Hallegraeff, G. M. 1993. The Transfer and Treatment of Shipping Ballast Waters to Reduce the Dispersal of Toxic Marine Dinoflagellates. In T. J. Smayda and Y. Shimizu (eds.), Toxic Phytoplankton Blooms in the Sea. Amsterdam: Elsevier, pp. 169–176.

Roether, W., Manca, B.B., Klein, B., et al. 1996. Recent changes in eastern Mediterranean deep waters. Science 271:333.

Sherman, K. 1993. Large Marine Ecosystems as Global Units for Management: An Ecological Perspective. In K. Sherman, L. M. Alexander, and B. D. Gold (eds.), Stress, Mitigation, and Sustainability of Large Marine Ecosystems. Washington, DC: AAAS Press, pp. 3–14.

Smayda, T. J. 1990. Novel and Nuisance Phytoplankton Blooms in the Sea: Evidence for a Global Epidemic. In E. Granéli, B. Sundström, L. Edler, and D. M. Anderson (eds.), Toxic Marine Phytoplankton. New York: Elsevier, pp. 29–40.

Smayda, T. J. 1991. Global Epidemic of Noxious Phytoplankton Blooms and Food Chain Consequences in Large Ecosystems. In K. Sherman, L. M. Alexander, and B. D. Gold (eds.), Food Chains, Yields, Models, and Management of Large Marine Ecosystems. Boulder, CO: Westview Press, pp. 275–308.

Tester, P., and Mahoney, B. 1993. Implication of the Diatom *Chaetoceros convulutus* in the Death of Red King Crabs, *Paralithodes camtschatica*, Captains Bay, Unalaska Island, Alaska, USA. Abstract presented at the Sixth International Conference on Toxic Marine Phytoplankton, 18–22 October 1993, Nantes, France.

Tester, P. A., Geesey, M. E., and Vukovich, F. M. 1993. *Gymnodinium breve* and Global Warming: What Are the Possibilities? In T. J. Smayda and Y. Shimizu (eds.), Toxic Phytoplankton Blooms in the Sea. Amsterdam: Elsevier, pp. 67–72.

Tester, P. A., Stumpf, R. P., Vukovich, F. M., et al. 1991. An expatriate red tide bloom, transport, distribution, and persistence. Limnol. Oceanogr. 36(5):1053–1061.

Thwaites, T. 1994. Are the antipodes in hot water? New Scientist 12 (Nov.):21.

Todd, E. C. D. 1989. Preliminary estimates of cost of foodborne disease in the United States. J. Food Protect. 52:595–560.

Todd, E. C. D. 1993a. Domoic acid and amnesic shellfish poisoning: A review. J. Food Protect. 56:69–83.

Todd, E. C. D. 1993b. Seafood-associated diseases in Canada. J. Assoc. Food Drug Officials 56(4):45–52.

Todd, E. C. D. 1993c. Costs of Diseases Associated with Seafood Toxins in Canada. Abstract presented at the Sixth International Conference on Toxic Marine Phytoplankton, 18–22 October 1993, Nantes, France.

Todd, E. C. D. 1994. Estimated Costs of Paralytic Shellfish, Diarrhetic Shellfish, and Ciguatera Poisoning in Canada. In M. E. Wilson, R. Levins, and A. Spielman (eds.), Disease in Evolution: Global Changes and Emergence of Infectious Diseases. New York: New York Academy of Science, pp. 831–834.

Todd, E. C. D., Brown, D. J., Grolla, A., et al. 1992. Scombroid poisoning: An outbreak in two Ontario communities. Can. Comm. Dis. Rept. 18(3):17–19.

Todd, E. C. D., and Holmes, C. F. B. 1993. Recent Illnesses from Seafood Toxins in Canada: Doses Relating to Fish Poisonings. In T. J. Smayda and Y. Shimizu (eds.), Toxic Phytoplankton Blooms in the Sea. Amsterdam: Elsevier, pp. 341–346.

Tomas, C. R., Steidinger, K. A., Garrett, J. K., and Baden, D. G. 1993. Toxic Algal Phytoplankton from Florida Coastal Waters. Abstract presented at the Sixth International Conference on Toxic Marine Phytoplankton, 18–22 October 1993, Nantes, France.

Travis, J. 1994. Taking a bottom-to-sky "slice" of the Arctic Ocean. Science 266:1947.

Trenberth, K.E., and Hoar, T. 1996. The 1990–1995 El Niño-Southern Oscillation event: Longest on record. Geophysical Research Letters 23:57.

Valiela, I. 1984. Marine Ecological Processes. New York: Springer-Verlag.

White, A. W. 1993. Biotoxins in Offshore Shellfish: A New Approach to Management of the Problem. Abstract presented at the Sixth International Conference on Toxic Marine Phytoplankton, 18–22 October 1993, Nantes, France.

Wooster, W. S., and Fluharty, D. L. (eds.). 1985. El Niño North: Niño Effects in the Eastern Subarctic Pacific Ocean. Seattle: Washington Sea Grant Program, University of Washington.

Worms, J., Bouchard, N., Cormier, R., et al. 1993. New Occurrences of Paralytic Shellfish Poisoning Toxins in the Southern Gulf of St. Lawrence, Canada. In T. J. Smayda and Y. Shimizu (eds.), Toxic Phytoplankton Blooms in the Sea. Amsterdam: Elsevier, pp. 353–358.

VI

Stress Mitigation: Environmental and Biological Considerations

24

Effects of Closure of a Continental Shelf Dumpsite*

Merton C. Ingham

In July 1986, the Environmental Processes Division (EPD) of the Northeast Fisheries Science Center (NEFSC), National Marine Fisheries Service (NMFS), National Oceanic and Atmospheric Administration (NOAA), proposed and initiated a study to document changes in living marine resources and their habitats during and after cessation of sewage sludge dumping at the 12-mile dumpsite in the New York Bight (NYB) (Environmental Processes Division 1988). The current report summarizes the preliminary findings from the second half of the field survey (January 1988 through September 1989), during which no sludge was dumped at the site.

Background

Since 1924, sewage sludge from some 200 sewage treatment plants in the New York Metropolitan Area was dumped at a site in the NYB approximately 12 nautical miles from Sandy Hook (Figure 24-1). Although the number of municipalities using this site declined over time, the volume of sludge dumped increased as waste treatment facilities were upgraded. In 1974, approximately 4.2×10^6 wet tons were dumped, whereas by 1983, the volume had almost doubled, reaching 8.3×10^6 wet tons (Figure 24-2). The volumes

---

*Excerpts taken from A. L. Studholme, M. C. Ingham, and A. Pacheco (eds.). 1991. Response of the Habitat and Biota of the Inner New York Bight to Abatement of Sewage Sludge Dumping. Third Annual Progress Report 1989. NOAA Tech. Mem. NMFS-F/NEC-82. February 1991. Washington, DC: U.S. Department of Commerce.

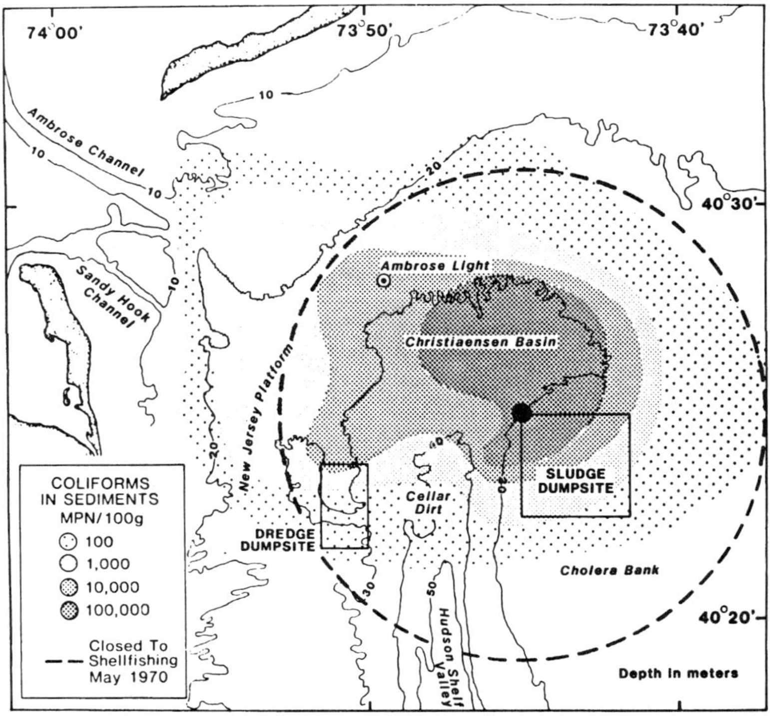

Figure 24-1. Location of the 12-mile dumpsite, dredged materials dumpsite, and area closed to commercial shellfishing in the NYB (after Verber 1976).

dumped during the early 1980s were larger than at any other sludge dumpsite in the world (Norton and Champ 1989).

Although the dumpsite is in a somewhat dispersive area with little evidence of sludge accumulation, in 1970 the U.S. Food and Drug Administration (FDA) closed an area within a radius of 11 km of the 12-mile dumpsite to commercial shellfish harvesting after finding elevated levels of coliform bacteria in sediment and shellfish (see Figure 24-1). Additionally, indications that sewage sludge dumped at the site was the major source of sewage-related contaminants in the adjacent Christiaensen Basin and Hudson Shelf Valley (HSV) led the U.S. Environmental Protection Agency (EPA) to deny further applications for dumping after December 1981 (Erdheim 1985; Santoro 1987).

After unsuccessful litigation by New York and New Jersey to continue use of the 12-mile dumpsite, the EPA essentially closed the site in April 1985 by denying requests for its redesignation (U.S. Environmental Protection Agency 1985)

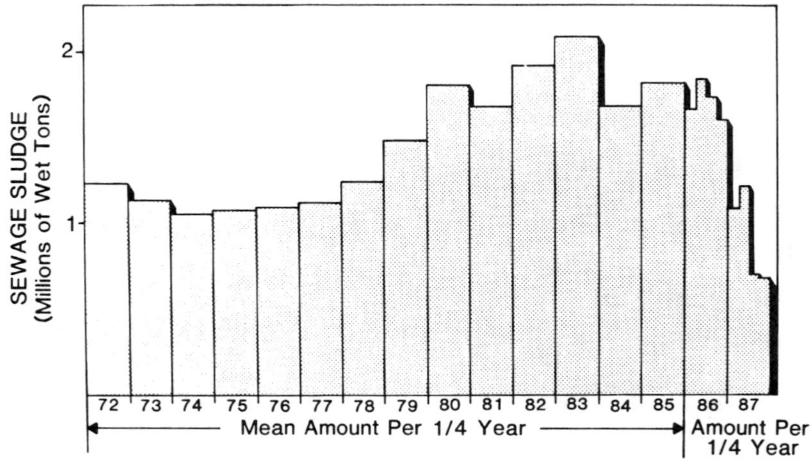

Sources: Suszkowski and Santoro, 1986; 1986–87 data are from EPA unpub.

Figure 24-2. Estimated volumes of sewage sludge dumped in the New York Bight. Annual means for 1972–1985; quarterly totals for 1986–1987 (after Swanson et al. 1985; Santoro 1987).

and selecting Deepwater Dumpsite 106 as the alternate location. The EPA adopted a schedule to phase out use of the 12-mile dumpsite beginning in early 1986, and all dumping was discontinued by the end of 1987 (Figure 24-3).

Since little is known about the recovery of dumpsites, discontinuation of the use of this site has provided marine scientists an opportunity to determine the response of this area to the removal of a major waste loading. Beginning in the summer of 1986, EPD developed and initiated a study in which biological, chemical, and physical oceanographic approaches were integrated to document changes in living marine resources and their habitats (Environmental Processes Division 1988). Results will document the response and extent of recovery of the site and its environs, providing information useful for:

1. Defining when and if the area can be reopened for shellfishing.
2. Assessing the changes in distribution and abundance of resource species in the vicinity of the site.
3. Determining probable sediment resuspension and transport of associated sludge components out of the Christiaensen Basin.
4. Quantifying the response of the area to the changes in pollutant loading in a way that will permit the use of this information for predicting response rates for other proposed temporary dumpsites.

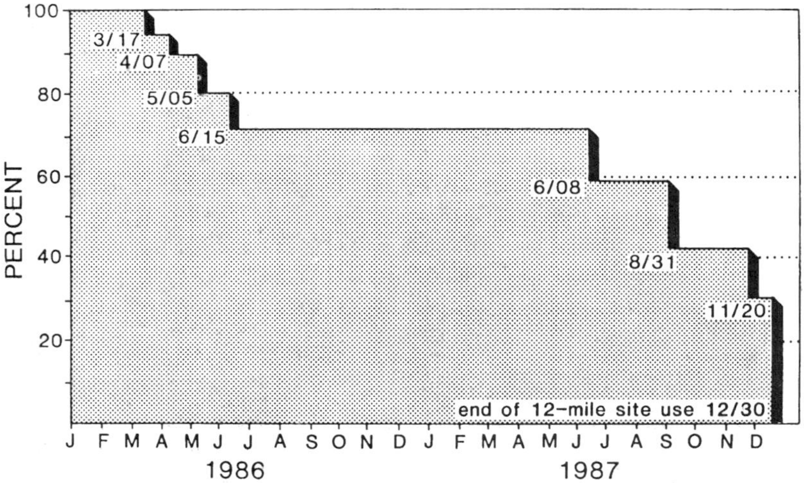

Figure 24-3. Phaseout schedule for sewage sludge dumping at the 12-mile dumpsite (after Santoro, personal communication).

5. Evaluating the management decision made by the EPA to divert the dumping of sewage sludge to Deepwater Dumpsite 106.

Development of the Study

Several factors had to be considered in the development of a practical and efficient study. The first relates to the long history of environmental degradation in the area that precludes the use of pre-dumping conditions as a baseline. Data from the first year's sampling, during which the volume of sludge dumped was reduced by 30% (see Figure 24-3), serve as a "baseline," and assessment of changes will be based on comparisons of this baseline with later sampling. Other information collected from as far back as the late 1960s does exist, as well, for several of the variables under study and will also be used to assess changes.

A second difficulty pertains to the separation of pollutant effects from various sources. It is estimated that, on average, 16% of the contaminant load in the NYB apex is traceable to sewage sludge, with most contributed by the Hudson-Raritan plume (27%), and to dredged material that is dumped on the western edge of the Christiaensen Basin (see Figure 24-1; Stanford and Young 1988). Nevertheless, by conducting synoptic surveys and initiating the study before a significant curtailment of dumping occurred, evaluation of the effects of abatement should be possible. Possible changes may include the following:

1. Dispersion of sewage sludge and cleansing of the Christiaensen Basin will be influenced by local windfield conditions and resultant changes in bottom-water circulation in the upper HSV.
2. Changes in water and sediment chemistry should occur, including reduction of sediment trace metals in sludge depositional areas. However, concentrations of certain organic compounds (e.g., polychlorinated biphenyls, or PCBs) may be more stable. Seabed oxygen consumption and nutrient regeneration rates should be reduced as benthic community metabolism and organic loading decrease.
3. Microbial concentrations should be reduced and bacteria indicative of sewage contamination should decrease to acceptable levels, permitting shellfish beds to be reopened for harvesting.
4. In areas that are presently heavily polluted, numbers of benthic macrofaunal species should increase; populations of *Capitella capitata*, an opportunistic polychaete found in disturbed areas, should decrease.
5. While dumping continues, abundance, distribution, and species composition of finfish and invertebrate communities should differ among areas that are bathymetrically similar but represent a gradient of sludge influence. After cessation of dumping and expected shifts in sediment contaminants and benthic forage species, these spatial differences should be reduced.

Sampling Strategy

The study was developed around two complementary sampling series: (1) a replicate survey; and (2) a broadscale survey, each conducted in alternate months, except for August when both were conducted to monitor rapid changes likely to occur at that time of year (see Figure 24-4; Environmental Processes Division 1988).

In the replicate survey, repeated measurements (n = 8) for most variables (Table 24-1) were made at three stations (Figure 24-4) that were similar bathymetrically and for which historical data exist, but that represent different levels of sewage sludge accumulation and effects. The most heavily degraded, NY6, was located approximately 1.6 km downslope from the corner of the dumpsite where the heaviest dumping occurred; it was thought to have the greatest accumulation of sludge constituents. A second station, R2, was about 3.4 km north of the dumpsite on the north edge of the Christiaensen Basin. Benthic communities in this area were presumably "enriched" with a high biomass of tolerant, although not necessarily pollutant-indicator, species. The

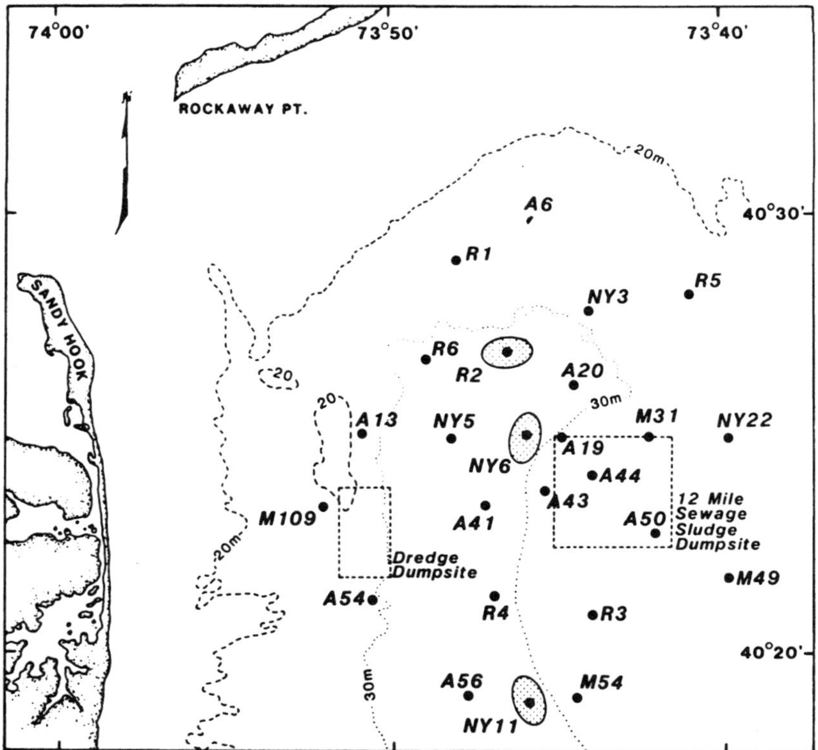

	1986 JAS	1986 OND	1987–1988 JFM	1987–1988 AMJ	1987–1988 JAS	1987–1988 OND	1989 JFM	1989 AMJ	1989 JAS
Replicate (◉)	x x x	x	x x	x	x x x	x	x x	x	x x x
Broadscale ●	x	x x	x	x x	x	x x	x	x x	x
Seabed metabolism	x x	x x x	x x x	x x x	x x x	x x x	x x x	x x x	x x x
Hydrographic surveys (with EPA)				x				x	
Seabed and water column rates, hydrographic surveys with EPA, NJ DEP, and others					x				x

Figure 24-4. Replicate, broadscale, and seabed metabolism stations and survey schedule of the 12-mile dumpsite study.

Table 24-1. Variables measured during the 12-mile dumpsite study

Habitat		
Water	Sediment	Biota
Bottom water	Heavy metals (R, B)	Resource species
Dissolved oxygen (R,B)	Organic contaminants (R, B)	Distribution/abundance (R, B)
Temperature (R, B)	Sulfide, pH profiles (R)	Diet (R)
Salinity (R, B)	Redox potential (R,B)	Winter flounder
pH (R, B)	Sediment BOD (R)	Red hake
Sulfide (R, B)	Chlorophyll pigments (R,B)	Silver hake
Nutrients (R, B)	Total organic carbon (R, B)	American lobster
Turbidity (R, B)	Characteristics	Gross pathology (R)
Water column	Grain size (R, B)	Winter flounder
Temperature	Erodibility	American lobster
Salinity (CTD)	Rates	Tissue organics (R)
Oxygen	Seabed oxygen consumption	Winter flounder
Current measurements (moored meters)	Sedimentation	American lobster
		Migration (tagging) (B)
		Winter flounder
		American lobster
		Benthos
		Macrofauna abundance/ diversity (R, B)
		Meiofauna abundance/ diversity (R, B)
		Bacteria—sediments
		Fecal and total coliform (R)
		C. perfringens (R)
		Vibrio spp. (R)
		Total count (R)
		Bacteria—shellfish

R = replicate survey; B = broadscale survey; BOD = biological oxygen demand.

third replicate station, NY11, was nearly 10 km south-southwest of the dumpsite center on the eastern shoulder of the HSV. It was considered to be the least polluted of the three sites, with the lowest concentrations of sediment contaminants and with benthic macrofauna typical of upper shelf valley sediments.

The broadscale survey consisted of single, nonreplicated bimonthly measurements at 25 stations covering most of the inner NYB and including all major habitat types. Station selection was based on considerations of bathymetry, known distribution of contaminants, patterns of environmental variables, dumping and dispersion patterns of sewage sludge, and existing historical benthic data. To add to our data base for interpreting the effects of sludge abatement, we collaborated with several agency and university researchers.

Executive Summary

In 1986, the Northeast Fisheries Science Center initiated a study to document changes in living marine resources and their habitats during and after cessation of sewage sludge dumping at the 12-mile dumpsite in the New York Bight (NYB). This summary highlights findings from the last 18 months of field work (1988–1989), focusing on changes at three intensively sampled stations: NY6—heavily degraded, R2—enriched, and NY11—a reference area assumed to be the least influenced by sludge dumping.

1. Current meters, deployed 1 m above the bottom in the NYB apex from May 1987 through June 1989, indicated that currents at subtidal frequencies are highly coherent with wind, especially at sites within the Hudson Shelf Valley (HSV). During the summer, the most intense bottom flow events (36 cm/sec maximum) were in a southward (down-valley) direction, driven by occasional episodes of east winds. During the winter, the persistent west winds often became strong enough to sustain bottom flow in a northward (up-valley) direction (54 cm/sec maximum).
2. Repeated hydrographic surveys indicated an exchange of bottom water in the dumpsite area in a one-week period, associated most likely with flow events up and down the HSV.
3. Erodibility and sludge-marker studies along the HSV indicated short-term storage of sludge in the Christiaensen Basin followed by resuspension events, probably coinciding with wind-driven currents. An exponential down-valley decline in the erodibility and sludge markers (chemical contaminants and *Clostridium* spores) was demonstrated, but station depth and HSV width governed deposition and subsequent resuspension potential. A predictable relationship between grain size and erodibility was demonstrated, permitting development of erosion models based on shear, grain size, and porosity values.
4. Measurement of fallout (cesium-137) and natural (beryllium-7) radionuclides in sediment cores indicated that the highest rates of net accumulation of fine-grained particles (a few centimeters per year) in the study area occur along the axis of the HSV up to several kilometers down valley from the dredge spoil and former sewage sludge disposal sites. This suggests resuspension of disposed material and down-valley transport.

   The dominance of DDT over DDD in surface particulate matter samples indicates that resuspension and Hudson-Raritan discharge were only minor sources of these compounds to the surface waters. Based on analysis of samples from the shelf break, it is conjectured that the DDT contamination of surface waters in the vicinity of the dumpsites is dominated by regional inputs derived from fine coastal aerosol transport during spraying of DDT before the domestic ban in 1972. The dominance of DDD over DDT in the near-bottom, suspended-particle samples is evidence of resuspension.

The high concentration of 2,3,7,8-TCDD (dioxin) found in sediment samples between the sludge and dredge dumpsites was probably derived from the disposal of material dredged from the lower Passaic River and Newark Bay.

5. Consumption of oxygen by the seabed is used as a measure of benthic community metabolism to understand energy flow and carbon cycling at the sediment–water interface in aquatic ecosystems. Stations not under the influence of sludge dumping maintained seabed oxygen consumption (SOC) rates of about 15 ml O_2 m^2 h^{-1}, and these rates did not change with the cessation of dumping. The stations under the influence of dumping maintained very high rates of SOC, especially during summer months. As dumping was phased out, the SOC rates at these stations declined to the "NYB apex background levels" of the stations not under the influence of dumping.
6. In 1989, dissolved oxygen (DO) concentrations in water near the seabed reached 2.5 mg/l (77 μM), the lowest value since the beginning of sludge volume reductions in 1986. Dissolved oxygen minima at NY6 had not been below about 4 mg/l (125 μM) during 1986–1988; however, before reductions in sludge dumping, values below 0.5 mg/l (15 μM) were observed in summers of 1983–1985. The 1989 minimum is within the range predicted for this study and probably reflects a general lowering of DO levels throughout the apex because of water-column processes that are unrelated to sludge dumping.
7. Biologically labile carbon concentrations in surface sediments in 1989 at NY6 and R2 were the lowest observed during the study. Sediment labile carbon content decreased from about 165 $\mu M/cm^3$ in 1984–1987 to about 55 $\mu M/cm^3$ in 1989. This is consistent with predictions from a model of sediment biogeochemistry. Concentrations at NY6 continued to be higher than those at NY11 and R2. However, the remaining difference between NY11 and R2 appeared to be largely the result of differential input of phytoplankton carbon.
8. Redox potential in surface sediments at NY6 and R2 generally increased since 1987, when sludge input was reduced to 30% of precessation dumping. At the same time, the amplitude of seasonal cycles diminished. This trend continued in 1989, with the observation of the highest values during

the study and a convergence in values among the three areas. These trends also followed predictions based on a model of sediment biochemistry.

9. Acid-soluble sulfide concentrations in surface sediments at NY6 and R2 were lower in 1989 than in previous years, continuing the decline first evident in 1988.
10. Metal contamination of surface sediments did not change appreciably at NY11 or R2 during the phaseout and for a year after cessation of dumping (July 1986–December 1988). Sediment contamination at NY6 was severe throughout the phaseout, but within a month of complete cessation, significant reductions occurred in surface–sediment metal levels.
11. Preliminary analyses of data on bottom-living invertebrates from the three replicate stations during the summers of 1986–1989 indicate that the numbers of crustacean, molluscan, and total species have increased. There were greater increases at those stations presumably influenced by sludge (NY6 and R2) than in the reference area (NY11); this may be an indication of the recovery of the altered areas. The pollution-indicator polychaete, *Capitella* sp., which had usually been very abundant while dumping was ongoing, was scarce in the summers of 1988 and 1989. Amphipod crustaceans, considered more pollution sensitive, were abundant at NY6 in June 1989 (18 months after cessation), but were rare again later in the summer.
12. Little skate and Atlantic rock crab ranked first in biomass of demersal fish and invertebrate megafauna, respectively, for all years at all replicate stations. Little skate constituted up to 92% of the summer demersal fish biomass. Winter flounder ranked second, and in the last year of study made up about one-fourth of the demersal fish biomass at R2 and NY6 from October to January. Increases in fish biomass during the post-dumping surveys were primarily the result of spiny dogfish, probably a reflection of a general offshore increase in numbers of elasmobranchs rather than to any changes in dumping. In part because of the transient nature of the finfish species and their seasonal occurrence, it appears that cessation of dumping has not had any measurable effect on finfish over the short term. Dominance of elasmobranchs has tended to mask any effects on teleosts; species analyses remain to be tested statistically.
13. From gross pathology observations of winter flounder, there were apparent reductions in the incidence of fin rot, cysts, *Glugea*, and lymphocystis at all three replicate stations after 1986–1987. No changes in the incidence of somatic anomalies, bentfin, or ambicoloration were detected. In the post-dumping period, a chitinoclasia incidence of about 20% occurred in American lobster; 5% were severe infections, whereas the rest were lighter infections.

14. Analyses for organic contaminants in hepatic tissues of winter flounder and American lobster collected from the vicinity of NY6 during and after the cessation of dumping revealed two statistically significant results. Within each sampling period: (a) concentrations of PCBs in both flounder and lobster from nearer NY6 were higher than in those animals collected from the reference area; and (b) lobster values were higher than those for flounder.
15. Over 7,000 winter flounder were tagged during the study. Some 3,000 were released in and around the dumpsite, and the remainder in Sandy Hook and Raritan Bays. About 3% were recovered, a few from as far as Nantucket Shoals. Results suggest the river systems associated with Sandy Hook Bay support a population that returns yearly during the spawning season. There was some indication that intermixing takes place among winter flounder populations in New Jersey, the 12-mile dumpsite, and Long Island, New York, indicating that these populations may be less discrete than described previously.

A summary of more recent results of the study was developed from the oral presentations given at a symposium held in Long Branch, New Jersey, on 18–19 June 1991 (Symposium on Effects of Cessation of Sewage Sludge Disposal at the 12-Mile Site, NOAA/NMFS/NEFSC) including the following:

Hydrography

- Water column variability in temperature and salinity occurs at all time scales.
- Hydrographic conditions observed during the 12-Mile Dumpsite study (12MDS) fell within the normal range for the inner NYB.

Bottom Currents

- Vector-mean bottom currents in the HSV were weak (1–5 cm/sec) and mostly up-valley, but the records were strongly event-dominated.
- Bottom current events in the upper HSV were strongly affected by local wind stress: up-valley with west winds, down-valley with east winds.
- The strongest bottom currents measured (>25 cm/sec) were up-valley and were in response to strong west-northwest winds in winter.
- Sediment transport was also event-dominated, and depended on the coincidence of strong bottom currents and wave action to provide resuspension energy.

- In the upper HSV about three strong transport events were identified per month; two were up-valley, one was down-valley. This ratio was typical for the period of record, and there was computed significant sediment transport 6% of the time: 4% up-valley and 2% down-valley.

Sludge Sediment Resuspension

- The sludge reservoir in the CB was temporally variable and consisted of an anoxic, silt-only cap (sometimes absent) and an underlying anoxic sand-silt matrix. The reservoirs in depressions in the upper HSV appeared to be less variable, and the silt deposits there were less likely to be sludge dominated.
- With cessation of dumping, sediments in the CB-HSV changed from anoxic silt-dominated to oxic sand-dominated and showed a reduction in concentration of sludge indicators.
- The peak in sediment erodibility shifted down the HSV about 50 km during 1987–1989, except for an unexpected peak that appeared in the CB-HSV transition in 1989. The unexpected increase in erodibility and *Clostridium perfringens* spores in the CB-HSV transition in June 1989 may be explained by the increase in dredge spoil dumping at the Dredge Dumpsite (see Figure 24-1) during the spring of 1989.
- Heavy metals found in upper HSV sediments are more likely to have come from the Dredge Dumpsite than the 12-MDS.

Sediment Deposition

- Sedimentation rates of fine-grained materials determined from radionuclide analysis were highest (cm/year) about 20 km down the HSV from the 12-mile dumpsite. This suggests transport of resuspended disposal material down the HSV.
- High concentrations of dioxin in sediments in the CB-HSV transition between the Dredge Dumpsite and the 12-mile dumpsite probably came from dredge spoils dumped at the Dredge Dumpsite.

Dissolved Oxygen and Bottom Oxygen Gradients

- The role played by the Christiaensen Basin in annual episodes of hypoxia in the New York Bight apex has decreased after abatement in sewage sludge dumping as suggested by various sediment changes such as:

1. Disappearance of high rates of seabed oxygen consumption in the basin,
2. Decrease in the area of strongly reducing surface sediments (redox),
3. Decreases in acid-soluble sulfide in surface sediments, and
4. Disappearance of strong near-bottom gradients in dissolved oxygen.

- It appears that the regulation of dissolved oxygen concentrations in bottom water in the Christiaensen Basin has shifted from sediment processes to water-column processes.

Organic Contaminants in Sediments

- The mean concentration of organic contaminants was lower in sediments collected in 1989 than those in 1986 during dumping, but the difference was not statistically significant. (Perhaps the completion of the remaining chemical analyses will improve the statistical contrast between dumping and post-dumping periods.)

Changes in the Abundance and Distribution of Clostridium perfringens

- Statistically significant decreases in concentration of *Clostridium perfringens* spores occurred at NY6 after abatement of sludge dumping.
- Concentrations of *C. perfringens* spores at NY6 before cessation were approximately 10 times concentrations at R2 and 100 times the concentrations at NY11.
- As found in earlier studies by Cabelli (1982) and others, the concentration of *C. perfringens* spores in the apex increased as bottom depth increased.
- In April 1991, sediment profiles revealed generally increasing densities of *C. perfringens* spores with sediment depth. Spore concentrations 2–10 cm below the sediment surface were comparable with concentrations observed in the surface layer during dumping.
- As recently as April 1991, spore concentrations in surface sediments at NY6 were still exceptionally high (1×10^4/g) and levels comparable with "New York Bight background" (approximately 10–100 spores/g) are not expected until 2000 A.D.

Changes in Sediment Biogeochemistry

- Two years after cessation of sludge dumping, the concentration of biologically labile carbon in the sediment at NY6 had decreased to one-third the levels observed during dumping, and redox potential in surface sediments responded to this decrease in biologically labile carbon.

- After cessation, the seasonal amplitude of the redox cycle at NY6 greatly diminished, converging with the pattern observed at R2 and NY11.
- The observed increase in the annual minimum redox in surface sediments at NY6 was approximately double that hypothesized before the study. During dumping, highly reducing conditions were observed in surface sediments throughout the Christiaensen Basin during summer. Approximately 200 km^2 of the 350 km^2 survey area had redox values less than 50 mV in surface sediments. Approximately two years after cessation, only 15 km^2 had values less than 50 mV.
- A model of sediment biogeochemistry that considers temperature, dissolved oxygen in overlying water, and labile carbon loading successfully emulated the record of redox potential at NY6, the most organically enriched site studied.
- A significant portion of the temporal variation in the number of benthic macrofauna species was correlated with sediment redox potential.

Changes in Sediment Oxygen Consumption

- Rates of sediment oxygen consumption (SOC) were responsive to increases and decreases in carbon loading to the Christiaensen Basin. Sediment oxygen consumption increased 57% in the sludge-affected area between 1974 and 1983, whereas sludge dumping increased 89%. During dumping, rates of SOC at station 33 (NY6) were highly elevated during summer (60 ml O_2 m^2 h^{-1}) and were high even when bottom water temperatures were low. After cessation, SOC rates decreased significantly, to 15–10 ml O_2 m^2 h^{-1}, and responded minimally to seasonal fluctuations in bottom water temperature.
- The observed changes in SOC were in good agreement with the anticipated changes hypothesized before the study.
- Station 30, 2.3 km east of the 12-mile dumpsite and outside the Christiaensen Basin, exhibited no significant response to changes in the quantity of sludge dumped.
- Sediment oxygen consumption at Station 35, adjacent to the Dredge Dumpsite, responded more to changes in dredged material input than to changes in sludge loading.

Changes in Sediment Metal Contamination

- Statistically significant decreases in lead and lead enrichment ([present lead/iron]/[pre-industrial lead/iron]) were found at the most contaminated replicate station (NY6) after the cessation of sludge dumping. These decreases occurred within one month after sludge dumping stopped.
- Metal concentrations after cessation decreased to about one-half to one-fourth the concentrations observed during dumping, **not** the order-of-magnitude decrease hypothesized before the study.
- No significant temporal changes were observed 4–5 cm below the sediment surface.
- A marked decrease in the area with lead enrichments greater than 20 (indicating heavy sludge pollution) was measured, from 105 km^2 during dumping to 25 km^2 about 1.7 years after dumping ceased. It is estimated that within 4 years the area of enrichment greater than 20 will disappear.

Biological Effects

- The only statistically significant finfish or shellfish biomass increase found after cessation of sludge dumping was an increase in lobster abundance. Lobster fishing effort has increased in the vicinity of the 12-mile dumpsite because of the elimination of the threat of pot fouling by sludge.
- Occurrence of skin nodules and fin rot in fish decreased after cessation, but these changes were only marginally significant.
- Studies of the diets of winter flounder, red hake, and lobster showed changes that were not statistically significant. There was, however, a striking decrease in sewage artifacts (hair, fibers) in the stomachs of red hake and lobster, and a decrease in *Capitella* from about 20% to less than 0.1% in winter flounder stomachs.
- There was a decrease in abundance of *Capitella* (pollution-tolerant) in the benthos and an increase in amphipods (pollution-sensitive). Significant increases occurred at NY6 in overall numbers of benthic species, including crustaceans, amphipods, and molluscs. Dominant species and species composition showed little change.

References

Cabelli, V. J., and Pedersen, D. 1982. The Movement of Sewage Sludge from the New York Bight Dumpsite as Seen from *Clostridium prefringens* Spore Densities. In Oceans '82 Conference Record. Washington, DC: Marine Technology Society, pp. 995–999.

Environmental Processes Division. 1988. A Plan for Study: Response of the Habitat and Biota of the Inner New York Bight to Abatement of Sewage Sludge Dumping. NOAA Tech. Mem. NMFS-F/NEC-55. Rockville, MD: National Oceanic and Atmospheric Administration.

Erdheim, E. 1985. United States Marine Waste Disposal Policy. In B. H. Ketchem et al. (eds.), Wastes in the Ocean, Vol. 6: Nearshore Waste Disposal. New York: Wiley-Interscience, pp. 421–460.

Norton, M. G., and Champ, M. A. 1989. The Influence of Site-specific Characteristics on the Effects of Sewage Sludge Dumping. In D. W. Hood et al. (eds.), Oceanic Processes in Marine Pollution, Vol. 4: Scientific Monitoring Strategies for Ocean Waste Disposal. Malabar, FL: Robert E. Krieger Co., pp. 161–183.

Santoro, E. 1987. Status report: Phaseout of ocean dumping of sewage sludge in the New York Bight apex. Mar. Pollut. Bull. 18(6):278–280.

Stanford, H. M., and Young, D. R. 1988. Pollutant Loadings to the New York Bight Apex. In Oceans '88 Conference Record. Washington, DC: Marine Technology Society, pp. 745–751.

Studholme, A. L., Ingham, M. C., and Pacheco, A. (eds.). 1991. Response of the Habitat and Biota of the Inner New York Bight to Abatement of Sewage Sludge Dumping. Third Annual Progress Report 1989. NOAA Tech. Mem. NMFS-F/NEC-82. Washington, DC: U.S. Department of Commerce.

Swanson, R. L., Champ, M. A., O'Connor, T., et al. 1985. Sewage Sludge Dumping in the New York Bight Apex: A Comparison with Other Proposed Ocean Dump Sites. In B. H. Ketchum et al. (eds.), Wastes in the Ocean, Vol. 6: Nearshore Waste Disposal. New York: Wiley-Interscience, pp. 461–488.

U.S. Environmental Protection Agency. 1985. Ocean Dumping Notice of Final Determination to Deny Petitioners to Redesignate the 12-Mile Site. Federal Register 50:14336–14340. Apr. 11, 1985.

Verber, J. L. 1976. Safe Shellfish from the Sea. In M. G. Gross (ed.), Middle Atlantic Continental Shelf and the New York Bight. Am. Soc. Limnol. Oceanogr. Spec. Symposium 2. American Society of Limnology and Oceanography, pp. 433–441.

25

Biological Effects of Contaminants on Shellfish Populations in Coastal Habitats: A Case History of New Bedford, Massachusetts

Judith E. McDowell Capuzzo

Abstract

The uptake and bioconcentration of organic contaminants by marine bivalve molluscs depend on the bioavailability of specific compounds, the duration of exposure, and the physiologic condition of populations. Species differ in their rate of uptake because of differences in filtration rates, lipid content, and habitat. Biological effects associated with bioconcentration of lipophilic contaminants have been attributed to the uptake of specific compounds and their metabolites rather than the total body burden of petroleum hydrocarbons or chlorinated hydrocarbons. Empirical data suggest that linkages exist between (1) developmental and reproductive abnormalities; (2) the physiologic and molecular processes involved in uptake, retention, and loss of contaminants; and (3) the toxicity and transformation of lipophilic contaminants. Biological responses that may contribute to the impairment of reproductive and developmental processes include responses that can be categorized as interfering with bioenergetic processes, such as feeding and nutrient allocation; biosynthetic processes, such as the synthesis of energy stores; and morphogenic processes, such as those involved in structural development. An understanding of the interrelationship of bioavailability, bioconcentration, and mechanisms of biological damage can lead to new insights into the resilience of shellfish populations to contaminated habitats. Case studies within the New Bedford Harbor (Massachusetts, USA) Superfund site are used to examine such relationships.

Bivalve molluscs have been used extensively during the past two decades as sentinel monitors of chemical contamination (Butler 1973; NRC 1980; Farrington et al. 1983) and more recently as organisms in biological effects monitoring (Bayne et al. 1988). As the relationships between levels of chemical contaminants and biological responses in bivalve molluscs continue to be explored, insights into the toxic action of specific compounds and groups of compounds have been elucidated. However, our knowledge of cause-and-effect relationships between tissue burdens of many contaminants and biological consequences in many species is still incomplete. Lipophilic organic contaminants such as polycyclic aromatic hydrocarbons (PAHs), polychlorinated biphenyls (PCBs), and other synthetic compounds for which PAHs and PCBs serve in part as model compounds, are generally resistant to degradation in the marine environment. Thus, such compounds or their metabolites may accumulate to high levels in animal tissues and interfere with normal metabolic processes that affect growth, development, and reproduction (Capuzzo et al. 1988). The bioavailability, bioconcentration, and toxic effects of lipophilic contaminants are related to their pharmacologic and toxicologic properties (Capuzzo 1987; Widdows et al. 1987; Abernathy et al. 1986; Donkin et al. 1990). The limited capacity of bivalve molluscs to detoxify organic contaminants (Stegeman 1985; Livingstone and Farrar 1984) results in the uptake and accumulation of high concentrations of organic contaminants.

The uptake and bioconcentration of organic contaminants by marine bivalves depend on the bioavailability of specific compounds, the duration of exposure, and the physiologic condition of populations. Species differ in their rate of uptake because of differences in filtration rates, lipid content, and habitat (Stegeman and Teal 1973; Clark and Finley 1974; Neff et al. 1976; Vandermeulen and Gordon 1976; Augenfeld et al. 1980; Farrington et al. 1982; Boehm et al. 1982). Bioconcentration patterns may be influenced by physicochemical properties such as molecular configuration or stearic properties of specific compounds that influence biotransformation and membrane transfer kinetics (Farrington et al. 1986; Shaw and Connell 1984) and biological factors such as partitioning between storage lipids and structural lipids (Neff 1979; Chiou 1985) and differential distribution of contaminants among different tissues (Widdows et al. 1982).

Steady-state pharmacokinetic models generally used to describe the kinetics of uptake and release of lipophilic organic contaminants consider that partitioning between the environment and the organism is controlled by steady-state uptake and depuration, and the organism is considered to be a single homogeneous unit (Moriarty 1975). Empirical data on the relationship of bioconcentration factors (K_B) to octanol-water partition coefficients (K_{OW}), however, often deviate from the predicted values derived from single compartment models, suggesting the interaction of several factors in determining bioconcentration with the potential for major seasonal differences to occur. Such factors may include differential uptake and release rates, dietary input, differential distribution within the organism, seasonal variability in lipid content, and metabolic transformations. Seasonal variation in biological responses to contaminant uptake and bioconcentration may be related to dependency on specific metabolic processes, such as storage and mobilization of lipid reserves during reproductive events (Capuzzo 1987; Capuzzo et al. 1989).

Biological effects associated with bioconcentration of lipophilic contaminants have been attributed to the uptake of specific compounds and their metabolites, rather than the total body burden of hydrocarbons or chlorinated hydrocarbons (Anderson et al. 1980; Malins and Hodgins 1981; Widdows et al. 1982, 1987; Capuzzo et al. 1984). Biological effects of organic contaminants have been observed at all levels of biological hierarchy (McIntyre and Pearce 1980; Capuzzo 1987; Moore et al. 1989). For bivalve molluscs, exposure to organic contaminants has resulted in impairment of physiologic mechanisms (Widdows 1985); histopathologic disorders (Moore 1988; Lowe 1988); and loss of reproductive potential (Berthou et al. 1987; Neff and Haensly 1982). Empirical data suggest that linkages exist between (l) histopathologic, bioenergetic, developmental, and reproductive abnormalities; (2) the physiologic and molecular processes involved in uptake, retention, and loss of contaminants; and (3) the toxicity or transformation of lipophilic contaminants (Capuzzo et al. 1988). Factors that may contribute to the impairment of reproductive and developmental processes in response to exposure to lipophilic contaminants include (1) deposition of contaminants in gametes and developing embryos; (2) lysosomal dysfunction associated with oocyte resorption; (3) interference with feeding mechanisms, such that exposure mimics starvation responses; (4) failure to incorporate sufficient yolk in oocytes; (5) morphologic abnormalities during embryogenesis resulting from failure of morphologic systems to develop properly; (6) limited capacity of developmental stages to metabolize or depurate contaminants; and (7) limited capacity of early developmental stages and reproducing adults to draw on excess energy reserves (Capuzzo et al. 1988). Thus, responses can be categorized as interfering with energetic processes (3, 4, and 7), biosynthetic processes (4), and structural development (2, 5) in addition to direct toxic effects (1, 6). Alterations in bioenergetics linked with observations of reduced fecundity and viability of larvae, abnormalities in gamete and embryologic development, and reduced reproductive success provide a strong empirical basis for examination of population responses. Incorporation of these responses in demographic models may lead to new insights into adaptations of specific life history stages to contaminant perturbations and the population consequences of stage- or age-specific effects of contaminants.

The effects of lipophilic organic contaminants on marine bivalve molluscs have been examined extensively during the past decade. The majority of the studies have been conducted on the blue mussel *Mytilus edulis* (e.g., Bayne et al. 1985, 1988) with an effort to integrate responses over several levels of biological hierarchy and to examine responses linked to specific classes of contaminants. Recent work has extended this approach to other species of bivalve molluscs, such as the subtropical turkey wing mussel, *Arca zebra* (Addison and Clarke 1990; Widdows et al. 1990). Differences in interspecific sensitivity and seasonal variation in intraspecific responses may be related to dependency on specific metabolic processes, such as storage and mobilization of lipid reserves during reproductive events (Capuzzo 1987). Alterations in lipid class distribution are consistent with observations of histopathologic damage and reductions in reproductive effort in bivalve molluscs in response to lipophilic contaminants (Capuzzo and Leavitt 1988; Moore 1988). These results suggest that alterations in lipid turnover may provide insight into the processes involved in both uptake and toxicity.

Case Study: New Bedford Harbor

Ongoing studies in New Bedford Harbor and adjacent Buzzards Bay, Massachusetts (USA), provide an excellent opportunity to investigate the bioavailability, bioconcentration, and biological effects of lipophilic organic contaminants in marine bivalve molluscs. The history of contaminant inputs to New Bedford Harbor is a carefully documented series of disturbances. Inputs of polychlorinated biphenyls (PCBs), polycyclic aromatic hydrocarbons (PAHs), and heavy metals are well described, and the fates of these contaminants have been examined in regards to chemical modifications, distributions, biogeochemical cycling, and physical transport of contaminants within the Acushnet River Estuary and in the outer regions of Buzzards Bay (Farrington et al. 1986; Pruell et al. 1990).

Uptake of both polycyclic aromatic hydrocarbons and polychlorinated biphenyls have been documented in native populations of bivalve molluscs, such as the soft shell clam *Mya arenaria,* and transplanted mussels *Mytilus edulis* (Table 25-1). Capuzzo et al. (1989) examined the distribution of chlorobiphenyl congeners in the mussel *Mytilus edulis L.* collected from a reference station in Sandwich, Massachusetts, and transferred to two stations in

Table 25-1. PCBs and PAHs in Sediments and Clams from New Bedford Harbor

Contaminant	Sediment[a] (μg/g dry wt.)	Mya arenaria[b] (ng/g dry wt.)	Mytilus edulis[b] (ng/g dry wt.)
PCB (IUPAC number)			
2,4,4' (28)	–	2,020	1,970
2,2',5,5' (52)	1.3	1,060	915
2,2',3,5',6 (95)	–	2,030	2,490
2,2',4,5,5' (101)	1.8	1,190	1,220
2,3',4,4',5 (118)	2.2	1,560	–
2,2',3,4,4',5' (138)	1.3	870	665
2,2',4,4',5,5' (153)	1.5	1,020	670
2,2',3,4,4',5,5' (180)	0.26	46	25
Total PCBs (as aroclor 1254)	30.0	20,100	9,350
PAH			
Phenanthrene	1.2	225	635
Fluoranthene	2.5	900	970
Pyrene	2.8	840	675
Chrysene	1.4	1,055	145
Benz[a]anthracene	1.2	2,500	635
Benzo[g,h,i]perylene	1.1	260	20
Total PAHs	26.0	10,090	1,940

[a]Pruell et al. (1990).
[b]Farrington, McDowell Capuzzo, and Leavitt (unpublished data).

Buzzards Bay (Cleveland Ledge and New Bedford Harbor) and one station in Nantucket Sound. Mussels were sampled biweekly and analyzed for concentrations of 20 chlorobiphenyl congeners or pairs of congeners in addition to changes in condition index and the distribution of neutral and polar lipid classes to examine differences in seasonal distribution of PCBs. Individual stations represented a gradient of chemical contamination; temperature profiles, chlorophyll *a* concentrations, and total suspended matter concentrations varied seasonally at all stations but were comparable at each of the stations during a specific sampling period. The sampling period extended over a complete annual cycle.

Mussel transplants in New Bedford Harbor showed uptake of chlorobiphenyl congeners to exceedingly high concentrations (10^{-6} g/g dry weight range). Fluctuation in the concentration of some chlorobiphenyl congeners (IUPAC No. 28/31 and 95) was apparent during the late spring and early summer with a marked decline occurring during autumn; this pattern was correlated with gametogenesis and spawning activity. Relative redistribution and release of individual chlorobiphenyl congeners associated with spawning is not consistent, suggesting differential partitioning of specific congeners in different tissues or different lipid pools. Similar seasonal patterns in the fluctuation of individual chlorobiphenyl congeners were observed among mussels at other stations, although concentrations at each of these stations were at least one to two orders of magnitude lower than those observed at New Bedford Harbor. Differential uptake and loss of lighter chlorobiphenyls (di- and trichlorobiphenyls) and heavier chlorinated biphenyls (hexa- and hepta-chlorinated biphenyls) have also been noted for the green-lipped mussel *Perna viridis* (Tanabe et al. 1987).

The relative distribution of specific chlorobiphenyls in mussels transplanted to New Bedford Harbor is markedly different from the distributions observed in mussels from other sites. For example, IUPAC 28/31 (a pair of trichlorobiphenyls) and IUPAC 95 (a pentachlorobiphenyl) are present in greater relative abundance than other chlorobiphenyls in mussels transplanted to New Bedford Harbor, whereas these compounds are at relatively low to intermediate concentrations at the other two sites. Although IUPAC 28/31 have lower K_{OW} than other chlorobiphenyls measured, their relatively high concentrations in harbor waters influence their rate of uptake.

Condition indices (measured as the ratio of dry weight to shell volume) for mussels at New Bedford Harbor are lower than values for mussels at other stations during the prespawning period (Table 25-2). Values for mussels from the Cleveland Ledge station are lower than values for mussels from the Nantucket Sound station during the first few months of the study but are equal during the period of gametogenesis. Accumulation and utilization of lipid reserves (primarily triacylglycerols) correlate with changes in condition index, and significant differences are evident between mussels from New Bedford Harbor and the other two stations. Maximum lipid accumulation among mussels from New Bedford Harbor is only 57.2% of values obtained at Nantucket Sound and Cleveland Ledge. As spawning proceeds, condition indices and lipid content are reduced in all populations.

Bioenergetic estimates of reproductive effort (measured as the proportion of energy allocated to reproduction through the collection of released gametes in relation to the total amount of energy assimilated and partitioned to growth and respiratory demands) was calculated for the entire spawning period (April to October). Reproductive effort (RE) was lowest for mussels from the New Bedford Harbor station (see Table 25-2).

Detailed analysis of the stereology and biochemical composition of gonadal tissues in mussels transplanted to New Bedford Harbor during the transplant experiment confirm findings of reduced reproductive effort in mussels associated with the accumulation of lipophilic organic contaminants. The percentage of developing follicles in female mussels from New Bedford

Table 25-2. Condition Indices and Reproductive Effort of Mussels Mytilus edulis Transplanted to New Bedford Harbor and Reference Sites

Station	Maximum Condition Index Prespawning (mg dry wt./shell volume[a])	Reproductive Effort (RE/total energy[b])
Nantucket Sound	335 ± 30	0.88
Cleveland Ledge	340 ± 20	0.71
New Bedford Harbor	230 ± 15	0.34

[a]Mean ± SE.
[b]Calculated from mean values from 8 individual animals at each site.

Harbor transplants is significantly reduced in comparison with control populations. This reduction in developing oocytes is correlated with a significant reduction in the triglyceride content of mantle/gonadal tissues. No aberrations in nutritive cells (adipogranular cells) or connective tissue (vesicular connective tissue cells) were noted, nor were there any aberrations in the follicle development of male mussels transplanted to New Bedford Harbor (McDowell Capuzzo et al., in preparation).

Other indicators of contaminant effects in bivalve molluscs from New Bedford Harbor include DNA strand breakage in gill tissues (Nacci et al. 1992) and alterations in gill function (Nelson and Gutjahr-Gobell 1990) in mussels; and high levels of hematopoietic neoplasia in the soft shell clam (*Mya arenaria) (*Reinisch et al. 1984; Leavitt et al. 1990). Both Reinisch et al. (1984) and Leavitt et al. (1990) reported high prevalences of hematopoietic neoplasia in soft shell clam (*Mya arenaria*) populations from a site in New Bedford Harbor in comparison with uncontaminated sites in Buzzards Bay. Leavitt et al. (1990) observed an overall prevalence of 39.3%; clams with advanced stages of the disease also showed a reduction in physiologic condition. The degree to which contaminant exposure influences the prevalence of hematopoietic neoplasia in clam populations is not fully understood but may be a function of increased susceptibility to the disease when animals are in poor physiologic condition.

Establishing a causal relationship between specific classes of contaminants and observed responses in bivalve mollusc populations from New Bedford Harbor, or from any contaminated habitat, is problematic because several classes of contaminants occur simultaneously, the responses observed may occur with exposure to several different contaminants, and synergistic or antagonistic effects of various classes of contaminants may occur. Specific questions that remain unanswered include: (1) To what extent does exposure to chlorinated biphenyls and polycyclic aromatic hydrocarbons influence neoplastic conditions of bivalve molluscs? (2) To what extent is biotransformation of contaminants involved in the responses of bivalve molluscs at different levels of biological hierarchy? (3) How does exposure to multiple classes of contaminants alter biological responses of bivalve molluscs? (4) Are responses in bivalve molluscs reversible with depuration of organic contaminants? Continuing studies during the planned dredging and clean-up of New Bedford Harbor may lead to new insights into these questions.

References

Abernathy, S., Bobra, A. M., Shiu, W. Y., et al. 1986. Acute lethal toxicity of hydrocarbons and chlorinated hydrocarbons to two planktonic crustaceans: The key role of organism-water partitioning. Aquat. Toxicol. 8:163–174.

Addison, R. F., and Clarke, K. R. 1990. The IOC/GEEP Bermuda Workshop. J. Exp. Mar. Biol. Ecol. 138:1–8.

Anderson, J. W., Kiesser, S. L., and Blaylock, J. W. 1980. The cumulative effect of petroleum hydrocarbons on marine crustaceans during constant exposure. Rapp. P.-V. Réun. Cons. Int. Explor. Mer 179:62–70.

Augenfeld, J., Anderson, J. W., Woodruff, D. L., and Webster, J. L. 1980. Effects of Prudhoe Bay crude oil-contaminated sediments on *Protothaca staminea* (Mollusca: Pelecypoda): Hydrocarbon content, condition index, free amino acid level. Mar. Environ. Res. 4:135–143.

Bayne, B. L., Addison, R. F., Capuzzo, J. M., et al. 1988. An overview of the GEEP workshop. Mar. Ecol. Prog. Ser. 46:235–243.

Bayne, B. L., Brown, D. A., Burns, K., et al. 1985. The Effects of Stress and Pollution on Marine Animals. New York: Praeger.

Berthou, F., Balouet, G., Bodennec, G., and Marchand, M. 1987. The occurrence of hydrocarbons and histopathological abnormalities in oysters for seven years following the wreck of the Amoco Cadiz in Brittany (France). Mar. Environ. Res. 23:103–133.

Boehm, P. D., Barak, J. E., Fiest, D. L., and Elskus, A. A. 1982. A chemical investigation of the transport and fate of petroleum hydrocarbons in littoral and benthic environments: The Tsesis oil spill. Mar. Environ. Res. 6:157–188.

Butler, P. A. 1973. Residue in fish, wildlife and estuaries. Organochlorine residues in estuarine mollusks, 1965–72. National Pesticide Monitoring Program. Pest. Monit. J. 6:238–246.

Capuzzo, J. M. 1987. Biological Effects of Petroleum Hydrocarbons: Assessments from Experimental Results. In D. F. Boesch and N. N. Rabalais (eds.), Long-term Environmental Effects of Offshore Oil and Gas Development. London: Elsevier, pp. 343–410.

Capuzzo, J. M., Lancaster, B. A., and Sasaki, G. C. 1984. The effects of petroleum hydrocarbons on lipid metabolism and energetics of larval development and metamorphosis in the American lobster (*Homarus americanus* Milne Edwards). Mar. Environ. Res. 14:201–228.

Capuzzo, J. M., and Leavitt, D. F. 1988. Lipid class composition of the digestive glands of *Mytilus edulis* and *Carcinus maenas* in response to contaminant gradients. Mar. Ecol. Prog. Ser. 46:139–145.

Capuzzo, J., McDowell, J. W., Farrington, P., et al. 1989. The relationship between lipid composition and seasonal differences in the distribution of PCBs in *Mytilus edulis* L. Mar. Environ. Res. 28:259–264.

Capuzzo, J., McDowell, J. W., Moore, M. N., and Widdows, J. 1988. Effects of toxic chemicals in the marine environment: Predictions of impacts from laboratory studies. Aquat. Toxicol. 11:303–311.

Chiou, C. T. 1985. Partition coefficients of organic compounds in lipid-water systems and correlations with fish bioconcentration factors. Environ. Sci. Technol. 19:57–62.

Clark, R.C., and Finley, J. S. 1974. Tidal aquarium for laboratory studies of environmental effects on marine organisms. Prog. Fish Cult. 36:134–137.

Donkin, P., Widdows, J., Evans, S. V., et al. 1990. Quantitative structure-activity relationships for the effect of hydrophobic chemicals on rate of feeding by mussels (*Mytilus edulis*). Aquat. Toxicol. 14:277–294.

Farrington, J. W., Davis, A. C., Brownawell, B. J., et al. 1986. Some Aspects of the Biogeochemistry of Polychlorinated Biphenyls in the Acushnet River Estuary, Massachusetts, U.S.A. In M. Sohn (ed.), Symposium on Estuarine and Marine Chemistry: Organic Geochemistry. American Chemistry Society Symposium Series 305. Washington, DC: American Chemical Society, pp. 174–197.

Farrington, J. W., Davis, A. C., Frew, N. M., and Rabin, K. S. 1982. No. 2 fuel oil compounds in *Mytilus edulis*: Retention and release after an oil spill. Mar. Biol. 66:15–26.

Farrington, J. W., Goldberg, E. D., Risebrough, R. W., et al. 1983. U.S. "Mussel Watch" 1976–1978: An overview of the trace metal, DDE, PCB, hydrocarbon and artificial radionuclide data. Environ. Sci. Technol. 17:490–496.

Leavitt, D. F., McDowell Capuzzo, J., Miosky, D., et al. 1990. Incidence of hematopoietic neoplasia in *Mya arenaria*: Monthly monitoring of prevalence and indices of physiological condition. Mar. Biol. 105:313–321.

Livingstone, D. R., and Farrar, S. V. 1984. Tissue and subcellular distribution of enzyme activities of mixed-function oxygenase and benzo[a]pyrene metabolism in the common mussel *Mytilus edulis* L. Sci. Tot. Environ. 39:209–235.

Lowe, D. M. 1988. Alterations in cellular structure of *Mytilus edulis* resulting from exposure to environmental contaminants under field and experimental conditions. Mar. Ecol. Prog. Ser. 46:91–100.

Malins, D. C., and Hodgins, H. O. 1981. Petroleum and marine fishes: A review of uptake, disposition, and effects. Environ. Sci. Technol. 15:1272–1280.

McIntyre, A.D., and Pearce, J. B. (eds.). 1980. Biological Effects of Marine Pollution and the Problems of Monitoring. Rapp. P.-V. Réun. Cons. Int. Explor. Mer 179:1–346.

Moore, M. N. 1988. Cytochemical responses of the lysosomal system and NADPH-ferrihemoprotein reductase in molluscan digestive cells to environmental and experimental exposure to xenobiotics. Mar. Ecol. Prog. Ser. 46:81–89.

Moore, M. N., Livingstone, D. R., and Widdows, J. 1989. Hydrocarbons in Marine Molluscs: Biological Effects and Ecological Consequences. In U. Varanasi (ed.), Metabolism of Polycyclic Aromatic Hydrocarbons in the Aquatic Environment. Boca Raton, FL: CRC Press, pp. 291–328.

Moriarty, F. 1975. Exposures and Residues. In F. Moriarty (ed.), Organochlorine Insecticides: Persistent Organic Pollutants. London: Academic Press, pp. 29–72.

Nacci, D., Nelson, S., Nelson, W., and Jackim, E. 1992. Application of the DNA alkaline unwinding assay to detect DNA strand breaks in marine bivalves. Mar. Environ. Res. 33:83–100.

National Research Council (NRC). 1980. The International Mussel Watch. Washington, DC: National Academy Press.

Neff, J. M. 1979. Polycyclic Aromatic Hydrocarbons in the Aquatic Environment: Sources, Fates and Biological Effects. London: Applied Science Publishers.

Neff, J. M., Cox, B. A., Dixit, D., and Anderson, J. W. 1976. Accumulation and release of petroleum-derived aromatic hydrocarbons by four species of marine animals. Mar. Biol. 38:279–289.

Neff, J. M., and Haensly, W. E. 1982. Long-term impact of the Amoco Cadiz oil spill on oysters *Crassostrea gigas* and plaice *Pleuronectes platessa* from Aber-Benoit and Aber-Wrach, Brittany, France. In Ecological Study of the Amoco Cadiz Oil Spill. NOAA-CNEXO Report. Washington, DC: National Oceanic and Atmospheric Administration, pp. 269–328.

Nelson, W., and Gutjahr-Gobell, R. 1990. New Bedford Harbor Pilot Study Progress Report: Biological Effects in *Mytilus edulis* after Pre-operational Deployments. Narragansett, RI: U.S. Environmental Protection Agency, Environmental Research Laboratory.

Pruell, R. J., Norwood, C. B., Bowen, R. D., et al. 1990. Geochemical study of sediment contamination in New Bedford Harbor, Massachusetts. Mar. Environ. Res. 29:77–101.

Reinisch, C. L., Charles, A. M., and Stone, A. M. 1984. Epizootic neoplasia in soft shell clams collected from New Bedford Harbor. Hazardous Waste 1:73–81.

Shaw, G. R., and Connell, D. W. 1984. Physicochemical properties controlling polychlorinated biphenyl (PCB) concentrations in aquatic organisms. Environ. Sci. Technol. 18:18–23.

Stegeman, J. J. 1985. Benzo(a)pyrene oxidation and microsomal enzyme activity in the mussel (*Mytilus edulis*) and other bivalve mollusc species from the Western North Atlantic. Mar. Biol. 89:21–30.

Stegeman, J. J., and Teal, J. M. 1973. Accumulation, release and retention of petroleum hydrocarbons by the oyster *Crassostrea virginica.* Mar. Biol. 22:37–44.

Tanabe, S., Tatsukawa, R., and Phillips, D. J. H. 1987. Mussels as bioindicators of PCB pollution: A case study on uptake and release of PCB isomers and congeners in green-lipped mussels (*Perna viridis*) in Hong Kong waters. Environ. Pollut. 47:41–62.

Vandermeulen, J. H., and Gordon, D. C., Jr. 1976. Re-entry of 5 year old stranded bunker C fuel oil from a low-energy beach into the water, sediments, and biota of Chedabucto Bay, Nova Scotia. J. Fish. Res. Board Can. 33:2002–2010.

Widdows, J. 1985. Physiological responses to pollution. Mar. Pollut. Bull. 16:129–134.

Widdows, J., Bakke, T., Bayne, B. L., et al. 1982. Responses of *Mytilus edulis* on exposure to the water accommodated fraction of North Sea oil. Mar. Biol. 67:15–31.

Widdows, J., Burns, K. A., Menon, N. R., et al. 1990. Measurement of physiological energetics (scope for growth) and chemical contaminants in mussels (*Arca zebra*) transplanted along a contamination gradient in Bermuda. J. Exp. Mar. Biol. Ecol. 138:99–117.

Widdows, J., Donkin, P., and Evans, S. V. 1987. Physiological responses of *Mytilus edulis* during chronic oil exposure and recovery. Mar. Environ. Res. 23:15–32.

26

Multispecies Approaches to Management of Large Marine Predators

Tim D. Smith, Robert B. Griffin, Gordon T. Waring, and John G. Casey

Abstract

Representatives of several groups of apex predators prey on fish and squid in North Atlantic temperate waters, including swordfish, oceanic sharks, tunas, toothed whales (sperm whales, pilot whales, ziphiid whales, and several species of pelagic dolphins) and baleen whales (fin whales, minke whales, sei whales, and humpback whales). These species appear to have similar habitat and ecological requirements in the North Atlantic temperate faunal region and are generally of high interest because of their economic value, ecological role, or perceived intrinsic worth. The management of fisheries in the offshore regions of the North Atlantic involves international arrangements for some species groups, such as cetaceans and bluefin tuna. International arrangements may be required in the future for others, such as oceanic sharks. An international research program to support resource management is proposed that will initially involve assembly and analysis of existing data sets from several disciplines and for several species. Analyses should be focused on generating and refining hypotheses about distribution and ecological relationships. Subsequently, multinational, multidisciplinary field studies should be designed and conducted to test the most promising hypotheses. The results of such studies would meet some of the information needs for existing international management regimes and would provide information relevant to other regimes that may be established.

In the region from Cape Hatteras to the Grand Banks, several species of large predators on fish and squid, termed here *large marine predators* (LMPs), occur seaward from the continental shelf break (depth >200 m). The distributions of LMPs extend into the northern margin of the Gulf Stream and eastward across the North Atlantic. East of the Grand Banks, the spatial distributions of several LMP species appear to spread northward toward Iceland and continue eastward past the Azores to the European and British coasts. The distributions of LMP species include the North Atlantic temperate faunal region, as defined by Backus et al. (1977) based on the distribution of myctophid species (Figure 26-1). In this region, LMP species of particular interest include oceanic sharks, swordfish, tunas, and several species of odontocete (toothed) cetaceans. These latter include sperm whales (*Physeter catadon*), pilot whales (*Globicephala* spp.), pelagic dolphins of the genera *Stenella, Delphinus, Grampus,* and *Tursiops* (referred to here as pelagic delphinids), and small whales of the family Ziphiidae (referred to here as ziphiid whales). Furthermore, several species of baleen whales (fin whale, *Balaenoptera physalus*, sei whale, *B. borealis*, minke whale, *B. acutorostrata*, and humpback whale, *Megaptera novaeangliae*) seasonally utilize continental shelf-edge and slope water habitats.

Although some LMP species occur primarily within the large marine ecosystems (LMEs) defined by Sherman and colleagues in continental shelf waters (Sherman 1986, 1990; Sherman et al. 1991), the focus here is on several LMP species that occur either regularly or exclusively seaward of the continental shelf regions included in the North Atlantic LMEs. The relationship of the oceanic areas defined by Backus and colleagues to concepts used in the definition of LMEs has been discussed (Sherman and Loughlin 1992), but formal definitions of LMEs in oceanic waters in the North Atlantic and elsewhere have not been proposed. Through feeding and seasonal migrations, many LMPs utilize both pelagic offshore and continental shelf habitats. A research program focusing on LMPs in oceanic waters would therefore aid in examining whether concepts used for defining LMEs in coastal waters can be extended to pelagic ecosystems.

Many LMP species in the North Atlantic are of particular interest because of historic and present international fisheries for them. For example, significant declines in abundances of large baleen whales, sperm whales, pilot whales, and other odontocete cetaceans have occurred over the past two centuries as a result of exploitation; in recent years, harvesting has been limited by the International Whaling Commission's (IWC) moratorium on commercial whaling. An ongoing fishery for pilot whales in the Faroe Islands has recently received increasing attention within the International Council for the Exploration of the Sea (ICES) (ICES 1992, 1993) and within the IWC (IWC 1992a). Fisheries for swordfish and several species of tunas and sharks have occurred across the North Atlantic in recent decades, and have expanded in some areas in recent years (Mejuto and Hoey 1991).

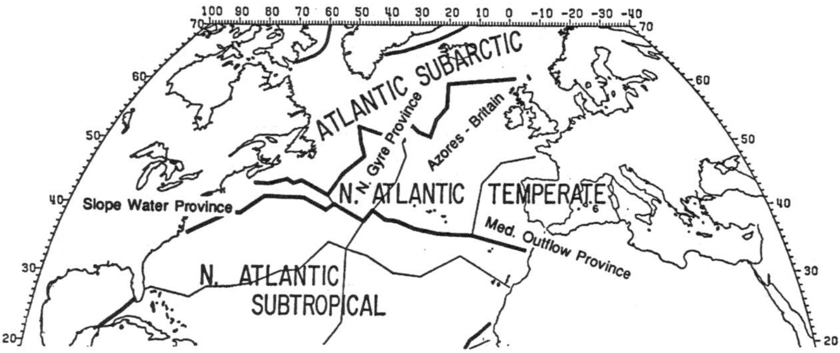

Figure 26-1. Map showing the location of the three faunal regions in the North Atlantic and four provinces within the North Atlantic Temperate Faunal Region, as defined by Backus et al. (1977) based on the distribution of mesopelagic fishes. (Adapted from R. Backus, J. Craddock, R. L. Haedrich, and B. H. Robison. 1977. Atlantic mesopelagic zoogeography. In Fishes of the Western North Atlantic, Part 7. New Haven, CT: Sears Foundation for Marine Research, Yale University, pp. 266–287; Fig. 1.)

For many LMP species, concerns have been raised about the ability of their populations to sustain past or present fishing levels. The U.S. National Research Council, in reviewing the scientific basis for management of bluefin tuna by the International Commission for the Conservation of Atlantic Tunas (ICCAT), identified major gaps in knowledge, including basic population structure (NRC 1994). Furthermore, fisheries for some of these species, such as swordfish and tunas, are known or suspected to interact directly with LMPs such as odontocete whales. For example, Japanese and American tuna and swordfish longline fisheries may have depredations of their catch by pilot whales and Risso's dolphins (*G. griseus*) (National Marine Fisheries Service, Northeast Fisheries Science Center, unpublished observer data); these fisheries, along with drift gillnet fisheries for the same species, are also known to have by-catches of oceanic sharks and cetaceans (NOAA 1991; Northridge 1992). Some LMP species attract interest because of high economic value, whereas others such as cetaceans are of interest because of possible ecological impact, perceived intrinsic value, or vulnerability to exploitation.

Fisheries for or incidentally involving LMP species of the North Atlantic faunal region (Backus et al. 1977) are difficult to regulate because the species' habitats tend to occur both within and outside of the exclusive economic zone of several nations, including the United States, Canada, Greenland, Iceland, Faroe Islands, Great Britain, Spain, and France. Cooperative international management is required for all of these species. In addition to the above-mentioned IWC, ICES, and ICCAT, the Northwest Atlantic Fisheries Organization

(NAFO), and, more recently, the North Atlantic Marine Mammal Commission (NAMMCO) are also involved in management of fisheries for LMP species.

Furthermore, the habitats occupied by these species are more extensive and less accessible to research than those of more coastally oriented species. As a result, some species, such as several of the ziphiid whales, which are generally considered rare, may in fact be rather abundant in offshore regions (J. Sigurjónsson, personal communication 1992).

Existing management approaches for LMPs are primarily single species in orientation and usually based on relatively simple density-dependent hypotheses. For example, within ICCAT, assessments of swordfish are based on virtual population analyses with an underlying assumption that productivity is directly related to abundance of the harvestable population (Mejuto and Hoey 1991). Within the IWC, the basis for resource assessments has shifted in recent decades from relying primarily on catch-based data and logistic density-dependent models to utilizing sighting-survey–based data and generalized (albeit single-species) density-dependent production models. There has been extensive development and computer-simulated testing of management procedures that better account for statistical uncertainty, possible misspecification of population models, and possible long-term environmental or ecological changes (IWC 1992b).

In neither ICCAT nor the IWC, however, has there been quantitative consideration of multispecies or environmental effects on population productivity relationships in actual management decisions. In ICCAT, for example, multispecies issues have been dealt with in a qualitative manner within working groups that deal with several species simultaneously. Within the IWC, there have been strong suggestions that such interactions may be more important than currently recognized (IWC 1992c; Lockyer and Sigurjónsson 1992). The likelihood of successful management of LMP species is reduced if potential multispecies interactions are not considered. Although the difficulties of developing sufficient ecological knowledge to address these issues are great, it is imperative to identify the types of relationships that may be relevant and to begin collecting and analyzing appropriate information.

Three theoretical approaches are of interest in considering the ecology of LMPs: oceanographic habitat determinants (Alexander 1990; Reilly 1990; Fiedler and Reilly 1994; Reilly and Fiedler 1994), trophic guilds (Root 1967; Adams 1985), and density-dependent habitat selection (Fretwell 1972; MacCall 1990). To explore the applicability of these theoretical approaches, we are proposing a cooperative multidisciplinary study involving researchers with interest in the ecology of several large marine predators in the North Atlantic temperate faunal region, and focusing on their prey, their population structure, and the oceanographic features defining their habitats. The initial objective of this study would be to identify and select among existing data sets pertinent to the ecology of large marine predator species. These several data sets would be used to characterize geographic and seasonal patterns of environmental processes and species distribution, to identify potential interactions and major information gaps, and to determine useful sampling methodologies. Sets of hypotheses about the ecology of LMP species, perhaps in the form of minimally complex ecological models, would be tested using existing data when possible, and would be used to guide future data collection activities.

Habitat Determinants in the Temperate North Atlantic

Oceanographic Structure

The North Atlantic temperate faunal region as defined by Backus et al. (1977) is bounded on the south by the northern edge of the Gulf Stream and on the north by a combination of the North American continental shelf, the northern boundary of the northern gyre just to the east of the Grand Banks, and the southern edge of the oceanic polar front (see Figure 26-1). Within the western portion of this region, the Gulf Stream and warm-core rings (meanders that have closed and detached from the Gulf Stream) form a mosaic of oceanic fronts that appear to attract several cetacean species, and are targeted by fishermen for several LMP species. Further to the east, the current and front structures become more variable, with a varying, branching current structure east of the Flemish Cap (Kraus 1986). Nearer Europe, there are several varying frontal structures of biological significance, as discussed by Holligan (1981).

Backus and colleagues defined four provinces within this faunal region, starting with the Slope Water Province in the west and continuing eastward with the Northern Gyre Province, the Azores–Britain Province, and the Mediterranean Outflow Province (see Figure 26-1). The boundary of the Slope Water Province is well defined by the continental shelf break on the north and the latitudinally fluctuating north wall of the Gulf Stream on the south. Within this province, the distribution of LMP species is correlated with the details of the continental shelf break, thermal fronts, and warm-core rings (Kenney and Winn 1987; Waring et al. 1992, 1993; Podesta et al. 1993; Waring

1995). The Northern Gyre Province immediately to the east is defined by the northern gyre, and is characterized by the complex current pattern resulting from the meeting of the Labrador Current and the Gulf Stream. The North Atlantic temperate faunal region expands northward into this province. The Azores–Britain Province occurs immediately to the east of the Northern Gyre Province and includes the region from the Azores on the south to Britain on the north. Its boundaries are defined by the oceanic polar front on the north and toward the west, and by the northern edge of the Gulf Stream (15°C at 200-m depth) on the south. Finally, the Mediterranean Outflow Province to the east represents the "areas in the open Atlantic in which the influence of the Mediterranean Water is strongest" (Backus et al. 1977), and is defined by those authors as the 35.7‰ isohaline.

Zoogeography

As mentioned earlier, the faunal provinces Backus and colleagues defined were based on the distribution of the several species of myctophids sampled during extensive transoceanic survey cruises. Five species of myctophids occur primarily in the North Atlantic Temperate Faunal Province, whereas five others occur in both this province and the one immediately to the north. Eight species of myctophids occur in this region and the one immediately to the south.

Mesopelagic fishes of the families Paralepididae (the barracudinas) and Myctophidae are known from the diets of swordfish and beaked whales in the Slope Water Province (Scott and Tibbo 1968; Stillwell and Kohler 1985; J. Nicolas, personal communication 1992). Other fishes eaten by LMP species vary by area. In the Slope Water Province, they include silver hake (*Merluccius bilinearis*), Atlantic herring (*Clupea harengus*), Atlantic mackerel (*Scomber scombrus*) and redfish (*Sebastes marinus*) (Scott and Tibbo 1968). Further east, little has been reported about predation on fishes by LMP species.

An analysis of squid distribution parallel to that by Backus and colleagues for myctophids does not appear to have been done. There are several squid species occurring in this region that are prey to LMP species. For example, stomach contents of pilot whales taken in the Faroe Islands fishery (Desportes 1990) included several families of squid, with the principal prey species being the ommastrephid squids *Todarodes sagittatus* and *Gonatus* spp. Several species of fishes are also consumed, but with markedly lower frequency. Blue sharks (*Prionace glauca*) also feed primarily on squid in oceanic environments (Kohler 1987), as do swordfish (Stillwell and Kohler 1985) and shortfin mako sharks (*Isurus oxyrinchus*) (Stillwell and Kohler 1982).

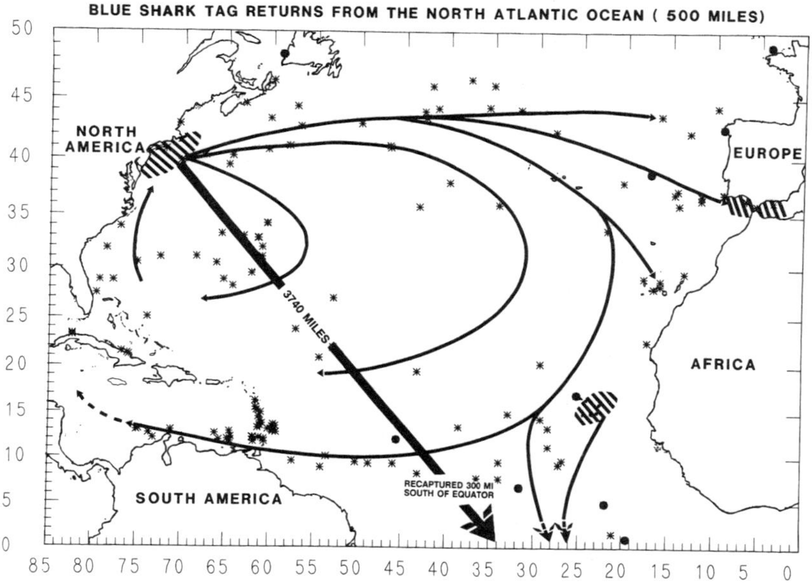

Figure 26-2. Tag returns from blue sharks traveling more than 500 miles between release and recapture; principal tagging areas are cross-hatched. (Reprinted from J. G. Casey. 1985. Transatlantic Migration of the Blue Shark: A Case History of Cooperative Shark Tagging. In R. H. Stroud [ed.], World Angling Resources and Challenges. Proceedings of the First World Angling Conference, Cap d'Adge, France, September 12–18, 1984. Ft. Lauderdale, FL: International Game Fish Association, pp. 253–268; Fig. 9. With permission.)

Principal Large Marine Predator Species in the Temperate North Atlantic

Based on fishery catch rates by area, a continuous distribution of swordfish (*Xiphias gladius*) across the North Atlantic has been demonstrated (Palko et al. 1981), primarily in the southern portion of the North Atlantic faunal region. This fishery has been pursued by several nations in recent years, and substantial information collected primarily during fishing operations allowed Mejuto and Hoey (1991) to develop hypotheses about habitat requirements and stock structure. These hypotheses identify differential migration patterns for larger and smaller swordfish, and trans-Atlantic movement patterns. Several species of sharks are caught with swordfish, and at least the blue shark appears to have similar movement patterns (Figure 26-2; Casey 1985). In contrast, the shortfin mako shark is also caught with swordfish but appears to have a more limited movement pattern (Figure 26-3; Casey and Kohler 1992). At least two

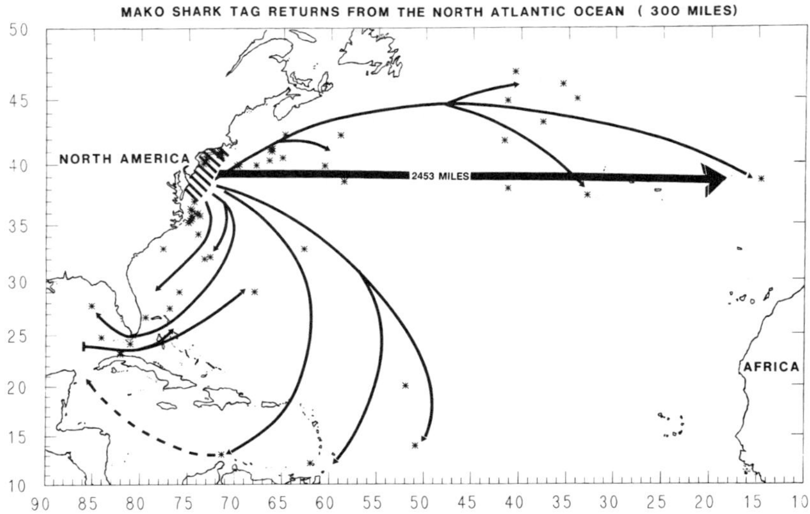

Figure 26-3. Tag returns from mako sharks traveling more than 300 nautical miles between release and recapture, as reported by Casey and Kohler (1992). (Reprinted from J. G. Casey and N. E. Kohler. 1992. Tagging studies on the shortfin mako shark (Isurus oxyrinchus) in the western North Atlantic. Aust. J. Mar. Freshw. Res. 43:45–60; Fig. 6. With permission.)

species of tuna found in these waters appear to have trans-Atlantic distributions; bigeye (*Thunnus obesus*) and yellow fin (*Thunnus albacares*) tunas are caught simultaneously with swordfish and sharks on longline and drift gillnet gear. Further study of their distribution and ecology is needed.

Baleen whales largely occupy continental shelf waters, although the migratory pathways to feeding sites for fin, minke, and humpback whales include shelf edge and slope waters and may include open ocean waters (Sutcliffe and Brodie 1977; CETAP 1982). For example, humpback whales were observed feeding in shelf edge (230-m) waters off the U.S. mid-Atlantic coast in the spring of 1992. A bottom trawl hauled through a biological hydroacoustic scattering layer at 50 m caught a large number of myctophids (ICES 1992). Sei whales are frequently seen in shelf edge waters or along edges of continental shelf banks (e.g., Browns Bank off Nova Scotia) that have steep relief features. Sei whales feed primarily on copepods, which are generally concentrated in areas of higher productivity.

Most species of odontocetes have trans-Atlantic distributions, but data to determine the degree of continuity of these distribution patterns, as seen for swordfish, are not generally available. In the western portion of the Slope

Water Province, the most frequently sighted species are sperm whales, long- and short-finned pilot whales, Risso's dolphins, bottlenose dolphins (*Tursiops truncatus*), and several species of ziphiid whales of the genera *Mesoplodon* and *Ziphius* (Hain et al. 1985). Spotted dolphins (*Stenella* spp.) are sighted with some frequency along the southern boundary of this province, whereas saddleback (common) dolphins (*Delphinus delphis*) are sighted along the northern boundary at the continental shelf break. These latter two species groups may be more representative of the adjacent oceanographic regions than of the Slope Water Province (Waring et al. 1992).

Several species of toothed whales in this province (sperm whales, pilot whales, bottlenose dolphins, *Stenella* spp., and ziphiid whales) are killed incidentally in the U.S. swordfish drift net fishery (Northridge 1992; Blaylock et al. 1995), which operates from south of Cape Hatteras north and east along the Slope Water Province to Georges Bank. This fishery focuses on areas of thermal gradients, particularly along the Gulf Stream wall and the associated warm-core rings. This suggests odontocete cetaceans, swordfish, and some species of sharks aggregate on thermal fronts in this area.

Cetacean sighting surveys have been conducted in the Slope Water Province east of Georges Bank much less frequently than they have been to the west. Some information on the distribution of sperm, sei, and fin whales has been obtained in this region from the Canadian fishery, which ended in 1972 (Sutcliffe and Brodie 1977). Also, repeated observations of recognizable individuals of one species of ziphiid whale, the bottlenose whale (*Hyperoodon ampullatus*), near the Laurentian Channel have suggested there may be resident populations in continental shelf canyons in this region (Faucher and Whitehead 1991).

Summer sighting surveys in the Central Atlantic subarctic faunal region and the Azores–Britain Province of the North Atlantic temperate faunal region have been conducted by Iceland, the Faroe Islands, and Spain. The Iceland surveys (Sigurjónsson et al. 1991) were conducted along the northwestern boundary of the Azores–Britain Province and in areas to the west. Sighting frequencies of sperm whales, pilot whales, and ziphiid whales (especially *Mesoplodon* spp.) were substantially higher along the boundary than immediately to the west in the Atlantic subarctic faunal region.

The Faroe Islands surveys were conducted north and south of the northern boundary of the Azores–Britain Province. Pilot whales were sighted much more frequently in the region southwest of the Faroe Islands (Desportes 1990). The Faroe Islands pilot whale fishery has been conducted for several centuries (Bloch et al. 1990) and throughout the year (Desportes 1990), confirming the consistent occurrence of the species in this region.

The Spanish survey included tracklines just east of the northwestern boundary of the Azores–Britain Province, during which several species of baleen whales as well as sperm whales and ziphiid whales were seen. During surveys on adjacent tracklines to the east, neither species group was seen (Lens 1991). This observation is consistent with the Icelandic survey, suggesting that these species may be associated with the northern boundary of the faunal region in this province. The Spanish survey also included tracklines in the region of the southeastern boundary of the Azores–Britain Province and in the Mediterranean Outflow Province to the east. High frequencies of sightings of sperm whales and pilot whales and some sightings of ziphiid species were recorded both near the province boundary and in the northern part of the Mediterranean Outflow Province.

Theoretical Approaches

As identified above, there are three theoretical approaches that may be useful in organizing research on LMP species in the North Atlantic: oceanographic habitat determinants, trophic guilds, and density-dependent habitat selection.

Oceanographic Habitat Determinants

Relationships of many LMP species to specific oceanographic features have been identified. Casey and Kohler (1992) found mako sharks preferred areas with water temperature in the range of 17–22°C. Cetacean distributions have been positively correlated with chlorophyll concentrations, and hence, primary production (Smith et al. 1986). Relative densities of different types of mixed-school cetaceans appear to be related to environmental variables, such as sea surface temperature, thermocline depth, and thickness of the oxygen minimum layer (Polacheck 1987). Sea surface temperature was important in explaining distribution patterns of dominant epipelagic fishes in the North Pacific (Brodeur 1988). However, the use of water temperature to describe tuna distributions may not be appropriate (Evans 1981). Tagging studies have shown that tunas routinely frequent water colder than was previously considered lethal.

Variables identified as important in some areas include surface and subsurface circulation, temperature and salinity patterns, frontal processes, and topographic features (e.g., seamounts, continental slope), which may relate to local productivity. Relationships to such variables can be pronounced and may have strong seasonal components. Furthermore, there may be year-to-year changes in many of the features affecting the distribution of LMP species. Because of the potential importance of frontal features, spatial distributions may have strong linear spatial components.

Several studies have noted movement patterns of tagged animals—e.g., mako sharks (Carey et al. 1978), swordfish (Carey and Robison 1981), hammerhead sharks (*Sphyrna* spp.) (Klimley and Butler 1988), skipjack tuna (*Katsuwonus pelamis*) (Rinaldo et al. 1981), and cetaceans (Mate 1989a, 1989b). Movement of hammerhead sharks into a region within the Gulf of California has been correlated with increase in surface water temperature and chlorophyll concentration (Klimley and Butler 1988).

General distribution patterns associated with continental shelf features have been identified for cetaceans (CETAP 1982), with distributions of some species paralleling the continental shelf edge. The importance of the shelf-break region for cetacean habitat has been identified (Kenney and Winn 1986); the shelf edge appears to be the most heavily utilized cetacean habitat off the northeastern United States after the western, inshore Gulf of Maine.

In the western North Atlantic, the shelf-break front closely follows the edge of the North American continental shelf (Mooers et al. 1978). Distribution patterns of several species of cetaceans that are delimited by the shelf edge may thus reflect an association with the shelf-break front (Kenney and Winn 1986). For example, the distribution of sightings of bottlenose dolphins and pilot whales appears to outline the shelf break and the shelf-break front. Furthermore, the distribution of pilot whales also outlines the 200-m contour of Georges Bank, presumably because of the association with the frontal system of the Georges Bank clockwise gyre. Other common shelf edge species include sperm whales, sei whales, pilot whales, Risso's dolphins, saddleback dolphins, bottlenose dolphins, and striped dolphins (*Stenella coeruleoalba*).

Many LMP species are cosmopolitan in distribution. The relevant geographic boundaries for some LMP species are therefore on the order of ocean basins, far larger than the LMEs defined by Sherman (1986). Individuals of some species may use all or seaward portions of several LMEs seasonally as "lunch counters." However, most work to date on the distribution of LMP species has focused on the smaller units.

Observations relevant to the possible correlation of distribution patterns of LMPs with oceanographic features in the North Atlantic temperate faunal region are contained in numerous data sets collected by scientists working in several disciplines and in many research institutions. Whereas analyses of many of these data sets have allowed identification of certain patterns and hypotheses, the data sets have generally not been examined thoroughly. As a result, maximum information has not been extracted from the data. For example, the available data have not been analyzed fully to assess the possible relationship of LMP species distributions to the faunal boundaries of Backus et al. (1977). Thorough analysis of all existing data sets will help answer questions pertaining to the oceanographic habitat determinants for LMPs.

Guild Theory

Despite wide taxonomic differences, LMPs share biological, ecological, and behavioral characteristics. Body size is large, requiring relatively large quantities of prey to sustain even basal metabolism. To meet these nutritional needs, many large marine predators feed on a wide variety of prey species. Sexual maturity is achieved relatively late in life, and although reproductive strategies differ greatly, the total fecundity is relatively limited. Specific genetic populations are frequently not well defined for LMP species, with clear cases such as the California gray whale (*Eschrichtius robustus*) being the exception (Rice et al. 1984).

Furthermore, LMP populations occupy spatially extensive ranges. For species restricted to nearshore areas, population ranges frequently include more than one LME. For example, harbor porpoise appear to move seasonally between the Gulf of Maine and areas around Cape Hatteras. In pelagic areas, population ranges can include more than one biological province and even significant fractions of entire ocean basins. Humpback whales move seasonally between Caribbean breeding grounds and feeding grounds across the North Atlantic temperate and the Atlantic subarctic faunal regions (Figure 26-1). Individual animals of many LMP species can move over large portions of the range occupied by their population over a lifetime. Even within areas that they may occupy seasonally, movements of individual animals over the average dimensions of LMEs can be made over periods of days to weeks.

A guild is defined as a group of organisms exploiting the same class of environmental resources in a similar fashion (Root 1967). The concept thus combines species that overlap to some degree in their habitat or trophic requirements. Through the nature of their interactions, member species of a guild must overcome problems of niche overlap, competitive exclusion, and resource partitioning. Assessments of community structure utilizing the concept of guilds may be useful in identifying important ecological principles (Karr 1987; Balon 1981).

Members of a guild need not be of closely related taxa. Criteria for defining a guild are arbitrary and may describe relationships that are trophic, reproductive, or behavioral. Furthermore, the same species may belong to more than one guild, each oriented along a different resource dimension. For example, a fish may belong with species group A in a foraging or trophic guild, while being associated with species group B in a nesting or reproductive guild.

Although guilds have not previously been suggested for LMP species, polytaxic associations of feeding tunas, dolphins, or large fish with sharks, whales, dolphins, and seabirds have been described in the Eastern tropical pacific (Au 1991) and the Gulf of Mexico (Hoffman et al. 1981), and have been suggested based on empirical observations in other areas, including the North Atlantic (Cort 1992). Associations of tunas with marine mammals and seabirds ("trophic shoals") have also been observed in the eastern tropical Atlantic (Ovchinnikov 1985).

Two faunal communities have been identified in separate regions of the eastern tropical Pacific (Au and Perryman 1985). "Species pairs" of small cetaceans serve as indicators of these communities. One community is delimited by a striped dolphin/common dolphin presence, and the other by a spotted dolphin/spinner dolphin (*Stenella attenuata/S. longirostris*) association. Tunas associate with dolphins only in the spotted/spinner community.

Although associations of fish and cetaceans may be an artifact of the patchiness of resources (Hoffman et al. 1981), behavioral evidence suggests some species may travel together for protective or foraging advantages (Au 1991). Evidence has been found for a specific feeding relationship between tunas and dolphins, and Edwards (1993) has suggested that these relationships can be explained on physiologic and trophic bases. Wolff (1982) suggested that deep diving by dolphins feeding on cephalopods drove cephalopods nearer the surface, where the more shallow-diving tuna could then feed on them.

To test the hypothesis of pelagic apex predator guilds, distributional records of LMP species (cetacean and fish) should be analyzed for distributional correlations between species, degree of similarity of prey utilization, and degree of niche overlap. By evaluating possible community associations, important resource axes (e.g., prey type, prey size) may be identified. Indices of niche breadth and niche overlap can then be calculated (Adams 1985; MacPherson 1981). In this manner, interaction among sympatric species may be analyzed. Through analysis of covariance for different niche dimensions, the degree of intra-guild competition may be evaluated (MacPherson 1981).

Several analytic methods have been used to measure resource utilization patterns. Gray (1979) used an adaptation of psychophysical unfolding theory to perform ecological analyses of resource utilization patterns. Adams (1985) used this adaptation to define a testable model of guilds, where guilds were described as "a class of sympatric species whose expressed preferences for a common set of key resources can be resolved to fit a single axis." His technique extracts guild structure from resource utilization matrices. Cardona (1991) modified the Gladfelter-Johnson index to better measure niche breadth within guilds.

Guild structure may also be investigated using MacCall's (1990) basin model. According to this model, habitat suitability for each species is assumed to vary spatially. A model might be constructed using environmental variables and prey species distributions as proxies for habitat basin definition. If a species group constitutes a guild, then multiple species should occupy similar basins. If an assembly of species does not constitute a guild, then species' basins should be independent of one another. Degree of "guildedness" might be estimated from degree of basin overlap.

Density-Dependent Habitat Selection

The extensive movements of LMP species and their highly patchy distribution patterns makes the consideration of habitat requirements complex. Models of the dynamics of LMP species do not usually contain explicit spatial components. For example, Hjort et al. (1933) suggested a simple sigmoid model for consideration of the dynamics of baleen whales harvested around Iceland that did not explicitly account for the then known spatial aspects of the biology of these species (Murray and Hjort 1912). More recently, the assessment of swordfish populations in the North Atlantic by ICCAT scientists has focused on questions of population discreetness rather than on spatial distribution. In these two contexts, the consideration of habitat selection scarcely arises.

Interestingly, within the International Whaling Commission, management approaches are being developed that address population discreetness concerns by defining a large number of "small" geographic areas. For example, in the North Atlantic, some nine areas have been defined for possible use in management of commercial harvesting of minke whales (IWC 1992b). These small areas are defined so that any one would contain animals from only one "biological stock" (IWC 1994).

Relative to interpreting the information that might result from such a management scheme, MacCall (1990) described several theoretical lines of reasoning relating to habitat selection. He suggested that density-dependent habitat selection theory could be applied as a basis for management strategy. In developing his theory, he identified several concepts of importance, including habitat suitability, habitat structure (e.g., mosaic versus homogeneity), and degrees of mobility. Visualizing habitat suitability spatially in terms of a basin defined by the maximum per capita growth rate that would be possible at different locations over the species' potential range provides a useful model from which to describe the major effects of such processes. In this context, the realized per capita growth rate for the population at any location declines as population size increases, and the differential growth rates determine the spatial distribution of a population. Depending on the geographic distribution of habitat quality, the area inhabited increases as the population approaches its carrying capacity and would be predicted to decrease as exploitation reduced abundance. That is, areas of marginal habitat tend to be occupied when a population is near its carrying capacity, whereas only optimal habitat areas are occupied at low population sizes.

MacCall (1990) suggests several possible implications of this modeling approach for large organisms. For example, the occupied ranges might be expected to vary more, and the densities within portions of the range with better suitability might be somewhat less variable, than is the case for smaller species. However, he points out the importance of extending this basic model to specific cases by accounting for the biology of an organism and the physical structure of the habitat it is occupying.

In the case of LMP species, several phenomena must be addressed. For example, recent successes in tracking tagged LMP species have suggested the spatial scale of within-season movement may be much larger than previously assumed (e.g., Carey and Robinson 1981; Mate 1989a, 1989b; Casey and Kohler 1992). However, evidence from individually identifiable cetaceans suggests at least some occur in the same specific locations from year to year. In the western North Atlantic, the potential habitat for several LMP species includes the continental shelf break and slope (roughly 200- to 2,000-m depth) and the northern edge of the Gulf Stream. These are strongly linear habitats, with unknown variability in suitability along them and within the shelf slope water between the two. Occupation of these habitats appears to differ seasonally.

Based on what is known about the mobility and basic habitat structure, it may be possible to extend MacCall's basin model suitably. One might explore the implications of the model for large, highly mobile animals moving seasonally along a mosaic of linear habitats, in lieu of MacCall's simple treatment of diffusion and advection within a continuous habitat organized from central optimum to peripheral marginal habitat.

More generally, MacCall also suggests the application of this approach to multispecies systems involving competition and predation. In the case of LMP species, it may be useful to consider simple guilds of species preying on seasonally available species such as squid and myctophids. An important consideration in making such extensions to MacCall's theory is how to parameterize them in a manner that information already available, and additional data collected in the future using sampling procedures described below, might be used to test predictions. Suitable predictions might be identified and designed into fisheries management schemes at the outset, for as MacCall suggests, the effects of varying exploitation may provide the opportunity for testing predictions from such models.

Field Data Collection and Analysis Methods

New field data collection programs can effectively draw on several scientific methodologies. Satellite telemetry, both existing capabilities for temperature measurement and color sensor capabilities that will soon come online, provide the ability to monitor within- and between-year variability in some oceanographic processes. Similarly, existing programs for monitoring the location of the Gulf Stream can help determine variability in at least part of the southern boundary of the North Atlantic temperate faunal region.

Multidisciplinary oceanographic and fishery resource survey cruises are necessary for in situ study of the distribution and ecological relationships of LMPs. Studies have demonstrated the utility, for example, of fishery trawl and cetacean sighting surveys in the western portion of the study area (Smith et al. 1988, 1990). For LMP species, longline fishing gear and line transect cetacean sighting surveys may be useful, as was recently demonstrated in a survey on the Blake Plateau just south of the western end of the faunal region (NOAA 1992). Conductivity, temperature, and depth measurements and access to satellite telemetry at sea using currently available commercial microcomputer software may prove useful for adjusting sampling to the actual boundaries of the faunal regions and provinces at the time of the survey. In addition, it may prove useful to conduct mid-water sampling for potential prey species, using both hydroacoustic monitoring and mid-water sampling trawls. Developments in undulating plankton recording devices may prove valuable for monitoring plankton distribution (Sherman and Loughlin 1992). Acoustic surveys using methods developed by Watkins and Moore (1982) and, more recently, Leaper et al. (1992) might expand cetacean survey effectiveness, especially for sperm whales and pilot whales. Finally, tissue samples for genetics/stock identification and ecological studies can be obtained from fish and sharks caught using longline fishing gear; tissues from cetaceans sighted might be collected using biopsy sampling methods developed in recent years (Baker et al. 1990).

Samples collected during multidisciplinary cruises should be analyzed using standard oceanographic and biological methods. Specific laboratory methods that may be useful for tissue samples from the LMP species include analysis of mitochondrial DNA to determine genetic relatedness (Baker et al.

1990), analysis of stable isotopes of nitrogen and carbon to determine trophic level (Fry 1988; ICES 1992; Abend 1993), and analysis of pollutants to determine contaminant levels (NOAA 1991).

Recently developed and new methods of statistical analysis will probably be required to adequately address the questions of interest. For example, Haney (1989) described iterative methods for characterizing transitory marine habitats. To address such questions, it will be necessary to assemble relevant data sets for simultaneous analysis, as was done for cetaceans in the western portion of the region by Waring et al. (1992, 1993) and for swordfish by Mejuto and Hoey (1991). Because many of the questions are primarily geographic, the use of geographic information systems could aid in simultaneous analysis (Waring et al. 1993). This is especially true when comparisons are made between dissimilar data such as satellite telemetry, sighting surveys, direct oceanographic measurements, current and water mass patterns, resource survey catches, and fishery logbook information on catch and effort.

A Research Program

Four aspects to a complete research program on LMP species are suggested: (1) assembly of existing data, (2) simultaneous analysis of assembled data and adaptation of existing theoretical approaches, (3) development of testable hypotheses, and (4) additional field sampling.

First, existing relevant data sources would be identified, and access arranged with the appropriate authorities and scientists. This would require substantial technical expertise because of the varying ways in which such data are recorded and stored. Some may be historic data sets not available in machine-readable form. An example is fisheries logbook data, which may have to be extracted by hand. It is extremely important that contents and collection methods of candidate data sets be reviewed to ensure they are appropriate and that the costs of retrieving them are reasonable in light of their potential value.

Such data sets must be prepared for input into a geographic information software system and modern database structures. This would involve preliminary summary and editing of the data, and manipulation of the form of the data to ensure comparability with other data sets. Substantial computer and subject matter expertise, and development of new methods of displaying such data (e.g., Waring et al. 1993) will be required.

Second, preliminary exploration of the assembled data would be done using correlation analysis and geographic and temporal overlay displays (Fiedler and Reilly 1994; Reilly and Fiedler 1994; Northridge and Smith

1994). At the same time, existing theoretical approaches involving oceanographic determinants or indicators of habitat, guild theory, and habitat selection theory would be adapted to the specifics of LMP species and available data. A strong interaction between the analyses of existing data and the extension and adaptation of theoretical approaches will be essential to ensure that the latter can be used given the types of information available.

Third, the necessary output of the second step will be the development of specific testable hypotheses. These may take many forms and would be tested using existing data. It may also be useful at this stage to encourage the development and testing of new and existing data collection methods—for example, individual animal telemetry, passive and active hydroacoustics, and satellite oceanographic measurements.

Fourth, as the above steps are completed, specific ideas for the types of sampling and the seasonal and spatial coverage required would emerge. It is apparent that multiple field operations would be required, perhaps over a period of 3–5 years. Plans for such sampling would be identified to the several national, governmental, and other organizations with interests in and responsibilities for the LMP species and other fisheries. Additionally, because some of the scientific issues involved bridge fishery management and ecological questions, it may be possible to secure support from agencies with primarily scientific responsibilities in addition to those agencies whose primary responsibility is fishery management.

This research program would clearly require the involvement of scientists from a variety of related disciplines. Additionally, and perhaps critically, it should interface with programs being conducted by several international organizations. For example, as noted above, the possible implementation of more sophisticated management approaches by the International Whaling Commission might eventually provide a stream of data on distribution and abundance as well as prey composition for selected baleen whales. Such management programs have collected a large portion of the data that we are proposing be assembled and analyzed, and a more interactive relationship with several of these organizations in the collection of data in the future would have valuable synergistic effects.

Conclusion

The ecology of any one species of large marine predator in the North Atlantic is not well known, but among the several species that appear to rely on similar prey species, some patterns emerge. Patterns of distribution of several LMP species are argued here to relate closely to a dominant physical factor, the Gulf Stream, and to relate to the patterns seen in the distribution of species of mesopelagic fishes and squids. Combining data from different disciplines and the several species involved will undoubtedly provide additional insight into the natural history and ecology of these species. To properly interpret existing data and, more important, to develop a sound basis for collecting additional data, existing theoretical bases must be further explored and adapted to LMP species. Future data collection activities would most usefully be based on testing hypotheses using existing data, and would be most effective if programs of different agencies could be linked.

The research program outlined here would support scientific and management questions that are primarily of single species orientation but will also provide support for more sophisticated questions about the relationships among LMP species and potentially about the underlying prey resources that support the LMP species. Both of these issues are arising in management considerations in the North Atlantic. For example, the adequacy of newer single-species management approaches based in part on spatial distribution of fishing is being evaluated within the International Whaling Commission. Similarly, questions concerning interactions between LMP species and their prey resources addressed for the Benguela Current system are also arising in the North Atlantic (NOAA 1991). Interest can also be seen in studies being conducted by Norway on minke whales (IWC 1992c) and the concerns raised during assessments of fin whales (NOAA 1992b). Although no specific actions toward multispecies approaches have been accepted within the IWC, informal consultations with scientists from ICES and other international organizations to develop revised proposals have been encouraged (IWC 1992b). Such discussions and initiative by other North Atlantic management organizations have resulted in a joint ICES and NAFO symposium in 1995 titled "Ecological Role of Marine Mammals." Furthermore, similar information needs about the feeding ecology of cetaceans that are reflected in these discussions have also been identified in ICES in the Pilot Whale Study Group (ICES 1992). The research program outlined here may serve to address some of the research needs for ecological and multispecies interactions in a more concrete manner.

ACKNOWLEDGMENTS

Several individuals assisted us by providing specific information relating to the several disciplines and species discussed here, including Kenneth Sherman (NEFSC, Narragansett, RI, USA), Richard Backus (Woods Hole, MA, USA), Ramon Conser, James Hain, Simon Northridge, John Nicolas, and Kathryn Bisack (all of NEFSC, Woods Hole, MA, USA), Johann Sigurjónsson (Marine Research Institute, Iceland), and Gary Stenson (DFO, St. John's, Newfoundland).

References

Abend, A. 1993. Distribution and diet of long-finned pilot whales in the North Atlantic using carbon and nitrogen stable isotopes. Masters thesis, University of Massachusetts.

Adams, J. 1985. The definition and interpretation of guild structure in ecological communities. J. Animal Ecol. 54:43–59.

Alexander, L. M. 1990. Geographic perspectives in the management of large marine ecosystems. In K. Sherman, L. M. Alexander, and B. D. Gold (eds.), Large Marine Ecosystems: Patterns, Processes, and Yields. Washington, DC: American Association for the Advancement of Science, pp. 220–223.

Au, D. W. 1991. Polyspecific nature of tuna schools: Shark, dolphin and seabird associations. Fish. Bull., U.S. 89:343–354.

Au, D. W., and Perryman, W. L. 1985. Dolphin habitats in the eastern tropical Pacific. Fish. Bull., U.S. 83:623–644.

Backus, R., Craddock, J., Haedrich, R. L., and Robison, B. H. 1977. Atlantic Mesopelagic Zoogeography. In R. H. Gibbs (ed.), Fishes of the Western North Atlantic, Part 7. New Haven, CT: Sears Foundation for Marine Research, Yale University, pp. 266–287.

Baker, C. S., Palumbi, S. R., Lambertsen, R. H., et al. 1990. Influence of seasonal migration on geographic distribution of mitochondrial DNA haplotypes in humpback whales. Nature 344:238–240.

Balon, E. K. 1981. Additions and amendments to the classification of reproductive styles in fishes. Environ. Biol. Fish. 6:377–390.

Blaylock, R. A., Hain, J., Hansen, L., et al. 1995. U.S. Atlantic and Gulf of Mexico Marine Mammal Stock Assessments. NOAA Techn. Mem. NMFS-SEFSC-363.

Bloch, D., Hoydal, K., Joensen, J. S., and Zachariassen, P. 1990. The Faroese catch of the long-finned pilot whale. N. Atlantic Studies 2:45–46.

Brodeur, R. D. 1988. Zoogeography and Trophic Ecology of the Dominant Epipelagic Fishes in the Northern North Pacific. In The Biology of the Subarctic Pacific (Part 2). Bull. Ocean Res. Inst. Univ. Tokyo, No. 26. Tokyo: University of Tokyo, pp. 1–28.

Cardona, L. 1991. Measurement of trophic niche breadth using occurrence frequencies. J. Fish Biol. 39:901–903.

Carey, F. G., Middleton, L., Stillwell, C. E., et al. 1978. Mako Shark Experiment on *Wieczno*, March 1978. Doc. No. 80-7. Woods Hole, MA: National Marine Fisheries Service, Northeast Fishery Science Center, Woods Hole Laboratory.

Carey, F. G., and Robison, B. H. 1981. Daily patterns in the activities of swordfish, *Xiphias gladius*, observed by acoustic telemetry. Fish. Bull., U.S. 79:277–292.

Casey, J. G. 1985. Transatlantic Migration of the Blue Shark: A Case History of Cooperative Shark Tagging. In R. H. Stroud (ed.), World Angling Resources and Challenges. Proceedings of the First World Angling Conference, Cap d'Adge, France, September 12–18, 1984. Ft. Lauderdale, FL: International Game Fish Association, pp. 253–268.

Casey, J. G., and Kohler, N. E. 1992. Tagging studies on the shortfin mako shark (*Isurus oxyrinchus*) in the western North Atlantic. Aust. J. Mar. Freshwater Res. 43:45–60.

Cetacean and Turtle Assessment Program (CETAP), University of Rhode Island. 1982. A Characterization of Marine Mammals and Turtles in the Mid- and North Atlantic Areas of the U.S. Outer Continental Shelf. Final report, Contract AA51-C78-48, Bureau of Land Management, Washington, D.C.

Cort, J. L. 1992. Study in Tuna Associations, Particularly Tuna–Dolphin Associations: Integration with the Biology of This Species of Migratory Fish. ICCAT Coll. Vol. Sci. Pap. 39:358–384.

Desportes, G. 1990. Pilot whale research in the Faroe Islands. N. Atlantic Studies 2:47–54.

Edwards, E. 1993. Energetics of associated tunas and dolphins in the eastern tropical Pacific Ocean: A basis for the bond. Fish. Bull., U.S. 90:678–690.

Evans, R. 1981. Comments on the use of water temperature to delimit tropical tuna distributions. ICCAT Coll. Vol. Sci. Pap. 15:49–56.

Faucher, A., and Whitehead, H. 1991. Population biology, social structure and movements of northern bottlenose whale (*Hyperoodon ampullatus*) in the Gully, Nova Scotia. Abstract presented at the Ninth Biennial Conference on the Biology of Marine Mammals, December 5–9, 1991, Chicago, IL.

Fiedler, P. C., and Reilly, S. B. 1994. Interannual variability of dolphin habitats in the eastern tropical Pacific. II: Effects on abundances estimated from tuna vessal sightings, 1975–1980. Fish. Bull., U.S. 92:451–463.

Fretwell, S. 1972. Populations in a Seasonal Environment. Princeton, NJ: Princeton University Press.

Fry, B. 1988. Food web structure on Georges Bank from stable C, N, and S isotopic compositions. Limnol. Oceanogr. 33:1182–1190.

Gray, L. 1979. The use of psychophysical unfolding theory to determine principal resource axes. Am. Nat. 114:695–706.

Hain, J. H., Hyman, M. A. M., Kenney, R. D., and Winn, H. W. 1985. The role of cetaceans in the shelf-edge region of the northeastern United States. Mar. Fish. Rev. 47:13–17.

Haney, J. C. 1989. Iterative techniques for characterizing marine bird habitats with time-series of satellite images. Colonial Waterbirds 12:78–89.

Hjort, J., Jahn, G., and Ottestad, P. 1933. The optimum catch. Hvalradets Skrifter 7:92–127.

Hoffman, W., Fritts, T. H., and Reynolds, R. P. 1981. Whale sharks associated with fish schools off South Texas. Northeast Gulf Sci. 5:55–57.

Holligan, P. M. 1981. Biological implications of fronts on the northwest European continental shelf. Phil. Trans. R. Soc. Lond. [A] 302:547–562.

International Council for the Exploration of the Sea (ICES). 1992. Study Group on Pilot Whales: Report of Meeting in Montreal, 3–4 December 1991. Copenhagen: ICES C.M. 1992/N:3.

International Council for the Exploration of the Sea (ICES). 1993. Report of the Study Group on Long-finned Pilot Whales, Copenhagen, 30 August–3 September 1993. Copenhagen: ICES C.M. 1993/N:5.

International Whaling Commission (IWC). 1992a. Report of the International Whaling Commission Meeting, July 1992, Glasgow. Rept. Int. Whaling Comm. 41:11–50.

International Whaling Commission (IWC). 1992b. Report of the Scientific Committee. Rept. Int. Whaling Comm. 41:51–86.

International Whaling Commission (IWC). 1992c. A Research Proposal to Evaluate the Ecological Importance of Minke Whales in the Northeast Atlantic. IWC Sci. Comm. Mtg. Doc. SC/44/NAB18.

International Whaling Commission (IWC). 1994. The revised management procedure (RMP) for baleen whales. Rept. Int. Whaling Comm. 44:145–152.

Karr, J.R. 1987. Biological monitoring and environmental assessment: A conceptual framework. Environ. Manag. 11:249–256.

Kenney, R. D., and Winn, H. E. 1986. Cetacean high-use habitats of the northeast United States continental shelf. Fish. Bull., U.S. 84:345–357.

Kenney, R. D., and Winn, H. E. 1987. Cetacean biomass densities near submarine canyons compared to adjacent shelf/slope areas. Continental Shelf Res. 7:107–114.

Klimley, A. P., and Butler, S. B. 1988. Immigration and emigration of a pelagic fish assemblage to seamounts in the Gulf of California related to water mass movements using satellite imagery. Mar. Biol. Prog. Ser. 49:11–20.

Kohler, N. E. 1987. Aspects of the Feeding Ecology of the Blue Shark (*Prionace glauca*) in the Western North Atlantic. Ph.D. Dissertation, University of Rhode Island, Kingston.

Kraus, W. 1986. The North Atlantic current. J. Geophys. Res. 91:5061–5074.

Leaper, R., Chappel, O., and Gordon, J. 1992. The development of practical techniques for surveying sperm whale populations acoustically. Rep. Int. Whaling Comm. 42:549–560.

Lens, S. 1991. North Atlantic sighting survey 1989: Report of the Spanish cruise. Rep. Int. Whaling Comm. 41:539–543.

Lockyer, C., and Sigurjónsson, J. 1991. The Icelandic Fin Whale (*Balaenoptera physalus*): Biological Parameters and Their Trends Over Time. IWC Scientific Committee Meeting, Document SC/F91/F8. International Whaling Commission.

MacCall, A. D. 1990. Dynamic Geography of Marine Fish Populations. Seattle: Washington Sea Grant, University of Washington Press.

MacPherson, D. 1981. Resource partitioning in a Mediterranean demersal fish community. Mar. Ecol. Prog. Ser. 4:183–193.

Mate, B. 1989a. Satellite-monitored radio tracking as a method for studying cetacean movements and behaviour. Rep. Int. Whal. Comm. 39:389–391.

Mate, B. 1989b. Watching habits and habitats from earth satellites. Oceanus 32:14–18.

Mejuto, J., and Hoey, J. 1991. An approach to a stock hypothesis for the swordfish (*Xiphias gladius*) of the Atlantic Ocean. ICCAT Coll. Vol. Sci. Pap. 35(2):482–501.

Mooers, C. N. K., Flagg, C. N., and Boicourt, W. C. 1978. Prograde and Retrograde Fronts. In M. J. Bowman and W. E. Esaias (eds.), Oceanic Fronts in Coastal Processes. New York: Springer-Verlag, pp. 43–58.

Murray, J., and Hjort, J. 1912. The Depths of the Ocean. London: Macmillan.

National Oceanic and Atmospheric Administration (NOAA). 1991. Fishery Management Plan for Sharks of the Atlantic Ocean. Silver Spring, MD: U.S. Department of Commerce, NMFS/NOAA.

National Oceanic and Atmospheric Administration (NOAA). 1992. Report, R/V *Oregon II*, Cruise 92-01. [Manuscript.] Miami, FL: NOAA, Southeast Fisheries Science Center.

National Research Council (NRC). 1994. An Assessment of Atlantic Blue Fin Tuna. Washington, DC: National Academy Press.

Northridge, S. 1992. Driftnet Fisheries and Their Impacts on Non-target Species: A Worldwide Review. FAO Fish. Tech. Pap. 320.

Northridge, S., and Smith, T. D. 1994. The use of a Geographical Information System in analyzing and representing catch and sightings data. Copenhagen: ICES C. M. 1994/N:6.

Ovchinnikov, V. V. 1985. Migrations and schooling behaviour of tunas. In Povedenie Promyslovykh Ryb (Behavior of Commercial Fishes) (in Russian). Moscow: Sbornik Navchnykh Trudov Vniro, pp. 101–111.

Palko, B. J., Beardsley, G. L., and Richards, W. J. 1981. Synopsis of the Biology of the Swordfish, *Xiphias gladius*. NOAA Tech. Rept. NMFS Circ. 441. Washington, DC: National Oceanic and Atmospheric Administration.

Podesta, G. P., Browder, J. A., and Hoey, J. J. 1993. Exploring the association between swordfish catch rates and thermal fronts on U.S. longline grounds in the Western North Atlantic. Continental Shelf Res. 13(2/3):253–277.

Polacheck, T. 1987. Relative abundance, distribution and interspecific relationship of cetacean schools in the eastern tropical Pacific. Mar. Mammal Sci. 3:54–77.

Reilly, S. B. 1990. Seasonal changes in distribution and habitat differences among dolphins in the eastern tropical Pacific. Mar. Ecol. Prog. Ser. 66:1–11.

Reilly, S. B., and Fiedler, P. C. 1994. Interannual variability of dolphin habitats in the eastern tropical Pacific. I: Research vessel surveys, 1986–1990. Fish. Bull. 92:434–450.

Rice, D. W., Wolman, A. A., and Braham, H.W. 1984. The gray whale, *Eschrichtius robustus*. Mar. Fish. Rev. 46:7–14.

Rinaldo, R., Evans, R., and Vergne, P. 1981. Preliminary results of a 1980 skipjack tuna tagging cruise in the Western Atlantic and Caribbean Sea. ICCAT Coll. Vol. Sci. Pap. 15:150–164.

Root, R. B. 1967. The niche exploitation pattern of the blue-gray gnatcatcher. Ecol. Monogr. 37:317–350.

Scott, W. B., and Tibbo, S. N. 1968. Food and feeding habits of swordfish, *Xiphias gladius*, in the western North Atlantic. J. Fish. Res. Board Can. 25:903–919.

Sherman, K. 1986. Measurement Strategies for Monitoring and Forecasting Variability in Large Marine Ecosystems. In K. Sherman and L. M. Alexander (eds.), Variability and Management of Large Marine Ecosystems. AAAS Selected Symposium 99. Boulder, CO: Westview Press, pp. 203–236.

Sherman, K. 1990. Productivity, Perturbations, and Options for Biomass Yields in Large Marine Ecosystems. In K. Sherman, L. M. Alexander, and B. D. Gold (eds.), Large Marine Ecosystems: Patterns, Processes, and Yields. Washington, DC: American Association for the Advancement of Science, pp. 206–219.

Sherman, K., Alexander, L. M., and Gold, B. D. (eds.). 1991. Food Chains, Yields, Models, and Management of Large Marine Ecosystems. Boulder, CO: Westview Press.

Sherman, K., and Loughlin, T. (eds.). 1992. Ad-hoc Committee on Large Marine Ecosystems Report. UNESCO Headquarters, Paris, France, 22–23 March 1991. NOAA Tech. Memo. NMFS/NEC-93. Washington, DC: National Oceanic and Atmospheric Administration.

Sigurjónsson, J., Gunnlaugsson, T., Ensor, P., et al. 1991. North Atlantic sighting survey 1989 (NASS-89): Shipboard surveys in Icelandic and adjacent waters, July–August 1989. Rept. Int. Whaling Comm. 41:559–572.

Smith, R. C., Dustan, P., Au, D., et al. 1986. Distribution of cetaceans and sea-surface chlorophyll concentrations in the California Current. Mar. Biol. 91:385–402.

Smith, T. D., Payne, P. M., Heinemann, D., et al. 1988. Utility of cetacean and seabird sighting surveys conducted during fishery surveys. Copenhagen: ICES C.M. 1988/N:5.

Smith, T. D., Payne, P. M., Heinemann, D., et al. 1990. Simultaneous fishery resource and seabird and cetacean sighting surveys. N. Atlantic Studies 2:90–101.

Stillwell, C. E., and Kohler, N. E. 1982. Food, feeding habits and estimates of daily ration of the shortfin mako (*Isurus oxyrinchus*) in the Northwest Atlantic. Can. J. Fish. Aquat. Sci. 39:407–414.

Stillwell, C. E., and Kohler, N. E. 1985. Food and feeding ecology of the swordfish, *Xiphias gladius*, in the western North Atlantic with estimates of daily ration. Mar. Ecol. Prog. Ser. 22:239–247.

Sutcliffe, W. H., and Brodie, P. F. 1977. Whale distributions in Nova Scotia waters. Fish. Mar. Serv. Tech. Rept. 722.

Waring, G. T. 1995. Fishery and Ecological Interactions for Selected Cetaceans off the Northeast USA. Ph.D. Dissertation, University of Massachusetts.

Waring, G. T., Fairfield, C. P., Ruhsam, C., and Sano, M. 1992. Cetaceans Associated with Gulf Stream Features off the Northeastern USA shelf. Copenhagen: ICES C.M. 1992/N:12.

Waring, G. T., Fairfield, C. P., Ruhsam, C., and Sano, M. 1993. Sperm whales associated with Gulf Stream features off the northeastern USA shelf. Fish. Oceanogr. 2:101–105.

Watkins, W., and Moore, K. 1982. An underwater acoustic survey for sperm whales (*Physeter catadon*) and other cetaceans in the southeast Caribbean. Cetology 46:1–7.

Wolff, G. A. 1982. A Study of Feeding Relationships in Tuna and Porpoise Through the Application of Cephalopod Beak Analysis. Dissertation Abstracts Int. Part B.—Science and Engineering: 43(6). Ph.D. Dissertation, Texas A & M University, Galveston.

27

Can We Manage Our Multispecies Fisheries?

Steven A. Murawski

Abstract

Fishery resources (fish, invertebrates, marine mammals, and reptiles) of the United States exclusive economic zone (EEZ, 3–200 nautical miles from the coast) are currently regulated under several pieces of key legislation enacted during the early 1970s. At that time, fishery science was only beginning to appreciate the interactions among species and fisheries as being potential impediments to simultaneous realization of competitive management goals. Many of the important management problems currently faced in the EEZ are exacerbated by the incompatibility of regulations promulgated separately under these statutes for ecologically or technologically related species. Reconciling the management of interacting species will require an institutional framework for evaluating the multispecies/multifishery consequences of management decisions and for articulating a clear set of compatible management goals for the various constituents. Traditional single-species biological reference points for fishery management must be re-evaluated to consider the effects of harvesting on ecosystem attributes such as stability, diversity, resistance, and resilience and on economic attributes, including optimization. Currently, dynamic mechanisms underlying species and fishery interactions are poorly understood. The scientific basis of decision making will increasingly have to come from research emphasizing the dynamic relationships of interacting species as well as the biological, economic, and technological factors contributing to fishery interactions. Such studies cannot replace traditional single-species, single-fishery analyses, but must complement and build on them.

Fishery resources in the United States exclusive economic zone (EEZ, 3–200 nautical miles from the coast) are primarily managed under statutory authorities developed in the early 1970s. The Marine Mammal Protection Act of 1972 (MMPA), the Endangered Species Act of 1973 (ESA), and the Magnuson Fishery Conservation and Management Act of 1976 (MFCMA) constitute the major regulatory authority for the conservation and utilization of living marine resources under the jurisdiction of the federal government. At the time Congress promulgated these acts, the scientific basis for resource management was only beginning to take into account the trophic and technological (mixed-species catch or multiple-fleet fisheries) interactions among resources (Larkin 1977). Far more prevalent a view at the time was the notion that each species existed more or less independently in a deterministic environment (Rodel 1975). Scientific inquiry mandated in support of these laws was primarily aimed at computing harvest effort levels corresponding to species-specific maximum sustainable yields (for fisheries), determining the status of marine mammal populations in relation to optimum sustainable population (OSP) sizes, or evaluating the feasibility of rehabilitating individual species threatened with extirpation. Optimality as a concept in fisheries management was primarily used in allocation of scarce resources and as justification for biological overfishing in the face of social or economic constraints on the management system.

Since these laws were passed, management plans for the conservation, protection, and recovery of living resources have been established for virtually all of the economically valuable or socially important species in the EEZ. Almost without exception, each of these plans has established management regulations for individual species or groups of technologically interrelated species. The implementation of these regulations has revealed numerous cases wherein the ecological interrelationships among species or the interjurisdictional nature of management problems have resulted in an infeasible set of management objectives. For example, fishing operations have resulted in unintentional marine mammal kills, impeding the recovery of mammal stocks to OSP goals (Smith 1983; Alaska Sea Grant 1984; Polacheck 1989). Such has also been the conflict in the southeastern United States between provisions of the Endangered Species Act, intended to conserve sea turtles, and the goal of fishery management to maximize catch of shrimp and associated species (Henwood and Stuntz 1987; Thompson 1988).

This chapter covers three areas. First, the institutional, ecological, and scientific impediments to simultaneous realization of multispecies management goals for marine fishery ecosystems are considered. Second, the feasibility of articulating and implementing ecosystem-level management objectives as an alternative to traditional biological reference points used in managing individual species or stocks is evaluated. Third, the scope of research programs necessary to support an ecological basis for management decisions is discussed.

Institutional Setting

Ecological relationships among various constituents of exploited marine communities have long been appreciated by scientists. By the turn of the twentieth century, the simultaneous management of interacting resources was being considered (Volterra 1926; Smith 1988). By the early 1970s, there was considerable recognition of the necessity for a systems approach to fisheries research and management (Dickie 1973; Regier and Henderson 1973; Andersen and Ursin 1973). Predator–prey analyses have subsequently formed an important component of fishery ecosystem studies throughout the world (see reviews in May et al. 1979; Beddington and May 1982; Mercer 1982). In a pair of articles separated by nearly 30 years, McHugh (1959, 1988) argued that mixed-species, multiple-gear fisheries obviated traditional species-oriented management programs. Yet resource management laws promulgated for U.S. coastal waters were primarily responsive to perceived problems in one or more important species or fisheries. The emphasis in implementing provisions of the MFCMA, the MMPA, and the ESA has been and remains primarily on single-species issues.

Interestingly, all three acts refer directly or allude to ecosystem or multispecies objectives as a basis for management action. The Endangered Species Act (PL 93-205) has as one of its stated purposes "to provide a means whereby the ecosystems upon which endangered species and threatened species depend may be conserved." In the findings and declaration of policy for the Marine Mammal Protection Act (PL 92-522), Congress noted that "species and population stocks should not be permitted to diminish beyond the point at which they cease to be a significant functioning element in the ecosystem of which they are a part, and, consistent with the major objective, they should not be permitted to diminish below their optimum sustainable population." The act further stipulates that marine mammal populations "should be protected and encouraged to develop to the greatest extent feasible commensurate with sound policies of resource management and that the primary objective of their management should be to maintain the health and stability of the marine ecosystem. Whenever consistent with this primary objective, it should be the goal to obtain an optimum sustainable population, keeping in mind the carrying capacity of the habitat." The Magnuson Fishery Conservation and Management Act (PL 94-265) notes in its National Standards for Fishery Conservation and Management that "to the extent practicable, an individual stock of fish shall be managed as a unit throughout its range, and interrelated stocks of fish shall be managed as a unit or in close coordination." The Magnuson Act further states that conservation and management measures are designed to ensure that "irreversible or long-term adverse effects on fishery resources and the marine environment are avoided; and there will be a multiplicity of options available with respect to future uses of these resources." The act also stipulates that a fishery research program shall be carried out to acquire knowledge and information "concerning the interdependence of fisheries or stocks of fish, the impact of pollution on fish, the impact of wetland and estuarine degradation, and other matters bearing upon the abundance and availability of fish."

What is not established in the current legislative and management structure is a mechanism for conflict resolution among competing resource interests. Perhaps more important, there is no formal requirement for evaluating the feasibility (or in fact the desirability) of attempting to simultaneously attain single-species objectives for various ecosystem constituents regulated under the acts. Fisheries goals that have been established are primarily those related to maximization (or at least optimization) of yields from relatively restricted species assemblages. Provisions of the MMPA and ESA seek to severely limit the kill of affected animals. A set of amendments to the MMPA allows for some harassment and incidental kill of marine mammals by fishing operations—for a 5-year period. The purpose of these exemptions is to gather information on the degree of interaction with marine mammals to be expected in various fisheries. Under these amendments, each fishery operating in the EEZ must be classified as to the likelihood of encounter with marine mammals. Those fisheries deemed to have significant encounters must submit to record-keeping requirements and embark scientific observers when asked to do so. What remains unclear is the basis for decision making once sufficient documentation of the fishery interactions with marine mammals has been gathered. Nevertheless, data on marine mammal–fishery interactions collected under this program will no doubt force a more integrative approach to conflict resolution in some cases.

Apart from the obvious cases of bycatch of species protected under one statute and taken in fisheries regulated under different laws, there are numerous cases wherein the multispecies interactions involve prey species that are regulated under different fishery management plans or, in fact, under the provisions of different statutes, than are their predators. Figure 27-1 depicts a simplified food web for some of the species constituting the pelagic fishery ecosystem off the northeast United States. There are three main prey species (herring, *Clupea harengus*; mackerel, *Scomber scombrus*; and sand lance, *Ammodytes dubius*) and a number of fish, marine mammal, and avian predators. Marine mammal and some avian predators are protected species. Some marine mammals are killed (pilot whale, *Globicephala* spp.; common dolphin,

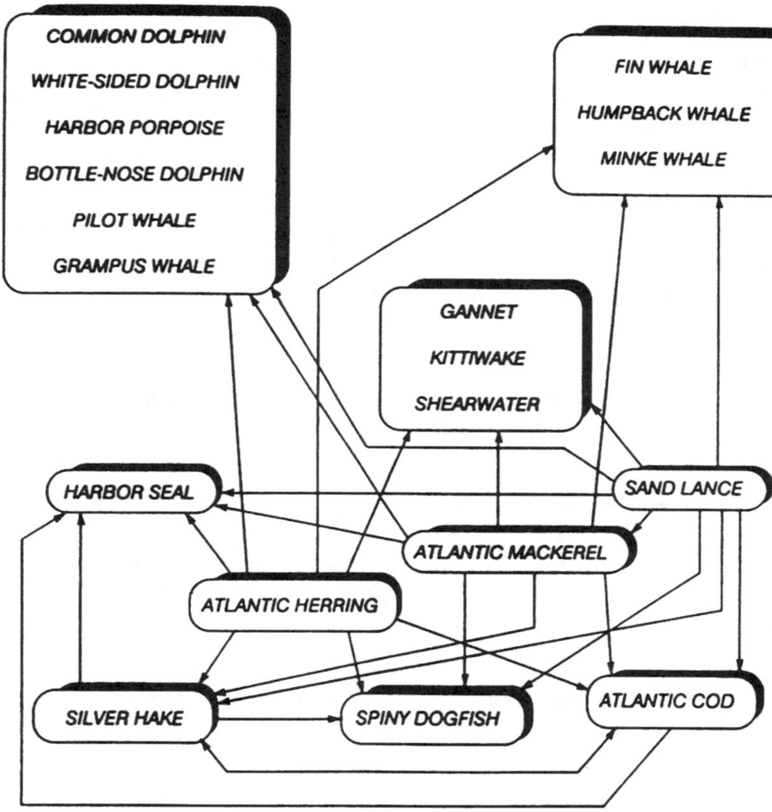

Figure 27-1. Predator–prey relationships among some fish, marine mammals, and seabirds of the pelagic ecosystem off the northeast United States. For the sake of clarity, not all predator and prey species are portrayed. Arrows indicate the direction of biomass flows, based on studies of feeding habits.

Delphinus delphis) during directed fishing on mackerel, primarily in joint venture fisheries with distant-water fleets. Mackerel fishing is regulated under a fishery management plan (FMP) developed by the Mid-Atlantic Fishery Management Council (under the MFCMA). Herring, sand lance, and spiny dogfish (*Squalus acanthias*) are currently not directly regulated in EEZ waters, except by a preliminary management plan (PMP) covering other finfish of the region. The PMP is promulgated primarily as a basis to exclude the declaration of harvestable surplus, which by statute must be considered for allocation to foreign fishermen. Atlantic cod (*Gadus morhua*) and silver hake (*Merluccius bilinearis*) are regulated under the Multispecies FMP for groundfish, developed by the New England Fishery Management Council. Although the FMP for groundfish implies consideration of the multispecies aspects of interacting species, the FMP has as its conservation goal the attainment of single-species equilibrium biological reference points for the 13 regulated species.

This relatively simple trophic system has undergone considerable change in fishery production, species dominance, and interspecies predation in the last two decades. Herring, mackerel, and silver hake stocks were intensively fished by distant-water fleets during the 1970s, resulting in very low stock sizes and ultimately the curtailment of directed fishing. Coincident with the decline of the dominant pelagic stocks of herring and mackerel, sand lance increased greatly in abundance in the late 1970s and early 1980s. Large whale species (humpback whales, *Megaptera novaeangliae,* and fin whales, *Balaenoptera physalus*) had utilized herring and (probably) mackerel as main forage items before this time. The demise of the herring population on Georges Bank brought with it an apparent shift in the distribution of whale predators to more nearshore areas, particularly those where concentrations of the more abundant sand lance could be found. An important whale-watch industry developed as it became more obvious that large whales occurred a few miles offshore.

More recently, stocks of mackerel and, to some extent, herring have rebounded, in part because of very low fishing mortality rates. Both of these species are currently in little demand by domestic markets. Thus, the impetus to catch them has primarily stemmed from foreign countries where domestic or third-country markets for these species are stronger. Fishing by foreign countries on these stocks has, however, been severely limited under various FMP and PMP provisions. The abundance of spiny dogfish has likewise increased dramatically in recent years. Fishing mortality on this small shark is currently quite low because of weak domestic market demand. Thus, the main species of the pelagic system have to some extent been allowed to proceed on trajectories primarily dictated by interspecies predation and competition. Sand lance has declined in abundance in recent years, coincident with increases in mackerel (a predator of larval and juvenile sand lance) and herring (a competitor with sand lance for zooplankton prey).

It is thus apparent that the productivity and species composition of fishery ecosystems are contingent not only on explicit management decisions, such as fishing certain stocks, but also on the implicit management decisions made for species of little market value (e.g., sand lance, spiny dogfish) or that are otherwise ignored. The ecosystem effects of specific management regulations are not generally evaluated in such decision making, and there is no mechanism for identifying a target ecosystem configuration or set of desirable attributes to be maintained within the ecosystems. If significant harvesting of mackerel, herring, and sand lance should occur, there would be serious concern about the ability of these prey resources to simultaneously support high levels of predatory fish and marine mammal stocks.

Even when there are limited trophic interactions among regulated species, the attainment of simultaneous goals for various species may be compromised by the interjurisdictional nature of fisheries. Figure 27-2 depicts the institutional and fishery interactions associated with demersal fisheries for groundfish and pandalid shrimp in the Gulf of Maine region. Four major categories of demersal fisheries can be defined, based primarily on gear: large-mesh trawls, small-mesh trawls, gill net, and shrimp trawling. These fisheries are interrelated by the mixed nature of the catch and the associated discards. Although each fishery targets a different set of species, interactions among fisheries occur when (1) small-mesh fisheries discard undersized individuals of target species for large-mesh fisheries, (2) large-mesh fisheries discard animals selected by the gear, but below the minimum landing sizes, and (3) the fisheries operate simultaneously or sequentially, the effort thus influencing the availability of shared species to competing fisheries. Discards of groundfish in small-mesh and shrimp trawls are considered particularly problematic because shrimp and other small-mesh trawl fisheries operate in nearshore areas that are the primary nurseries for groundfish species.

There are no less than eight separate regulatory entities (including the three states) that are directly involved in managing components of the demersal fishery complex in the Gulf of Maine. The New England Fishery Management Council has developed an FMP for 13 of the most valuable groundfish species. The goals of this FMP are to attain, on a species-by-species basis, the target levels of the percentage of virgin spawning stock biomass per recruitment thought to provide adequate protection to spawning fish. Primary regulations implemented under the groundfish FMP are minimum mesh size regulations (for large-mesh fisheries) and minimum landing lengths for eight of the species. The shrimp fishery is regulated separately by a subcommittee of the Atlantic States Marine Fisheries Commission (a compact of Atlantic coastal states). Shrimping regulations specify a closed season (May–November) and a bycatch limit restricting landings of groundfish to a percentage of shrimp landings. Small-mesh fisheries for silver and red hake (*Urophycis chuss*) are regulated under the multispecies FMP. Other small-mesh target species, including spiny dogfish, are nominally managed as "other finfish" by a preliminary management plan promulgated directly by the National Marine Fisheries Service. The three coastal states (Maine, New Hampshire, and Massachusetts) control territorial seas (0–3 nautical miles)

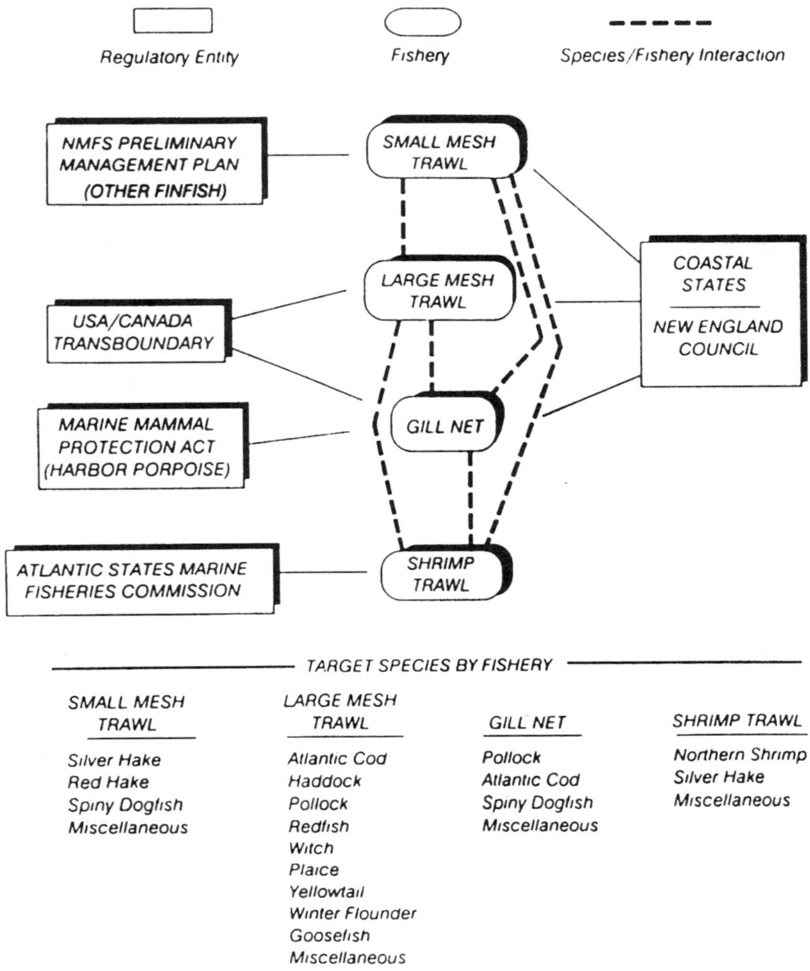

Figure 27-2. Schematic of institutional and fishery interactions among Gulf of Maine demersal fisheries. Dashed lines indicate fishery interactions resulting from multiple fisheries catching the same species but with different partial recruitment-at-age patterns.

and impose separate management programs that regulate portions of the contiguous groundfish and shrimp stocks. Some of the groundfish stocks of the region are distributed or migrate across the United States–Canada maritime boundary and are thus subjected to vastly differing management schemes in each country's waters. Finally, gill-net fishing in the Gulf of Maine has been classified as a fishery in which there may be frequent takes of marine mammals (harbor porpoise, harbor seal, and others). In the next several years, logbook data and at-sea observer coverage will further document the extent of marine mammal take. Regulation of this fishery will depend on the exploitation rate caused by incidental kills in relation to the trajectory of the marine mammal populations with respect to OSP goals.

Target spawning biomass-per-recruit levels for most Gulf of Maine groundfish species regulated under the multispecies FMP are currently not being achieved. Stock sizes and catches of the groundfish stocks have generally declined in response to significant increases in overall fishing effort in recent years. Shrimp and hake stocks have fared better than groundfish stocks in recent years and landings have been good. There has, however, been considerable debate over the potential negative effects of catch and discard patterns within the small-mesh fisheries, because such discarding may affect the attainment of groundfish management targets. The situation is complicated because individual vessels usually participate in several of both the large- and small-mesh fisheries on a seasonal basis. In winter months, effort is transferred from the large-mesh trawl and gill-net fisheries primarily to the shrimp fishery. Curtailment of the shrimp fishery would result in a significant short-term revenue loss to fishermen and would increase effort in alternative fisheries (such as large-mesh trawling) that already have excess effort. Likewise, further regulation of the gill-net fishery to reduce harbor porpoise bycatch may drive effort into other fleet sectors (small- and large-mesh trawling), thereby potentially exacerbating overfishing of groundfish stocks.

Some debates over the merits of regulating small-mesh fisheries of the Gulf of Maine center on potential technological innovation to minimize the bycatch of sublegal groundfish in the shrimp fishery (with so-called separator trawls) and around reductions in the geographical distribution and overall level of effort in small-mesh fisheries (Murawski et al. 1991). What has not developed is a clear consensus on an overall objective function from the several demersal fisheries as an interacting system rather than a series of nonrelated fisheries. Clearly, the attainment of percentage of spawning biomass-per-recruit targets for all important species simultaneously is highly improbable because of the technological interactions among fisheries. Potential criteria for developing an overall strategy include revenue and profit maximization, species diversity of landings and stock abundance, stability in overall catches or catch rates, employment maximization, and minimization of marine mammal kills. Currently, however, most debate focuses on addressing single-species issues. Without an overall perspective and management approach, it is doubtful that a consensus for resolution of fishery interaction issues can be developed. No single regulatory entity has the mission to do so.

Ecological Realities

The ecological determinants of multispecies/multifishery interaction result from three basic processes: (1) co-occurrence of animals in space and time; (2) simultaneous capture because of indiscriminate harvesting technique; and (3) interdependencies among species because of predator–prey or competitive interactions. Co-occurrence is a requisite for simultaneous capture, but the selection of co-occurring species can be influenced by choice of gear (e.g., trawls, set lines, gill nets). Similarly, co-occurrence is necessary for predation, but predation effects are dictated by numerous other processes including predator–prey size compatibility and behavioral characteristics.

Some of the most vexing of current multispecies management problems arise from what are scientifically the most tractable of the above determinants of species interaction. Two examples of the conflict between resource protection statutes and fishery management occur in the controversies surrounding the capture of protected sea turtles in shrimp fishing and the unintentional killing of marine mammals in fishing operations. In both of these cases, the conflict arises because of the inadvertent killing of the protected species, which are not allowed to be marketed. Attempts to minimize such negative interactions have had as their scientific goal understanding the ecological and behavioral attributes of the species involved, so as to determine the finest time and space scales at which interactions occur. These studies have been the basis for technological development (conservation fishing gear and procedures) as well as area–time management of the fisheries aimed at minimizing interactions. Research on mixed-species assemblages has as its basis the need to evaluate spatial and temporal patterns in resource co-occurrence as a basis for optimizing fisheries on mixed resources (Murawski and Finn 1988). Environmental determinants of species abundance and distribution are thus important ecological relationships to be considered in multispecies studies.

Species interactions and community structures are influenced in part by what explicit management decisions are made (extraction or protection of particular components), along with an array of implicit management decisions for unregulated resources. Fishery ecosystems will act to integrate these decisions within the constraints imposed by primary production, trophic efficiency, and such factors as ecosystem stability, resilience (return rate after per-

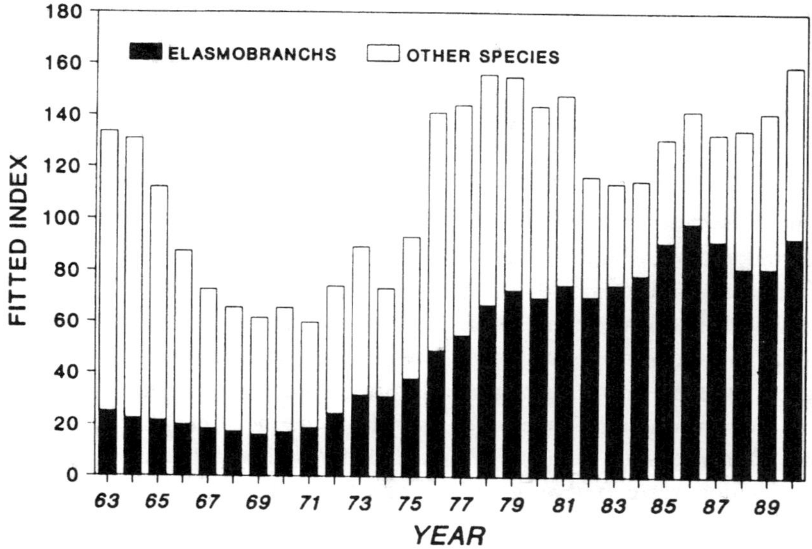

Figure 27-3. Bottom trawl survey abundance indices (in kilograms per tow, smoothed) for total mixed-species catches on Georges Bank, 1963–1990. Species are aggregated into two categories: small elasmobranch species (primarily spiny dogfish and seven species of skates) and other species (dominated by commercially valuable gadoid and flounder species).

turbation), and resistance [production flux related to species change (May 1974; Pimm 1982)]. The end result of the integration of these factors may be fishery ecosystems that are perfectly functional (and balanced), but that exhibit a number of undesirable characteristics from a biological or economic perspective.

An example of system state changes in reaction to alternative harvest policies is illustrated by groundfish fisheries from the Georges Bank region off the northeast United States. Figure 27-3 presents trawl survey data collected by the Northeast Fisheries Center. Trawl survey data are separated into two groups: small elasmobranchs (spiny dogfish and skates), and others (primarily commercially valuable gadoids and flounders). Abundance of both categories decreased precipitously during the 1960s, primarily in response to pulse-fishing by distant-water fleets. The proportion of elasmobranchs to other species remained relatively constant through this period, because distant-water fleets used virtually all trawlable species for human consumption or industrial purposes. Since the late 1960s, total trawlable biomass has increased to levels exhibited before intensive distant-water fisheries, but the proportion of elasmobranch biomass to that of other species has generally reversed. Marketable species have declined steadily in abundance since 1977, with spiny dogfish and skates constituting the bulk of research vessel catches in recent years. These patterns are corroborated by the species composition of commercial catches monitored at sea (Figure 27-4). Interestingly, multispecies size composition (relative numbers by length interval) has remained comparatively stable over the period, notwithstanding the dramatic changes in species composition (Murawski and Idoine 1989). This may imply that energy transfer among various functional trophic levels is a conservative process of such ecosystems and is thus preserved over a wide range of species compositional states. Whether species changes on Georges Bank are the result of biological interactions among species or are simply the result of differential fishing mortality rates remains conjectural. However, total biomass in the system does seem to have again reached a threshold. The ability to increase the abundance of marketed species may thus be limited by predation from or competition with the elasmobranch species.

Scientific Challenges

Management goals of preserving rare or otherwise protected species while allowing interacting fisheries to occur will test not only the ability of managers to effect appropriate system-based management policies, but also the ability of scientists to refine their understanding of the ecological determinants of species co-occurrence. Similarly, the goals of developing engineering solutions to reduce inadvertent captures while minimizing impact on otherwise legitimate fishing operations represents a very rigid set of specifications that bound a management resolution. The relative importance of the various criteria rests in the institutional and political arenas.

Our ability to direct fishery ecosystems toward a set of specific management goals is fundamentally dependent on our understanding of how such systems operate and how they respond to a complex set of simultaneous perturbations. The most robust set of predictions of community change are those based on empirical observation of the system at previous times, all else being equal (Sainsbury 1988). One problem with this approach is that even when time-series of data are sufficiently long, they usually encompass only a few selected ecosystem components (e.g., economic resources). Even then, our knowledge of dynamic relationships among components is usually fragmentary (Daan 1980).

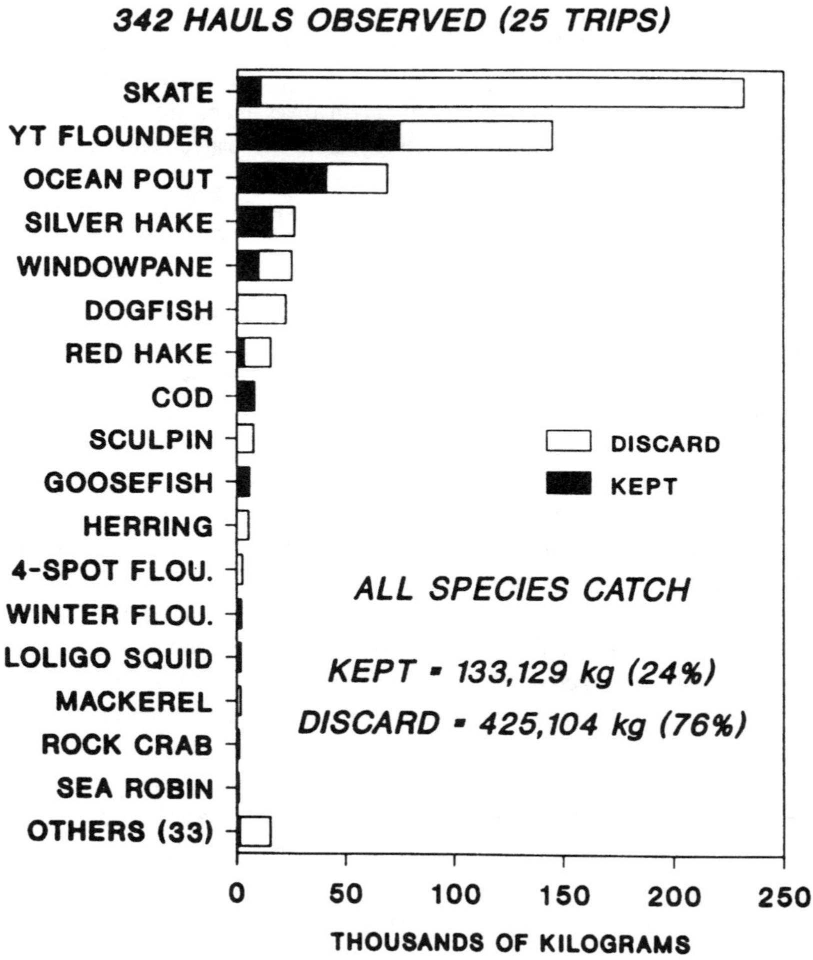

Figure 27-4. Species composition and disposition of the catch from otter trawl fishing for yellowtail flounder off the northeast United States in 1990. Data were obtained from trips on which scientific observers sampled catches at sea on a tow-by-tow basis. Landings from the trips are represented by the black bars and discards by open bars. Primary reasons for discards are the lack of markets for large quantities of some species (skates, dogfish, sculpin, etc.) and fish below legal minimum sizes (yellowtail flounder, Atlantic cod).

In the absence of historical analogy with which to predict ecosystem change, we must try to decipher the dynamic mechanisms underlying species and fisheries interactions. The basis for predicting changes in yield to a particular single- or mixed-species fishery (Y_i) involves analyzing the individual and cumulative effects of effort changes (*df*) in all associated fisheries (*j*):

$$dY_i = \frac{\delta Y_i}{\delta f_1} df_1 + \frac{\delta Y_i}{\delta f_2} df_2 + \ldots \frac{\delta Y_i}{\delta f_j} df_j$$

The matrix of partial derivatives of yield contingent on change of effort in each fishery (the Jacobian) specifies the relative magnitude and sign of yield change (tonnage or value) as a function of effort in the individual fisheries (ICES 1989). The individual matrix elements subsume the species interrelationships (predator–prey and competition) and the partial fishing mortality rates (F) generated on the mix of species taken by competing fisheries. Estimation of the elements of the matrix is an important requisite for management since it allows one to evaluate alternative management scenarios with respect to single- and multispecies goals.

The estimation of the Jacobian for technologically interrelated fisheries (such as the open-access Gulf of Maine system) may be every bit as complex as that for systems linked by biological interactions (ICES 1989, 1991). This is because of the complex set of factors determining fleet interactions, such as those controlling effort transfer among competing fisheries, the effects of catch and discards for mixed-species fisheries, and (in the case of economic yields) the economics of fishery production in relation to species abundance.

When fisheries target predators as well as prey, the Jacobian includes the feedback loops of instantaneous predation mortality rates (M2) on each prey species (i) and age group (j) caused by each predator species (k) and age group (l):

$$M2_{ijkl} = f[N_{ij}, N_{kl}] \mid N_{ijkl} \; \forall \, \text{i, j, k, l}$$

where $M2_{ijkl}$ is a function of the stock abundance (N) of prey species/age ij and predator species/age kl, and is contingent on the abundances of all potential predator and prey species (N_{ijkl}). The functional relationships between predation mortality rates and predator abundance in complex, many-species ecosystems depend on the availability and vulnerability of various alternative prey resources that are suitable for each predator group. Predation rates on particular species/ages of prey are thus likely to be highly dynamic. Abundance of each predator group is, of course, partially determined by the level of fishing mortality on them and by non-predation related sources of natural death (M1 + M2 = M). These equations apply as well to predator species that are not caught in directed fisheries but that may suffer mortality because of inadvertent bycatch, such as marine mammals (Harwood and Croxall 1988).

The estimation of dynamic feeding relationships depends on considerable knowledge of prey selectivity among the various predator species and ages, as well as the functional relationships of predation mortality to prey abundance (Holling 1965; Murdoch et al. 1975). Although estimation of such a set of variables or of functions is indeed a complex undertaking, it is within the capacity of current technical expertise and abilities to sample it in situ.

Such a project is ongoing for the North Sea fishery ecosystem (ICES 1989, 1991). A nine-species system including gadoids [cod; haddock (*Melanogrammus aeglefinus*); whiting (*Merlangius merlangus*); saithe (*Pollachius virens*); Norway pout (*Trisopterus esmarkii*)], herrings [Atlantic herring; sprat (*Sprattus sprattus*)], and others (mackerel; sand eel, Ammodytidae) is being modeled to evaluate the system-level effects of manipulating the harvesting intensity on fisheries for both predators and prey. The impetus to model the North Sea as a multispecies system stems from the long-term decline in stock biomasses of commercially important predator species (e.g., cod, haddock, whiting) combined with the development of large-scale industrial fisheries for prey species including sand eel, sprat, and Norway pout (Figure 27-5). Are these industrial fisheries for prey negatively influencing the human consumption fisheries for predators? A key element of the modeling has been an intensive stomach sampling program conducted in 1981 (the "Year of the Stomach" program), supplemented by additional years of feeding data for selected predators (cod, whiting).

The North Sea modeling approach has included a combination of retrospective analyses to tune the system to known fluctuations in catches of the various species as a basis of a forecast model for predicting changes in predation mortality rates in response to alternative fishing scenarios and fluctuations in recruitment. Results so far have challenged the scientific basis of some of even the most fundamental of single-species calculations supporting the fishery management decision process. For example, traditional yield-per-recruit analysis suggests that higher equilibrium cod yields would result from an increase in the trawl mesh size used when fishing for cod (see Figure 27-5). Multispecies calculations indicate, however, that because of the proportionally higher predation rate of larger cod on their young, and the release by the larger mesh of other species preying on young cod, the effect of increasing mesh size is, in fact, to reduce not only long-term cod yields but also yields for a variety of other species and the multispecies aggregate

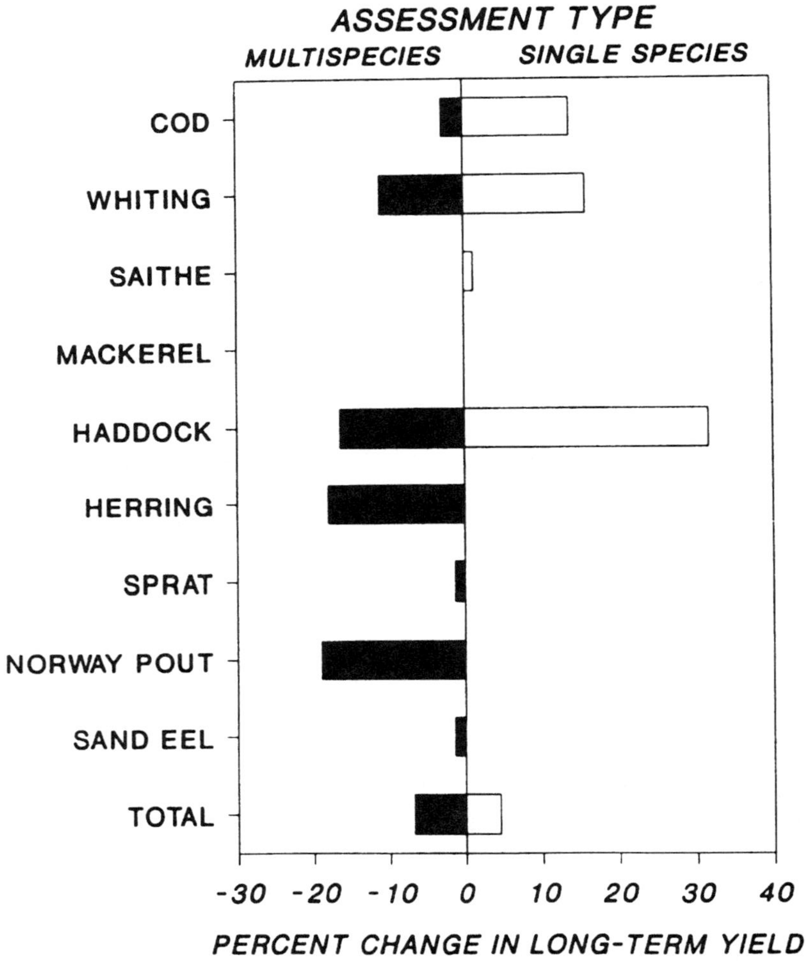

Figure 27-5. Percentage changes in long-term fishery yields for North Sea stocks resulting from an increase in trawl mesh size from 85 to 120 mm for the directed fishery for cod. Results are presented for traditional single-species/multi-fleet analyses, and for multispecies/multi-fleet analyses including interspecies predation rates. Lower yields in the multispecies cases result from greater predation rates from larger predatory fish (cod, whiting, haddock, saithe) released by the larger mesh (ICES 1989).

(see Figure 27-5). An ecosystem-level approach to this system thus indicates that even simple manipulations of the fishery may have complex interaction effects. Given the importance of these specific conclusions regarding the North Sea system, a repeat of the intensive 1981 stomach sampling project will be conducted in 1991 to better refine and validate the dynamic fishery models being constructed. An ecosystem-level approach thus need not entail a bottom-up understanding of all trophic levels contributing to fishery yields. As in the above example, fishery interaction studies are compatible with, and in fact require a basis in, traditional single-species population dynamics studies.

The challenges of providing a sound scientific basis for multispecies management relate both to understanding dynamic processes governing interactions and the precision of measurement of ecosystem change relative to the hypothesized process at work. It is possible, and indeed likely, that several alternative models of processes governing ecosystem change may be equally plausible, especially when beginning to analyze such problems (such as the Georges Bank example; see Figure 27-3). Sainsbury (1988) noted that three alternative models of fishery community change include (1) multiple single-species responses, assuming little interaction among species or fisheries; (2) biological interactions (predation and competition) governing system outcomes; and (3) changes in the physical environment (climate or habitat) that may alter community structure. Given our uncertainty about which of these mechanisms may dictate community dynamics, management of such systems must be regarded as experimental. Conducting fishery management programs as semi-controlled experiments implies that the system will be monitored with sufficient precision and scope to allow for testing of the original hypotheses of mechanisms governing the community. The scientific challenge then is to develop alternative hypotheses of mechanisms likely to govern community changes, and to put in place sampling programs that will ultimately allow rejection of untenable hypotheses.

Discussion

Can we manage our multispecies fisheries? Clearly, the operational word here is "manage." We can and are now manipulating our fishery ecosystems through a combination of explicit regimes (e.g., fishery regulations and protection statutes) and a series of implicit decisions for components for which there is little commercial or societal interest. The totality of such explicit and implicit manipulations determines current ecosystem structures and productivity. A far more compelling question is: Can we identify a target ecosystem structure and, through a series of management actions, manipulate the ecosystem toward that desired structure? A major impediment to achieving such a scenario is the lack of an institutional mechanism for doing so. Surprisingly, ecosystem goals can be and have been the subject of a number of international conservation agreements. Conservation objectives adopted under the Convention on the Conservation of Antarctic Marine Living Resources (CCAMLR) include (1) prevention of any harvested populations from falling below levels ensuring greatest net increment to stable populations, (2) maintaining ecological relationships between harvested, dependent, and related populations, (3) restoring depleted populations, and (4) preventing or minimizing risks of ecosystem change that are not reversible over two or three decades (Sherman and Ryan 1988). The International Commission for the Northwest Atlantic Fisheries (ICNAF) adopted a "second tier" quota system (less than the sum of individual species quotas) expressly incorporating the interdependence of harvested populations and the mixed nature of fisheries into management regulations (Anthony and Murawski 1986). Implementing objectives similar to those of ICNAF and CCAMLR for the U.S. EEZ resources will necessarily engender contentious debate among various segments of the harvesting industries and those interested in protection of certain species in the ecosystems. The issues will be politically difficult to resolve and will tax the ability of current science to provide adequate bases for decision making. However, the consequences of failing to resolve such issues are that the ecosystems will integrate whatever explicit and implicit decisions are made, whether desirable or not. We will simply have to accept the level of and variability in production and species mix dictated by the cumulative effects of external forces (fishing, pollution, and physical environmental variability) and internal structuring mechanisms that regulate energy compartmentalization and flux. As a society, we will have to be more cognizant that our consumer preferences for types and sizes of living marine resources and our decisions regarding which species to exclude from exploitation may have complex feedbacks into the ecosystems that produce these resources (see Figure 27-4). Management approaches to resolving multispecies conflicts have generally been based on systems of disincentives (e.g., bycatch quotas, area and fishery closures). More effective management of these conflicts might result from positive incentive plans, but they have not been widely explored. As we more fully utilize the diverse resources of our marine fishery ecosystems and extend protected status to a wider array of endangered and threatened species, the necessity to consider multispecies effects and to find more workable approaches to their mitigation becomes ever more urgent.

ACKNOWLEDGMENTS

Much of the content of this article is based on the work and experiences of my colleagues at the Northeast Fisheries Center and in the International Council for the Exploration of the Sea. In particular, I acknowledge the contributions of Tim Smith, Bill Overholtz, Mike Fogarty, and especially John Pope, in preparation of the manuscript. Dr. Robert Hofman of the Marine Mammal Commission provided a wealth of research material, as well as perspective on ecosystem aspects of marine mammal research and management. Finally, I appreciate the careful manuscript reviews of Dr. Brad Brown and an anonymous referee.

This article appeared in "Fisheries" 16(5):5–13.

References

Alaska Sea Grant. 1984. Proceedings of the Workshop on Biological Interactions Among Marine Mammals and Commercial Fisheries in the Southeastern Bering Sea. Alaska Sea Grant Report 84-1. Anchorage: Alaska Sea Grant.

Andersen, K. P., and Ursin, E. 1977. A multispecies extension to the Beverton and Holt theory of fishing, with accounts of phosphorous circulation and primary production. Medd. Dan. Fisk. Havunders. 7:319–435.

Anthony, V. C., and Murawski, S. A. 1986. Managing Multispecies Fisheries with Catch Quota Regulations—The ICNAF Experience. International Council for Exploration of the Sea. Coop. Res. Rept. 139, Annex 7:42–57. Copenhagen: ICES.

Beddington, J. R., and May, R. M. 1982. The harvesting of interacting species in a natural ecosystem. Sci. Am. 247(5):62–69.

Daan, N. 1980. A review of replacement of depleted stocks by other species and the mechanisms underlying such replacements. Rapp. P.-V. Réun. Cons. Int. Explor. Mer 177:405–421.

Dickie, L. M. 1973. Management of fisheries: Ecological subsystems. Trans. Am. Fish. Soc. 102:470–480.

Harwood, J., and Croxall, J. P. 1988. The assessment of competition between seals and commercial fisheries in the North Sea and the Antarctic. Mar. Mammal Sci. 4(1):13–33.

Henwood, T. A., and Stuntz, W. E. 1987. Analysis of sea turtle captures and mortalities during commercial shrimp trawling. Fish. Bull. 85(4):813–817.

Holling, C. S. 1965. The functional response of predators to prey density and its role in mimicry and population regulation. Mem. Entomol. Soc. Can. 91:293–320.

International Council for Exploration of the Sea (ICES). 1989. Report of the Multispecies Assessment Working Group. ICES C.M. 1989/Assess. 20. Copenhagen: ICES.

International Council for Exploration of the Sea (ICES). 1991. Report of the Multispecies Assessment Working Group. ICES C.M. 1991/Assess. 7. Copenhagen: ICES.

Larkin, P. 1977. An epitaph for the concept of maximum sustained yield. Trans. Am. Fish. Soc. 106:1–11.

May, R. M. 1974. Stability and Complexity in Model Ecosystems. Monographs in Population Biology 6. Princeton, NJ: Princeton University Press.

May, R. M., Beddington, J. R., Clark, C. W., et al. 1979. Management of multispecies fisheries. Science 205(4403):267–277.

McHugh, J. L. 1959. Can we manage our coastal fishery resources? Trans. Am. Fish. Soc. 88(2):105–110.

McHugh, J. L. 1988. Can we manage our Atlantic coastal fishery resources II? Mar. Fish. Rev. 50(4):41–45.

Mercer, M. C. (ed.). 1982. Multispecies Approaches to Fisheries Management Advice. Can. Spec. Publ. Fish. Aquat. Sci. 59.

Murawski, S. A., and Finn, J. T. 1988. Biological bases for mixed-species fisheries: Species co-distribution in relation to environmental and biotic variables. Can. J. Fish. Aquat. Sci. 45(10):1720–1735.

Murawski, S. A., and Idoine, J. S. 1989. Multispecies Size Composition: A Conservative Property of Exploited Fishery Systems? Northwest Atlantic Fish. Org. Sci. Counc. Rept. Doc. 89/76. Northwest Atlantic Fisheries Organization Scientific Council.

Murawski, S. A., Lange, A. M., and Idoine, J. S. 1991. An Analysis of Technological Interactions Among Gulf of Maine Mixed-species Fisheries. ICES Mar. Sci. Symp. 193. Copenhagen: ICES.

Murdoch, W. W., Avery, S., and Smyth, M. E. B. 1975. Switching in predatory fish. Ecology 56:1094–1105.

Pimm, S. L. 1982. Food Webs. London: Chapman and Hall.

Polacheck, T. 1989. Harbor porpoises and the gill net fishery. Oceanus 32(1):63–70.

Regier, H. A., and Henderson, H. F. 1973. Towards a broad ecological model of fish communities and fisheries. Trans. Am. Fish. Soc. 103:56–72.

Rodel, P. M. (ed.). 1975. Optimum Sustainable Yield as a Concept in Fisheries Management. Am. Fish. Soc. Spec. Publ. 9. American Fisheries Society.

Sainsbury, K. J. 1988. The Ecological Basis of Multispecies Fisheries, and Management of a Demersal Fishery in Tropical Australia. In J. A. Gulland (ed.), Fish Population Dynamics (2nd ed.). New York: John Wiley, pp. 349–382.

Sherman, K., and Ryan, A. F. 1988. Antarctic marine living resources. Oceanus 31(2):59–62.

Smith, T. D. 1983. Changes in size of three dolphin (*Stenella* spp.) populations in the eastern tropical Pacific. Fish. Bull. 81(1):1–13.

Smith, T. D. 1988. Stock Assessment Methods: The First 50 Years. In J. A. Gulland (ed.), Fish Population Dynamics (2nd ed.). New York: John Wiley, pp. 1–33.

Thompson, N. 1988. The status of loggerhead, *Caretta caretta*; Kemp's ridley, *Lepidochelys kempi*; and green, *Chelonia mydas*, sea turtles in U.S. waters. Mar. Fish. Rev. 50(3):16–23.

Volterra, V. 1926. Variations and fluctuations of the number of individuals in animal species living together. J. Cons. Int. Explor. Mer 3:3–51.

28

Potential Benefits from Efficient Harvest of New England Groundfish

Steven F. Edwards and Steven A. Murawski

Abstract

The dissipation of food and economic value from commercial harvest of the multispecies groundfish resource in U.S. waters off New England is reported. Net economic value to the nation would be maximized by an estimated 70% reduction in fishing effort, resulting in a sevenfold increase in the size of the harvestable resource and a threefold increase in sustainable yield. Under these conditions, the fishing industry and consumers would benefit from an estimated \$150 million increase in sustainable net economic value each year. New England consumers could profit further from the nutritional and health benefits of an additional 6 pounds of fresh fish per capita. Private property and common property may be the most effective controlled access options to achieve the estimated food and economic values. These options could evolve from market transfers of individual landings or effort quotas for self-sustaining assemblages of groundfish species.

H. Scott Gordon's (1954) seminal economic theory of open access and the dissipation of the value of an open access resource is perfectly illustrated by events in the traditional multispecies groundfish fishery in U.S. waters off New England. Despite nearly four decades of management by the International Commission for Northwest Atlantic Fisheries and, since 1977, the New England Fishery Management Council, virtual open access to commercially vital resources of Atlantic cod (*Gadus morhua*), haddock (*Melanogrammus aeglefinus*), and yellowtail flounder (*Limanda ferruginea*) has led to record high levels of fishing effort and mortality and record low levels of resource abundance and landings (Anthony 1990). The fishery now survives on juveniles of these species and landings of previously underutilized resources of American plaice (*Hippoglossides platessoides*), witch flounder (*Glyptocephalus cynoglossus*), and other groundfish, many of which have also become overfished recently (NEFC 1991).

In this chapter, estimates of the dissipation of the New England groundfish food resource and its economic value to the fishing industry and consumers are reported. Also, different controlled access policies to rationalize the fishery are contrasted.

Economic Efficiency as a Fishery Management Policy

In section 1801.(c) of the Magnuson Fishery Conservation and Management Act of 1976 (Magnuson Act), Congress identified economic efficiency as a matter of policy for fishery management in the United States. Economic efficiency is maximized when the net economic value (NEV), or the difference between the total social utility of seafood consumption and the total social opportunity costs of producing seafood, is greatest. Economists measure social utility in terms of the most that people are willing to spend for a particular good or service (including an environmental good or service) rather than do without it; the amount is greater than actual expenditures. Social opportunity costs are defined, in part, as the value foregone when labor and capital are used to produce seafood (i.e., catch, process, distribute, wholesale, retail, and manage fish) instead of something else valued by consumers. In addition, the foregone potential growth of harvested fish is part of the total social opportunity costs of using a renewable resource.

The Magnuson Act also stipulates that goals other than economic efficiency are important in fisheries management. However, one such possible goal, maximizing food production, is difficult to justify in the United States where, overall, there is no shortage of dietary protein. Whether markets would deliver additional seafood to the malnourished at prices that they could afford is speculative. Furthermore, landings that are significantly greater than those that correspond to maximum economic efficiency would lower the price of seafood, causing a reduction in the demand for and consequent production of chicken and other meats that consumers consider to be substitutes for seafood. Thus, production at maximum sustained yield would not necessarily increase the nation's total supply of food.

The components of NEV that could be measured in our study are consumer surplus and resource rent. Consumer surplus measures that fraction of total social utility that consumers enjoy at no cost—i.e., a "profit" for consumers. The estimates of potential gains in consumer surplus in our study can be thought of in terms of being able to buy greater amounts of seafood at lower prices. Resource rent is the economic value of a fish resource that, like labor and capital, is a factor of production in the harvest of groundfish. The rental value of a fish resource is analogous to the value of farmland, forest land, or residential land. Although harvesters pay for use of all other publicly owned natural resources—including energy resources and hard minerals on the continental shelf (Mead et al. 1985)—the Magnuson Act probably precludes charging fishermen for use of marine fish stocks. Unless the Magnuson Act is amended, resource rents would accrue to fishermen in the form of profits.

Compared with NEV under maximum economic efficiency, NEV from harvest of a depleted fish resource is expected to be relatively small. Economists maintain that without property rights to a fish resource, no one fisherman has an incentive to conserve it. Fishermen seek to maximize their return in a competitive industry only with respect to the costs of labor, capital, and material factors of production. Consequently, fishing effort expands until the profits attributable to the resource are dissipated (Gordon 1954). Also, harvest of a depleted biological resource and, therefore, consumer surplus are smaller than when NEV is maximized.

Methodology

Mathematical Model

Models that represent and relate the behaviors of ecological and economic systems are needed to estimate the waste associated with exploitation of common property resources, such as New England groundfish. Accordingly, the most efficient fishery management policy can be expressed as follows:

$$\text{maximize}_{X,Q,T}\ J\{Q\} = \int_{o}^{\tau} e^{-\delta \cdot T}[U(Q,T) - C(X,Q,T)]dT \qquad (1)$$

subject to the state equation

$$X_T = F(X) - Q(T),\ X(0) = X_0$$

and to the constraints

$$X(T) \geq 0,\ Q(T) \geq 0$$

Above, X is the biomass of the harvestable resource; Q is harvest; T is the time period $(0, \ldots \tau)$; $U(Q,T)$ is the total social utility of seafood consumption; $C(X,Q,T)$ is the total social opportunity costs of seafood production; δ is the social discount rate; $U(Q,T) - C(X,Q,T)$ is NEV; and subscripts denote partial differentiation. Social utility is written as a function of time to allow for changes in consumers' preferences for seafood. Costs are written as a function of time because technological change lowers harvesting and other production costs.

The dynamic nature of the optimization can be simplified with resort to the current value Hamiltonian:

$$H = e^{\delta \cdot T}\left[U(Q,T) - C(X,Q,T)\right] + \mu\left[F(X) - Q\right] \tag{2}$$

where μ is a Lagrangian multiplier representing the marginal value of a fish resource during a time period. Maximization of functionale (2) results in the following optimum rule:

$$\delta = F_x - \frac{C_x + C_{QX}\left[F(X) - Q\right] - \left[U_{QT} - C_{QT}\right]}{U_Q - C_Q} \tag{3}$$

as reported by Clark (1976; Ch. 5). The subscripts denote first and second partial differentiation. U_Q is the demand for fish in the landings market, and U_{QT} is the rate of change of demand over time. C_Q is the marginal cost of supplying fish to the landings market, and C_{QT} is the rate of change of marginal costs over time.

Empirical Models

To estimate the waste of food and economic value caused by depletion of the New England groundfish resource, empirical models were substituted for the generic functions in rule (3). Growth of the groundfish resource was modeled with the Generalized Stock Production Model (GSPM) of Pella and Tomlinson (1969):

$$F(X) = H \cdot X^m - K \cdot X \tag{4}$$

where *H*, m, and *K* are variables that characterize growth.

As in other empirical bioeconomic analyses of fisheries (e.g., Bell 1972; Cook 1988), the dearth of microeconomic time-series data on crew size and vessel technology necessitates use of the following linear technological production function:

$$Q = q(T) \cdot E \cdot X \tag{5}$$

where *E* is a composite index of fishing effort (i.e., a combination of crew, capital, and material inputs, such as days fished), and $q(T)$ is the catchability coefficient, or the percentage of a resource that is harvested by a unit of fishing effort. Catchability is modeled as a function of time to reflect advances in fishing technology. Judging from nominal catchabilities since the year 1976, a 2% rate of increase compounded annually was a reasonable first approximation for the rate of technological change. Therefore,

$$q(T) = q \cdot [1.02]^T$$

where q is a constant catchability variable, and $T = 1, \ldots 14$ during 1976–1989.

The cost function,

$$C = c\left[\frac{Q}{q(T) \cdot X}\right] = C(X,T) \cdot Q \tag{6}$$

is derived by minimizing the costs of fishing effort subject to the harvesting constraint imposed by the technological production function. The variable c is the unit cost of fishing effort (i.e., the market costs of labor, capital, and material inputs, such as fuel and ice).

The absence of vessel cost data necessitates use of a behavioral assumption to derive the unit-cost variable. As noted above, one expects fishing effort to expand until the average cost of harvesting fish is equal to landings price. Therefore, from equation (6):

$$AC = \frac{c}{q(T) \cdot X} = P$$

where AC is average harvesting costs, and P is landings price. On transposition,

$$c = P \cdot q(T) \cdot X \tag{7}$$

Information on landings price, catchability, and resource size allows calculation of unit-cost variables from equation (7).

Finally, the demand for groundfish landings in New England is modeled as follows:

$$P = f + g \cdot Q + h \cdot T \tag{8}$$

where f, g, and h are variables that reflect consumer preferences. The time period ($T = 1, \ldots 28$ during 1962–1989) serves to account for all factors correlated with the passage of time, including trends in imports, growth in the population of consumers, and strengthened preferences of consumers for seafood (Edwards 1992).

After appropriate differentiation of models (4), (6), and (8), rule (3) can be solved numerically for the resource size that maximizes NEV. Optimum harvest and fishing effort can then be derived from models (4) and (5), respectively, thereby allowing calculation of potential increases in NEV relative to an open access situation.

Data and Variable Estimates

Data on total U.S. and Canadian groundfish landings and fishing effort by otter trawl fleets on Georges Bank and in the Gulf of Maine (NAFO statistical areas 5Y and 5Ze) during 1976–1989 are from the National Marine Fisheries Service (NMFS) and the Northwest Atlantic Fisheries Organization. As noted above, otter trawlers land various demersal species in addition to Atlantic cod, haddock, and yellowtail flounder.* Nominal fishing effort (days fished) is standardized by vessel length and adjusted by the catchability factor, $[1.02]^T$, to reflect advances in fishing technology. Additional data from NMFS on landings and nominal prices in New England during 1962–1989 are used to estimate the landings demand model with the Cochrane-Orcutt iterative least squares procedure in the econometrics package SHAZAM (White 1987). Landings prices are adjusted for inflation and expressed in 1989 U.S. dollars. The average of the annual values of the unit cost during 1976–1989 is used. Finally, based on a recent evaluation of social discounting (Howe 1990), a 2% social discount rate was applied. Estimates of variables are reported in Table 28-1.

To the extent possible, the period 1976–1989 is used to control for management, ecological, and fishery factors exogenous to our simple model. This period corresponds with the Magnuson Act era and therefore is relevant to the current management regime. In addition, an implicit assumption of the GSPM and other single-species stock production models is that all factors other than the population being studied are constant. However, community structure is known to have shifted from dominance by gadoids before the 1970s to dominance by elasmobranchs currently (Sissenwine and Cohen 1991). Finally, it is not yet possible to harmonize data on foreign fishing effort with domestic effort before the Magnuson Act (Mayo et al. 1992).

---

*The Northeast Multispecies Fishery Management Plan (NEFMC 1991) manages American plaice (*Hippoglossoides platessoides*), witch flounder (*Glyptocephalus cynoglossus*), winter flounder (*Pseudopleuronectes americanus*), windowpane flounder (*Lophopsetta maculata*), pollock (*Pollachius virens*), white hake (*Urophycis tenuis*), silver hake (*Merluccius bilinearis*), red hake (*Urophycis chuss*), ocean pout (*Macrozoarces americanus*), and redfish (*Sebastes fasciatus*). Other groundfish species landed by the New England otter trawl fleet include cusk (*Brosme brosme*), and Atlantic wolffish (*Anarhichas lupus*). In addition, negligible amounts of shellfish are landed (about 5% of the total harvest and revenues), particularly pink shrimp (*Pandalus borealis*), northern lobsters (*Homarus americanus*), and sea scallops (*Placopecten magellanicus*).

Table 28-1. Estimates of variables of the bioeconomic model

Model component	Variable estimate
Stock production, $F(X) = H \cdot X^m - K \cdot X$	
H	−0.0005
m	1.9
K	−0.40
$N = 14; R^2 = 0.41$	
Harvest technology, $Q = q \cdot [1 + \delta]^T \cdot E \cdot X$	
q	0.007
Ex-vessel demand, $P = f + g \cdot Q + h \cdot T$	
f	0.44 (8.7)
g	−0.0010 (−4.7)
h	0.0121 (10.1)
$N = 28; R^2 = 0.94$	

Note: F(X) is surplus production (thousand metric tons); X is harvestable biomass (thousand metric tons); Q is harvest (thousand metric tons); E is fishing effort (thousand standardized days fished); δ is the social discount rate (i.e., 0.02 here); P is landings price ($ U.S. per pound in 1989 dollars); T is the time period; N is number of observations; R^2 is coefficient of determination; and t statistics are reported in parentheses for the demand model.

Potential Increases in Net Economic Value

Economic Overfishing

Figures 28-1 and 28-2 portray the extent of economic overfishing of the New England groundfish resource. Notice that the GSPM software adjusts for disequilibrium in the fishery. That is, observed landings during the study period could not be sustained by the extant resources and fishing effort.*

The size of the harvestable resource estimated to maximize NEV is about seven times greater than the 1989 resource (Figure 28-1). Optimum fishing effort is estimated to be 70% less than during 1989 (Figure 28-2).† Landings by an otter trawl fleet that maximizes NEV are estimated to be about 40% greater than landings in 1989 and three times greater than that estimated to be sustainable by the 1989 resource (see Figures 28-1 and 28-2).

Since the World Court divided Georges Bank between Canada and the United States in November 1984, U.S. landings of groundfish from the com-

*By sustainable, we do not imply constant harvest year after year. These estimates are best understood as being average conditions.

†Gates and Norton (1974) reported that the domestic fleet fishing for yellowtail flounder on Georges Bank and the Southern New England grounds during the early 1970s would have to be reduced by nearly 70% in order to maximize economic efficiency.

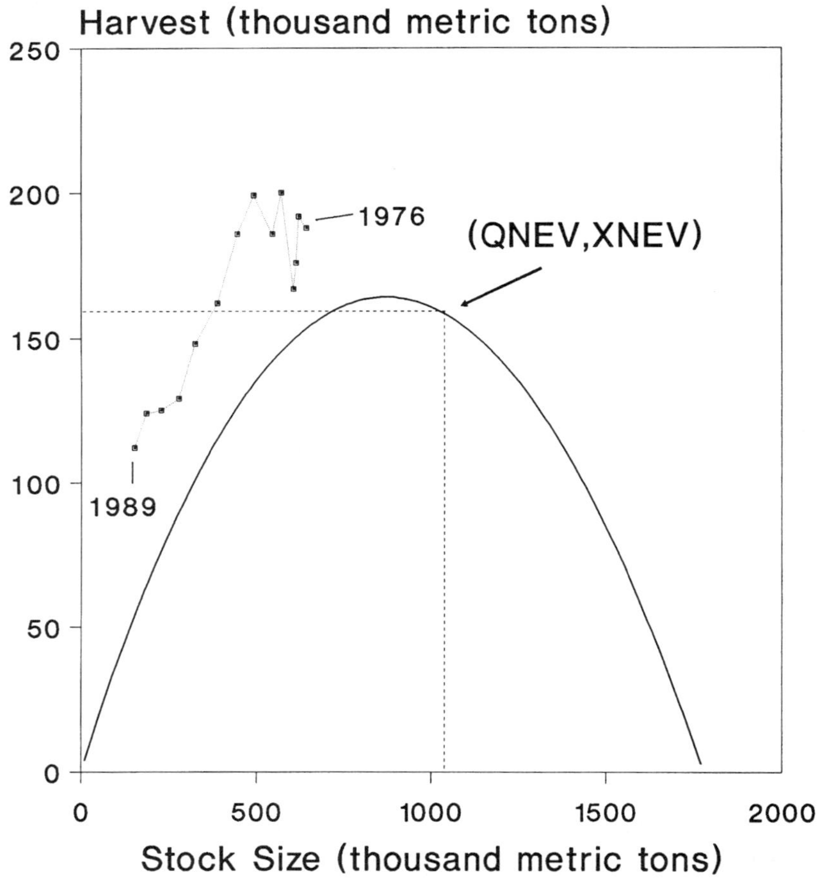

Figure 28-1. Bioeconomic equilibria relative to the size of the harvestable New England groundfish resource. Net economic value is maximized at (QNEV, XNEV). Data are also shown.

bined areas of Georges Bank and the Gulf of Maine have averaged 86% of total U.S. and Canadian landings. Using this percentage to prorate optimum yield, per capita consumption of groundfish by New Englanders could be increased by 2.5 lb when compared with 1989 landings, or 6.3 lb when compared with the harvest that could be sustained by the 1989 resource. These amounts compare with a national average of 10.2 lb of fresh or frozen seafood per capita in 1989 (USDOC 1991) and therefore represent a significant nutritional and health benefit.

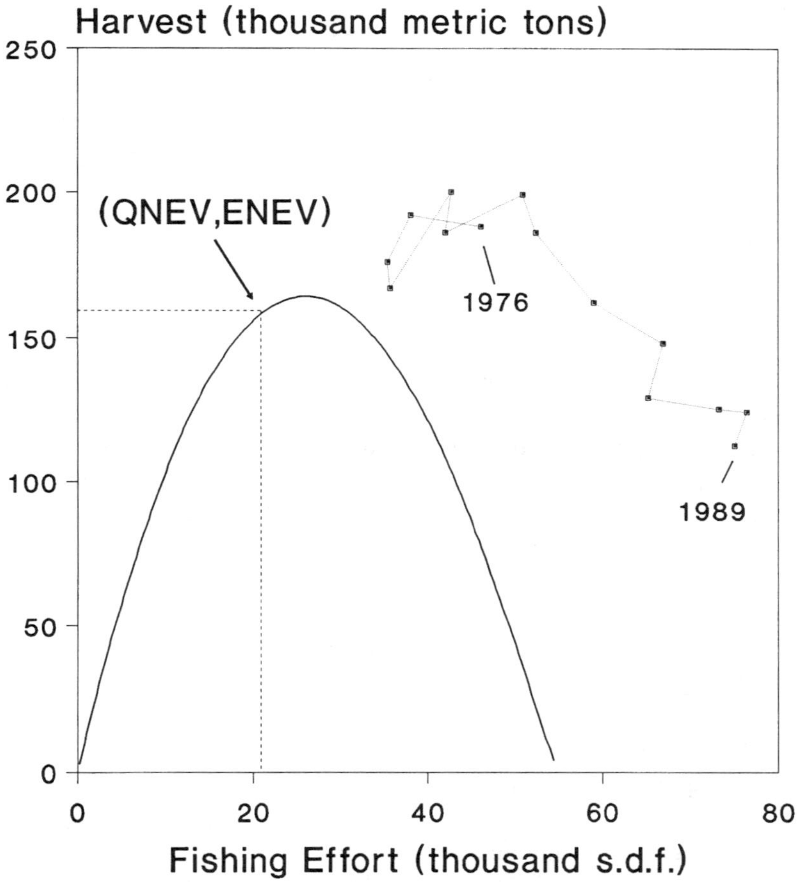

Figure 28-2. Bioeconomic equilibria relative to fishing effort by the otter trawl fleet. Net economic value is maximized at (QNEV, ENEV). Data are also shown. (s.d.f.= standardized days fished [see text].)

Dissipated Economic Value

The potential sustainable increase in NEV in the United States is estimated to be about $150 million a year (Table 28-2). About 90% of this value is resource rent, or what would be profit in the fishing industry. However, the estimated gain in consumer surplus of about $20 million a year is not negligible.

Table 28-2. Estimates of annual net economic value from commercial harvest of groundfish in U.S. waters off New England

Value*	Open access in 1989	Sustainable by 1989 resource	Optimum (MEE)	Sustainable net gain
Resource rent	0	0	128	128
Consumer surplus	10	2	21	19
Total net economic value	10	2	149	147

MEE = maximum economic efficiency.
*Value in $ U.S. million in 1989 dollars.

Mapping the optimum time path to maximum NEV was beyond the scope of this study. However, the summation of NEVs over time gives an indication of the magnitude of potential economic value (or waste, if open access continues). Using 10 years as an upper bound for recovery of the groundfish resources (NEFC 1991) and assuming further that no benefits would accrue during the interim, the net present value of $150 million over a 30-year period is about $2 billion when a 2% social discount rate is applied. Over a 50-year period, the total potential gain in net present value is over $3 billion. Despite the rough nature of these calculations, it should be clear that the fishing industry and consumers risk losing—and probably have already foregone since the Magnuson Act took effect—billions of dollars of economic value from the groundfish resource.

Although large, our estimates of potential gains in NEV may be conservative. The growth potential of the groundfish resource is underestimated because of unknown amounts of discards. For example, the Northeast Fisheries Center in Woods Hole, Massachusetts, estimates that discards of undersized yellowtail flounder in the Gulf of Maine and Georges Bank otter trawl fishery amounted to over 20% of sea-sampled catches by weight during 1989. Also, a return to a gadoid- and flounder-dominated community structure could increase the production potential of groundfish well beyond that which could be modeled here. In addition, an increase in resource size would be characterized by an increase in the age and size structures of the harvestable fish. However, the higher value of larger fish is not taken into account when estimating demand (see Gates and Norton 1974). Finally, without microeconomic data on vessel costs and productivities, it is not possible to assess either the cost savings from relaxing regulation-induced inefficiencies or the economic profit earned by inframarginal fishermen, such as highliners.

Management Alternatives

Maximizing NEV through efficient use of the New England groundfish resource is neither inexpensive nor easy. Rationalizing the fishery would be costly in terms of the labor and capital used in research, management, and enforcement—costs that must be subtracted from the estimate of potential gain in NEV reported in Table 28-2.

In addition, not all management options would be equally effective at maximizing NEV. Since the Magnuson Act banned foreign fleets from freely fishing within 200 miles of the U.S. coastline, the New England Fishery Management Council (the New England Council) has attempted to micromanage fishing mortality by instituting mesh-size restrictions, minimum sizes of fish, and trip limits. The failure of this policy is apparent from the status of the groundfish resources and fishery.

Depletion of the once productive New England groundfish resource and of other fish resources throughout the world has helped focus attention on the fundamental cause of overfishing—i.e., a lack of property rights to fish resources. This problem is differentiated from a preoccupation with symptoms of inadequate property rights, such as high fishing mortality and excess capacity and effort.

Limited entry and individual transferable quotas (ITQs) have been applied in Australia, Canada, Iceland, New Zealand, and the United States to rationalize harvest of fish resources (Muse and Schelle 1989). Sales of licenses and of vessel quotas reveal that property rights confer value to fish resources (see Meany 1979; Geen and Nayar 1988; Muse and Schelle 1989; Wilen 1989; Ackroyd et al. 1990). However, neither limited entry nor ITQs comprehensively address the problems inherent with common property and government regulation. Limited entry would "close" the New England groundfish fishery to a net increase in the number of vessels, but the excess effort already in place would perpetuate common property losses.* Most likely, managers and the fishing industry would enter a wasteful and futile cycle of effort controls and "capital stuffing" that would make fishing technology increasingly inefficient and would fail to control fishing mortality (Wilen 1989).

An ITQ system would be superior to limited entry by providing fishermen incentives to economize on fishing effort. However, landings quotas do not influence fishermen to husband a fish resource (Keen 1983; Copes 1986), par-

*Note that property has different dimensions. Limiting the number of fishermen will add a degree of necessary exclusivity to the fishery. However, without further controls on access by participants, common property and associated wastes persist.

ticularly a multispecies resource. Consequently, the sixfold differential in landings prices for different species and sizes of New England groundfish would induce fishermen to "highgrade" catches in order to maximize profit from landings quotas, rather than to maximize the NEV of the resource for the benefit of the nation.* Discarding bycatch when one lacks a landings quota or when the aggregate fishery quota is filled can be a related problem in a multispecies ITQ fishery, such as in New Zealand (Ackroyd et al. 1990). The waste associated with these specific forms of discarding should not plague an effort quota system, but an effort quota system would be susceptible to technology creep and the tendency to micromanage fishing effort. Furthermore, managing and enforcing an effort quota system would be expensive to taxpayers and to gross regional and national products.

In addition to these and other drawbacks of ITQs that Copes (1986) has already detailed, there is the potentially significant, albeit less appreciated, issue of rent dissipation "on land." With ITQs, the distribution of value would become the salient concern of special interest groups, which would align themselves by state, vessel classes, and type of use, including recreational fishing and marine mammal protection. The time and effort used to lobby Congress, the New England Council, and the NMFS for favorable groundfish allocations could dissipate much of the potential resource rent and thereby diminish regional production. Ackroyd et al. (1990) and Hide and Ackroyd (1990) characterized such dissipation of resource rents in the multispecies fisheries in New Zealand.

Sole ownership shows more promise for maximizing NEV from use of fish resources, including New England groundfish. A policy of sole ownership was recently resurrected by Keen (1983, 1988) in the fishery economics literature. Based on their observations of the performance of ITQs in New Zealand, Ackroyd et al. (1990) and Hide and Ackroyd (1990) also have promoted private sole ownership.

Sole ownership is distinguished from other rights-based policies by the incentive it provides harvesters to husband a fish resource (Keen 1988). There are other potential advantages of sole ownership, however. Given the owner's vested interest, and depending on the need for state control, the social and financial costs of managing the groundfish resource could be reduced substantially. That is, there should be less need to monitor and police the sole

*Highgrading occurs when catch is sorted by the value of its contents and the lesser valued portion is discarded. A quantitative understanding of highgrading in ITQ fisheries is lacking except for anecdotal reports and one study that places it at 30–50% of total catch in the Lake Superior lake trout fishery (Muse and Schelle 1989).

owner than the many fishermen competing for a resource in a limited entry or ITQ fishery. In addition, if resource rights are transferable, an owner would consider the value of alternative uses of his or her quota. Markets would allocate exclusive rights to the highest valued use, including, possibly, recreational fishing or marine mammal protection as well as commercial fishing. The inefficiencies and costs associated with political allocations would be largely overcome.

Despite these potential comparative advantages, sole ownership suffers from being misunderstood and ignored. Concerns of sole ownership leading to a monopoly or it being unconstitutional may be overstated. Taking consumer preferences into account, a sole owner of the New England groundfish resource would compete with substantial imports of groundfish from Canada, domestic landings of other fish, and, increasingly, the aquaculture industry. Also, regarding the legality of sole ownership, it may appear from the actions of state and federal governments that all citizens have a right to access fish stocks. Where fishermen unions and trade associations have agreed to informal controls on effort and catch, their actions have been judged as violations of the Sherman Antitrust Act (Johnson and Libecap 1982). However, sole ownership is not necessarily unconstitutional provided that it is established for the benefit of the nation and is not discriminatory (Fletcher 1965). Furthermore, there are already many cases where fishing grounds are privately owned, particularly oyster beds along the coasts of the Atlantic Ocean and Gulf of Mexico (Agnello and Donnelley 1979).

The various possible forms of sole ownership apparently are not recognized either. Centralized federal ownership (vis-à-vis the current system of eight regional councils and NMFS) is one possibility, although the public sector has failed to allocate other natural resources on federal lands efficiently, particularly forests, range, and water (Deacon and Johnson 1985; Libecap 1989; Anderson and Leal 1991). Private ownership may not be a satisfactory alternative, either, because a profit-oriented corporation would not seek to maximize NEV, although maximum NEV could result depending on competition from other product sources and on the corporation's private discount rate (Copes 1972). A third possibility is common property (vis-à-vis open access, or "no property") whereby a group of people—not government—determines harvest policy.

Libecap (1989) explains that history determines the kind of property rights to natural resources that can be adopted at any time. Accordingly, the Magnuson Act, the fishery management community's apparent reluctance to adopt property rights, and the political influence of special interest groups make private or common ownership of fish resources in U.S. waters an impractical and unlikely prospect for the near future. However, momentum may be building for adoption of ITQs in fisheries throughout the United States. The ITQs for groundfish could be designed to facilitate the eventual evolution of such ownership. In particular, quotas for landings or fishing effort could be defined for self-sustaining assemblages of groundfish species that are combined in terms of close ecological and technological interactions over time and space. Research by Gabriel (1989) suggests that groundfish occurring on Georges Bank and in the Gulf of Maine may qualify as separate sole ownership resources, but the single, larger scale of the Northeast Shelf ecosystem is also a possibility.

For private or common ownership to evolve from an ITQ system, quotas for groundfish resources would have to be exclusive, secure, and transferable. Concentration may be complete, or a manageable number of quota holders could be "unitized." Under unitization, harvesting rights would be delegated through negotiation among quota holders to a single resource harvester, with net revenues being apportioned according to the negotiated value of separate quotas. Unitization is used by oil companies that own leases to a common petroleum reservoir on outer continental shelf lands (Libecap 1989; Mead et al. 1985).

As noted above, private or common ownership of the New England groundfish resource could also marshall the unappreciated ability of markets to allocate the environment to recreation or conservation.* Associations of party or party boat owners or of recreational fishermen could purchase or lease rights to fish resources. This is practiced in Great Britain and by Trout Unlimited in the United States. Alternatively, conservation organizations could purchase and not use the right to fish during certain times of the year and in specific areas in order to protect marine mammals from fishing gear. The success of conservation organizations such as the Nature Conservancy and the Audubon Society reveals how markets can be used to preserve wildlife.†

Summary and Conclusions

In this chapter, a bioeconomic model of the multispecies groundfish fishery in U.S. waters off New England was reported and used to estimate the waste of food and economic value resulting from open access exploitation of the common property resource. The fishing industry and consumers could profit from an estimated sustainable increase of $150 million in NEV a year. In addition,

*See Anderson and Leal (1991) for a wide ranging discussion of "free market environmentalism."
†A report on sole ownership is available from the first author on request.

the quantity of fresh seafood available to consumers in New England could increase by up to about 6 lb per capita. However, the failure of effort controls and the evidence that demand has increased over time portends further waste and the possible demise of the fishery.

We argued that limited entry and ITQ policies do not comprehensively address the problems inherent with open access or management by the public sector. However, private or common ownership could evolve from an appropriately configured ITQ or a transferable license system that assigns exclusive and secure property rights to self-sustaining assemblages of the groundfish resources.

References

Ackroyd, P., Hide, R. P., and Sharp, B. M. H. 1990. New Zealand's ITQ System: Prospects for the Evolution of Sole Ownership Corporations. Unpublished Report. New Zealand: Centre for Resource Management, Lincoln University.

Agnello, R. J., and Donnelley, L. P. 1979. Prices and property rights in the fisheries. South. Econ. J. 42:253–262.

Anderson, T. L., and Leal, D. R. 1991. Free Market Environmentalism. Boulder, CO: Westview Press.

Anthony, V. C. 1990. The New England groundfish fishery after 10 years under the Magnuson Fishery Conservation and Management Act. North Am. J. Fish. Manage. 10:175–184.

Bell, F. W. 1972. Technological externalities and common-property resources: An empirical study of the U.S. northern lobster fishery. J. Polit. Econ. 80:148–158.

Clark, C. W. 1976. Mathematical Bioeconomics: The Optimal Management of Renewable Resources. New York: John Wiley.

Cook, B. A. 1988. Discount effects and Canada's Pacific halibut fishery. Mar. Res. Econ. 5:71–77.

Copes, P. 1972. Factor rents, sole ownership and the optimum levels of fisheries exploitation. The Manchester School of Economics 40:145–163.

Copes, P. 1986. A critical review of the individual quota as a devise in fishery management. Land Econ. pp. 278–291.

Deacon, R. T., and Johnson, M. B. 1985. Forestlands Public and Private. San Francisco: Pacific Research Institute for Public Policy.

Edwards, S. F. 1992. Evidence of structural change in preferences for seafood. Mar. Res. Econ. 7:141–151.

Fletcher, R. L. 1965. Constitutionality of Limiting the Number of Fishermen in a Commercial Fishery. In J. A. Crutchfield (ed.), The Fisheries: Problems in Resource Management. Seattle: University of Washington Press, pp. 95–119.

Gabriel, W. G. 1989. Persistence in Northwestern Demersal Fish Assemblages. NAFO SCR Doc. 89/77.

Gates, J. M., and Norton, V. J. 1974. The Benefits of Fisheries Regulation: A Case Study of the New England Yellowtail Flounder Fishery. Mar. Tech. Rept. No. 21. Kingston, RI: University of Rhode Island.

Geen, G., and Nayar, M. 1988. Individual transferable quotas in the southern bluefin tuna fishery: An economic appraisal. Mar. Res. Econ. 5:365–387.

Gordon, H. S. 1954. The economic theory of a common property resource. J. Polit. Econ. 62:124–142.

Hide, R. P., and Ackroyd, P. 1990. Depoliticising Fisheries Management: Chatham Islands' Paua (Abalone) as a Case Study. Unpublished report. New Zealand: Centre for Resource Management, Lincoln University.

Howe, C. W. 1990. Introduction: The social discount rate. J. Environ. Econ. Manage. 18(S-2):S-1–S-2.

Johnson, R. N., and Libecap, G. D. 1982. Contracting problems and regulation: The case of the fishery. Am. Econ. Rev. 72:1005–1022.

Keen, E. A. 1983. Common property in fisheries: Is sole ownership an option? Mar. Pol. 7:197–211.

Keen, E. A. 1988. Ownership and Productivity of Marine Fishery Resources: An Essay on the Resolution of Conflict in the Use of the Ocean Pastures. Blacksburg, VA: McDonald and Woodward.

Libecap, G. D. 1989. Contracting for Property Rights. New York: Cambridge University Press.

Mayo, R. K., Fogarty, M. J., and Serchuk, F. M. 1992. Aggregate fish biomass and yield on Georges Bank, 1960–1987. J. Northwest Atlantic Fish. Sci. 14:59–78.

Mead, W. J., Moseidjord, A., Muraoka, D. D., and Sorenson, P. E. 1985. Offshore Lands: Oil and Gas Leasing and Conservation on the Outer Continental Shelf. San Francisco: Pacific Institute for Public Policy Research.

Meany, T. F. 1979. Limited entry in the western Australia rock lobster and prawn fisheries: An economic evaluation. J. Fish. Res. Board Can. 36:789–798.

Muse, B., and Schelle, K. 1989. Individual Fisherman's Quotas: A Preliminary Review of Some Recent Programs. CFEC 89-1. Juneau, AK: Alaska Commercial Fisheries Entry Commission.

Northeast Fisheries Center (NEFC). 1991. Status of the Fishery Resources Off the Northeastern United States for 1990. NOAA Tech. Mem. NMFS-F/NEC-81. Woods Hole, MA: National Oceanic and Atmospheric Administration.

New England Fishery Management Council (NEFMC). 1991. Amendment #4 to the Fishery Management Plan for the Northeast Multispecies Fishery. Saugus, MA: New England Fishery Management Council.

Pella, J. J., and Tomlinson, P. K. 1969. A generalized stock production model. Bull. Int. Am. Trop. Tuna Comm. 13:421–458.

Sissenwine, M. P., and Cohen, E. B. 1991. Resource productivity and fisheries management of the Northeast Shelf Ecosystem. In K. Sherman, L. M. Alexander, and B. D. Gold (eds.), Food Chains, Yields, Models and Management of Large Marine Ecosystems. Boulder, CO: Westview Press, pp. 107–123.

United States Department of Commerce (USDOC). 1991. Fisheries of the United States, 1990. Current Fishery Statistics No. 9000. Washington, DC: National Marine Fisheries Service.

White, K. J. 1987. SHAZAM: A general computer program for econometric methods (version 5). Am. Statis. 41:80.

Wilen, J. E. 1989. Rent Generation in Limited Entry Fisheries. In P. A. Neher, R. Arnason, and N. Mollett (eds.), Rights-based Fishing. Dordrecht, The Netherlands: Kluwer Academic, pp. 249–262.

29

Developing International Law and Ecosystem-Based Fisheries Management

Lawrence Juda

Pressure on world fisheries has grown enormously through the past century and a half as a consequence of a combination of factors. Perhaps the two most important are the rapid increase in world population and a host of improvements in the technology associated with fisheries operations.

In terms of human population, the start of the nineteenth century was roughly the period in which population crossed the one billion mark; by the close of the century it will stand at some six billion.[1] Associated with this veritable population explosion is much greater demand for the living resources of the sea. On the technological side, developments that made it possible to locate fish, catch ever larger amounts, and better preserve those catches, together with the increased availability of modern transport facilities, which opened up larger markets for the products of the sea, have also served to magnify the pressure on the limited resources of the oceans.

Increasingly, threats to the commercial viability of fish stocks have become evident. Despite greater effort, catch of desired species seems to be in decline and a larger proportion of the total catch is composed of lower value species. The need to limit catch to preserve valuable fish stocks has become apparent. But this reality has been difficult to act on in logical fashion since, traditionally, marine fish stocks have had the character of a common proper-

[1]On population trends and the effect of larger populations on the environment, see World Resources Institute. 1994. World Resources 1994–1995: A Guide to the Global Environment. New York: Oxford University Press, pp. 27–42.

ty resource and have thus been subject to the phenomenon of the "tragedy of the commons."[2] In a perversion of Adam Smith's "invisible hand," everyone following his or her own best interest contributes to the inevitable collapse of fisheries consequent to the cumulative impact of individual actions.

Over time, the international legal system sought to respond to the emerging problem of stock decimation by shrinking the ocean commons—that is, the area designated as high seas and in which traditional freedom of fishing was operative. It did this by legitimizing wider limits to the portions of ocean space subject to the control of the coastal state. Under the regime of the 1982 United Nations Convention on the Law of the Sea, for example, territorial seas were recognized to a maximum extent of 12 miles rather than the more traditional 3-mile limit.[3] But most important, this convention sanctioned the creation of the exclusive economic zone (EEZ),[4] a development marking a major change in the international law of fisheries.

The EEZ, extending to a maximum distance of 200 miles from the baselines used to measure the territorial sea, encompasses the area from which comes an estimated 90–95% of the world marine fish catch.[5] In this area the coastal state has sovereign rights to and jurisdiction over living resources and has the authority and responsibility to set total allowable catches (TACs).[6] Nationals of non-coastal states may fish in this zone only with the consent of the coastal state.[7] Given such developments and taking into account the geographic distribution of fish, it might at first consideration appear that the predicament of the tragedy of the commons finally had been overcome, since there was now some recognized and legitimate authority with the legal capacity to limit catch and to regulate behavior in those areas where the vast bulk of fish was located.

It was quickly evident, however, that the EEZ had its limits as a management unit for world fisheries.[8] In particular, at least two difficult problems were obvious: those presented by transboundary stocks and those occasioned

[2]Garrett Hardin. 1968. The tragedy of the commons. Science 162:1243–1248.

[3]The text of this convention appears in International Legal Materials 21:1261–1354 (1982). Hereinafter cited as 1982 Convention. See Article 3 on the 12-mile limit for territorial seas.

[4]1982 Convention, Chapter V.

[5]Food and Agricultural Organization (FAO). 1984. Report of the FAO World Conference on Fisheries Management and Development. Rome: FAO. This report estimates that 95% of the marine fish catch comes from within 200 miles of the coast. The International Union for the Conservation of Nature (IUCN) uses the 90% figure. International Union for the Conservation of Nature (IUCN). 1995. The Law of the Sea: Priorities and Responsibilities in Implementing the Convention. Gland: IUCN, p. 71.

[6]1982 Convention, Article 61 (1).

[7]1982 Convention, Article 62 (3)(4).

[8]On some of the problems of the use of the EEZ as a management unit, see Lawrence Juda. 1987. The Exclusive Economic Zone and ocean management. Ocean Develop. Int. Law 18:305–331.

by straddling stocks. In the former case, fish move through the jurisdictional space of adjacent countries, as in the example of the Gulf of Guinea, and are thus subject to the authority of different states at different times. In the latter case, as evidenced in the recent clash between Canada and Spain in the Northwest Atlantic, fish move from the area under the authority of the coastal state into the high seas, where stocks become subject to capture by anyone based on the traditional concept of freedom of the seas. In neither situation are stocks subject to one uniform management regime.

Boundaries drawn by people to mark off territories are important; through history, they have been the subject of countless conflicts and international arbitrations. Fish pay little attention to human sensitivities, however, and go where they find what they need to survive. The connection between fish stocks and continental shelf areas has been noted for many years. It is in the waters over the world's continental shelves that the fish find what they need: nutrients and appropriate habitat. Continental shelf areas, however, do not coincide with the 200-mile limit, nor do they stop at the boundaries of adjacent states. The fish are not deterred by the boundaries so carefully drawn by government officials; they follow their own imperatives. Effective fishery management therefore requires consideration of the perspective of the fish and the increasing body of scientific data concerning fish and their interaction with the environment that supports them.

The ongoing study of large marine ecosystems (LMEs), the documentation of which appears in many publications,[9] suggests strongly the validity of the need to manage stocks on the basis of the ecosystems. If the following propositions are valid:

- EEZs in most parts of the world do not coincide with identifiable LMEs, and
- Fish will not change their habits to suit human needs or to conform to human-made management schemes, then it would appear that management regimes must better be brought into accord with the relevant units of nature if maximum yields are to be sustained.

What are the possibilities and problems of such a reconciliation of international law and nature? The 1982 United Nations Convention on the Law of

[9]See, for example, the contributions of many authors in this volume and in K. Sherman, L. Alexander, and B. Gold [eds.], Food Chains, Yields, Models, and the Management of Large Marine Ecosystems (Boulder, CO: Westview Press, 1992); Large Marine Ecosystems: Patterns, Processes, and Yields (Washington, D.C.: American Association for the Advancement of Science, 1990); and Large Marine Ecosystems: Stress, Mitigation, and Sustainability (Washington, D.C.: American Association for the Advancement of Science, 1993).

the Sea provides the basic framework for the international legal system relevant to the world's oceans. Whereas its provisions indicate an awareness of ecosystem concepts, as in regard to species interdependence,[10] and stipulate unequivocally that states "have the obligation to protect and preserve the marine environment,"[11] its approach to fisheries management retains the use of artificial zones encompassing areas that do not correspond to those of identified LMEs. Furthermore, it treats separately those portions of the ocean coming under the jurisdictional limits of coastal states and those areas beyond them that maintain their character as high seas.

Whereas Chapter V of the 1982 Law of the Sea Convention establishes the authority of the coastal state over fisheries within its EEZ, article 116 of that convention reiterates the right of all states for their nationals to fish on the high seas. Significantly, however, this right is made subject to the rights, duties, and interests of coastal states. The relationship between stocks within the EEZ and in adjacent areas of high seas is thus acknowledged, as is the need to consider the implications of high seas fisheries for those within the EEZs. However, the exact nature of legal obligations of states under the convention, with respect to straddling stocks, remains ambiguous and has been the subject of dispute. With regard to both straddling and transboundary stocks, states are subject to the general obligation to "seek" to agree with one another on necessary measures to ensure stock conservation either directly or through appropriate regional or subregional organizations.[12] Yet different states that share or have access to stocks may well have different goals, interests, and priorities and their consequent behavior may well serve to undercut attempts at effective long-term management. Achievement of needed cooperation may well prove to be difficult, especially given the possible restrictions on the use of the convention's binding third party dispute settlement machinery in fisheries disputes.[13]

Like biological systems, however, legal norms continue to evolve. Among the important events pertinent to this discussion are the 1972 Stockholm Conference on the Human Environment, the 1980 Convention for the Conservation of Antarctic Marine Living Resources, the 1992 United Nations Conference on Environment and Development, and the United Nations Conference on Straddling Fish Stocks and Highly Migratory Fish Stocks, which met in the period of 1993–95. All have made significant contributions to linking international law to environmental perspectives in general and to ecosystems in particular.

---

[10]1982 Convention, Article 61(4).

[11]1982 Convention, Article 192.

[12]1982 Convention, Article 63.

[13]1982 Convention, Article 297 (3).

Although the Stockholm Conference did not directly result in the adoption of any binding conventions, it had a major effect in shaping the perception of important decision-makers. Serving to emphasize and to make salient to the politicians and diplomats who make the law the need to comprehend ecological systems and balances, it thereby affected the future directions of international environmental and resource management law. It was noted in the Conference Declaration of Principles that natural ecosystems "must be safeguarded for the benefit of present and future generations."[14] Moreover, if renewable resources were to be protected, the natural systems that sustained them must also be safeguarded. Thus, the conference underscored the relationship between the health of fish stocks and the environmental condition of estuaries, intertidal marshes, and other near or inshore environments.[15]

The adoption in 1980 of the Convention on Conservation of the Antarctic Marine Living Resources marked a major step forward in international resource management based on ecosystem approaches. Most particularly, by its definition of the treaty area as encompassing the ecosystem delineated by the Antarctic Convergence rather than a simple boundary based on a line of latitude, as had been stipulated in the 1959 Antarctic Treaty,[16] the diplomats were acknowledging what was long known to the scientific community: political boundaries did not, per se, serve the needs of effective resource management. The operation of natural systems had to be taken into account in the Antarctic and, implicitly, elsewhere as well.

Meeting in Rio de Janeiro 20 years after the Stockholm Conference, the United Nations Conference on Environment and Development (UNCED) demonstrated the growing interest in ecosystem-based management. Integrated management of coastal and marine areas was the central theme of Chapter 17 of Agenda 21, which addressed ocean and coastal protection and management. In the perspective of UNCED, the marine environment formed "an integrated whole," a realization that necessitated new approaches to management.[17] The Food and Agricultural Organization secretariat, in support of such a view, noted the need for "alternative geographic frames of reference" in

---

[14]Declaration of the United Nations Conference on the Human Environment, principle 2. In United Nations. Report of the United Nations Conference on the Human Environment, A/Conf. 48/14/Rev. 1 (1973).

[15]Ibid. Action Plan for the Human Environment, recommendation 48.

[16]The 1959 Antarctic Convention used the 60-degree south parallel of latitude to demarcate the area under its jurisdiction (402 U.N.T.S. 71, article 6). On the use of the Antarctic Convergence, see the Convention on the Conservation of Antarctic Marine Living Resources, article 1 (1). International Legal Materials 19:841–859 (1980).

[17]UNCED, Agenda 21, 17.1.

which to consider the marine environment and its resources.[18] Yet despite the concern for a more "holistic" approach to fisheries management, Agenda 21 continued the pattern reflected in the 1982 United Nations Convention on the Law of the Sea of treating living resources separately with regard to their location within or beyond areas of national jurisdiction.[19]

Recognizing the inadequacy of international fishery management schemes and specifically concerned with the problem of straddling and highly migratory stocks, UNCED recommended and the United Nations General Assembly called for an international conference on that subject.[20] The Straddling Stock Conference provides the latest opportunity to advance the concept of ecosystem-based fisheries management through global negotiations. The concept that fish stocks must be managed in some consistent fashion throughout their migratory range appears to be widely accepted, as is recognition of the need for compatibility of management efforts and rules inside and outside of the EEZ.

Achieving consensus on how such compatibility is to be achieved, however, has proven more problematic. Should the dominant position of the coastal state in the EEZ, recognized in the 1982 Law of the Sea Convention, be extended out to adjacent high seas because of some coastal state "special interest"? Or, rather, should the right of all states to have their nationals fish on the high seas compel coastal states to give "due regard" to the high seas fishing interests of distant-water fishing states even though the exercise of that right might damage stocks found predominantly in the jurisdictional waters of the coastal state?

The conference, in the international agreement it adopted in August 1995, provides the potential for more effective management on an ecosystem basis through the use of regional bodies, such as the Northwest Atlantic Fisheries Organization, which would coordinate conservation measures for high seas fisheries with those taken by the coastal state in its jurisdictional waters.[21] Significantly, the new agreement would limit fishing rights in high seas areas to nationals of those states that participate in or abide by regional manage-

---

[18]Food and Agricultural Organization (FAO). 1991. Living Marine Resources. Research Paper No. 5. Rome: FAO, pp. 9–16, 31–32.

[19]UNCED, Agenda 21, Chapter 17. Section C addresses "Sustainable Use and Conservation of Marine Living Resources of the High Seas," whereas Section D considers "Sustainable Use and Conservation of Marine Living Resources Under National Jurisdiction." On the deliberations leading to this result, see William T. Burke. 1993. UNCED and the oceans. Marine Policy 17:519–533 at 521–525.

[20]United Nations General Assembly A/Res/47/192 (22 December 1992).

[21]See the Agreement for the Implementation of the Provisions of the United Nations Convention on the Law of the Sea of 10 December 1982 Relating to the Conservation and Management of Straddling Fish Stocks and Highly Migratory Fish Stocks. International Legal Materials 34:1547–1580 (1995). Articles 8–11.

ment systems.[22] However, would such bodies be able to reach agreement with coastal states on consistent management policies and practices? If the will to provide for ecosystem-based fisheries management is present, the law may be changed to accommodate it, as international law, like all law, must have the capacity to adapt to new conditions so as to remain relevant and meaningful.[23]

For marine scientists, the interactions and connections observed in what are termed large marine ecosystems are becoming better known and understood through the collection of data and the continuing conduct of studies. This knowledge is now increasingly recognized as relevant to the policymakers who shape the law and rules under which the uses of the oceans proceed. The process of incorporation of knowledge of ecosystems and their operation into the thinking and perspective of lawmakers is of great significance because it will affect the nature of emerging law. If the concept of LMEs is to become an organizing principle in fisheries management, then the case for it will have to be made effectively by the scientific community to the broader public.

Finally, although it appears that the idea of ecosystem-based management is gaining acceptability, the fact remains that conservation of resources is but one major issue in resource and environmental management. In the real world of politics and diplomacy, conservation is linked with the issue of allocation of resources. If agreement cannot be reached on the equity of the allocation of resources and on the distribution of costs and benefits among states, then the prospects for achieving effective ecosystem-based management will be greatly hindered.

[22]Ibid., article 8 (4).

[23]For a detailed examination of the development of international ocean law in response to changes in technology and human understanding of natural systems, see L. Juda, International Law and Ocean Use Management: The Evolution of Ocean Governance (London: Routledge, 1996).

30

Summary and Recommendations for the Mitigation of Stress

Kenneth Sherman, Norbert A. Jaworski, and Theodore J. Smayda

Based on the peer-reviewed reports, the editors, who also served as conveners of the Symposium on the Northeast Shelf ecosystem, conclude that:

1. The Northeast Shelf ecosystem is stressed in the nearshore areas by growing eutrophication and pollution. Relatively small areas considered toxicologic "hot spots" have been identified from the monitoring efforts of the National Oceanic and Atmospheric Adminstration (NOAA) and the Environmental Protection Agency (EPA). In these hot spots, contaminant loadings of the sediments are quite high. Toxicologic effects are expressed in pathologic conditions among fish and shellfish. However, no persistent population mortalities are associated with present "toxic pollution" levels.
2. Signs of increasing levels of nutrient overload or eutrophication, resulting in depletion of dissolved oxygen, are evident in estuaries and coastal estuarine plumes. The cumulative effects on the nearshore ecology of the Northeast Shelf ecosystem are unknown.
3. The increasing incidence of biotoxins causing deaths among marine fish, shellfish, and mammals appears to be associated with an increased frequency and extent of phytoplankton blooms.
4. Loss of estuarine habitat for spawning and feeding areas of several species continues to be of concern. It is a problem that can be addressed by judicious application of funds accrued from fines levied on polluters toward support of ecological restoration projects. Several restoration projects to reclaim loss of habitat because of pollution have been initiated.

5. The more open water areas of the Northeast Shelf ecosystem are "healthy." The structure and function of the lower end of the food chain is highly productive.
6. Major shifts in the dominance of the finfish community from gadoids (Atlantic cod, haddock, pollock, hakes) in the 1960s to a predominately elasmobranch (spiny dogfish and skates) and pelagic biomass in 1991 are attributed to perturbations caused by excessive fishing mortality of gadoids and flatfish. The loss of consistently high recruitment of high-value gadoids and flounders contributes to the depleted state of the economically desirable fish stocks.
7. Consideration is being given to "adaptive management" strategies focused on the effects of directed removals of low-economic-yield predators (e.g., spiny dogfish and skates) to enhance recovery of the depleted gadoid and flounder stocks.
8. Marine mammal populations at risk from fishery-caused deaths are being subjected to increasing research and monitoring efforts by NOAA/National Marine Fisheries Service and other scientific groups and institutions. This effort is to ensure recovery of depleted populations and to eliminate fishery bycatches that could be detrimental to reproductive and recruitment success and the rebuilding of depleted populations.
9. As a result of the Symposium, coordinated programs of principal federal agencies are under way in collaboration with the states, academic institutions, and the private sector to improve monitoring strategies aimed at supporting efforts to mitigate detrimental effects of habitat loss, coastal pollution, eutrophication, and overexploitation of marine resources. These include NOAA's Status and Trends Program and its Marine Resources Monitoring, Assessment, and Prediction (MARMAP) program, and the EPA's Environmental Monitoring and Prediction (EMAP) program. In addition, the National Science Foundation's ecosystem dynamics program (GLOBEC), which is designed to study the effects of physical variability on biological populations, has a Northwest Atlantic component and is funding studies on Georges Bank.
10. Pending legislation to provide funds for additional waste-treatment facilities for municipalities, and best management practices for non-point sources of pollution along the coast of the Northeast Shelf ecosystem, will aid in reducing the loadings of nutrients that lead to eutrophication within the estuaries of the region.
11. Dumping urban sludge within the EEZ has been legislated as unlawful; illegal dumping is now subject to federal penalties. As a result, all sludge dumping has been terminated within the EEZ.

12. Symposiums or workshops on the changing states of health of the Northeast Shelf ecosystem should be scheduled at 3-year intervals.
13. Consideration should be given to the development of more integrative governance of the ecosystem's capital assets (e.g., fish, water quality, habitat, other renewable and nonrenewable resources); this should include participation by the public and the municipal, state, regional, and national levels of government serving as resource stewards.

Index

In this index, the abbreviation t stands for table and f stands for figure.